RESEARCH METHODS FOR INFORMATION SYSTEMS

Research Methods for Information Systems

Ronald S. King

Mercury Learning and Information
Dulles, Virginia
Boston, Massachusetts
New Delhi

Publisher: David Pallai
Mercury Learning and Information
22841 Quicksilver Drive
Dulles, VA 20166
info@merclearning.com
www.merclearning.com
1-800-758-3756

This book is printed on acid-free paper.

Ronald S. King. *Research Methods for Information Systems.*
ISBN: 978-1-936420-12-4

Library of Congress Control Number: 2011931635

121314321

Our titles are available for adoption, license, or bulk purchase by institutions, corporations, etc. For additional information, please contact the Customer Service Dept. at 1-800-758-3756 (toll free).

CONTENTS

PREFACE

Research Methods for Information Systems is intended to provide a simple and practical introduction to an area that many students and professionals find difficult. The book takes a non-threatening approach to the subject, avoiding excessive mathematics and abstract theory. It shows how to apply research and statistical methods to the current problems faced in a variety of fields, but with special emphasis on computer science and information systems. The book includes numerous exercises and examples that help students understand the relevance of research methodology applications. Assuming no previous knowledge, the text provides complete coverage for a first course in research and statistical methods in computing, or "on the job" self-training.

The text was designed to be an academic book that explains the "why" and the "how" of practical ways to conduct research in the computing field. In computer science, research methods have historically been passed from advisor to student via apprenticeship. At the same time, a bewildering diversity of research methods have been employed within the field of computing including: implementation driven research, mathematical proof techniques, empiricism, and observational studies. Additionally, research methods texts, from other fields, are inadequate for the field of computing research methods. Fortunately, a growing number of degree programs, in the computing field, have been exploring models and the content for computing research methods courses. In 2005, the SIGCSE Committee on Teaching Computer Science Research Methods [SIGCSE-CSRM] was founded. *Research Methods for Information Systems* exposes the reader to SIGCSE-CSRM's complete taxonomy for computing research methods: case studies, conceptual analysis, field experiments, field studies, instrument development, laboratory experiments, literature reviews, simulation, and exploratory surveys.

A research model applicable to applied research is proposed and discussed throughout the text. This model accommodates scientific methods of research, including empirical, quantitative, qualitative, case study and mixed methods. The pedagogical approach is described in terms of thematic areas of scholarship, practice, and intended outcomes. Research method topics covered and summarized are: proposal formulation, research design, methods of investigation, methods of demonstrating concepts; approaches to research validation and documentation of research results.

Throughout the text lead articles, examples and exercises provide the reader with actual instances of: implementation driven research, mathematical proof techniques, empiricism, and observational research. This diversity in research methods in the computing fields is due to the fact that topics in the disciplines of computer science and information systems are technologically based rather than theory-driven logic. The reader is also exposed to qualitative and mixed research methods. This is extremely important since within information systems there has been a general shift in information system research away from technological studies, toward more managerial and organizational issues.

Throughout this text the concern is to help the student understand quantitative reasoning and the research process; therefore, there are exercises at the end of every chapter. It has probably already occurred to you that it is one thing to understand the research process and quite another to actually be able to perform in a research setting.

You already know that constructive progress in quantitative reasoning abilities are not easy to accomplish. You learn through active participation rather than simply by listening and absorbing ideas. Like a finely tuned runner, you enhance your competitive chances in the future race through many practice sessions.

The purpose of the following discussion is to provide students and teachers with advice on structuring the upcoming class and how to properly approach the subject. The teacher's responsibility is merely to *guide* class discussion and *stimulate* interaction within the class during classroom problem solving. The teacher should *encourage* all students to discover and implement problem solving methods during class, at least one day each week.

Quantitative reasoning is a never-ending task that reaches into most areas of business, education and the behavioral sciences. Develop a solid foundation, practice diligently, and you will receive many benefits in the future.

Suggestions for the Student

To help you have the best opportunities to succeed in this class, we suggest you follow a few simple rules. They can be classified under three headings: *Prepare, Participate,* and *Apply.*

Prepare:

1. Read the text completely and carefully. Underline or highlight sections that you feel are important. Put a question mark next to those ideas or concepts you don't understand or with which you find it difficult to agree.
2. Work the exercises! Learn by doing! Answer each question thoughtfully and critically.

Participate:

1. Don't be afraid to ask questions. Your questions may voice the questions of other class members. Your courage to speak out will give those people permission to talk and may encourage more stimulating discussion.
2. Don't hesitate to share your ideas. Abstract thinking has its place, but personal thoughts and illustrations will help you and others remember the material much better.
3. Be open to others. Listen to other members' approaches to solving problems.

Apply:

1. Commit yourself to be an active participant in problem solving. Involve yourself in what is happening around you during the class problem solving sessions.
2. Identify the keys and highlights that enable the problems to be solved.

Suggestions for the Teacher

Here are several guidelines that will help you encourage discussion, facilitate learning, and implement the statistical methodologies:

1. Encourage the students to prepare thoroughly and bring all necessary resources to the weekly problem solving class sessions.
2. Discuss each question, or case study, individually. Ask for several strategies and encourage students to react to the comments made by others.
3. Provide visualization, with charts and tables, to enhance the ideas being presented. Outline major concepts.
4. Always invite concrete illustrations.
5. Look for ways to practically apply the methods studied in class and the suggestions offered.

Unit 1 introduces the ideas of statistics and statistical inference and describes the types of data analyzed with these techniques. The remaining chapters in the unit define and analyze the usual measures of central tendency, variability, and distributional shape. Chapter 6 provides an introduction to data mining, or data exploratory analysis.

Fundamental concepts of probability—events, random variables, probability distributions, expected value and variance of a probability distribution—are covered in Unit 2. Chapters 9 through 11 examine the most important discrete and continuous distributions. Chapter 12 presents the use of one distribution to approximate another.

This discussion leads into Unit 3, which considers the sampling distributions of the sample mean and other sample statistics. The Central Limit Theorem is stated and discussed. Chapter 15 describes point estimation, interval estimation, and hypothesis testing.

Investigation of hypothesis testing with one sample is continued in Unit 4, with discussion of the Type II error, one-tail and two-tail tests, small-sample tests for the mean using the t distribution and chi-square tests of variance. One-sample tests and two-sample tests, beginning with the F test for the ratio of two variances, since the outcome of that test determines how to test the difference of two means when using independent samples. Matched-pair tests are also considered. Unit 5 describes the use of the chi-square statistic to perform tests of multinomial data, independence, goodness-of-fit, and missing data.

Unit 6 explores correlation, defining not only the Pearson correlation coefficient but other coefficients which apply to situations in which the Pearson R may not be appropriate. Chapter 14 discusses linear regression, both simple and multiple. Development of the multiple linear regression equation is accomplished through an iterative matrix technique rather than the cumbersome normal equations. The final section of the chapter describes transformations which can be used to develop nonlinear regression equations with the techniques of linear regression. As an alternative to using multiple linear regression for this situation, neural nets are introduced An example based on judgment analysis (JAN) is included in the homework exercises, as well as in Unit 9.

Units 6 through 10 are a primary text for the first course in research design and methodology. There is no single, best way to teach such a course—except, of course, the way each of us does it. This text has structured the order of presentation in the way the author finds most effective when teaching the course. The organization flows from an overview of what research is all about in Chapter 40, to specific instruction and examples of writing a research article. The major sections include Unit 9 and Unit 10.

Realizing that choice of one's presentation order, which may be different from mine, prompted the author to make the chapters within each section to be fairly independent. The reader should encounter little difficulty using the chapters in whatever order they prefer. The major exception to this is Unit 6; it really should be read first. A key feature of this book is its unified approach to the application of linear statistical models in regression, analysis of variance, and experimental designs. Instead of treating these areas in isolated fashion, emphasis is on seeking to show the interrelationships between them. Use of a common notation for regression on the one hand and analysis of variance and experimental designs on the other facilitates a unified view. The notion of a general linear statistical model, which arises naturally in the context of regression models, is carried over to analysis of variance and experimental design models to bring out their relation to regression models. This unified approach also has the advantage of simplified presentation.

Applications of linear statistical models frequently require extensive computations. Explanations of the basic mathematical steps in fitting a linear statistical model are provided, but discussions do not dwell on computational details. This approach permits one to avoid many complex formulas and enables emphasis to focus on basic principles. Extensive use of computer capabilities for performing computations, and illustrating a variety of computer printouts helps in explaining how these are used for analysis. A wide variety of case examples is presented, both to illustrate the great diversity of applications of linear statistical models and to show how analyses are carried out for different problems. Theoretical ideas are presented to the degree needed for good understanding in making sound applications. Emphasis is placed on a thorough understanding of the models, particularly the meaning of the model parameters, since such understanding is basic to proper applications.

Calculus is not required for reading this text. In a number of instances, use of calculus to demonstrate how some important results are obtained, but these are confined to supplementary comments or notes and can be omitted without any loss of continuity. Readers who do know calculus will find these comments and notes in natural sequence so that the benefits of the mathematical developments are obtained in their immediate context. Some basic elements of matrix algebra are needed for multiple regression and related areas. Chapter 16 introduces these elements of matrix algebra in the context of simple regression for easy learning.

A secondary purpose of this volume is its use as a reference after the student completes the course. Unit 7 discusses one-way and two-way analysis of variance, using completely randomized and randomized-block designs, and concludes with a discussion of the analysis of covariance. Chapter 34 is a discussion of the design of experiments.

Unit 8 explores nonparametric hypothesis tests, beginning with tests of randomness and goodness-of-fit, and concluding with nonparametric substitutes for parametric tests. This material should be covered before studying Monte Carlo simulation in Unit 10.

In the final two Units, the tools of statistical analysis are described and discussed. Chapter 43 examines Excel, which is used throughout the book, and commonly employed as statistical software packages. Statistical tables and data files for examples and exercises are contained in the appendices.

In each chapter, the discussion of a statistical procedure is accompanied by an example of Microsoft's Excel software to perform that procedure. For example, pivot table-tabulations are introduced in Chapter 6. These examples are drawn from Excel, but this book can be used successfully with any statistical software package.

This book differs from other applied statistics and research methods texts in several important ways:

a. Except in simple examples, computations are left to the computer system. This frees the student to concentrate on selecting the correct procedure and interpreting the results.

b. The student learns statistics by doing statistical analyses.

c. The use of a particular program package is not mandatory; analyses can be performed with any statistical software package. The packages used in the text (Excel or SPSS) are present primarily for illustration.

d. To begin performing statistical analyses immediately, one primary data file is used in most discussions in the text.

e. Surveys of currently available software and guidelines for the evaluation of these packages are included, as is a short discussion of future trends.

f. Complicated concepts and relationships are explained verbally as well as mathematically.

g. The key to thinking like a statistician is the ability to visualize sampling distributions. These are explored and illustrated in detail.

h. The notion of fitting a linear statistical model to experimental data is easy for the beginner to visualize because he has considered the simple problem of fitting a straight line in an introductory course. Thus the approach is familiar, intuitively appealing, and very easily extended to a multidimensional space of independent variables. Introducing the sums of squares for the analysis of variance as the difference in sums of squares for error for two linear models is meaningful and provides an easily understood intuitive justification for the F-test. Once the student sees how the sums of squares are obtained for a few examples, he or she is content to memorize the formulas for various types of designs. In contrast, the

> sums of squares for the analysis of variance are often presented in a cookbook manner and appear to most beginners to have been acquired out of the blue. This fact is not difficult to explain because proof of expectations is usually omitted. Some authors give the tedious algebraic proof of additively of sums of squares, but this by itself does not give intuitive justification to the F-test. Other advantages to the least-squares approach are numerous. It forces the student to think about the probabilistic model for his conceptual population when he designs his experiment, not after. Thus he or she realizes that he must in some way relate the practical question that he or she wishes to answer to an inference about one or more parameters in the probabilistic model. He or she early achieves a single and powerful method of analysis that, unlike the analysis of variance, can be applied to data from undersigned (or badly designed) experiments. It leads early and easily into the analysis of variance, and from that point on the student possesses two powerful methods of analysis. This, along with unity of presentation and intuitive appeal, is perhaps the most important advantage to the approach. It is not proposed that least squares be *substituted* for the analysis of variance. Rather, the least-squares approach can be used as a powerful tool for estimation that will unify and *supplement* the analysis of variance

Research Methods for Information Systems is designed to be used in a one-year graduate course in research methods, or in a one-semester corresponding undergraduate course. The first five units, which cover the basic ideas of descriptive statistics, probability, and statistical inference, and introduce Excel/SPSS, are intended for a rapid review of statistical methods. The instructor may then select material from the remaining Units, Unit 6 through Unit 9 in particular, without disturbing the continuity of the course. Unit 10 covers Monte Carlo and discrete event simulation studies and a research study vignette is presented. The purpose of the vignette is to identify the various types of information system methodologies which an information system researcher should maintain in the research toolbox and to interweave these methods with current practices in project management.

UNIT 1

DESCRIPTIVE STATISTICS

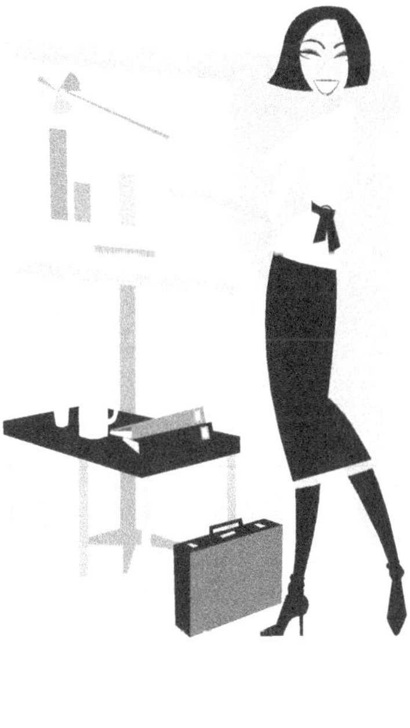

"The CIO profession: driving innovation and competitive advantage," at *www.935.ibm.com/services/us/cio/pdf/2007_cio_leadership_survey_white_paper.pdf*, was a survey conducted by the Center for CIO Leadership, in collaboration with Harvard Business School and MIT Sloan Center for Information Systems Research (CISR). The respondents included 175 CIOs from leading companies around the world. Sentences were given to the respondents to rate their level of agreement with on a scale of 1 to 5 with 1 = "not at all" and 5 = "to a great extent." Many charts, tables, and other visual aids are provided for helping the reader discern an understanding of the relationships among the CIO role, different aspects of IT performance, and organizational performance.

After studying this article, discuss how the visual aids enhance the reader's understanding of the concepts presented:

1. What type of relationship appears to exist between the senior executive teams and CIOs?
2. What type of involvements do CIOs experience within their organizations?
3. Do CIOs have an influence over their organization's strategic decisions?
4. Which strategic decision making do CIOs participant in and to what degree?

This lead article demonstrates, over and over again, that pictures are worth a great deal. The illustrations enable us to make sense of all the data. Computing power allows the data to be rapidly processed, summarized, and analyzed. Analyzed data encourages the production and storage of more data. Data such as stock quotes is brought to our fingertips. Clearly, everyone needs to properly analyze and interpret all of this available data.

For the lead article you should ask:

1. Why was this study done?
2. How was the study done?
3. What were the findings?
4. What was the selection process for the participants in the study?
5. Is the sample taken representative of the national population?

But first you must be able to read with understanding and then determine the impact of the results of the study. This unit will aid in these pursuits.

CHAPTER 1

INTRODUCTION TO STATISTICS

OVERVIEW AND LEARNING OBJECTIVES

In This Chapter

1.1 Introduction
1.2 What Is Statistics?
1.3 The Role of Probability in Statistical Inference
1.4 Types of Data
1.5 How to Perform a Statistical Study
1.6 Exercises

1.1 Introduction

What is statistics? Why should a person study statistics? How does one perform a statistical study? We consider these questions in this chapter.

For an example application, consider "Winning Database Configurations: An IBM Informix Database Survey" by Marty Lurie of the IBM Software Group, available at *www.ibm.com/developerworks/db2/zones/informix/library/techarticle/lurie/0201lurie.html*. This article written for database administrators evaluates the performance of the IBM® Informix® database management system. How are database servers used and configured? How much hardware does it take to process a large amount of data? The answers often lie with the database administrators (DBAs), who, through many hard-earned lessons, have found configurations that work well. In this article, Lurie presented the results of a survey on more than 60 IBM Informix servers deployed at over 40 organizations.

Why do a database survey?

There are two primary reasons for performing the survey:

- To develop sizing metrics to define how many CPUs, how much memory, and how many disks are needed for a given workload
- To find out what people are actually doing with the database, including which features are being used, such as mirroring

One of the most common questions that database administrators ask is "How many CPUs do I need for this server?" Another question is "What is the best way to back up the server?" Consultants and pre-sales tech support staff base their responses on what has succeeded at other accounts, with a liberal dose of what the product design team recommends. By examining the existing configurations, we can get a good idea of what a system is capable of handling. Using statistical methods, Lurie provides a formula that one can use to determine how many CPUs are needed, based on the amount of data and the version of the IBM Informix server one is using.

Answer the following questions:

1. How did the author achieve the "statistical significance" for the survey?
2. How was the survey conducted?
3. Clearly a vast amount of data was collected. How was control achieved for this process?
4. What appears to be the major limitation in the study design?

1.2 What Is Statistics?

According to Webster's *New Collegiate Dictionary*, statistics is

> The science of the collection and classification of facts on the basis of relative numbers of occurrence as a ground for induction; systematic compilation of instances for the inference of general truth.

Several aspects of statistics are brought out by the second part of this definition. The first is the need for a well-defined method of summarizing the observations in an experiment in order to make the information contained in the observations easier to understand. For example, a businessperson might summarize the sales per month of a particular item by calculating the average sales per month, thus reducing a group of numbers to a single value that tells the reader about a particular characteristic of the original values. The measure of dispersion would describe how sales of this item are distributed across the months. A manager might inquire as to whether sales of the item are steady, with all monthly sales figures near the average, or whether sales are unusually high in certain months. Particularly in the latter case, the manager would rank the months in order of decreasing sales to obtain each month's position with respect to the others. Finally, the manager could summarize monthly sales of the item by constructing a graph or table.

Each of these techniques facilitates the communication and interpretation of a large mass of data (a *population*), producing from that data information that can be used. This step leads us to the following definition:

> *Descriptive statistics* is a collection of methodologies used to describe a population's central tendencies, dispersion or variability, distribution, and the relative positions of the data in the population. Included in these methodologies are quantitative, graphical, and tabular techniques.

A manager might also be concerned with more complicated questions, such as the interdepartmental allocation of resources within a large corporation. Is it a major objective of the organization to maximize product output for a given level of resource input? Or is some other goal more important to the corporation? What do the various resources contribute to the outputs and products of the organization? What are the economic implications of changing the system of resource allocation? Answers to these and similar questions involve large data sets. Time, personnel required, cost, and legal restrictions involved in answering the latter questions prevent their being answered simply by applying descriptive statistical techniques to entire populations of values.

Instead, managers will examine subsets of the population called *samples* to which descriptive statistics can be reasonably applied. From these samples, a manager can make inferences and generalizations about the population from which they came. Subsets of this kind are the second major aspect of statistics.

> *Statistical inference* is a collection of methodologies in which decisions about a population are made by examining a sample of the population.

In *statistical inference*, a *parameter* represents some numerical property of a population. A *statistic* is a numerical property of a sample, and is generally used to *estimate* the value of the corresponding population parameter. For example, a manager might take a sample from the population of all monthly sales figures of the item in which he is interested, and use the average of the sample to estimate the average monthly sales of the item since its introduction.

1.3 The Role of Probability in Statistical Inference

When the manager in our example is using a sample statistic as an estimate of a population parameter, he needs to know how accurate an estimate he is obtaining. That is, the manager needs a measure of the confidence that he has in the results. *Probability* is the link between the characteristics of the sample and those of the population; it is the key to that measure of confidence.

> *Probability* is reasoning used to deduce characteristics of a sample from the characteristics of the population from which the sample was taken.

For example, if a researcher is familiar with the properties of several different populations, he could take the results of a given sample and determine from which of the populations, if any, the sample was taken. In this book, we will detour into the study of probability before pursuing *statistical inference*.

Note that *statistical inference*, in which conclusions are drawn about a population based on a sample from the population, is *inductive reasoning*. This is the reverse of probability, in which we make statements about a sample based on the properties of the population from which it came.

1.4 Types of Data

The general area of statistics may also be divided according to the types of data being examined, and data can be classified according to two general schemes. The first scheme classifies data by *measurement scales* and the second by the number of values that the data may have. We usually think of measurement as the assignment of numbers to objects or observations, as when we measure the length of a piece of lumber. Such measurements, however, constitute just one in a range of *levels* or *scales* of measurements.

The lowest of these levels is *nominal measurement*, the classification of observations into categories. These categories must be mutually exclusive and collectively exhaustive; that is, they must not overlap and must include all possible categories of the characteristic being observed. Examples of nominal variables are sex, type of automobile, and job classification.

An *ordinal scale* is distinguished from a nominal scale by having equal intervals between the units of measure by the property of order among the categories, as in the rank of a contestant in a competition. We know that “first” is above “second,” but we do not know how far above.

An *interval scale* is distinguished from an ordinal scale by having equal intervals between the units of measure. Scores on an exam are an example of values on an interval scale. However, though a person may score zero on an exam, this does not demonstrate that the person has none of the knowledge or traits that the exam intended to measure. *Ratio scales* have the properties of interval scales but with a true zero. Age, height, and weight are all measured on ratio scales.

This classification of scales of measurement has historically divided statisticians into two camps. The first holds that using the common arithmetic operations on nominal or ordinal data will distort the meaning of the data. Members of this camp have developed procedures intended to be used with nominal and ordinal data, and these procedures constitute the field of *nonparametric statistics*. The second camp believes that since statistical analyses are performed on the numbers yielded by the measures rather than on the measures themselves, they may apply the same *parametric* statistical procedures to all the above types of data. They hold that common arithmetic operations performed on values produced by any measurement scale will not affect the validity of the results. Mathematical and empirical studies are available in the literature to support both contentions.

Data may also be classified according to the range of values available for the variable of interest. If the variable can take on only a finite or countable number of values, it is said to be *discrete*. Job classification, sex, and number of cars sold in a month are examples of discrete variables. If a quantitative variable can take on any value over a range of values, it is called *continuous*. Height, weight, distance, and temperature are thought of as continuous variables.

The classification schemes discussed so far have dealt with quantitative variables. *Quantitative data* requires that we describe the characteristics of the objects being studied numerically. *Qualitative data* is concerned with traits that are not numerically coded. A qualitative variable is called an *attribute*; for example, in a data file of information on licensed drivers, "blue" would be a value of the attribute "eye color." Figure 1.1 shows the data classification schemes discussed thus far.

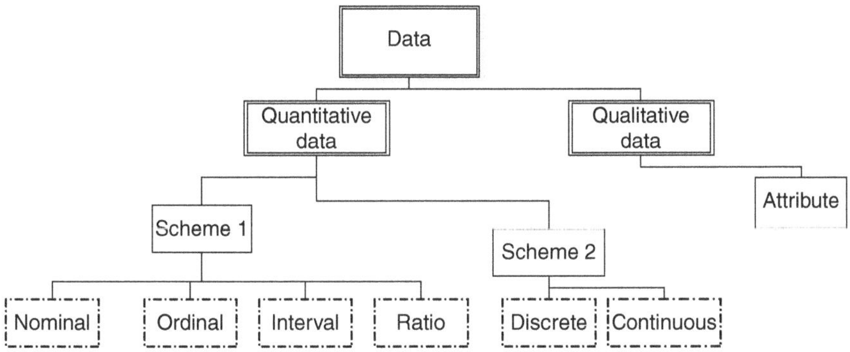

FIGURE 1.1 Data classification schemes.

All figures and tables in this chapter appear on the companion DVD. ON DVD

1.5 How to Perform a Statistical Study

A statistician, like a scientist, deals in probabilities. It is a common misconception that science deals in certainties. In the language of statistics, the best a scientist can hope to do is:

a. to specify the levels of the independent quantities in the research study

b. to control the effects of extraneous quantities

c. to determine the probable effects to be observed on the dependent quantity being examined

Once the scientist has formulated (c) from (a) and (b), the scientist states the hypotheses to be tested, then states a decision rule to which the sample data, the results of the study, can be compared. Based on this comparison, the scientist decides whether to accept the hypotheses, or to reject them and begin the process again. Those accepted hypotheses become theories, not proven beyond all doubt, but accepted as indicating and describing the probable behavior of the system under study.

Throughout this process of theoretical development, scientists strive for theories that:

a. are consistent with known facts

b. do not contradict one another

c. can be tested experimentally

d. generate explanations for a wider variety of phenomena

e. generate useful predictions

The same guidelines determine statistical methods for decision making under conditions of uncertainty. The decision maker must choose among alternative actions. The decision maker is uncertain about the possible results of these actions, since they also depend on conditions beyond a person's control. The probabilities with which these other conditions will occur can often be estimated. Applying statistics to sample values of the independent and control variables, the decision maker calculates values of the dependent variables. These values are used in a previously formulated decision rule to decide which of the alternative actions to take. This approach to problem solving is called *decision analysis*. Most statistics texts devote a separate chapter to this topic, setting it apart from the rest of statistics. The topic is called *Bayesian statistics*, as opposed to *traditional* or *classical statistics*. Here, since decision analysis is a philosophical approach to the question of decision making under uncertainty, it will be included throughout the book.

1.6 Exercises

1. Classify the following measures as nominal, ordinal, interval, and/or ratio.
   a. Plant maintenance expenditure
   b. Educational classification
   c. Percentage of workers with MBA degrees
   d. Building age
   e. Rating the flavor of a soft drink on a scale of 1 to 9
   f. Church affiliation
2. Repeat Exercise 1 with respect to whether the variables are continuous or discrete.
3. For a recent article from *The Wall Street Journal*, or another comparable business resource:
   a. Formalize a set of hypotheses that could be researched to resolve the answer.
   b. State the dependent variables of interest.
   c. State the independent variables of interest.
   d. State the control variables of interest.
   e. Discuss how you might draw the sample and describe a decision rule you might use.
4. Describe the construction of a spreadsheet to help the concession stand at the football stadium maintain a high degree of cost-effectiveness:
   a. Identify the data.
   b. Identify the fields of interest for the data.
5. Consult your instructor or another advisor about journals in your field of interest that apply statistical analyses to databases to extract informational content. Begin reading articles of interest from these sources.
6. Indicate which of the following terms are associated with a sample, S, or the population, P:
   a. Parameter
   b. Statistic
   c. Inductive reasoning is applied to the ______________________ to draw inferences about the ________________________.

CHAPTER 2

MEASURES OF CENTRAL TENDENCY

OVERVIEW AND LEARNING OBJECTIVES

In This Chapter

2.1 Sample Study

Winning Database Configurations: An IBM Informix Database Survey by Marty Lurie of the IBM Software Group, available at *www.ibm.com/developerworks/db2/zones/informix/library/techarticle/lurie/0201lurie.html*, employs various statistical methods to conduct a performance analysis of the Informix database management system. After studying this lead article, answer the following questions:

1. Which measures of central tendency were used for deciding the size factors for the Informix system? Why?
2. What sample sizes were used? Are these values sufficient?
3. How can the amount of RAM be modeled? Why?
4. Which server clearly has larger data volume capabilities?
5. Explain what a TPC-H benchmark is.

One could easily size a system using the above average values, but we would miss the opportunity to provide a much better estimating tool, presented in Chapter 25, "Introduction to Simple Linear Regression."

2.2 Introduction

In descriptive statistics we are concerned with finding measures of central tendency. From a large group of values, we want to extract numbers that will characterize certain qualities of the group and distill information that we can more easily understand and communicate.

2.3 Measures of Central Tendency

When we are confronted by a long list of values, we often ask, "What is a typical value?" or "What is an average value?" We are asking for a measure of the *center* of the values, for which there are several commonly used statistics.

2.3.1 Mode

The simplest measure of the central tendency of a group of values is the *mode*, the most common value.

The *mode* is the value with the largest frequency.

In the group of values 5, 7, 6, 7, and 10, the mode is 7, since it appears more often than any of the others. If several values share the largest frequency, then all of those values are modes.

As a measure of the center of a distribution of values, the mode has several characteristics that should be noted:

1. It is the fastest and roughest measure of central tendency.
2. It is not necessarily unique, since a given group of values may have more than one mode.
3. It does not necessarily exist; if all values occur only once, there is no mode.
4. It is determined by the most common value or values, and does not consider any others.

When giving the mode, it is wise to state the frequency of the mode and the total number of values.

2.3.2 Median

A more useful measure of a "typical" value in a group of values is the *median*, which is defined in this way:

> The *median* is that value which divides all the values of the variable so that half are larger and half are smaller than the median.

To find the median, arrange the values in order (they are then called *order statistics*), and select the one in the middle. If the number of values n is odd, the median is the $[(n+1)/2]$th value. For example, the median of the values 2, 7, 5, 4, 6, 9, 6 is 6:

$$2, 4, 5, 6, 6, 7, 9$$

If the number of values is even, the median is halfway between the $(n/2)$th and $[(n/2)+1]$th values:

$$10, 12, 16, 17, 20, 23, 25, 26$$

$$\text{median} = 18.5 = \frac{20+17}{2}$$

Relevant characteristics of the median are these:

1. It always exists; for any group of values, the median can be computed.
2. It is unique; any group of values has only one median.
3. It is not greatly affected by extreme values; in the second example in this section, if we include the value 1643 in the list, the median moves up to 20.

Note that the median does take into account the *relative positions* of all the values and is more sensitive to *all* the values than is the mode.

2.3.3 Mean

The measure of the middle of a group of values that considers the *magnitudes* of all the values is the *mean* or *arithmetic mean*, the quantity most often associated with the word "average."

> The *mean* is the sum of all the values of the variable divided by the number of values.

Stated more precisely, if $X_1, X_2, \ldots, X_N$ are a *population of values*, their *mean* is

$$\frac{X_1+X_2+\ldots+X_N}{N} = \frac{\sum_1^N X_i}{N},$$ and is denoted μ (mu).

For example, the mean of the values 9, 15, 23, 21, 12, and 16 is

$$\frac{9+15+23+21+12+16}{6} = 16$$

If the values are elements of a *sample*, the notation is different, but the process is the same: If $X_1, X_2, \ldots, X_N$ are a sample of values, their *mean* is

$$\frac{X_1+X_2+\ldots+X_n}{n} = \frac{\sum_1^n X_i}{n} = \frac{1}{n}\sum_1^n X_i,$$ and is denoted $\overline{X}$ (X-bar).

The mean has several important characteristics:

1. Like the median, it always exists and is unique.
2. It is a good estimator; if we take repeated samples from the same population, the means of the samples will tend to cluster around the population mean.
3. It is sensitive to extreme values; consider the previous example, with the value 1920 included. The mean is then:

$$\frac{9+15+23+21+12+16+1920}{7} = 288.$$

2.4 Measures of the Middle and Distributional Shape

When a distribution of values is symmetrical and unimodal (has one mode), the mean is the best measure of central tendency. In fact, in such a case, the mean, median, and mode will be approximately equal, as shown in Figure 2.1.

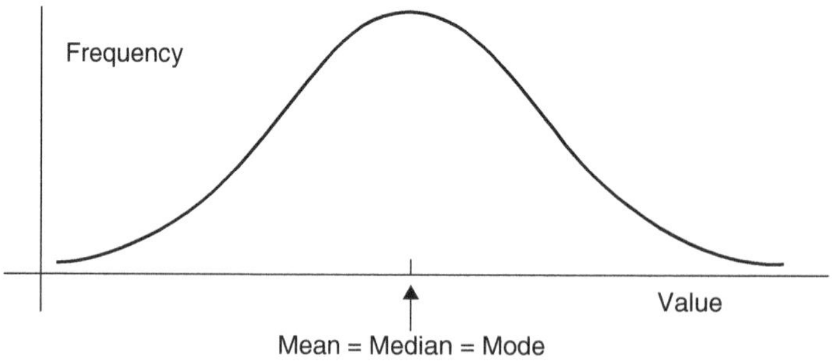

FIGURE 2.1 Measures of central tendency of a symmetrical distribution.

ON DVD All figures and tables in this chapter appear on the companion DVD.

If a distribution of values is unimodal but not symmetrical, we say that it is *skewed*, and since the mean, median, and mode differ in sensitivity to extreme values, they will be different, as shown in Figure 2.2.

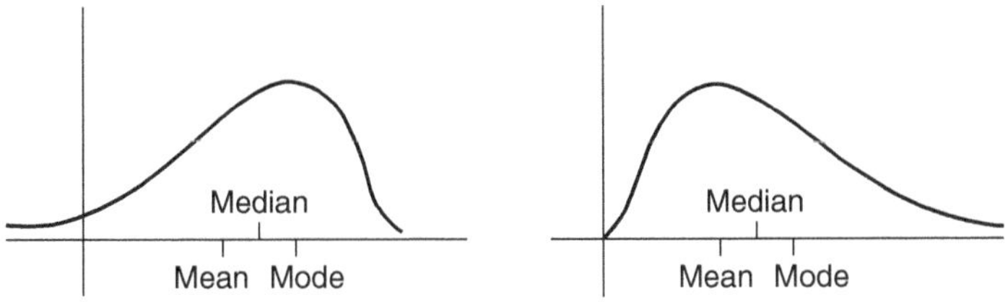

FIGURE 2.2 Measures of central tendency of skewed distributions.

On the left side of Figure 2.2, we see a distribution that is *skewed to the left*, or *negatively skewed*. The mean is most affected by the low extreme values, and is less than the median, which is less than the mode. In a distribution that is *skewed to the right (positively skewed)*, this relationship is reversed. Such distributions arise in situations where all the extreme values lie on the same side of the bulk of the data, as with salaries in a business, where a few of the values are larger than the majority.

EXCEL

USING EXCEL TO COMPUTE THE MEAN, MEDIAN, AND MODE

Functions for computing the mean, median, and mode in Excel are: AVERAGE(<CELL RANGE>), MEDIAN(<CELL RANGE>), and MODE(<CELL RANGE>).

Data can be sorted by doing the following:

1. Select the **Data** menu.
2. Choose the **Sort** option.
3. When the Sort Dialog box appears:
   a. In the Sort by box, make sure the variable label appears and that either **Ascending** or **Descending** is selected.
   b. Click **OK**.

2.5 Exercises

1. Find the mean, median, and mode of this sample of values: 4, 6, 12, 6, 9, 5.
2. Find the mean, median, and mode of this population of values: 26, 35, 29, 27, 35, 35, 29.

3. Over the past 10 years, mutual fund A has paid a mean return of 12%, with a median return of 7%; mutual fund B has paid a mean return of 10%, with a median return of 9%. You plan to invest a sum of money for two years. Based on this information, would you choose to invest in fund A or fund B? Why?

4. Ten workers are paid the following hourly wages: $10.10, $10.75, $15.55, $17.50, $17.70, $10.80, $16.15, $7.75, $5.30, and $16.20. What is the mean wage paid to the workers? The median? Is this group of values skewed? Answer this question both manually and with Excel.

5. A fleet of 30 cars achieves mean mileage of 22.2 mpg, while another fleet of 50 cars gets 20.5 mpg. What is the mean mileage of both fleets together? Does this suggest any generalization about the mean of two groups of values whose means are known?

6. Use Excel to find the mean, median, and mode of the variable X7 (change in construction activity) from the American cities database. What do these values tell you?

7. Would it make sense to find the mean and median of the values of X1 (city region) in the American cities database? Why or why not? What does this tell you about these measurements?

CHAPTER 3

MEASURES OF DISPERSION

OVERVIEW AND LEARNING OBJECTIVES

In This Chapter

3.1 Introduction

In descriptive statistics not only are we concerned with finding measures of central tendency, but we are also interested in variation, distributional shape, and relative location of individual values for the values of a variable. From a large group of values, we want to extract numbers that will characterize certain qualities of the group and to distill information that we can more easily understand and communicate.

3.2 Measures of Variation

After graduation, you are offered two jobs. Employer A tells you that the median salary at his organization is $39,500, while employer B's company pays a mean salary of $41,500. If the salary were your only consideration, would you accept B's offer over A's? Do you have enough information here to make an informed decision?

Probably not. You know nothing of the variability of the salaries in the two companies; that is, whether they are bunched tightly around their respective centers, suggesting a good starting salary but little chance for increase, or widely varying, suggesting that you might work toward pleasantly high levels. What is needed is a measure of the *variation* or *dispersion* of a group of values.

3.2.1 Range

The simplest and quickest measure of dispersion is the *range*. As its name implies, it is the difference between the maximum and minimum values of the variable.

Range = maximum value – minimum value.

Clearly the range will tend to be larger if the values are more varied and smaller if the values are more uniform, but the range is an unreliable measure of variation since it is entirely determined by only two values out of the entire sample or population.

3.2.2 Variance and standard deviation

We would like a value that indicates variability and takes into account *all* the values in the sample or population.

Suppose that $X_1, X_2, \ldots, X_N$ are a population of values. We begin our search for a better measure of variability by considering the *deviations from the mean,* $(X_i - \mu)$, for i = 1, 2, ..., N. These will be larger if the values (the X_i's) are more varied, so we might consider the mean of these deviations. However,

$$\frac{\sum_1^N (X_i - \mu)}{N} = \frac{1}{N}\sum_1^N X_i - \frac{1}{N}\sum_1^N \mu = \mu - \frac{N\mu}{N} = 0$$

no matter what the values might be, so this quantity is not a measure of anything.

The above result is due to the canceling out of positive and negative deviations from the mean. This can be prevented by using the *absolute values* of the deviations. Thus we can define the *mean deviation* as:

$$\frac{1}{N}\sum_{1}^{N}|X_i - \mu|$$

This is a valid measure of variability, which increases as the values of the population become more dispersed, but it is rarely used. Instead, statisticians have taken advantage of the fact that the square of any real number is nonnegative. That is, $(X_i - \mu)^2$ is a positive number whose magnitude depends on the difference between X_i and μ. From this, we make the following definition.

> The *variance* of a population of values is the mean of the squared deviations from the mean, and is denoted σ^2 (read "sigma squared"). That is,
>
> $$\sigma^2 = \frac{1}{N}\sum_{1}^{N}(X_i - \mu)^2.$$

Using calculus, it can be shown that for any value *d*, the sum of the squared deviations from *d* of the values in a population,

$$\sum_{1}^{N}(X_i - d)^2,$$

achieves its minimum value when $d = \mu$. This lends a geometric as well as an arithmetic meaning to the idea of the variance.

There are, however, two difficulties with the variance. First, when the values themselves are large, then the variance can become huge, and it is difficult to relate the variance to the original values. Second, the units of measurement for the population of values can cause trouble. If the values are measured in "dollars," then the units of the variance are "dollars squared," creatures never seen outside the laboratory.

To solve both of these problems, we define a related measure of variability.

> The *standard deviation* of a population, denoted σ, is the square root of the variance
>
> $$\sigma = \sqrt{\sigma^2}.$$

The standard deviation is in the same units as the original values, and its magnitude can be more easily related to the dispersion of those values.

For example, if 9, 15, 23, 21, 12, and 16 are a population of values,

$$\mu = \frac{9+15+23+21+12+16}{6} = 16,$$

$$\sigma^2 = \frac{(9-16)^2 + (15-16)^2 + (23-16)^2 + (21-16)^2 + (12-16)^2 + (16-16)^2}{6} = 23.33,$$

and

$$\sigma = \sqrt{23.33} = 4.83.$$

In the definitions of variance and standard deviation, we specified that the values form a *population*. When the values are elements of a *sample*, we compute the variance in a slightly different way:

If $X_1, X_2, \ldots, X_n$ are a sample of values, their *sample variance*, denoted s^2, is

$$\frac{1}{n-1}\sum_{1}^{n}(X_i - \mu)^2.$$

We divide by $(n - 1)$, one less than the number of values, to make s^2 a better estimator of the variance of the population from which the sample was taken. This procedure will be explained and justified in Chapter 24, "Correlation Analysis." Also,

> The *sample standard deviation* of a population, denoted σ, is the square root of the variance:
>
> $$s = \sqrt{s^2}.$$

3.3 Chebyshev's Inequality and the Empirical Rule

We know that the variance and standard deviation of a population or sample of values measures the variability of the population or sample, but we can be more precise in our understanding of the relationship between these measures and the locations of the values through a result called Chebyshev's *inequality*:

> If a population of values has mean μ and variance σ^2, then for any number $k \geq 1$, at least $\left(1 - \frac{1}{(k^2)}\right) * 100\%$ of the values must lie within k standard deviations of μ; alternatively, at most $\left(\frac{1}{k^2}\right) * 100\%$ of the values lie above $\mu + k\sigma$ or below
>
> $$\mu - k\sigma.$$

This result also holds for a sample with mean $\overline{X}$ and variance s^2.

Suppose that the mean and standard deviation for the salaries of factory workers at U.S. plants are reported to be \$15,066 and \$2,315. Thus, by Chebyshev's inequality we can say that in at least $\left(1 - \frac{1}{3^2}\right) * 100\% \Rightarrow 89\%$ of the U.S. plants in our sample, the average salary of factory workers is between $15066 - 3 * 2315 = \$8{,}121$ and $15066 + 3 * 2315 = \$22{,}011$.

The proportion between these limits may be greater, but Chebyshev's inequality assures us that it is at least => 89%.

Alternatively, we can say that at most $\frac{1}{3^2} * 100\% = 11.11\%$ of the values lie below \$8,121 or above \$22,011. Again, the proportion may be less, but we are assured that it is no more than 11.11%.

Chebyshev's inequality tells us nothing about the shape of the distribution of values, though if the distribution is unimodal and symmetrical, we can apply the *empirical rule*, a stronger statement than Chebyshev's inequality.

> The *empirical rule*: For a unimodal, symmetrical distribution of values, Figure 3.1 shows the approximate percentage of all the values that lie within specific intervals centered on the mean.

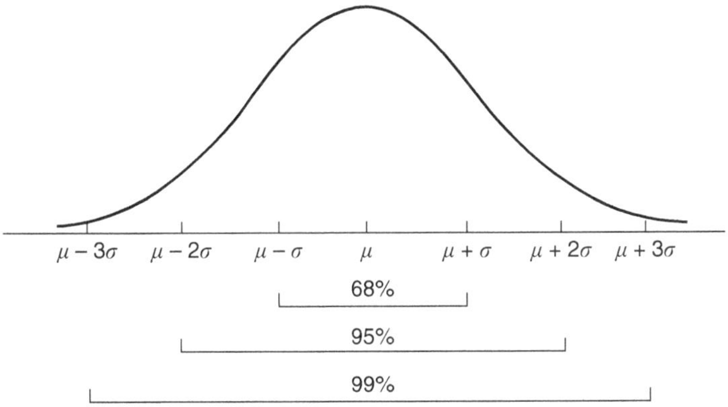

FIGURE 3.1 The empirical rule.

All figures and tables in this chapter appear on the companion DVD. ON DVD

Therefore, in the previous example on factory workers' salaries, we could say that in more than two-thirds of the U.S. plants in our survey, the average salary of factory workers is between \$15,066 – \$2,315 = \$12,751 and \$15,066 + \$2,315 = \$17,381; more than 95% of these values lie between \$15066 – 2*\$2315 = \$10,426 and \$15,066 + 2*\$2,315 = \$19,696; and nearly all the values lie between \$8,121 and \$22,011.

3.4 Comparing Variability

Often, the researcher will want to compare the variability of one population with that of another. If the means of the two populations are different, it would be risky to compare their variances or standard deviations directly. We require a measure of variability that is independent of the magnitudes of the values in the sample or population. This definition provides such a measure:

> The *coefficient of variation* of a group of values is their standard deviation expressed as a percentage of their mean; that is,
>
> $$\frac{s}{\overline{X}} * 100\% \quad \text{or} \quad \frac{\sigma}{\mu} * 100\%.$$

3.5 Measures of Distributional Shape: Skewness

We have mentioned that a distribution of values may be symmetrical or skewed. A measure of this characteristic of distributional shape is defined in this way:

> The *skewness* of a sample of values $X_1, X_2, \ldots, X_n$ is:
>
> $$\frac{\sum_{1}^{n}(X_i - \overline{X})^3}{ns^3}.$$

If the skewness is near 0, the distribution is symmetrical. Positive and negative values of skewness indicate positively and negatively skewed distributions, respectively, though skewness is not considered extreme unless it is less than –1 or greater than +1. See Figure 3.2.

This statistic should be applied only to unimodal distributions composed of interval- or ratio-level data, and is referred to in some texts as *relative skewness*.

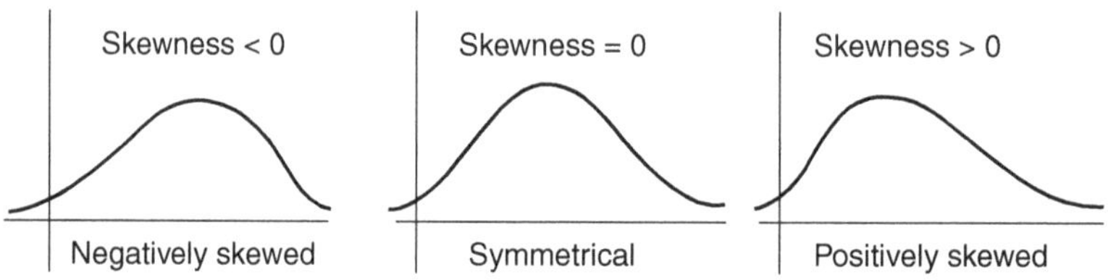

FIGURE 3.2 Skewness.

3.6 Measures of Distributional Shape: Kurtosis

The *kurtosis* of a sample of values $X_1, X_2, \ldots, X_n$ is defined to be

$$\frac{\sum_{1}^{n}(X_i - \overline{X})^4}{n^4} - 3.$$

This statistic is a measure of the relative variation of a symmetrical, unimodal distribution. urtosis measures the "pointedness" or "flatness" of a distribution of values, as shown in Figure 3.3.

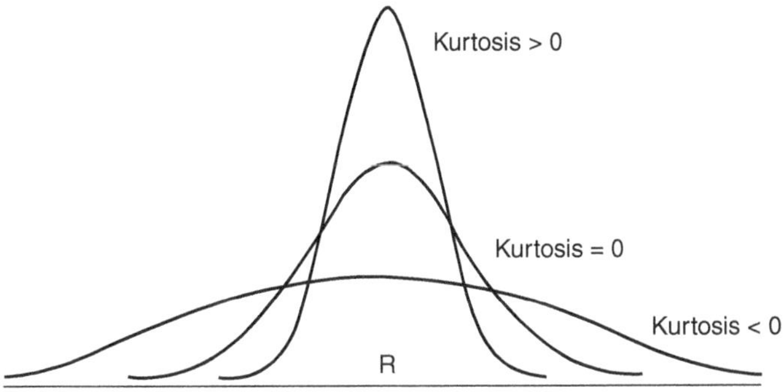

FIGURE 3.3 Kurtosis.

EXCEL

USING EXCEL TO COMPUTE THE MEASURES OF VARIATION

Functions for computing the mean, median, and mode in Excel are VAR(<CELL RANGE>) and STDEV(<CELL RANGE>).

Follow these steps to analyze data using Excel's Descriptive Statistics tool:

1. Select the **Tools** menu.
2. Choose the **Data Analysis** option.
3. Choose **Descriptive Statistics** from the list of analysis tools.
4. When the Descriptive Statistics dialog box appears:
    a. Enter the appropriate input range.
    b. Select **Group by Columns**.
    c. Select **Labels in XX Row**.
    d. Select the appropriate output range.
    e. Enter where results are to be displayed in the **Output Range** box.
    f. Select **Summary Statistics**.
    g. Click **OK**.

SAMPLE EXCEL COMPUTATIONS FOR GENERIC MEASURES OF VARIATION

Given the following Auto file:

TABLE 3.1 Auto file descriptive statistics.

	A	B	C	D	E	F
1	Car	Basic Price	Displacement cc	Horsepower	Curb Weight lb.	MPG
2	Maserti Merak	$31,000	2965	182	3185	14
3	Maserti Bore	$39,927	4931	315	3540	12
4	Maserti Khamsin	$43,587	4931	315	3800	12
5	Mazda GLC	$3,895	1415	65	1995	30
6	Mazda RX-7	$6,395	1146	100	2350	17
7	Mercedes-Benz 240D	$20,775	2746	142	3560	14
8	Mercedes-Benz 300TD	$25,000	2998	77	3780	23
9	Mercedes-Benz 280CE	$22,481	2998	77	3475	23
10	Mercedes-Benz 450SL	$34,760	4520	180	3795	12
11	MG Midget	$5,200	1493	50	1850	23
12	MGB	$6,550	1798	67	2415	16
13	Peugeot 504	$7,922	1971	88	2905	22

generation of the following output is possible:

TABLE 3.2 Sports car statistics.

	A	B	C	D	E	F
1	Car	Basic Price	Displacement cc	Horsepower	Curb Weight lb.	MPG
2	Maserti Merak	$31,000	2965	182	3185	14
3	Maserti Bore	$39,927	4931	315	3540	12
4	Maserti Khamsin	$43,587	4931	315	3800	12
5	Mazda GLC	$3,895	1415	65	1995	30
6	Mazda RX-7	$6,395	1146	100	2350	17
7	Mercedes-Benz 240D	$20,775	2746	142	3560	14
8	Mercedes-Benz 300TD	$25,000	2998	77	3780	23
9	Mercedes-Benz 280CE	$22,481	2998	77	3475	23
10	Mercedes-Benz 450SL	$34,760	4520	180	3795	12
11	MG Midget	$5,200	1493	50	1850	23
12	MGB	$6,550	1798	67	2415	16
13	Peugeot 504	$7,922	1971	88	2905	22
14						
15						
16						
17						
18	*Column1*					
19						
20	Mean	$20,624				
21	Standard Error	4182.988287				
22	Median	$21,628				
23	Mode	#N/A				
24	Standard Deviation	$14,490.30				
25	Sample Variance	$209,968,692.06				
26	Kurtosis	1.50				
27	Skewness	0.260362702				
28	Range	$39,693				
29	Minimum	$3,895				
30	Maximum	$43,587				
31	Sum	$247,492				
32	Count	12				
33						

The latter output is generated by using the formulas shown in Table 3.3:

TABLE 3.3 Formulas to generate auto file descriptive statistics.

	A	B
1	**Car**	**Basic Price**
2	Maserti Merak	31000
3	Maserti Bore	39927
4	Maserti Khamsin	43587
5	Mazda GLC	3895
6	Mazda RX-7	6395
7	Mercedes-Benz 240D	20775
8	Mercedes-Benz 300TD	25000
9	Mercedes-Benz 280CE	22481
10	Mercedes-Benz 450SL	34760
11	MG Midget	5200
12	MGB	6550
13	Peugeot 504	7922
14		
15		
16		
17		
18	*Column1*	
19		
20	Mean	=AVERAGE(B2:B13)
21	Standard Error	4182.98828650649
22	Median	=MEDIAN(B2:B13)
23	Mode	=MODE(B2:B13)
24	Standard Deviation	=STDEV(B2:B13)
25	Sample Variance	=VAR(B2:B13)
26	Kurtosis	=-KURT(B2:B13)
27	Skewness	=SKEW(B2:B13)
28	Range	=MAX(B2:B13)-MIN(B2:B13)+1
29	Minimum	=MIN(B2:B13)
30	Maximum	=MAX(B2:B13)
31	Sum	=SUM(B2:B13)
32	Count	=COUNT(B2:B13)
33		

3.7 Exercises

Computational Exercises

1. Find the range, variance, and standard deviation of this sample of values: 4, 6, 12, 6, 9, 5.

2. Find the range, variance, and standard deviation of the population of values: 26, 35, 29, 27, 33, 35, 29.

3. The formulas given by the definitions of population and sample variance are cumbersome, but we can develop shortcut formulas that involve fewer operations. Show that these statements are true:

a. $\sigma^2 = \frac{1}{N}\sum_{i=1}^{n}(X_i - \mu)^2 = \left[\frac{1}{N}\sum_{i=1}^{n} X_i^2\right] - \mu^2.$

b. $s^2 = \frac{1}{n-1}\sum_{i=1}^{n}\left(X_i - \overline{X}\right)^2 = \frac{\sum_{i=1}^{n}\left(X_i^2\right) - n\overline{X}^2}{n-1}.$

4. Ten workers are paid the following hourly wage: $10.10, $10.75, $15.55, $17.50, $17.70, $10.80, $16.15, $17.74, $15.30, and $6.20. Find the range, variance, and standard deviation for these wages. Are these values most appropriately considered a population or a sample?

Excel Exercises

ON DVD

5. Using Excel, find the range, variance, and standard deviation of the values of the variable X7 in the American Cities database in Appendix B. What are the units of these statistics?

6. A hypothetical city constructed in the 1970s includes a subdivision called New Suburb. The homes have been grouped into continuous neighborhoods, and data about the households are listed in Table 3.4.

TABLE 3.4 New suburb household data.

Household	Household Address	Income	Household	Address	Income
	East Boondocks			Main Street	
1	22	$15,772	29	1	4,676
2	24	14,667	30	2	6,778
3	26	21,539	31	4	5,558
4	28	11,814	32	3	8,905
5	30	7,644	33	4	5,731
6	32	12,888	34	5	7,088
7	34	11,119	35	6	6,775
8	36	10,024	36	7	9,222
	West Boondocks		37	8	9,776
9	1	9,836	38	9	5,783
10	2	8,448	39	108-01	14,453
11	3	10,887	40	108-02	10,113
12	4	13,464	41	108-03	8,985
13	5	11,113	42	108-04	21,119
14	6	12,747	43	108-05	16,668
15	7	10,777	44	108-06	10,569
13	8	9,007	45	108-07	14,554
17	9	12,225	46	108-08	11,800

(Continued)

TABLE 3.4 Continued.

Household	Household Address	Income	Household	Address	Income
18	10	12,345	Tranquil Court		
19	11	10,554	47	1	24,776
20	12	13,098	48	2	26,123
21	13	10,567	49	3	30,001
22	14	8,553	50	4	28,888
23	15	11,363	51	5	28,998
24	16	13,119	52	6	23,556
25	17	12,225	56	7	27,956
26	18	10,887	54	8	24,665
27	19	11,008	55	9	29,545
28	20	11,080	56	10	$26,997
East Court					
57	3	18,988	79	3	42,735
58	4	13,556	80	4	60,600
59	7	17,956	81	5	38,887
60	8	14,665	82	6	71,775
61	11	19,545	83	8	31,119
62	12	16,997	84	10	40,000
63	15	15,305	85	12	56,337
64	16	15,555	West Court		
65	19	16,885	86	1	12,223
66	20	17,554	87	5	10,678
67	23	21,115	88	9	14,556
68	24	20,997	89	11	13,665
69	27	16,666	90	17	15,997
70	28	17,007	91	21	14,555
71	31	15,155	92	25	16,664
72	32	18,444	93	26	22,115
73	35	15,876	94	29	19,997
74	36	16,123	95	30	17,666
75	39	20,001	96	33	16,002
76	40	18,888	97	34	16,155
Hillcrest			98	37	17,444
77	1	7,845	99	28	16,876
78	2	28,553			

7. Using Excel, find:

   a. The average income for families living in East Boondocks.

   b. The variance and standard deviation of the incomes of families living in East Boondocks.

Interpretation Exercises

8. What is the smallest value a population variance can ever have? Under what conditions will this value occur?

9. Use Chebyshev's inequality and your answers to 7a and 7b to express family income ranges that are present within East Boondocks.

10. Redo problem 9 using the empirical rule.

11. Plan a survey of jobs available during the summer near your campus for high school and college students. Develop a spreadsheet for this purpose in Excel. For this data, find descriptive statistics for all the variables. Interpret your results with both the empirical rule and Chebyshev's inequality. Additionally, analyze the skewness and kurtosis of the sample of values.

CHAPTER 4

FREQUENCY DISTRIBUTIONS

OVERVIEW AND LEARNING OBJECTIVES

In This Chapter

4.1 Introduction

4.2 Sample Study

4.3 Presenting Qualitative Data

4.4 Exercises

4.1 Introduction

Tables and graphs commonly are used to summarize both quantitative and qualitative data. When examining newspaper articles, annual reports, and research studies, one will often encounter tabular and graphical summaries. This chapter is an introduction to the preparation and interpretation of one such method of visualization—frequency distributions. Clearly, a picture is often worth a thousand words.

4.2 Sample Study

Winning Database Configurations: An IBM Informix Database Survey by Marty Lurie of the IBM Software Group, available at *www.ibm.com/developerworks/db2/zones/informix/library/techarticle/lurie/0201lurie.html*, employs pie charts to analyze workload for the Informix database system.

1. Discuss all surprises that were encountered when analyzing the workload.
2. Is a pie chart the best chart choice to use for the analysis? Why or why not? Which other chart types could be used effectively?
3. Redo questions 1 and 2 for backup procedures.
4. Redo questions 1 and 2 for servers used.
5. How could questions 1 through 4 be answered when comparing the Informix database system to other database systems?

Notice how large the data is for this opening article, yet the data is summarized in such a way that a clear and accurate picture emerges. Experimenters need to reduce a mass of data as much as possible, but at the same time guard against the possibility of obscuring important features due to the reduction process. We all must use proper analysis and interpretation when utilizing charts and graphs.

4.3 Presenting Qualitative Data

We begin by considering tabular and graphical representations of the following data set, in Excel on the companion DVD:

ON DVD All figures and tables in this chapter appear on the companion DVD.

TABLE 4.1a Products purchased by customers within specific regions.

First Name	Age	Occupation	State/Province	Product
Brenden	14	Child	Washington	Galaxia
Dawn	28	Teacher	Nevada	Galaxia
Doris	22	Student	Washington	Galaxia
Franz	32	Teacher	New York	Voyage to Saturn
Gary	36	Consultant	California	Galaxia
Henrietta	19	Student	Maryland	Galaxia
Hiromi	15	Child	California	Galaxia
Hugh	29	Teacher	Alberta	Voyage to Saturn
Jonathon	13	Child	Louisiana	Knight Time
Manjit	12	Child	Alberta	Knight Time
Marilyn	42	Teacher	Oregon	Voyage to Saturn
Mario	12	Child	California	Galaxia
Michelle	40	Consultant	California	Galaxia
William	13	Child	Oklahoma	Knight Time
Zahra	11	Child	New York	Knight Time

TABLE 4.1b Frequency distribution for customers by occupation.

Occupation	Absolute Frequency	Relative Frequency	Cumulative Frequency
Consultant	2	0.13	0.13
Student	2	0.13	0.27
Teacher	4	0.27	0.53
Child	7	0.47	1.00

Note that the frequency distribution is a table for the specified variable listing its values in order, with the *absolute frequency* of each value being the number of times it occurs. In Table 4.1b, we see that four of the purchasers in the survey are teachers.

The third column of the table forms, with the first, a *relative frequency distribution*, whose entries give the *proportion* of all the values equal to each particular value. In Table 4.1b, we see that 13%, $\left(\frac{2}{2+2+4+7}=\frac{2}{15}\right)*100\%$, of the occupations in the survey are students. The first and last columns of the table form a *cumulative relative frequency distribution* in which the entries give the percentage of all the values less than or equal to each value of the variable. We see that $(2 + 2)/15 * 100\%$ = approximately 27% of the individuals are either consultants or students.

EXCEL

USING EXCEL'S COUNTIF FUNCTION TO CONSTRUCT A FREQUENCY DISTRIBUTION

The COUNTIF function can be used to construct a frequency distribution. First enter data and related functions and formulas for the original spreadsheet. Elsewhere on the spreadsheet, enter the tables shown in Tables 4.2a and 4.2b:

TABLE 4.2a Excel data sheet.

	A	B	C
1	Last Name	First Name	Age
2	Tanaka	Brenden	11
3	Amin	Dawn	12
4	Pirelli	Doris	12
5	Dashwood	Franz	13
6	Morris	Gary	13
7	Singh	Henrietta	14
8	Calhoun	Hiromi	15
9	Grant	Hugh	19
10	Leblanc	Jonathon	22
11	Russell	Manjit	28
12	Leung	Marilyn	29
13	Schuler	Mario	32
14	McAdams	Michelle	36
15	French	William	40
16	Van Home	Zahra	42

TABLE 4.2b Related Excel frequency distribution (formulas displayed).

	K	L	M
1			
2			
3			
4		**relative**	**cumulative**
5	**frequency**	**frequency**	**frequency**
6	=COUNTIF($D$2:$D$16,J6)	=K6/SUM($K$6:$K$9)	=L6
7	=COUNTIF($D$2:$D$16,J7)	=K7/SUM($K$6:$K$9)	=L7+M6
8	=COUNTIF($D$2:$D$16,J8)	=K8/SUM($K$6:$K$9)	=L8+M7
9	=COUNTIF($D$2:$D$16,J9)	=K9/SUM($K$6:$K$9)	=L9+M8
10			
11		=SUM(L6:L9)	
12			
13			
14			
15			
16			

Notice that the COUNTIF statements in cells K6 through K9 specify the cell ranges in the original data using absolute addresses and the search value uses a relative address. Similarly, the denominator of the fractions to calculate the relative frequencies specifies the cell range of the SUM function in absolute addresses.

To display the data in a way that is clear and easily understood, one can construct a histogram. Figure 4.1 is the histogram for the frequency distribution illustrated in Table 4.1b.

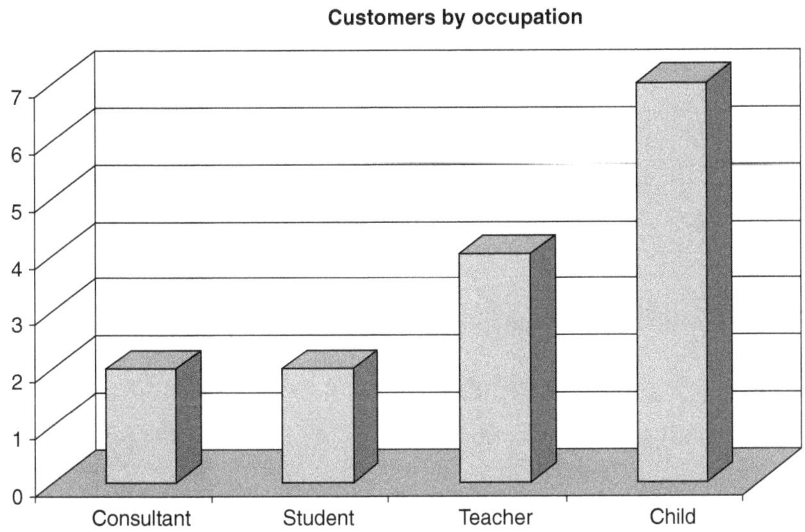

FIGURE 4.1 Frequency distribution histogram.

The primary task is to produce *frequency distributions* of the values of the variables named in the specified field of interest. In Figure 4.1, we can see that seven of the purchasers were children, which was the most common occupation.

EXCEL

USING EXCEL'S CHART WIZARD TO CONSTRUCT HISTOGRAMS

1. Select the input cell range.
2. Click the **Chart Wizard** and choose the appropriate chart type.
3. Choose **Column** in the Chart Type list.
4. Choose **Clustered Column** from the Chart sub-type display list and click **Next**.
5. When the Chart Source Data dialog box appears, click **Next**.
6. When the Chart Options dialog box appears, input the title (and legend, if desired).
7. Click **Finish**.

Sometimes, it is even more informative to sort the data before graphing the results, as shown in Figure 4.2.

SORTING IN EXCEL

EXCEL

1. Select the entire table as the input cell range.
2. Select the **Data** menu.
3. Choose the **Sort** list item.

NOTE

All graphs on the spreadsheet will automatically generate the modifications needed to reflect the newly sorted data.

This sorted frequency distribution enables the reader to easily answer queries such as:

- Which is the most common occupation?
- Which is the least common occupation?
- Are all the occupations fairly common or is there a predominant occupation?

Alternative diagrams are also available for presentation of the frequency distributions, as shown in Figures 4.2 and 4.3.

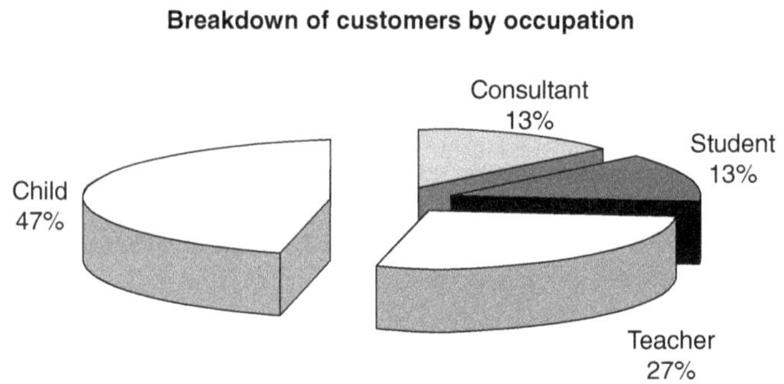

FIGURE 4.2 Pie chart example.

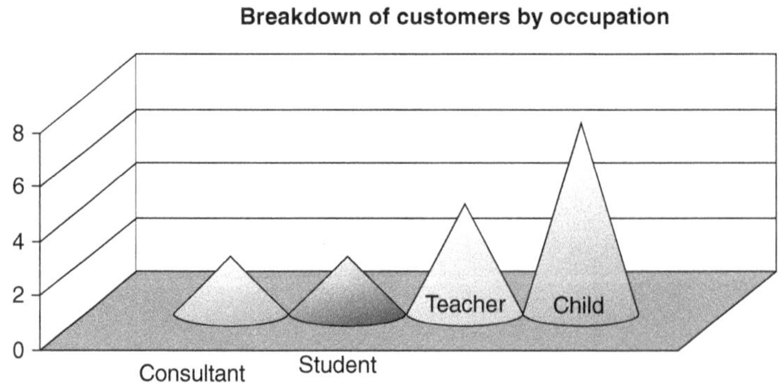

FIGURE 4.3 Cone chart example.

In frequency and cumulative frequency distributions, we should be cautious about giving percentages for particular values when the total number of cases is small. Percentages carry a ring of precision, yet in our example the addition to or removal from a category of only one value would change the corresponding relative frequency by $\frac{1}{15} * 100\% = 6.67\%$. In a study with only 10 cases, shifting one entry would change the corresponding relative frequency by 10%.

4.4 Exercises

Computational Exercises

1. Use the following data on the historical aspects of investments, as shown in Table 4.3:

TABLE 4.3 Investment analysis.

	Annual Returns on Investments in		
Year	**Stocks**	**T-bills**	**Bonds**
1928	43.81%	3.08%	0.84%
1929	-8.30%	3.16%	4.20%
1930	-25.12%	4.55%	4.54%
1931	-43.84%	2.31%	-2.56%
1932	-8.64%	1.07%	8.79%
1933	49.98%	0.96%	1.86%
1934	-1.19%	0.30%	7.96%
1997	31.86%	4.91%	9.94%
1998	28.34%	5.16%	14.92%
1999	20.89%	4.39%	-8.25%
2000	-9.03%	5.37%	16.66%
2001	-11.85%	5.73%	5.57%
Averages	12.05%	3.96%	5.21%

Create histograms for annual returns on stocks and bonds. Then compare the annual returns on stocks and bonds.

Excel Exercises

2. Using the data in the StockPrices file in Appendix B, construct histograms for monthly returns on GE and Intel on the companion DVD.

ON DVD

Interpretation Exercises

3. Defend and illustrate the following statements when constructing histograms:

a. "Inappropriate bucket sizes can result in a loss of information or in too much detail."

b. "Histograms with a broken scale are often used to exaggerate small differences."

CHAPTER 5

GROUPED FREQUENCY DISTRIBUTIONS

OVERVIEW AND LEARNING OBJECTIVES

In This Chapter

5.1 Introduction

When examining newspaper articles, annual reports, and research studies, one will often encounter tabular and graphical summaries derived from quantitative data. This chapter is an introduction to the preparation and interpretation of grouped frequency distributions. As stated before, a picture is often worth a thousand words.

5.2 Summarizing Quantitative Data

We have seen in the previous chapter that simple frequency distributions or histograms provide very little information when the number of individual values is large. In situations like this, we can create more intelligible frequency distributions and graphs by grouping values into *intervals* or *classes*, then counting the number of values in each interval. Such a distribution is called a *grouped frequency distribution*.

To choose the number of intervals into which the values will be divided and the widths of those intervals, two conventions are useful:

- Use not less than 5 nor more than 15 class intervals.
- The interval widths should be 1, 2, 3, 5, or 10, or some multiple of 2, 3, 5, or 10, and should be equal.

(An exception: The highest and lowest intervals may be unbounded.)

Violation of these rules will tend to produce tables and graphs that are difficult to read.

To create a grouped frequency distribution by hand, simply choose the class intervals and count the number of values in each. With a large number of cases, this becomes tedious, so we again turn to the computer. Use of the Excel If statement can be used to accomplish the latter feat.

Given the following Auto file:

ON DVD

TABLE 5.1 Sports car statistics.

	Displacement		Weight		
Car	**Basic Price**	**Cc**	**Horsepower**	**lb**	**MPG**
Maserati Merak	$31,000	2965	182	3185	14
Maserati Bore	$39,927	4931	315	3540	12
Maserati Khamsin	$43,587	4931	315	3800	12
Mazda GLC	$3,895	1415	65	1995	30
Mazda RX-7	$6,395	1146	100	2350	17
Mercedes-Benz 240D	$20,775	2746	142	3560	14
Mercedes-Benz 300TD	$25,000	2998	77	3780	23
Mercedes-Benz 280CE	$22,481	2998	77	3475	23

(Continued)

TABLE 5.1 Continued.

Car	Displacement Basic Price	Cc	Weight Horsepower	lb	MPG
Mercedes-Benz 450SL	$34,760	4520	180	3795	12
MG Midget	$5,200	1493	50	1850	23
MGB	$6,550	1798	67	2415	16
Peugeot 504	$7,922	1971	88	2905	22

EXCEL

USING EXCEL TO GENERATE A GROUPED FREQUENCY

Assuming that the Auto file is stored in cells A1 through J41, labels included, then Table 5.2 can be set up to generate the grouped frequency distribution:

TABLE 5.2 SPSS commands to generate a grouped frequency distribution.

	D	E
18	**MPG**	**Frequency**
19	10-14	=FREQUENCY(F2:F13,{14,19,24,29,34})
20	15-19	=FREQUENCY(F2:F13,{14,19,24,29,34})
21	20-24	=FREQUENCY(F2:F13,{14,19,24,29,34})
22	25-29	=FREQUENCY(F2:F13,{14,19,24,29,34})
23	30-34	=FREQUENCY(F2:F13,{14,19,24,29,34})

The first argument for the Frequency function is the input cell range being searched. The second argument, called a *bin range*, is a list of the upper bounds for the class intervals. Be sure to notice that { }'s are used to set off the bin range.

You can generate the frequency distribution shown in Table 5.3.

TABLE 5.3 Sports car vital statistics.

	A	B	C	D	E	F	G	H	I	J
1	Car	Basic Price	Displacement cc	Horsepower	Curb Weight lb.	MPG				
2	Maserti Merak	$31.000	2965	182	3185	14				
3	Maserti Bore	$39.927	4931	315	3540	12				
4	Maserti Khamsin	$43.587	4931	315	3800	12				
5	Mazda GLC	$3.895	1415	65	1995	30				
6	Mazda RX-7	$6.395	1146	100	2350	17				
7	Mercedes-Benz 240D	$20.775	2746	142	3560	14				
8	Mercedes-Benz 300TD	$25.000	2998	77	3780	23				
9	Mercedes-Benz 280CE	$22.481	2998	77	3475	23				
10	Mercedes-Benz 450SL	$34.760	4520	180	3795	12				
11	MG Midget	$5.200	1493	50	1850	23				
12	MGB	$6.550	1798	67	2415	16				
13	Peugeot 504	$7.922	1971	88	2905	22				
14										
15	MPG	Frequency	Relative	Frequency		low	high	midpoint	fM	dev.
16	10-14	4	4.00	4.00		10	14	12	48	100.00
17	15-19	2	2.00	6.00		15	19	17	34	547.56
18	20-24	4	4.00	10.00		20	24	22	88	244.73
19	25-29	0	0.00	10.00		25	29	27	0	583.58
20	30-34	1	1.00	11.00		30	34	32	32	196.00
21	Total	11	11.00					Total	202	1671.87
22							N	5		
23							X Bar	22.00		
24							Var	40.40		
25	Descriptive Stats for MPG						Std. Dev.	6.36		
26	*Column1*						Std. Error	2.84		
27							Median	18		
28	Mean	18.17					Mode 1	12		
29	Standard Error	1.00					Mode 2	22		
30	Median	16.50								
31	Mode	12.00								
32	Standard Devia	5.87								
33	Sample Varian	34.52								
34	Kurtosis	0.54								
35	Skewness	0.64								
36	Range	19.00								
37	Minimum	12.00								
38	Maximum	30.00								
39	Sum	218.00								
40	Count	12.00								
41										

USING EXCEL TO CREATE A GROUP FREQUENCY DISTRIBUTIONAL ANALYSIS

TABLE 5.4 SPSS commands for generating the sports car vital statistics.

	A	B	C	D	E	F	G	H	I	J
1	Car	Basic Price	Displacement cc	Horsepower	Curb Weight lb.	MPG				
2	Maserti Merak	31000	2965	182	3185	14				
3	Maserti Bore	39927	4931	315	3540	12				
4	Maserti Khamsin	43587	4931	315	3800	12				
5	Mazda GLC	3895	1415	65	1995	30				
6	Mazda RX-7	6395	1146	100	2350	17				
7	Mercedes-Benz 240D	20775	2746	142	3560	14				
8	Mercedes-Benz 300TD	25000	2998	77	3780	23				
9	Mercedes-Benz 280CE	22481	2998	77	3475	23				
10	Mercedes-Benz 450SL	34760	4520	180	3795	12				
11	MG Midget	5200	1493	50	1850	23				
12	MGB	6550	1798	67	2415	16				
13	Peugeot 504	7922	1971	88	2905	22				
14										
15	MPG	Frequency	Relative Frequen	Frequency		low	high	midpoint	fM	dev.
16	10-14	=FREQUENCY(F3 F14,{14,19,24,29,34})	=B16/$B$20	=C16		10	14	=(F16+G16)/2	=B16*H16	=(H16-H23)^2
17	15-19	=FREQUENCY(F3 F14,{14,19,24,29,34})	=B17/$B$20	=D16+C17		15	19	=(F17+G17)/2	=B17*H17	=(H17-H24)^2
18	20-24	=FREQUENCY(F3 F14,{14,19,24,29,34})	=B18/$B$20	=D17+C18		20	24	=(F18+G18)/2	=B18*H18	=(H18-H25)^2
19	25-29	=FREQUENCY(F3 F14,{14,19,24,29,34})	=B19/$B$20	=D18+C19		25	29	=(F19+G19)/2	=B19*H19	=(H19-H26)^2
20	30-34	=FREQUENCY(F3 F14,{14,19,24,29,34})	=B20/$B$20	=D19+C20		30	34	=(F20+G20)/2	=B20*H20	=(H20-H27)^2
21	Total	=SUM(B16 B20)	=SUM(C16 C20)					Total	=SUM(I16 I20)	=SUM(J16 J20)
22							N	=COUNT(H16 H20)		
23							X Bar	=SUM(H16 H20) / COUNT(H16 H20)		
24							Var	= SUM(I16 I20)/H22		
25	Descriptive Stats for MPG						Std. Dev.	=SQRT(H24)		
26		*Column1*					Std. Error	=H25/SQRT(H22)		
27							Median	=F18+((B21/2)-(B16+B17))*(G18-F18)		
28	Mean	=AVERAGE(F2 F13)					Mode 1	12		
29	Standard Error	=B32 / B32					Mode 2	22		
30	Median	=MEDIAN(F2 F13)								
31	Mode	=MODE(F2 F13)								
32	Standard Deviation	=STDEV(F2 F13)								
33	Sample Variance	=VAR(F2 F13)								
34	Kurtosis	=-KURT(F2 F13)								
35	Skewness	=SKEW(F2 F13)								
36	Range	=MAX(F2 F13)-MIN(F2 F13)+1								
37	Minimum	=MIN(F2 F13)								
38	Maximum	=MAX(F2 F13)								
39	Sum	=SUM(F2 F13)								
40	Count	=COUNT(F2 F13)								
41										

To approximate the arithmetic mean of data organized into a grouped frequency distribution, we begin by assuming the observations in each class are represented by the *midpoint* of the class. The mean of a sample of data organized in a grouped frequency distribution can be computed by:

Sample mean for a grouped frequency distribution

$$\overline{X} = \frac{\sum fM}{n}$$

Where:

$\overline{X}$ = sample mean

M = midpoint in each class

f = frequency in each class

fM = frequency in each class times the midpoint of the class

$\sum fM$ = sum of fM products

n = total number of frequencies

The midpoint, M, is simply the sum of the lower and upper bounds for the interval divided by 2.

To calculate the standard deviation of data grouped into a frequency distribution, we need to adjust the common formula for the standard deviation for ungrouped data. We weight each of the squared differences by the number of frequencies in each class. This formula is:

Sample standard definition for a grouped frequency distribution

$$s = \sqrt{\frac{\sum f(M - \overline{X})^2}{n-1}}$$

Where:

s = sample standard deviation

M = midpoint of the class

f = class frequency

n = number of observations in the sample

$\overline{X}$ = sample mean

The median was found using an interpolation procedure that assumes that the values in each class are evenly distributed through the class. To find the median, we proceed through the class in which the median must lie until we reach the hypothetical middle value.

$$\text{median} = L + \frac{j}{f} * c,$$

Where L = lower endpoint of the median class, f = frequency of median class, j = (number of values/2) – (number of values ≤ L), and c = width of median class.

Likewise, the mode is the midpoint of the class with the largest frequency.

Notice that the grouped data can generate statistical values unequal to the same statistical indices computed on the ungrouped data. Therefore, to obtain maximum accuracy, use the original data values when computing statistics.

5.3 Exercises

Computational Exercises

1. A food processing plant wants to test the shelf life of a new product. 50 items were randomly selected and tested under identical conditions. These are the shelf lives in weeks for the items tested:

18.4	28.3	14.4	24.2	14.4	16.3	12.2	18.8	8.8	22.3
13.6	23.9	4.7	32.1	18.5	27.7	19.6	26.8	19.0	11.0
2.7	26.1	19.7	26.7	23.9	20.3	34.4	23.1	14.4	26.0
6.9	17.7	2.2	22.6	9.3	19.1	7.6	18.9	17.2	15.9
24.0	12.5	11.4	8.6	13.2	8.5	17.9	15.6	30.1	24.0

a. Find the mean, median, mode, variance, and standard deviation of this data.

b. Group the data in 7 classes with interval widths of 5. Let the lower bound of the first interval be 2.

I. Find the mean, variance, and standard deviation of the grouped data.

II. Construct tables of the absolute frequency, the relative frequency, and the cumulative frequency distributions of this data.

III. Construct a histogram and a frequency polygon of the absolute frequencies of this data.

Excel Exercises

2. Redo problem 1 using Excel.

Interpretation Exercises

3. For problem 1, which set of summary values is more accurate, those in 1a or 1b? Why?

CHAPTER 6

DATA MINING

OVERVIEW AND LEARNING OBJECTIVES

In This Chapter

6.1 Introduction

Simple arithmetic and easy-to-draw graphs can be helpful when summarizing data sets. This chapter looks at several such techniques for discovering relationships present in data sets, often referred to as *exploratory data analysis*, or *data mining*.

6.2 Single Variate Exploratory Data Analysis

Data mining, often referred to as *exploratory data analysis*, is the nontrivial extraction of implicit, previously unknown, and potentially useful information from data. It is usually associated with a business or other organization's need to identify trends. Data mining involves the process of analyzing data to show patterns or relationships, sorting through large amounts of data, and picking out pieces of relative information or patterns that reoccur in datafiles.

A simple example of data mining is its use in a particular retail sales department. If a store tracks the purchases of a customer and notices that a customer buys a lot of silk shirts, the data mining system will make a correlation between that customer and silk shirts. The sales department will look at that information and may begin direct mail marketing of silk shirts to that customer, or it may alternatively attempt to get the customer to buy a wider range of products. In this case, the data mining system used by the retail store discovered new information about the customer that was previously unknown to the company.

Another widely used (though hypothetical) example is that of a very large North American chain of supermarkets. Through intensive analysis of the transactions and the goods bought over a period of time, analysts find that beer and diapers were often bought together. Though explaining this interrelationship might be difficult, taking advantage of it, on the other hand, should not be hard (e.g., placing the high-profit diapers next to the high-profit beer). This technique is often referred to as *market basket analysis*.

We first will look at data exploration focused on tabular and graphical methods used to summarize the data for one variable at a time. Some of the methods for data exploration include stem-and-leaf plots, box plots, and time plots.

6.2.1 Stem-and-leaf plots

In previous chapters, we illustrated how to summarize quantitative data into a meaningful format. Frequency distributions quickly generate a visual presentation of the shape of a distribution without doing any further calculations. The reader is able to determine where the data is concentrated and whether there are any extreme values. But frequency distributions lose the exact identity of each datum. Additionally, with a frequency distribution the reader is at a loss as to how the values within each class are distributed. While overcoming the latter disadvantages of frequency distributions in a condensed form, one method that is employed to display quantitative information is the *stem-and-leaf display*.

To make a stem-and-leaf display:

1. Separate each observation into a stem, which consists of all but the final (rightmost) digit, and a leaf, which contains the final digit.
2. Write the stems vertically in increasing order from top to bottom and draw a vertical line to the right of the stems.
3. Go through the data, writing each leaf to the right of its stem.
4. Write them again, and rearrange the leaves in increasing order out from each stem.

Consider the following example.

EXAMPLE 6.1

A STEM-AND-LEAF DISPLAY

Given: The number of home runs that Babe Ruth hit in each of his 15 years with the New York Yankees from 1920 to 1934 were: 54, 59, 35, 41, 46, 25, 47, 60, 54, 46, 49, 46, 41, 34, 22.

Solution:

Stem	Leaf
2	25
3	45
4	1166679
5	449
6	0

Sometimes it is convenient to round (or even truncate) the data so that the final digit after rounding is suitable for a leaf. Do this when the data has many digits.

You can also split stems to double the number of stems when all the leaves would otherwise fall on just a few stems. Each stem then appears twice. Leaves 0 to 4 go on the upper stem and leaves 5 to 9 go on the lower stem.

Stem	Leaf
2	2
2	5
3	4
3	5
4	11
4	66679
5	44
5	9
6	0

Further improvement to the organization of stem-and-leaf displays can be achieved by sorting the digits on each line into rank order.

Although the stem-and-leaf display may appear to offer the same information as a histogram, it has two primary advantages:

- The stem-and-leaf display is easier to construct by hand.
- Within a class interval, the display provides information that is unavailable in a histogram because the display shows the actual values.

6.2.2 Quartiles, deciles, and percentiles

When analyzing ordinal data, one must use the median for describing the central tendency of a data set. Therefore, a new measure of dispersion needs to be defined. One method is to determine the location of values that divide a set of observations into equal parts. These measures include *deciles, quartiles,* and *percentiles*.

> The *Pth percentile* is a value that at least P percent of the data set are less than or equal to and at least $(100 - P)$ percent of the data set are greater than or equal to.

$$L_p = (n+1)\frac{P}{100}.$$

Where:

n = total number of values in the data set

P = the desired percentile

L_p = location of the desired percentile

Note that the median has the position at $(n+1)/2$, or at $(n+1)\frac{P}{100}$. The *median* value is the observation in the center. *Quartiles* divide a set of data into four equal parts: the 25th percentile, 50th percentile, and 75th percentile, which are referred to as the 1st, 2nd, and 3rd quartiles. *Deciles* divide the set of data into 10 equal parts.

A common measure of dispersion for ordinal data is the *inter quartile range.*

> The *inter quartile range* for an ordinal data set is the difference between the third and second quartiles.

$$\mathbf{Q} = L_{75} - L_{25}$$

where:

Q = inter quartile range

L_p = location of the desired percentile; $P = 25$ and $P = 75$

6.2.3 Box plots

A *box plot* is an appropriate graphical method to employ for ordinal data. Constructing a box plot requires:

- The minimum value, m
- The first quartile, L_{25}
- The median, L_{50}
- The third quartile, L_{75}
- The maximum value, M

Consider an auto rental service that is trying to audit the miles per gallon of its rentals. For a sample of 20 of the rentals, the owner has discovered that:

$$\mathrm{m} = 13 \text{ mpg}$$

$$L_{25} = 15 \text{ mpg}$$

$$L_{50} = 18 \text{ mpg}$$

$$L_{75} = 22 \text{ mpg}$$

$$\mathrm{M} = 30 \text{ mpg}$$

To make the box plot, follow these steps:

1. Create an appropriate scale along the horizontal axis.
2. Draw a box that starts at L_{25} and ends at L_{75}.
3. Inside the box place a vertical line to represent L_{50}.
4. Extend horizontal lines from the box out to m and M. (These lines are referred to as *whiskers*.)

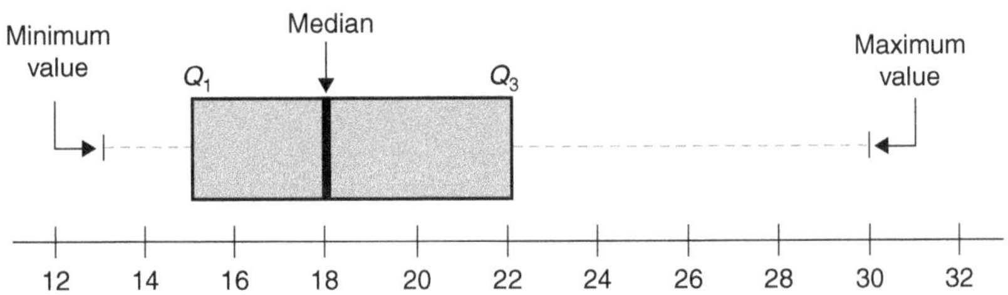

FIGURE 6.1 Box plot.

The box illustrates that the middle achieved mpg is between 15 mpg and 22 mpg. The distance between the end points of the box is the inter quartile range, which is the dispersion for the majority of the mpg for the rentals.

Notice that the box plot also reveals that the distribution of mpg for the rentals is positively skewed, since the dashed lines are of unequal lengths and the median is not in the middle of the axis line.

6.2.4 Time plots

When a variable is measured over time, we can depict changes over time if we plot each observation against the time it was measured. The resulting graph is called a *time plot*.

We will take another looks at Example 6.1 where the number of home runs that Babe Ruth hit in each of his 15 years with the Yankees from 1920 to 1934 were: 54, 59, 35, 41, 46, 25, 47, 60, 54, 46, 49, 46, 41, 34, 22.

The time plot for Babe Ruth's career performance with the Yankees is:

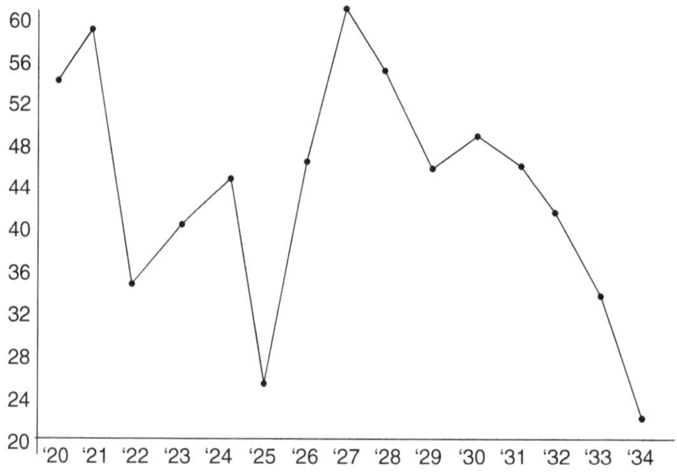

FIGURE 6.2 Time plot.

6.3 Bivariate Exploratory Data Analysis

Sometimes a manager desires to discover the relationship between two variables. Two methods utilized for this purpose are pivot tables and scatter diagrams.

6.3.1 Pivot tables and pivot charts

A pivot table provides the ultimate flexibility in data analysis. It divides the records in a list into categories, then computes summary statistics for those categories. Pivot tables are illustrated in this section in conjunction with the data in Table 6.1 that displays sales information for a hypothetical advertising agency. Each record in the list in Table 6.1 displays the name of the sales representative, the quarter in which the sale was recorded, the type of media, and the amount of the sale.

TABLE 6.1 Sales data for an advertising agency.

	A	B	C	D
1	Sales Rep	Quarter	Media	Avg Sales
2	Alice	1st quarter	TV	$15,000
3	Alice	1st quarter	Radio	$4,000
4	John Doe	1st quarter	Magazine	$2,000
5	John Doe	1st quarter	Radio	$1,000
6	Carol	1st quarter	Radio	$4,000
7	Ted	1st quarter	Radio	$2,000
8	Alice	2nd quarter	Magazine	$2,000
9	Alice	2nd quarter	Radio	$4,000
10	John Doe	2nd quarter	Radio	$4,000
11	Carol	2nd quarter	Magazine	$2,000
12	Carol	2nd quarter	Magazine	$1,000
13	Carol	2nd quarter	TV	$10,000
14	Ted	2nd quarter	TV	$6,000
15	Ted	2nd quarter	TV	$6,000
16	Alice	2nd quarter	Radio	$2,000
17	John Doe	3rd quarter	TV	$10,000
18	Carol	3rd quarter	TV	$8,000
19	Carol	3rd quarter	TV	$18,000
20	Ted	3rd quarter	TV	$20,000
21	Ted	3rd quarter	Magazine	$15,000
22	Ted	3rd quarter	Magazine	$2,000
23	Alice	4th quarter	Radio	$4,000
24	Alice	4th quarter	Radio	$1,000
25	John Doe	4th quarter	Magazine	$10,000
26	John Doe	4th quarter	Magazine	$12,000
27	John Doe	4th quarter	Radio	$1,000
28	John Doe	4th quarter	Magazine	$7,000
29	Carol	4th quarter	TV	$13,000
30	Ted	4th quarter	TV	$13,000
31	Ted	4th quarter	TV	$15,000

The pivot table in Figure 6.3 shows the total sales for each media type-sales representative combination. Look closely and you will see four shaded buttons, each of which corresponds to a different area in the table. The Media and Sales Rep buttons are in the row and column areas, respectively. Thus, each row in the pivot table displays the data for a different media type (magazine, radio, or TV), whereas each column displays data for a different sales representative. The Quarter button in the page area provides a third dimension. The value in the drop-down list box indicates that all of the records in the underlying worksheet are used to compute the totals in the first, second, third, or fourth quarters. You can also click the arrows next to the other buttons to display selected values for the media type or sales representative.

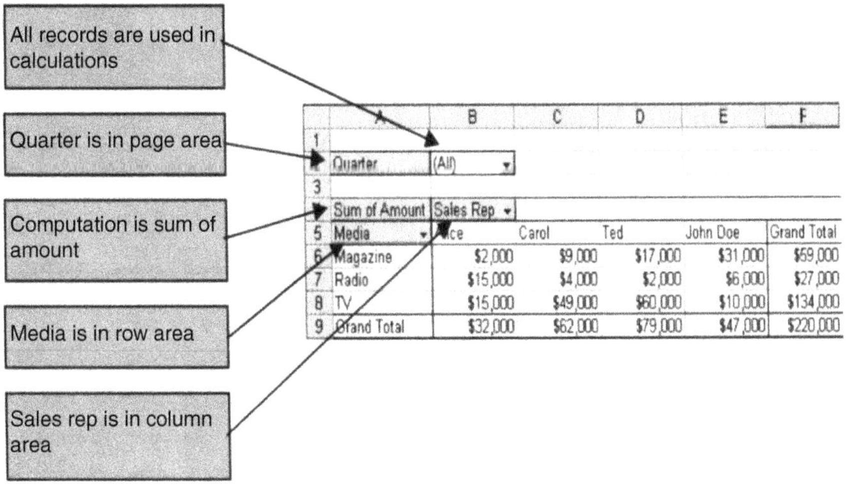

FIGURE 6.3 Microsoft window components of a pivot table.

The best feature about a pivot table is its flexibility; you can change the orientation to provide a different analysis of the associated data. Figure 6.4, for example, displays an alternate version of the pivot table shown in Figure 6.3 in which the fields have been rearranged to show the total for each combination of quarter and sales representative. You can go from one pivot table to another simply by clicking and dragging the buttons corresponding to the field names to different positions.

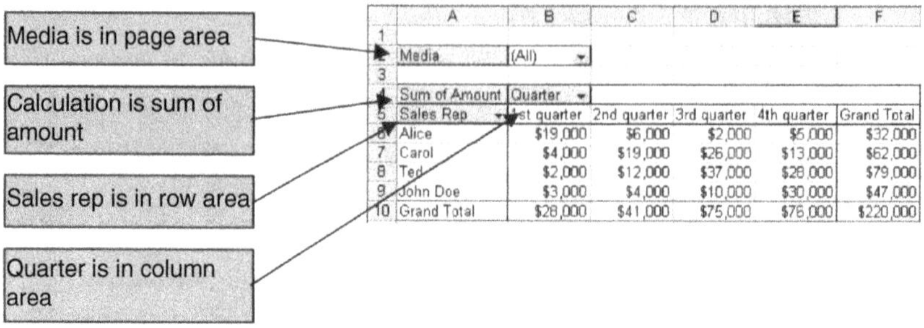

FIGURE 6.4 Window structure of a pivot table.

You can also change the means of computation within the data area. Both of the pivot tables in Figures 6.3 and 6.4 use the Sum function, but you can choose other functions such as Average, Minimum, Maximum, or Count. You can also change the formatting of any element in the table. More importantly, pivot tables are dynamic in that they reflect changes to the underlying worksheet. Thus, you can add, edit, or delete records in the associated list and see the results in the pivot table, provided you execute the Refresh command to update the pivot table.

The Pivot Table Wizard is used to create the initial pivot table in conjunction with an optional pivot chart. The pivot chart in Figure 6.5, for example, corresponds to the pivot table in Figure 6.3 and at first glance, it resembles any other Excel chart. Look closely, however, and you will see shaded buttons similar to those in the pivot table that enable you to make changes to the chart by dragging the buttons to different areas. Reverse the position of the Media and Sales Rep buttons, for example, and you have a completely different chart. Any changes to the chart are reflected in the underlying pivot table and vice versa.

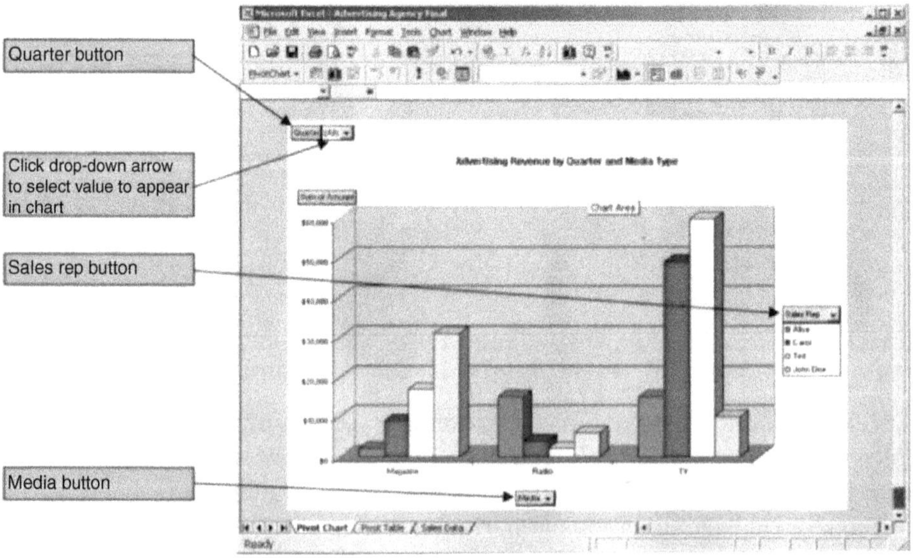

FIGURE 6.5 Graphical Display for a Pivot Table.

Drop-down arrows next to each button on the pivot chart let you display selected values. Click either arrow to display a drop-down list in which you select the values you want to appear in the chart. You could, for example, click the drop-down arrow next to the Sales Rep field and clear the name of any sales rep to remove his data from the chart.

Pivot tables are one of the best-kept secrets in Excel, even though they have been available in the last several releases of Excel. (Pivot charts were introduced in Excel 2000.)

EXCEL

CREATING A PIVOT TABLE AND A PIVOT CHART IN EXCEL

Phase 1: Start the Pivot Table Wizard

Step 1. Open the original spreadsheet.

Step 2. Click anywhere on the spreadsheet.

Pull down the **Data** menu.

Click **PivotTable and PivotChart** report.

Close the Office Assistant, if necessary.

Step 3. Choose the same options as shown in Figure 6.6.

The pivot table will be created from data in an Excel list or a database.

Click **Next**.

In this example, cells A1 through D31 have been selected as the basis of the pivot table.

Step 4. Click **Next**.

The option button to put the pivot table into a new worksheet is already selected!

Click **Finish**.

Two additional sheets have been added to the workbook, but the pivot table and chart area are not yet complete.

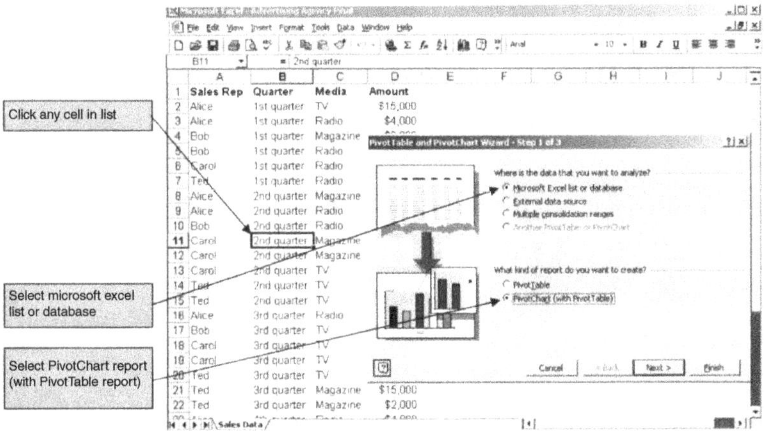

FIGURE 6.6 Wizard generation of a pivot table.

Phase 2: Complete the Pivot Table

Step 1. Click the tab that takes you to the new worksheet (Sheet 1 in this example).

Click the **Media** field button and drag it to the row area.

Click the **Sales Rep** button and drag it to the column area.

Click the **Quarter** field button and drag it to the page area.

Click the **Amount** field button and drag it to the data area.

See Figure 6.7 to check the placement of your elements. You should see the total sales for each sales representative and for each type of media in a pivot table. Rename the worksheets so that they are more descriptive of their contents.

Step 2. Double-click the **Sheet1** tab to select the name of the sheet.

Step 3. Double-click the **Chart 1** worksheet and change its name to **Pivot Chart** in a similar fashion. Save the workbook.

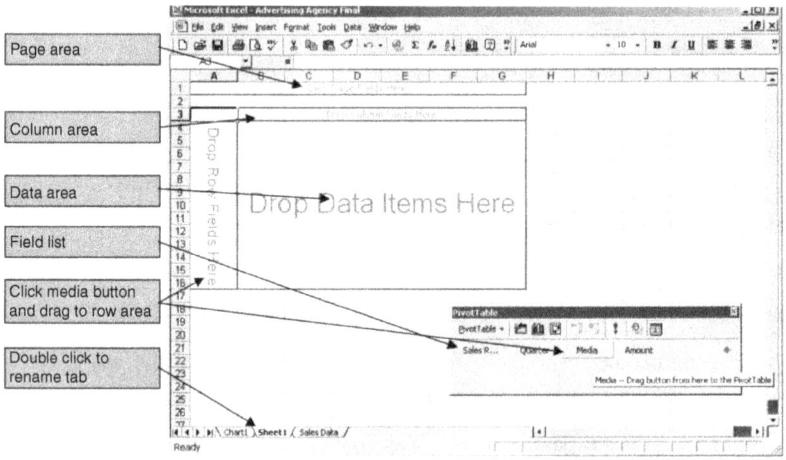

FIGURE 6.7 Work spaces for pivot table construction.

Phase 3: Modify the Sales Data

You will replace Bob's name with your own name in the pivot table.

Step 1. Click the **Sales Data** tab to return to the worksheet.

Pull down the **Edit** menu.

Click the **Replace** command to display the Find and Replace dialog box.

Step 2. Enter **Bob** in the Find What dialog box.

Type your name in the Replace With dialog box.

Click the **Replace All** button.

Click **OK** after the replacements have been made.

Close the Find and Replace dialog box.

Step 3. Click the **Pivot Table** tab to return to the pivot table. The name change is not yet reflected in the pivot table because the table must be manually refreshed whenever the underlying data changes.

Step 4. Click anywhere in the pivot table, then click the **Refresh Data** button on the Pivot Table toolbar to update the pivot table. You should see your name as one of the sales representatives, similar to what is shown in Figure 6.8. (Note that "Bob" was replaced by "John Doe.")

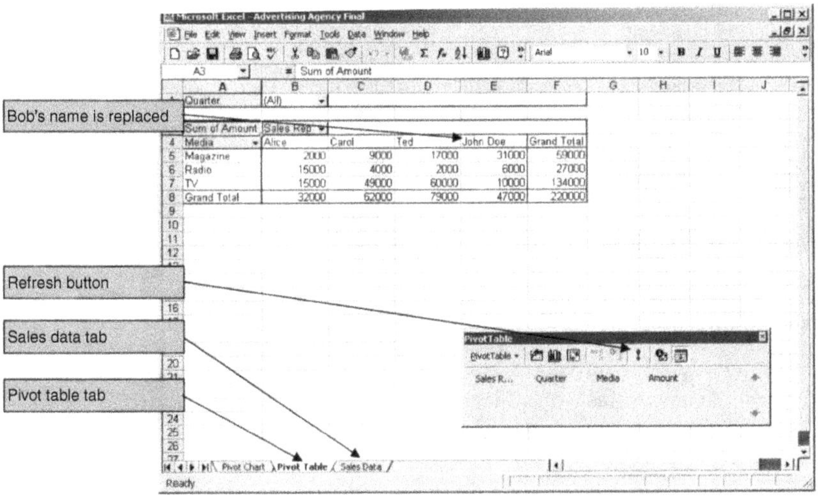

FIGURE 6.8 Pivot table window navigation.

Phase 4: Pivot the Table

You can change the contents of a pivot table simply by dragging fields from one area to another. Click and drag the Quarter field to the row area. The page field is now empty and you can see the breakdown of sales by quarter and media type.

Step 1. Click and drag the **Media** field to the column area, then drag the **Sales Rep** field to the page area. Your pivot table should match the one in Figure 6.9.

Step 2. Click anywhere in the pivot table, then click **Field Settings** button on the Pivot Table toolbar to display the Pivot Table Field dialog box.

Step 3. Click the **Number** button, and choose **Currency format** (with zero decimals). Click **OK** to close the Format Cells dialog box. Click **OK** a second time to close the Pivot Table Field dialog box.

Step 4. Save the workbook.

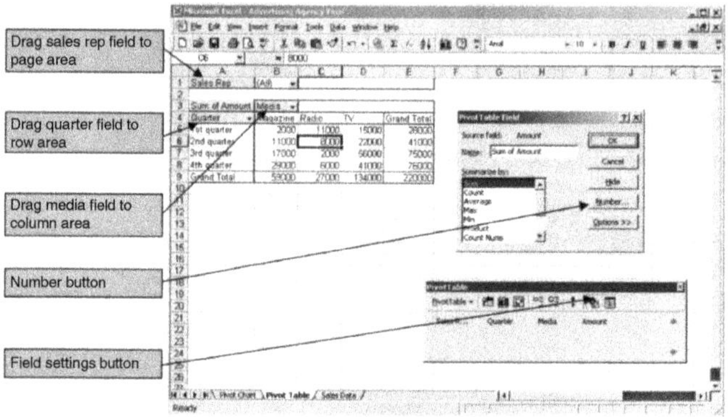

FIGURE 6.9 Pivot table field editing.

Phase 5: Change the Chart Type

Step 1. Click the **Pivot Chart** tab to view the default pivot chart as shown in Figure 6.10.

Step 2. Pull down the **Chart** menu and click the **Chart Type** command to display the Chart Type dialog box.

Select **Clustered column with a 3-D visual effect**.

Step 3. Check the **Default formatting** box. This is a very important option, because without it, the chart is rotated in an awkward fashion. Click **OK**.

Step 4. Save the workbook.

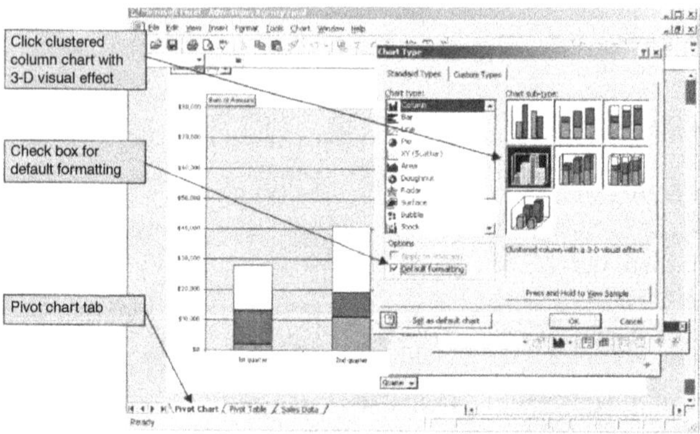

FIGURE 6.10 Control of visual effects for pivot table editing.

Phase 6: Complete the Chart

Step 1. Pull down the **Chart** menu.

Click **Chart Options**.

Click the **Titles** tab and enter the title.

Click **OK**.

Step 2. Click the **Sales Data** tab to select the worksheet.

Press and hold the **Ctrl** key as you click the **Pivot Table** tab to select the worksheet containing the pivot table.

Step 3. Pull down the **File** menu.

Click the **Page Setup** command and click the **Sheet** tab.

Check the boxes to print Gridlines and Row/Col headings.

Click the **Margins** tab and check the box to center the worksheet horizontally.

Click **OK**.

Save the workbook.

Step 4. Pull down the **File** menu and click the **Print** command to display the Print dialog box.

Click the option button to print the entire workbook.

Click **OK**.

Your printed pivot chart should look like Figure 6.11.

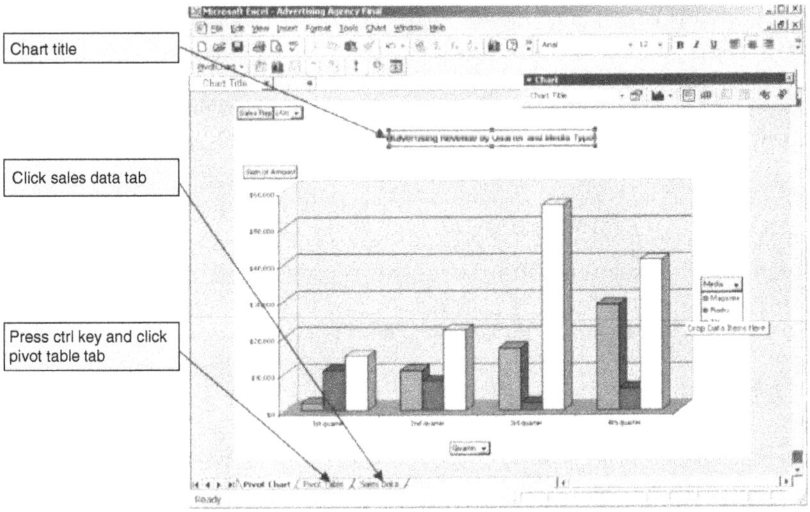

FIGURE 6.11 Generating graphics for a pivot table.

6.3.2 Scatter diagrams

Scatter diagrams are used to study possible relationships between two variables. Although these diagrams do not prove a cause and effect relationship between the variables, they indicate the possible existence and strength of a relationship. Relationships between variables exist when one variable depends on the other, and changing one variable affects the other variable.

A scatter diagram is composed of a horizontal axis and a vertical axis containing the measured values of one of the variables on each of them. The purpose of the scatter diagram is to display what happens to one variable when the other one changes. The diagram is used to test a theory that the two variables are related, with the slope of the diagram indicating the type of relationship between variables. An analysis method used to decide whether there is a statistically significant relationship between two variables is called *correlation*. Correlation may be positive, negative, or display no relationship. A positive correlation is indicated by an ellipse of points that slopes upward, demonstrating that an increase in the key variable also increases the effect variable. A negative correlation is indicated by an ellipse of points that slopes downward, demonstrating that an increase in the key variable results in a decrease in the effect variable.

Scatter diagrams can be used for several purposes:

- As a method of validating hunches about a cause-and-effect relationship between types of variables

Examples: Do students who spend more time watching TV have higher or lower average GPAs? Is there a relationship between the production speed of an operator and the number of defective parts made? Is there a relationship between typing speed and number of errors made?

- To display the direction of the relationship (positive, negative, etc.)

Examples: Will test scores increase or decrease if students spend more time in study hall? Will increasing assembly line speed increase or decrease the number of defective parts made? Do faster typists make more or fewer typing errors?

- To display the strength of the relationship

Examples: How strong is the relationship between measured IQ and grades earned in chemistry? How strong is the relationship between assembly line speed and the number of defective parts produced? How strong is the relationship between typing faster and the number of errors made?

Scatter diagrams can be used in a variety of situations not only in business but also in education (e.g., finding a possible relation between time spent watching TV and grades in school), sociology (finding a possible relation between education level and income), and even chemistry (finding a possible relation between temperature and strength of chemical reaction or physics.)

In business, scatter graphs can be useful in almost every type of service or manufacturing company. In manufacturing, scatter diagrams can be used to analyze the performance of equipment and workers, the relation of temperature and frequency of equipment breakdowns, or the relation of equipment age and breakdown frequency. Later, this data is valid for deciding, for example, whether there is any relation between temperature and performance of equipment. Additionally, analyzing the experience of workers and number of nonconformities occurring can help in deciding whether it is important for the company to spend extra money to keep experienced workers.

Sales companies can try to find a relationship between expenditures for advertising and number of clients or an increase in payment to sales personnel and number of satisfied customers. Especially when dealing with labor and customers, analysts have to take into account that the relationship of the variables is strongly influenced by other factors. For example, the number of clients cannot increase solely due to the amount of money the company spends on advertising, but the number of clients can decrease because of advertisements by competitors, improved quality of products sold, constant expansion to new markets, and other influencing factors.

In finance, scatter diagrams can be used to measure the relationship between various statistical data, such as growth of GDP in various countries with respect to growth of share indexes.

Although scatter diagrams are quite commonly used, it should always be remembered that even if they show a positive correlation of variables, this just indicates a possible relationship that still has to be examined and proved. A positive correlation cannot only be a coincidence (if we are taking a small number of variables), but it is also possible that variables are supposed to influence some other variables.

EXCEL

GENERATION OF SCATTER DIAGRAMS IN EXCEL

Assume that we have the data set shown in Table 6.2 stored in cells A1 through C21 on a spreadsheet. We want to study the relationship between income and home price. In fact, it would be great if income could be used to predict home price.

TABLE 6.2 Home Price - Income Data.

City	Income	Home Price
Bismarck, ND	62.8	92.8
Columbia, SC	66.8	116.7
Savannah, GA	67.8	108.1
Birmingham, AL	71.2	130.9
Toledo, OH	71.2	101.1
Akron, OH	74.1	114.9
Lancaster, PA	75.2	125.9
Fort Lauderdale, FL	75.8	145.3

(Continued)

TABLE 6.2 Continued.

City	Income	Home Price
Nashville, TN	77.3	125.9
Madison, WI	78.8	145.2
Cleveland, OH	79.2	135.8
Atlanta, GA	82.4	126.9
Denver, CO	82.6	161.9
Detroit, MI	85.3	145
Philadelphia, PA	87	151.5
Hartford, CT	89.1	162.1
Washington, DC	97.4	191.9
Naples, FL	100	173.6
Trenton, NJ	106.4	168.1
Danbury, CT	132.3	234.1

Step 1. Select cells B2:C21

Step 2. Click the **Chart Wizard** button.

Choose **XY (Scatter)** in the **Chart Type** list.

Choose **Scatter** from the **Chart** sub-type display.

Click **Next**.

Step 3. Click **Next**.

Step 4. Select the **Titles** tab and enter the title.

Click **Next**.

Step 5. Click **Finish**.

You should have a scatter diagram similar to Figure 6.12.

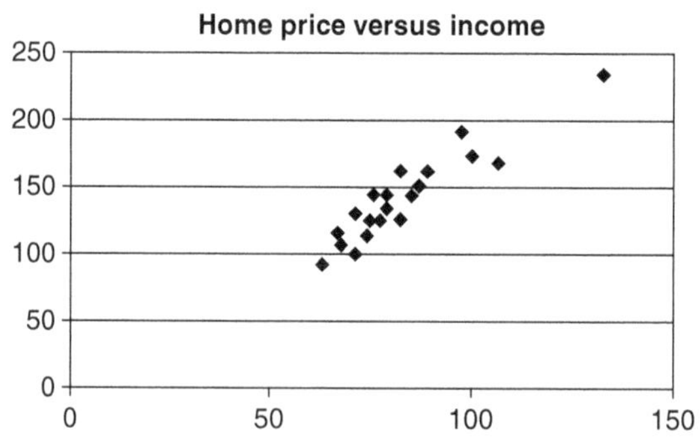

FIGURE 6.12 Scattergram for home price by income.

The following steps illustrate how to add a trend line.

Step 1. Position the mouse pointer over any data point in the original scatter diagram and Right-click to display a list of options.

Step 2. Choose **Add Trend line**.

Step 3. When the **Add Trend line** dialog box appears:

Select the **Type** tab.

Choose **Linear** from the **Trend/Regression type** display.

Click **OK**.

Your diagram should look like Figure 6.13.

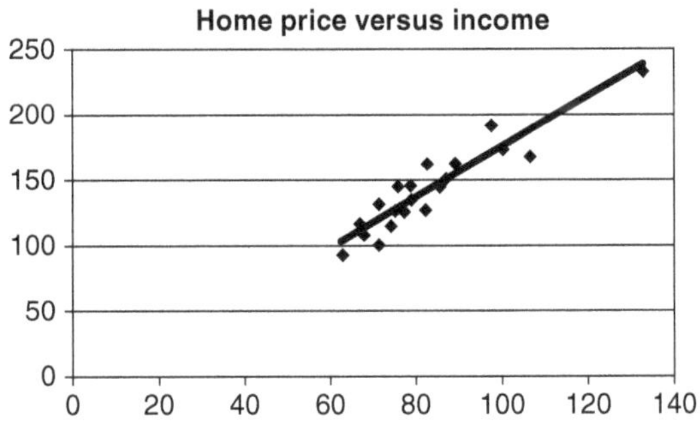

FIGURE 6.13 Regression line for home price by income.

We observe that there is a linear relationship between income and home price. As the income, the horizontal axis, increases, the home price, the vertical axis, also increases. The fact that the spread of the actual observations is tightly packed around the trend line indicates that the trend line is a valid representation of the linear relationship.

6.4 Exercises

Excel Exercises

1. You own a fast food restaurant and have done some market research in an attempt to better understand your customers. For a random sample of customers, you are given the income, gender, and number of days per week that the residents go out for fast food, found in the Fast Food datafile in Appendix B. ON DVD

With the aid of a pivot table, use the information to determine how gender and income influence the frequency with which a person goes out to eat fast food.

2. For the years 1985–1992, you are given monthly interest rates on bonds that pay money one year after the day they are bought. It is often suggested that the interest rates are more volatile—tend to change more—when interest rates are high. Does the data in the Interest Rate Volatility datafile, found in Appendix B, support this statement? ON DVD

Pivot tables can display standard deviations.

Interpretation Exercises

3. The Excel file Makeup Inform contains information about the sales of makeup products, as shown in the partial listing in Table 6.3:

TABLE 6.3 Makeup product data.

Trans Number	Name	Date	Product	Units	Dollars	Location
1	Betsy	4/1/2004	lip gloss	45	$137.20	South
2	Hallagan	3/10/2004	foundation	50	$152.01	Midwest
3	Ashley	2/25/2005	lipstick	9	$28.72	Midwest
4	Hal	5/22/2006	lip gloss	55	$167.08	West
5	Caret	6/17/2004	lip gloss	43	$130.60	Midwest
6	Colleen	11/27/2005	eye liner	58	$175.99	Midwest
7	Cristina	3/21/2004	eye liner	8	$25.80	Midwest

Information is maintained on:

- Name of salesperson
- Date of sale
- Product sold
- Units sold
- Transaction revenue
- Location

Tables 6.4 through 6.7 are pivot tables created from this spreadsheet. For each pivot table, describe and interpret the information captured.

TABLE 6.4 Pivot Table 1.

Count of Trans Number	
Name	**Total**
Ashley	197
Betsy	217
Cici	230
Colleen	206
Cristina	207
Emilee	203
Hal	200
Jen	217
Zaret	214
Grand Total	1891

TABLE 6.5 Pivot Table 2.

	Product					Summary
Name	**Eye Liner**	**Foundation**	**Lip Gloss**	**Lipstick**	**Mascara**	**Grand Total**
Ashley	5844.948744	4186.058628	6053.684565	3245.442978	6617.100349	25947.23526
Betsy	6046.534282	8043.486462	5675.650045	3968.605496	4827.253996	28561.53028
Cici	5982.823291	6198.248632	5199.949201	3148.84065	7060.711397	27590.57317
Colleen	3389.625314	6834.767608	5573.323725	2346.413777	6746.525368	24890.65579
Cristina	5397.273636	5290.989935	5297.97981	2401.668343	5461.646997	23849.55872
Emilee	7587.38898	5313.787561	5270.250313	2189.137568	4719.299731	25079.86415
Hal	6964.621074	6985.734333	5603.119378	3177.871325	5703.34667	28434.69278
Jen	7010.440514	5628.643036	5461.61479	3953.300132	6887.17495	28941.17842
Zaret	8166.749063	6451.650057	5670.329329	2448.707163	3879.949944	26617.38556
Grand Total	56390.4049	54933.37125	49805.90116	26879.98743	51903.0094	239912.6741
Jen's						
Lipstick	3953.300132					
Sales						

TABLE 6.6 Pivot Table 3.

Sum of Dollars		
Name	**Location**	**Total**
Ashley	East	7772.704761
	Midwest	4985.896509
	South	7398.565792
	West	5790.068203
Ashley Total		25947.23526
Betsy	East	8767.431725
	Midwest	4878.085848
	South	7732.05698
	West	7183.955727
Betsy Total		28561.53028
Cici	East	5956.320446
	Midwest	8129.619289
	South	7174.448975
	West	6330.184462
Cici Total		27590.57317
Colleen	East	5713.069445
	Midwest	6586.142169
	South	7785.632708
	West	4805.811471
Colleen Total		24890.65579
Cristina	East	4126.268644
	Midwest	5870.034488
	South	5964.158473
	West	7889.097115
Cristina Total		23849.55872
Emilee	East	6295.472056
	Midwest	5642.196163
	South	6050.594346
	West	7091.601589
Emilee Total		25079.86415
Hal	East	4965.615813
	Midwest	7378.321391
	South	8210.814251
	West	7879.941325
Hal Total		28434.69278

TABLE 6.6 Continued.

Sum of Dollars		
Name	**Location**	**Total**
Jen	East	6949.209483
	Midwest	6381.320681
	South	7116.016774
	West	8494.631484
Jen Total		28941.17842
Zaret	East	6825.995148
	Midwest	4953.797616
	South	6864.065862
	West	7973.52693
Zaret Total		26617.38556
Grand Total		239912.6741

TABLE 6.7 Pivot Table 4.

Sum of Dollars		
Years	**Name**	**Total**
2004	Ashley	9495.068134
	Betsy	9420.270725
	Cici	8965.262077
	Colleen	9361.385804
	Cristina	9132.086152
	Emilee	7805.647572
	Hallagan	10676.87903
	Jen	9049.299912
	Zaret	9078.507356
2005	Ashley	9547.543701
	Betsy	9788.728323
	Cici	9024.965709
	Colleen	7996.802973
	Cristina	7976.353025
	Emilee	9326.418545
	Hallagan	9102.484269
	Jen	8920.272064
	Zaret	8639.703793
2006	Ashley	6904.623429
	Betsy	9352.531233
	Cici	9600.345387
	Colleen	7532.467015
	Cristina	6741.119544
	Emilee	7947.798037
	Hallagan	8655.329478
	Jen	10971.60645
	Zaret	8899.174407
Grand Total		239912.6741

UNIT 2

ELEMENTARY PROBABILITY

Peter Drucker, in a article in **The Wall Street Journal** (12/1/92), http://www.theacagroup.com/performancemeasureforcustomers.htm, stated that information technology has provided the means of collecting vast amounts of data, but, in order for data to be converted into information it must be organized for the task, directed toward specific performance and applied to a decision. Many managers don't know what information they need to do their job and how to get that information. Others don't understand how the availability of that information has changed their management task. Finally, few managers know what information they owe to the organization to insure its success.

- What are the characteristics of performance measurement systems?
- Explain the difference between hard versus soft measures.
- How can probability measures be used for performance measurement systems?
- Explain the difference between probability and reliability measures.

This lead file illustrates that probability models are used to study the variation in observed data so that inferences about the underlying process can be developed. The mission of this unit is to understand probabilities and how they are determined.

Through the knowledge of the parameters associated with probability distributions one can construct probability models for the various statistics computed from sample data. These probability models are referred to as sampling distributions. In fact theory, in particular the Central Limit Theorem, uses these sampling distributions to develop procedures for statistical inference.

CHAPTER 7

RANDOM EXPERIMENTS, COUNTING TECHNIQUES, AND PROBABILITY

OVERVIEW AND LEARNING OBJECTIVES

In This Chapter

7.1 Introduction

Having defined several descriptive statistics, our next objective is the development of the tools of statistical inference. To do this, we must first study probability, and this topic, useful and interesting in its own right, is pursued in the next five chapters.

Many impressions you may have about what probability is and how it behaves are likely to be true. If we flip a coin, it seems reasonable to say that the probability that it will land with the "heads" side up is equal to $\frac{1}{2}$. This simple example illustrates the major goal of this chapter: to develop a set of processes and rules for assigning to an event that might occur a number (its *probability*) that is proportional to its likelihood.

7.2 Random Experiments

We begin by defining the situation to which we apply our methods:

> A *random experiment* is any well-defined situation whose outcome is uncertain and in which we make an observation or take a measurement.

The word "random" indicates the element of chance; the experiment may result in any one of several possible outcomes, and we do not know which one will occur. Flipping a coin is a random experiment; it will result in either heads or tails, but we cannot know which. Other examples of random experiments are these:

1. Roll two dice. How many dots are on the two upper faces?
2. Deal a hand of five cards. Which particular group of five cards is dealt?
3. Choose one city at random from all the cities in Texas. Which city was chosen?
4. Observe the closing Dow Jones average.
5. Count the number of email messages that are processed daily on a specific client machine.

7.3 Sample Spaces and Events

In each of these experiments, the precise result is unknown, but we can list all the possibilities. Each of these possibilities is called an *outcome*, or *elementary event*, and they cannot be subdivided. That is, outcomes are the smallest units of what might happen. For a given experiment, the set of all outcomes is called the *sample space*, *event space*, or *outcome space*, and it is indicated by the letter S.

If we flip one coin, the two outcomes are "heads" and "tails." We can represent this as H and T, and we write $S = \{H, T\}$. If we roll two dice, the total number of dots showing on the upper faces might be any integer from 2 to 12; S = {2, 3, 4, ..., 12}. In Example 2 above, the sample space is the set of all 5-card hands that can be created from the usual deck of 52 cards. In Example 3, S is the set of all cities in the state of Texas. How would you describe the sample spaces for Examples 4 and 5?

Note that the performance of a random experiment always results in one and only one outcome; when a coin is flipped, it must fall either heads or tails. Also, no two outcomes can ever occur simultaneously.

There are many experiments in which we are concerned with the occurrence or nonoccurrence of a set of outcomes, rather than just one result. For example, we choose one city from the state of Texas at random. Is it in Central Texas? Here we are asking if the experiment resulted in one of a set of outcomes, and from this idea we extract the following definition:

> An *event* is any subset of the sample space S.

Since the set of cities in Central Texas is a subset of S, the set of all cities in Texas, it is an event.

Further:

> If A is an event in S, written $A \subseteq S$, we say that A *occurs* if the experiment results in an outcome that is in A.

If A is the event that we pick a city in Central Texas, then when we choose one city at random, for instance, Austin, we say that A occurs. If El Paso is chosen, then A does not occur.

Two events deserve special attention. First, the sample space S is a subset of itself, so S is an event. Since S contains all the outcomes associated with the experiment, the experiment always results in an outcome that is in S, S *always* occurs. The *null set*, or *empty set* (ϕ), is a subset of every other set, so it is a subset of S, and, therefore, an event, the *null event*. Since it contains no outcomes, it *never* occurs. Second, each outcome by itself is a subset of S, so each outcome is also an event.

7.4 What Probability Means

An experiment is a situation whose result is uncertain. We can list, in the sample space, all the possible outcomes of the experiment, but we do not know which will occur. We want to assign numbers (probabilities) to events in the sample space indicating how likely they are to occur. What will these numbers mean?

Consider the simple experiment of rolling a single die. The event A is the appearance of six dots on its upper face. We used a computer to simulate the repetitions and counted the number of times A occurred, or the *frequency* of A., was counted, as illustrated The results are shown in Table 7.1.

TABLE 7.1 Relative frequency distribution.

Number of Trials	Frequency of A	Relative Frequency of A
50	7	0.14
100	19	0.19
500	77	0.154
1000	176	0.176
5000	871	0.1742
10000	1692	0.1692

ON DVD All figures and tables in this chapter appear on the companion DVD.

Also calculated was the *relative frequency* of A, the proportion of all the trials in which A occurred. It is clear that as the number of trials increased, the relative frequency of A appeared about 17 times out of every 100 repetitions of the experiment, and that the *probability* of A might then be near 0.17.

> As the number of repetitions of an experiment increases, the relative frequency of an event A will appear to stabilize around some value p. We call p the *probability* of A, written P(A).

In our example, we would conclude that P(A) = 0.17. Later, we will see that $P(A) = \frac{1}{6} = 0.1667$.

7.5 Equally Likely Outcomes

Now that we have a definition of the probability of an event, we must develop ways to assign probabilities to events. The simplest case is that in which all the outcomes in the sample space of the experiment are assumed to be *equally likely*. In situations of this kind, we make the following definition:

> Let the sample space S contain k equally likely outcomes. Then each outcome in S is assigned probability 1/k.

An example is the rolling of a single die. The die may come to rest with any face on top, and it seems reasonable to believe that the six faces are equally likely. Therefore, the probability that any one face will come to rest on top is $\frac{1}{6}$. In particular, the probability of rolling 6 is $\frac{1}{6} = 0.1667$, verifying the experiment discussed in the previous section.

We are also interested in the probabilities of *events* in the sample spaces in which the outcomes are equally likely. The definition of the probability of such an event is a reasonable extension of the previous definition:

> Let A be an event in S, a sample space in which the outcomes are equally likely. Then
>
> $$P(A) = \frac{number-of-outcomes-in-A}{number-of-outcomes-in-S}.$$

If we roll a single die, the probability of rolling 5 or 6 is $\frac{2}{6} = \frac{1}{3}$.

7.6 Putting Events Together: Union, Intersection, and Complement

We know that given several sets, we can apply the set operations of union, intersection, and complement to them to produce other sets. Events are subsets of the sample set S, so we can apply set operations to sets. What kinds of things emerge?

The *union* of two sets is that set containing all the elements that are in either or both of the original sets. The union of two sets A and B contains all the outcomes that are in A, in B, or in both. Thus:

> If A and B are events in S, the event $A \cup B$, read "A union B" or "A or B," occurs when A occurs or B occurs or both occur.

Similarly, the *intersection* of two sets contains the elements that are simultaneously in both. The intersection of two events contains the outcomes in both events:

> If A and B are events in S, the event $A \cap B$, read "A intersect B" or "A and B," occurs when both A and B occur.

Finally, the *complement* of a set is the set of all elements not in the set. The complement of an event is composed of all the outcomes in the sample space that are not in the event:

> If A is an event in S, the event A', read "A complement" or "not A," occurs when A does not occur.

Consider the experiment of drawing one card at random from an ordinary deck of 52 cards, with A the event that we draw a club and B the event that we draw a face card (jack, queen, or king).

Then $A \cup B$ is the event that we draw either a club or a face card; $A \cap B$ occurs if the card drawn is both a club and a face card; and A' is the event that the card is a diamond, heart, or spade, but not a club.

Consider A and the event C that a club is drawn. It is impossible for A and C to occur simultaneously, since they have no outcomes in common. In such a case, we say that A and C are *mutually exclusive* and write $A \cap C = \phi$.

Two useful results from set theory are DeMorgan's laws, which relate the operations of union, intersection, and complement:

$$(A \cap B)' = A' \cup B'$$
$$(A \cup B)' = A' \cap B'.$$

For events A and B in a sample space S, consider the interpretations of these expressions. For example, $(A \cup B)'$ is the event that neither A nor B occurs.

7.7 Venn Diagrams

Sets and their relationships can be depicted schematically with Venn diagrams, in which interiors of circles represent the elements of sets. Events and their interactions are often illustrated with Venn diagrams, with circles representing events and the enclosing rectangle corresponding to the sample space.

Figure 7.1 illustrates $A \cap B$, $A \cup B$, and A'.

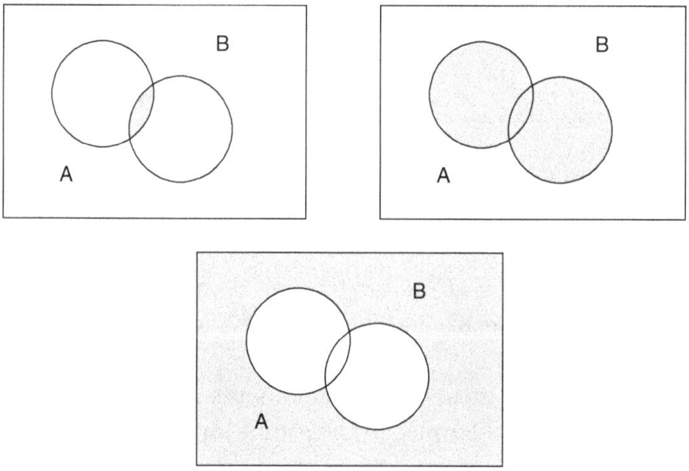

FIGURE 7.1 Venn diagrams for intersection, union, and complement.

7.8 The Axioms of Probability

We have discussed the idea of probability, and have defined a method for assigning probabilities to events in experiments where the outcomes in the sample space are equally likely.

We have also outlined the results of applying set theoretical operations to events. It is in investigating the probabilities of events made with the set operations that we begin to build the mathematical structure of probability theory. That is, A', $A \cup B$, and $A \cap B$ are events; what are their probabilities?

The construction of a mathematical system begins with a statement of *axioms*, assumptions from which the structure will be derived. In probability, we state three axioms.

> Let A be an event in the sample space S. Then:
>
> Axiom 1: $P(A) \geq 0$.
>
> Axiom 2: $P(S) = 1$.
>
> Axiom 3: If B is another event in S, with $A \cap B = \phi$
>
> Then $P(A \cup B) = P(A) + P(B)$.

These are reasonable assumptions. It would not make sense to assign an event a negative probability, and since S contains all the outcomes for the experiment and therefore must occur, it has probability 1. To see the validity of Axiom 3, consider the act of rolling a die and A' is the event where the top face is a two and B the event where the top face is a five. $P(A') = \frac{1}{6}$, $P(B) = \frac{1}{6}$, and $A' \cap B = \phi$. Then $P(A' \cup B) = \frac{2}{6} = 0.33 = \frac{1}{3} + \frac{1}{3} = P(A) + P(B)$.

7.8.1 Laws, or theorems, derived from the axioms

From the axioms of probability, we can develop theorems that tell us more about the probabilities of events in S.

THEOREM 7.1

For any event $A \subseteq S$, $P(A) = 1 - P(A')$.

Proof:

$S = A \cup A'$, and $A \cap A' = \phi$.

By Axiom 3,

$$P(S) = P(A) + P(A').$$

But $P(S) = 1$, so $1 = P(A) + P(A')$.

Then $P(A) = 1 - P(A')$.

The result of Theorem 7.1 will be useful in finding the probabilities of complicated events where complements can be more easily examined.

THEOREM 7.2

For any event $A \subset S$, $P(A) \leq 1$.
Proof:

From Theorem 7.1,

$P(A) = 1 - P(A')$.

By Axiom 1, $P(A') \geq 0$, so $P(A) \leq 1$.

This result verifies a statement that is intuitively reasonable, that the probability of an event cannot exceed 1. *For any event A, $0 \leq P(A) \leq 1$.*

THEOREM 7.3

$P(\phi) = 0$.
Proof:

Let $A = \phi$ in the statement of Theorem 7.1.

Then $P(\phi) = 1 - P(\phi')$.

But $\phi' = S$, so $P(\phi) = 1 - P(S) = 1 - 1 = 0$.

This again is a reasonable result: The null event contains no outcomes; therefore it cannot occur and has probability 0.

THEOREM 7.4

For any events A and B in S, $P(A \cup B) = P(A) + P(B) - P(A \cap B)$.

Proof:

$A \cup B = A \cup (A' \cap B)$, and $A \cap (A' \cap B) = \phi$, so by Axiom 3,

$P(A \cup B) = P(A) + P(A' \cap B)$,

Also, $B = (A \cap B) \cup (A' \cap B)$ with $(A' \cap B) \cap (A \cap B) = \phi$,

So $P(B) = P(A \cap B) + P(A' \cap B)$.

Subtracting $P(A \cap B)$ from both sides of this equation,

$P(B) - P(A \cap B) = P(A' \cap B)$.

Substituting into the second line of this proof we obtain

$P(A \cup B) = P(A) + P(B) - P(A \cap B)$.

Theorem 7.4 is an important one, and more subtle than the previous three. To visualize what is happening, it is useful to turn to Venn diagrams in which the *area* of the part of the diagram that represents an event corresponds to the *probability* of the event. The entire area $A \cup B$ is the area of A plus the area of B (see Figure 7.2), but this would include the area $A \cap B$ twice. To avoid this situation, we subtract that area once. Area corresponds to probability, so $P(A \cup B) = P(A) + P(B) - P(A \cap B)$.

As an example of this result, consider rolling a die and observing the top face. Let E be the event that the face is an even number, and let B be the event that the top face is a multiple of 6. Then $P(E) = \frac{3}{6} = \frac{1}{2}$, and $P(B) = \frac{2}{6} = \frac{1}{3}$. Note that $E \cap B = \{6\}$, thus $P(E \cap B) = \frac{1}{6}$. Then, by an indirect computation using the result of Theorem 7.4, we have

$$\begin{aligned} P(E \cup B) &= P(E) + P(B) - P(E \cap B) \\ &= \frac{1}{2} + \frac{1}{3} - \frac{1}{6} = \frac{4}{6} \end{aligned}$$

But by noting that $E \cup B = \{2, 4, 3, 6\}$ and by direct computation, we find that $P(E \cup B) = \frac{4}{6}$. Whether through the application of Theorem 7.4 or by direct computation, we arrive at the same result.

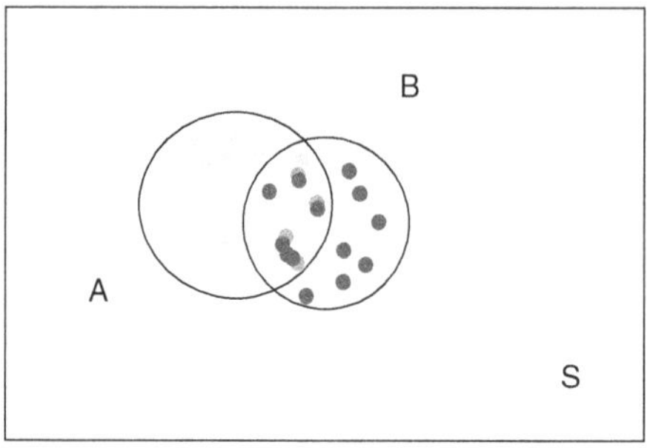

FIGURE 7.2 Venn diagram illustrating Theorem 7.4.

7.9 Counting Techniques: Permutations and Combinations

In the examples so far, we have used experiments whose sample spaces contained equally likely outcomes, and we have assigned probabilities by counting outcomes in events and in sample spaces. It might seem that this technique is limited to simple situations, but we can increase its usefulness by extending our ability to count.

ON DVD

Consider the Cost of Living table in Appendix B that lists hypothetical data on the cost of living for 45 cities in the U.S.:

Suppose we decide to classify the cities into three regions, three levels of health care costs (high to low), and five levels of housing costs (cheap to extremely expensive). Then an interesting question to ask is: If by "different" we mean "different in any detail," in how many different ways can a city be classified?

Given the number of alternatives at each step of the classifications given above, it seems reasonable that there are $3 * 3 * 5 = 45$ possible city classifications. This is an example of the *multiplication principle*.

> If a selection consists of k steps, each with n alternatives (i = 1, 2, …, k), then the entire selection can be made in $n_1 * n_2 * \ldots * n_k$ different ways.

If we extend the classification scheme of the example above to include three categories of grocery costs and four categories of transportation costs, then a city could be classified in $3 * 3 * 5 * 3 * 4 = 540$ different ways.

Similarly, consider how many different nonsense sequences of four letters can be made from the letters A, B, C, D, and E, if letters can be repeated. (These nonsense words would be things like ABAD, CACB, etc.) We assemble these words by making a four-step selection, and at each step, we have five alternatives. Thus, there are

$$5 * 5 * 5 * 5 = 625$$

possible nonsense words. We can think of this as selecting a letter four times from a hat containing five letters, each time replacing the letter chosen.

If we are not allowed to repeat letters, and if we select without replacement, the number of remaining alternatives decreases by one at each step of the selection, so there are only possible words.

$$5 * 4 * 3 * 2 = 120$$

In general, consider the number of ways r objects might be selected *in order* from n objects. For the first selection, we have n alternatives; for the second selection, since one object has been used, $n - 1$ alternatives; and so on. At the last selection, there remain $n - r + 1$ objects from which to choose (after all r objects have been selected and lined up, there remain $n - r$ objects not chosen), so the number of such arrangements is

$$n * (n - 1) * (n - 2) * \dots * (n - r + 1).$$

> Each of these ordered arrangements is called a *permutation* of n objects taken r at a time.

For example, of the 45 cities in the cost of living study, you plan to visit 5. If the order in which you visit the cities matters, then each possible trip is a permutation of the 45 cities taken 5 at a time, and there are such permutations.

$$45 * 44 * 43 * 42 * 41 = 146{,}611{,}080$$

Expressions and formulas involving products like these can be written more efficiently using this notation:

> For any positive integer n, the product $n * (n{-}1) * (n{-}2) * \dots * 2 * 1$ is called n *factorial*, and is written $n!$ or $0!{=}1$.

For example, $5! = 5 * 4 * 3 * 2 * 1 = 120$. This notation lets us write the number of permutations of n objects taken r at a time as

$$\frac{n!}{(n-r)!}$$

At this point, reconsider the trip described above, and assume that the order in which the cities are visited does not matter. When order does matter, there are possible trips.

$$\frac{45!}{(45-5)!}$$

Each unordered group of 5 cities could be ordered in $5!{=}120$ ways, and these have all been counted separately by the number

$$\frac{45!}{(45-5)!}$$

The number of permutations has counted each group of 5 cities 5! times, so that the number of different *unordered* groups of 5 cities is

$$\frac{45!}{(45-5)!}/5! = \frac{45!}{(45-5)!5!} = 1{,}221{,}759$$

This brings us to the following definition:

> The number of *unordered groups* of r objects that can be selected out of n objects is
>
> $$\frac{n!}{(n-r)!r!}, \quad \text{often written} \quad \binom{n}{r}$$
>
> And sometimes read "n choose r." Each of these unordered groups is called a *combination* of n objects taken r at a time.

A classic example of this concept comes from card playing. How many 5-card poker hands are possible from an ordinary deck of 52 cards? The order of cards in a hand does not matter, so each hand is a combination of 5 cards selected from 52 cards. The total number of such hands is

$$\binom{52}{5} = \frac{52!}{(52-5)!5!} = \frac{52*51*50*49*48}{5*4*3*2*1} = 2{,}598{,}960.$$

7.10 Counting Techniques and Probability

We can use these counting techniques to expand the range of events to which we can assign probabilities. For example, we select three cities at random from the data set of the cost of living in U.S. cities. What is the probability that all 3 have a composite index of at least 100? Each possible group of three cities is one outcome in the sample space of this experiment, and there are:

$$\binom{45}{3} = \frac{45!}{(45-3)!3!} = 14{,}190.$$

Of these:

$$\binom{21}{3} = \frac{21!}{(21-3)!3!} = 1{,}330.$$

have a composite index of at least 100. Therefore, the probability that all 3 cities have a composite index of at least 100 is:

$$\frac{\binom{21}{3}}{\binom{45}{3}} = \frac{1{,}330}{14{,}190} = 0.0937.$$

In later chapters we will see other examples of counting techniques being applied to questions of probability.

7.11 Conditional Probabilities

In many random experiments, knowledge of the occurrence or nonoccurrence of one event may change our estimate of the probability of another event. We construct an example of such a situation from the cost of living in U.S. cities file.

Our experiment will be the random selection of one city from the 45 in our study, and we will consider the interaction of events based on the variables X1 and X3, composite index and transportation component index. There are many distinct values for X1, however, and our illustration will be clearer if we group these values into *low, medium,* and *high* classes. Low will be an index value ≤100, medium an index value between 100 and 120, and high an index value ≥120. The same reclassification is also applied to the transportation index data. This can be achieved as shown in Table 7.2:

TABLE 7.2 Composite index and transportation component index frequency distribution.

Component Index Weights	**100%**	**9%**		
	Composite Index[2]	**Transportation**	**Comp Idx Class**	**Trans Idx Class**
Montgomery, Ala.	96.1	98.1	Low	low
Juneau, Alaska	131.6	117.5	high	medium
Phoenix, Ariz.	98.2	107.6	low	medium
Los Angeles, Calif.	153.1	116.5	high	medium
San Diego, Calif.	141	119.7	high	medium
San Francisco, Calif.	177	125.9	high	medium
Colorado Springs, Colo.	96.5	103.6	low	medium
Denver, Colo.	103.5	96.7	medium	low
Washington, DC	137.8	112.8	high	medium
Jacksonville, Fla.	92.4	100.2	low	medium
Atlanta, Ga.	97.2	101.4	low	medium
Honolulu, Hawaii	162.4	130.1	high	high

Table 7.3 shows the frequency of each of these classifications:

TABLE 7.3 Composite index frequency distribution.

Comp Idx Class	Frequency
low	21
medium	16
high	8
Trans Idx Class	**Frequency**
low	18
medium	26
high	1

The Excel functions to achieve this feat are listed in Tables 7.4 and 7.5:

TABLE 7.4 Excel generation tables for composite index frequency distribution.

Component Index Weights	1	0.09	
	Composite Index[2]	Transportation	Comp Idx Class
Montgomery, Ala.	96.1	98.1	=IF(B4<=100,“low”,IF(B4<=130,“medium”,“high”))
Juneau, Alaska	131.6	117.5	=IF(B5<=100,“low”,IF(B5<=130,“medium”,“high”))
Phoenix, Ariz.	98.2	107.6	=IF(B6<=100,“low”,IF(B6<=130,“medium”,“high”))
Los Angeles, Calif.	153.1	116.5	=IF(B7<=100,“low”,IF(B7<=130,“medium”,“high”))

TABLE 7.5 Excel frequency command application.

Trans Idx Class	Frequency
Low	=COUNTIF($E$4:$E$48,I13)
Medium	=COUNTIF($E$4:$E$48,I14)
High	=COUNTIF($E$4:$E$48,I15)

If C was selected as the event that the chosen city had a low composite index, we can see that $P(C)=\frac{21}{45}=0.4667$.

Suppose that T is the event that the chosen city has a low transportation index, then $P(T)=\frac{18}{45}=0.4$. Next a table is constructed for the low transportation index instances only, as shown in Table 7.6, and from this table we find those instances with a low composite index, as shown in Table 7.7.

TABLE 7.6 Excel frequency command application.

Component Index Weights	**100%**	**9%**			
	Composite Index[2]	**Trans-portation**	**Comp Idx Class**	**Trans Idx Class**	**Comp Low within Low Tran**
Montgomery, Ala.	96.1	98.1	low	low	TRUE
Denver, Colo.	103.5	96.7	medium	low	FALSE
Springfield, Ill.	92.1	99.7	low	low	TRUE
Des Moines, Iowa	94.4	94.2	low	low	TRUE
Minneapolis, Minn.	102.6	95.5	medium	low	FALSE
Jackson, Miss.	92.3	93.4	low	low	TRUE
St. Louis, Mo.	99.9	96.9	low	low	TRUE
Billings, Mont.	97.5	99.5	low	low	TRUE
Omaha, Neb.	92.3	99.4	low	low	TRUE
Albuquerque, N.M.	103.3	100	medium	low	FALSE
Charlotte, N.C.	93.5	99.4	low	low	TRUE
Oklahoma City, Okla.	92	97.4	low	low	TRUE
Memphis, Tenn.	89.9	94	low	low	TRUE
El Paso, Tex.	91.5	95	low	low	TRUE
San Antonio, Tex.	94.2	89.1	low	low	TRUE
Salt Lake City, Utah	97.3	97.5	low	low	TRUE
Richmond, Va.	101	95	medium	low	FALSE
Cheyenne, Wyo.	107.7	95.2	medium	low	FALSE

TABLE 7.7 Resultant composite index frequency distributions by levels.

	Conditional Counts		
Trans Idx Class	**Frequency**	**Low Comp within Low Trans**	
Low	18	low/low Tran	13
Medium	26	low/not low Tran	5
High	1		

Since we know that T occurred, we can then use this information to revise the probability of C by finding how many cities with a low composite index also have a low transportation index: $P(C \text{ given } T) = P(C \mid T) = \frac{13}{18} = 0.7222.$

We have seen that P(C)=0.4, but if we know that L occurred, we know $P(C \mid T) = 0.7222$. Therefore, knowledge of whether or not L has already occurred does make a difference in the final results, $P(C) \neq P(C \mid T)$.

Table 7.8 shows that $P(C \text{ and } T) = \frac{13}{45} = 0.2889$. Notice that $P(C \mid L) = \frac{13}{18}$ and $\frac{\left(\frac{13}{45}\right)}{\left(\frac{18}{45}\right)} = \frac{13}{18}$.

TABLE 7.8 Frequency distribution table.

AND Idx Counts	
Comp Low/Trans Low	**Frequency**
low/low	13
Other	32

Thus we have the following definition:

Let A and B be events in S, with P(B) > 0. The *conditional probability* of A given B, written $P(A \mid B)$, is

$$\frac{P(A \cap B)}{P(B)}.$$

Probabilities of the form $P(A \cap B)$ are called *joint probabilities*. *Marginal probabilities*, are of the form P(L) and P(C). A conditional probability is a joint probability divided by a marginal probability.

Another example is to suppose that A and B are events in S with P(A) = 0.46 and $P(A \cap B) =$ 0.14. Then

$$P(B|A) = \frac{P(A \cap B)}{P(B)} = \frac{0.14}{0.46} = 0.304.$$

7.12 Independent Events

Consider the simple experiment of rolling two dice. Let A be the event that either one or two dots appear on the first die, and let B be the event that an even number of dots appear on the second die. Then

$$P(A) = \frac{12}{36} = \frac{1}{3};\ P(B) = \frac{18}{36} = \frac{1}{2}; \quad \text{and} \quad P(A \cap B) = \frac{6}{36} = \frac{1}{6}.$$

With this information, we can compute conditional probabilities:

$$P(A|B) = \frac{P(A \cap B)}{P(B)} = \frac{\left(\frac{1}{6}\right)}{\left(\frac{1}{2}\right)} = \frac{1}{3} = P(A).$$

$$P(B|A) = \frac{P(A \cap B)}{P(A)} = \frac{\left(\frac{1}{6}\right)}{\left(\frac{1}{3}\right)} = \frac{1}{2} = P(B).$$

Note that the conditional and unconditional probabilities are the same; knowing that one of these events has occurred does not affect the probability of the other. In fact, the above two statements are equivalent:

$$\begin{aligned}
&P(A|B) = P(A) \Leftrightarrow \\
&\frac{P(A \cap B)}{P(B)} = P(A) \Leftrightarrow \\
&P(A \cap B) = P(A) * P(B) \Leftrightarrow \\
&\frac{P(A \cap B)}{P(A)} = P(B) \Leftrightarrow \\
&P(B|A) = P(B)
\end{aligned}$$

Both statements are equivalent to the third statement above, from which we obtain this definition:

> Let A and B be events in S; A and B are *independent events* if $P(A \cap B) = P(A) * P(B)$.

Independent events do not interact in the sense that the occurrence of one has no effect on the probability of the other. This is true of the events in the dice example, where

$$P(A \cap B) = \frac{1}{6} = \frac{1}{3} * \frac{1}{2} = P(A) * P(B).$$

As another example: C and D are events in S, with P(C) = 0.6, P(D) = 0.5, and $P(C \cup D)$ = 0.8. Are C and D independent?

We must find $P(C \cap D)$, and we know that $P(C \cup D) = P(C) + P(D) - P(C \cup D)$. Therefore, $0.8 = 0.6 + 0.5 - P(C \cap D)$, and $P(C \cap D) = 0.3$. Then $P(C \cap D) = 0.3 = 0.6 * 0.5 = P(C) * P(D)$, so the events C and D are independent.

7.13 Exercises

Interpretation Exercises

1. Describe the sample spaces for these random experiments:

   a. Roll one die and observe the number of dots on its upper face.

   b. Flip a coin until the first head appears.

   c. Choose a group of five people from your class.

   d. Select a person at random and measure his or her height.

2. Let C be the event that tomorrow's weather is hot, while D is the event that it rains. Describe these events:

   a. $C \cap D$

   b. $C \cup D$

   c. D'

   d. $C \cap D'$

   e. $C' \cup D'$

   f. $(C \cup D)'$

3. A and B are events in S. Indicate the portions of Venn diagrams corresponding to:

   a. A'

   b. $A \cup B$

   c. $A' \cap B$

   d. $(A \cap B)'$

   e. $A' \cup B$

4. Use Venn diagrams to verify DeMorgan's laws.

5. If events A and B are independent, are A′ and B? A and B′? A′ and B′? Why or why not?

6. Are there any circumstances in which mutually exclusive events will be independent?

Computational Exercises

7. A and B are events in S, with $P(A) = 0.25$, $P(B) = 0.55$, and $P(A \cap B) = 0.10$. Find the following probabilities.

   a. $P(A')$

   b. $P(A \cup B)$

c. $P(A' \cup B')$

d. $P(A' \cap B)$

e. $P((A \cap B)')$

8. E and F are mutually exclusive events in S, with P(E) = 0.33 and P(F) = 0.45. Find these probabilities:

   a. $P(E \cap F)$

   b. $P(E \cup F)$

   c. $P(E \cap F')$

   d. $P(E \cup F')$

9. In a marketing survey, the people interviewed are classified according to five education categories, seven job categories, four income categories, and three marital status categories. In how many ways can a person be classified?

10. How many four-letter patterns can be made from the letters a, b, c, d, e, f, and g

    a. if letters can be repeated?

    b. if letters cannot be repeated?

11. How many different ways can a group of 10 people be divided into

    a. two committees of five each?

    b. a committee of four and another of six?

12. You plan a trip to visit seven cities; in how many orders might this be done?

13. At Acme Mail Order Specialties, 80% of all orders are shipped within 24 hours, and 50% are shipped within 24 hours and delivered within a week.

    a. What is the probability that an order was shipped within 24 hours and will arrive within a week?

    b. If 60% of all orders are delivered in a week, what is the probability that an order delivered in a week was shipped within 24 hours?

14. C and D are events in S, with P(C) = 0.4, P(D) = 0.5, and $P(C \cap D') = 0.2$. Show that C and D are or are not independent.

Excel Exercises

15. Create a cross-tabulation of X3 by X1 for the American Cities database (in Appendix B) using Excel. Find these probabilities: ON DVD

    a. P(high unemployment rate)

    b. P(medium unemployment rate)

c. P(high unemployment rate $\cap$ medium unemployment rate)

d. P(medium unemployment rate | Western city)

e. P(low unemployment rate | Western city)

f. P(low or high unemployment rate | Central city)

g. Are any pairs of events among those in the cross-tabulation independent?

CHAPTER 8

PROBABILITY TOOLKIT

OVERVIEW AND LEARNING OBJECTIVES

In This Chapter

8.1 Introduction

Having defined elementary operations and laws for probability, our next objective is the development of the tools for describing and studying different types of experiments. Before embarking on a study of various types of experiments, however, we must first study the generic concept of a probability distribution and the related toolkit; this topic is pursued in this and the next several chapters.

8.2 Random Variables

Consider the experiment of flipping three coins and counting the number of heads that appear. Equally likely outcomes for this experiment are triples of H's and T's, and from them we can see that these probabilities hold:

$$\begin{array}{lll} S = \{HHH, HHT, & P(\text{no heads}) = \frac{1}{8} & P(1 \text{ head}) = \frac{3}{8} \\ \quad HTH, HTT, & & \\ \quad THH, THT, & P(2 \text{ heads}) = \frac{3}{8} & P(3 \text{ heads}) = \frac{1}{8} \\ \quad TTH, TTT\} & & \end{array}$$

What we have done is associate a number (0, 1, 2, or 3) with each outcome in the sample space. HHH has "3" attached to it. When we perform the experiment, an outcome will occur, and with it, the number of that outcome. This is an example of a *random variable*:

> A *random variable* is a numerical quantity whose value is determined by a random experiment.

That is, a random variable is a mapping from the sample space S to the set of real numbers R.

When an outcome occurs, the random variable takes on the value attached to that outcome. Here are some examples.

1. In the above experiment, let the random variable X be the number of heads.
2. Roll two dice; let the random variable Y be the total number of dots on the upper faces.
3. Inspect 20 items out of a production lot of 1,000; let Z be the number of items that fail the inspection.
4. Select a city at random from the Cost of Living file discussed in the previous chapter; let X be the number of cities with low transportation index values.
5. Flip a coin until the first head appears; let Y be the number of trials before the first head.

In each of these examples, the random variable can take on only certain exact values. Such random variables are called *discrete*, and are defined in this way:

> A random variable is *discrete* if the set of its possible values, called the *range* or *space* of the random variable, is finite or countably infinite.

In Example 5, the range of the random variable is 0, 1, 2, 3, 4, ..., and is countably infinite.

> A random variable is *continuous* if it can assume any real value in an interval; that is, its range is an interval.

These are examples of continuous random variables:

1. Let X be the height of a randomly chosen person.
2. Let Y be the time required to perform some task.
3. Let Z be the distance required to stop a car traveling at 50 mph.

In these experiments, the random variables can take on any value within an interval of values, though in practice we might approximate them with discrete random variables.

For now, we will consider discrete random variables, then extend the concepts we develop to the continuous case.

8.3 Probability Distributions

We have seen that frequency distributions are tabular displays of the values of a variable that help to illustrate the shape of the distribution, its central tendency, variability, and other characteristics. Since probabilities can be interpreted as relative frequencies when the number of trials is large, it is natural to use the idea of a distribution to describe the allocation of probability among the possible values of a random variable.

Return to the experiment in which we flip three coins, and let the random variable X be the number of heads that appear. The range of X is {0, 1, 2, 3}, and we can give the probabilities associated with each possible value of X:

$$P(X=0)=\frac{1}{8} \qquad P(X=2)=\frac{3}{8}$$

$$P(X=1)=\frac{3}{8} \qquad P(X=3)=\frac{1}{8}$$

This display resembles a relative frequency distribution, and is one way of describing the *probability distribution* of X. It is also an example of this definition:

> Let X be a discrete random variable. The *probability function* f associated with X is defined to be:
>
> $$f(x) = P(X = x).$$

f is also called the *probability density function (PDF)*, or *probability mass function (PMF)* of X.

In this example:

$$f(0) = P(X = 0) = \frac{1}{8},\ f(1) = \frac{3}{8},\ f(2) = \frac{3}{8},\ f(3) = \frac{1}{8}.$$

f(x) = 0 for all other values of x. This can be most economically written as:

$$f(x) = \frac{1}{8},\ x = 0,3,$$

$$f(x) = \frac{3}{8},\ x = 1,2,$$

$$f(x) = \text{all other } x,$$

and is another way of describing the probability distribution of X.

Two observations should be made about discrete probability functions:

- The value of a probability function is a probability, f(x) = P(X = x), so for any value of x, $0 \le f(x) \le 1$.
- A random variable divides the sample space into distinct events (as shown in Figure 8.1), so the sum of the values of f(x) over all the possible values of x must be one: $\sum_x f(x) = 1$.

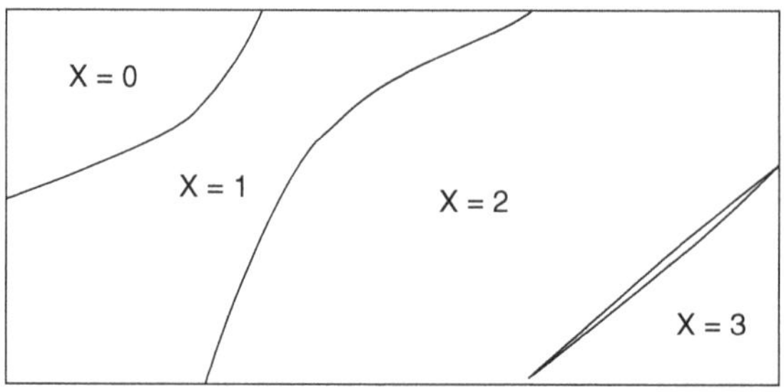

FIGURE 8.1 The random variable X divides the sample space into separate events.

ON DVD All figures and tables in this chapter appear on the companion DVD.

Note that both of these conditions are met by the probability function in the example.

Just as we can construct histograms corresponding to a frequency distribution, so can we draw the graph of a probability function. The probability associated with each value in the range of the random variable is represented by a vertical line above that value, as shown in Figure 8.2:

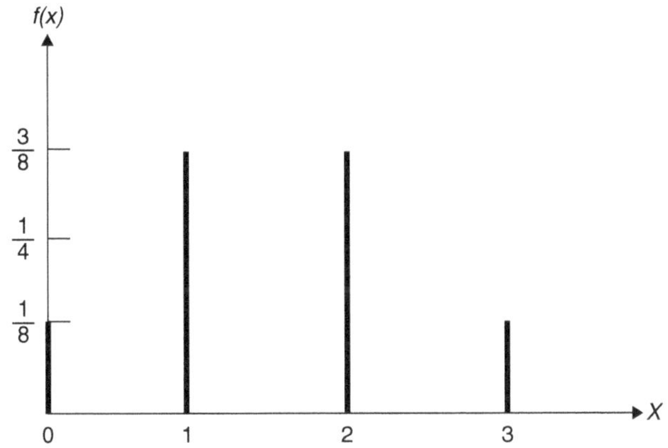

FIGURE 8.2 The probability function f(x) of X.

Finally, another function can be defined that corresponds to the cumulative relative frequency distribution.

> Let X be a discrete random variable. The *cumulative density function* F associated with X is defined to be $F(x) = P(X \le x) = \sum_{y \le x} P(X = y)$.

That is, while f(x) gives the probability that the random variable X will take on the value x, F(x) gives the probability that X will take on a value less than or equal to x.

This function can be graphed as in Figure 8.3, which shows the cumulative density function and its graph for our example. Note that each step in the value of F is equal to the value of f at that step.

$$F(x) = \begin{cases} 0, & x < 0 \\ \frac{1}{8}, & 0 \le x < 1 \\ \frac{4}{8}, & 1 \le x < 2 \\ \frac{7}{8}, & 2 \le x < 3 \\ 1, & x \ge 3 \end{cases}$$

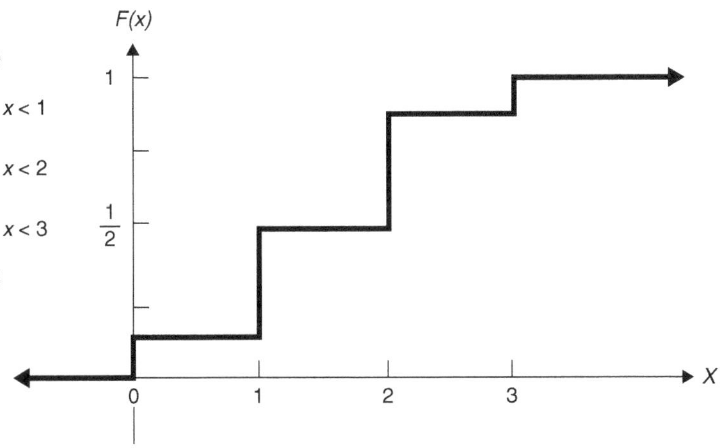

FIGURE 8.3 The cumulative density function.

8.4 Expected Value

Just as we can indicate the center of a distribution of values by its mean, so can we indicate the center of the probability of a random variable. This center is an average of the possible values of the random variable, weighted by their respective probabilities and corresponding to the center of the graph of the probability function. More precisely:

> If X is a random variable with probability function f, the *expected value* or mean of X, indicated by E(X) or μ, is defined to be
>
> $$\sum_x (x * f(x)) = \sum_x (x * P(X = x)),$$
>
> where the sum is taken over all possible values of X

In the example developed in the previous section, we found a probability function and graph, shown in Figure 8.2. The expected value or mean of X is

$$E(X) = \mu = \sum_x (x * f(x)) = \sum_x (x * P(X = x)) = (0 * \frac{1}{8}) + (1 * \frac{3}{8}) + (2 * \frac{3}{8}) + (3 * \frac{1}{8}) = \frac{3}{2}.$$

Note that this value corresponds to the center of the distribution as shown in Figure 8.2, and that there is one term in the sum for each value in the range of X; each term is the product of a possible value of the random variable with the probability that X takes on that value.

The random variable of the example can never take on the value $\frac{3}{2}$; we can never "expect" this to happen. Rather, an expected value can be thought of as the *average* (mean) value of the random variable over many repetitions of the experiment. If we repeated the tossing of the three coins 1,000 or 10,000 times, the mean of the values of X from all the trials would be near $\frac{3}{2}$.

Several results about expected value should be established for later use:

- **For any constant c, E(c) = c.**

  That is, if a random variable always takes on the same value, that value is the expected value of the random variable. Though this result may seem trivial, it will be useful in simplifying complicated expressions that involve expected values.

- **For any constant c, E(cX) = cE(X).**

  If we create a new random variable from another by multiplying by a constant, the mean of the new random variable is the constant times the mean of the old random variable.

- **For any random variables X and Y, E(X + Y) = E(X) + E(Y).**

  The expected value of the sum of two (or more) random variables is the sum of their expected values.

8.5 Variance

Having established a measure of the middle of a probability distribution, we then inquire into its *variability*. Are the possible values of the random variable concentrated around the mean, or are they widely scattered? We answer this question as we did when we investigated the variability of a frequency distribution, by using the squared deviations of the values from their mean:

> Let X be a random variable with probability function f and mean μ. The *variance* of X, indicated by Var(X) or σ^2, is
>
> $$E[(X-\mu)^2] = \sum_x (x-\mu)^2 f(x).$$

Notice how similar this is to the definition of the variance of a sample or population.

The definition of variance leads to another familiar definition:

> The *standard deviation* of X, indicated by σ, is $\sqrt{\sigma^2} = \sqrt{Var(X)}$.

Again, using the example of the three coins, with X the number of heads appearing when the coins are tossed:

$$\sigma^2 = \text{Var}(X) = \left(0-\frac{3}{2}\right)^2 * \frac{1}{8} + \left(1-\frac{3}{2}\right)^2 * \frac{3}{8} + \left(2-\frac{3}{2}\right)^2 * \frac{3}{8} + \left(3-\frac{3}{2}\right)^2 * \frac{1}{8}$$

$$= \frac{24}{32} = \frac{3}{4} = 0.75, \text{ and}$$

$$\sigma = \sqrt{\sigma^2} = \sqrt{0.75} = 0.866.$$

As with population and sample variances, we can develop formula for Var(X), using the properties of the expected value given in the previous section:

$$\begin{aligned}
\sigma^2 &= \text{Var}(X) = E[(X-\mu)^2] = E[X^2 - 2X\mu + \mu^2] \\
&= E[X^2] + E[-2X\mu] + E[\mu^2] \\
&= E[X^2] - 2\mu E[X] + \mu^2 \\
&= E[X^2] - 2\mu^2 + \mu^2 \\
&= E[X^2] - E[X]^2 \\
&= \sum_x (x^2 f(x)) - \mu^2 \text{ Therefore,}
\end{aligned}$$

$$\sigma^2 = E(X^2) - E(X)^2$$

These formulas are very similar to those developed in previous exercises. We use them to find the variance of the random variable in the example of the three coins:

$$\sigma^2 = \mathrm{Var}(X) = (0^2 * \frac{1}{8}) + (1^2 * \frac{3}{8}) + (2^2 * \frac{3}{8}) + (3^2 * \frac{1}{8}) - \left(\frac{3}{2}\right)^2$$

$$= \frac{24}{8} - \frac{9}{4} = \frac{3}{4} 0.75.$$

We can state two results about the variance of a random variable:

- **For any constant c, Var(cX) = c^2Var(X).** If we multiply a random variable by a constant, the variance of the new random variable is the constant squared times the variance of the old random variable.
- **For any constant c, Var(c + X) = Var(X).** Adding a constant to a random variable does not change its variance.

8.6 Independent Random Variables

We have seen that when two events A and B are independent, $P(A \cap B) = P(A) * P(B)$, and the occurrence of one does not change the probability of the other. We can extend this concept to include random variables:

> The random variables X and Y are *independent* if, for all possible values of X and Y, P(X = x and Y = y) = P(X = x) * P(Y = y).

That is, any pair of events (X = x) and (Y = y) is independent; the value taken on by one of the random variables has no effect on the probability of the other.

Rolling two dice provides us with an example of a pair of independent random variables. Let X be the number of dots on the upper face of the first die, and let Y be the number of dots on the upper face of the second. Since there is no physical connection between the two dice, it seems reasonable that these random variables should be independent, and a check of the 36 pairs of nontrivial events of the form "(X = x) and (Y = y)" will show that this is so.

You may have noticed that in the previous section we said nothing about Var(X + Y). We will make no statement about Var(X + Y) in general, unless X and Y are independent. To do that, we must first prove this result:

THEOREM 8.1

If X and Y are independent random variables, then E[XY] = E[X] * E[Y].
Proof:

$$
\begin{aligned}
\mathrm{E}[\mathrm{XY}] &= \sum_{x,y} xyP((X=x)and(Y=y)) \\
&= \sum_{x,y} xyP(X=x)P(Y=y) \\
&= \sum_{x,y} xP(X=x)yP(Y=y) \\
&= \sum_{x,y} xf(x)yg(y) \\
&= \sum_{x}[\sum_{y} xf(x)yg(y)] \\
&= \sum_{x}[xf(x)\sum_{y} yg(y)] \\
&= [\sum_{y} yg(y)][\sum_{x} xf(x)] \\
&= \mathrm{E}[\mathrm{X}]*\mathrm{E}[\mathrm{Y}]
\end{aligned}
$$

Using this result, we can prove the following theorem.

THEOREM 8.2

If X and Y are independent random variables, then Var(X + Y) = Var(X) + Var(Y).
Proof:

$$
\begin{aligned}
\mathrm{Var}(\mathrm{X}+\mathrm{Y}) &= \mathrm{E}[(\mathrm{X}+\mathrm{Y})^2] - [\mathrm{E}(\mathrm{X}+\mathrm{Y})]^2 \\
&= \mathrm{E}(\mathrm{X}^2 + 2\mathrm{XY} + \mathrm{Y}^2) - [\mathrm{E}(\mathrm{X}) + \mathrm{E}(\mathrm{Y})]^2 \\
&= \mathrm{E}(\mathrm{X}^2) + 2\mathrm{E}(\mathrm{XY}) + \mathrm{E}(\mathrm{Y}^2) - [\mathrm{E}(\mathrm{X}^2) + 2\mathrm{E}(\mathrm{X})\mathrm{E}(\mathrm{Y}) + \mathrm{E}(\mathrm{Y})^2] \\
&= [\mathrm{E}(\mathrm{X}^2) - \mathrm{E}(\mathrm{X}^2)] - [\mathrm{E}(\mathrm{Y}^2) - \mathrm{E}(\mathrm{Y})^2] + [2\mathrm{E}(\mathrm{XY}) - 2\mathrm{E}(\mathrm{X})\mathrm{E}(\mathrm{Y})] \\
&= [\mathrm{E}(\mathrm{X}^2) - \mathrm{E}(\mathrm{X})^2] + [\mathrm{E}(\mathrm{Y}^2) - \mathrm{E}(\mathrm{Y})^2] + 0 \\
&= \mathrm{Var}(\mathrm{X}) + \mathrm{Var}(\mathrm{Y}) + 0
\end{aligned}
$$

It does not follow from this result that Var(X – Y) = Var(X) – Var(Y). In fact, if X and Y are independent:

$$\begin{aligned} \mathrm{Var}(X-Y) &= \mathrm{Var}(X+(-Y)) \\ &= \mathrm{Var}(X)+\mathrm{Var}(-Y) \\ &= \mathrm{Var}(X)+(-1)^2\mathrm{Var}(Y) \\ &= \mathrm{Var}(X)+\mathrm{Var}(Y). \end{aligned}$$

Also,

$$\sigma(X+Y)=[Var(X)+Var(Y)]^{\frac{1}{2}} \neq \sigma(X)+\sigma(Y), \text{ and}$$

$$\sigma(X-Y)=[Var(X)+Var(Y)]^{\frac{1}{2}}.$$

8.7 Exercises

Computational Exercises

1. Let X be a random variable with this probability function: f(x) = x/6; x = 1, 2, 3.

a. Show that f is acceptable as a probability function.

b. Find E(x) and Var(x), (μ and σ^2).

2. Y and Z are random variables with E(Y) = 10, Var(Y) = 9, E(Z) = 15, Var(Z) = 16. Find the following:

a. E(2Y + Z)

b. Var(2Y) and σ (2Y)

c. E(Z/4)

d. Var(Z/4) and σ (Z/4)

e. E(Y – 3Z)

f. Var(Y + 5) and σ (Y + 5)

Excel Exercises

ON DVD

3. In the American Cities database, recode X3, the unemployment rate, and use the output to construct the probability function of X and draw its graph. Construct the cumulative function associated with X and chart the graph.

4. Find the mean and variance of the random variable in problem 2.

5. The file HousePriceData.xls contains the following information for selected homes: ON DVD

- Square footage
- Price
- Number of bathrooms
- Number of bedrooms

Use this information to answer the following questions.

a. What is the average price of all homes having 2 1/2 bathrooms and 4 bedrooms?

b. What is the variance in the prices of all homes having 3 1/2 bathrooms + 4 bedrooms?

Interpretation Exercises

6. Prove or demonstrate the following results about the expected value for a discrete random variable:

a. $E(c) = c$, where c is a constant.

b. $E(cx) = c\,E(x)$.

c. $E(x + c) = E(x) + c$.

7. Consider the random variables in problem 2. Can we say anything about $Var(Y + Z)$? Why or why not?

CHAPTER 9

DISCRETE PROBABILITY DISTRIBUTIONS

OVERVIEW AND LEARNING OBJECTIVES

In This Chapter

9.1 Introduction

Probability distributions arise in a variety of situations in the real world, and many of these distributions fall into a relatively small number of categories. Investigations of these distributions will provide powerful tools for analyzing the natural systems in which they correspond. We will be able to construct mathematical models of real-world systems incorporating these probability distributions, and to describe and predict the behavior of the systems through analysis of the models. In this chapter, we will provide an overview of discrete probability distributions.

9.2 Problem Solving through Modeling

In this chapter, we will use *deductive* logic, reasoning from the whole to a part, rather than the *inductive* logic we apply in statistical inference. Probability distributions will be constructed as models of populations, and we will find probabilities of events within those populations. Corresponding to each population, the model will consist of a type of probability distribution with the mean and variance of the associated random variable.

Figure 9.1 illustrates the process of solving a real-world problem through the application of such a probability model. This process is composed of these steps:

1. From the real-world problem, identify the appropriate type of probability model.
2. Using values from the real-world problem, describe the parameters of the probability.
3. Find a solution to the model, and examine its plausibility in terms of the model. If the solution is unreasonable, repeat this step.
4. From the solution of the model, find the corresponding solution to the original problem and consider its plausibility. If the real-world solution is unreasonable, reconsider step 2 and continue.

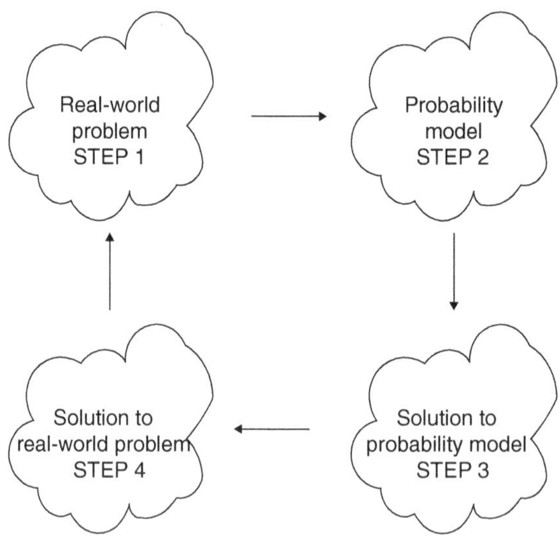

FIGURE 9.1 Process of solving a real-world problem.

There are two ways in which this process can fail. First, the solution of the model may be incorrect; this type of error generally is easily found and easily corrected. Second, the model itself may not correspond sufficiently to the real-world problem to provide an adequate representation of its relevant features. This will render incorrect any solution based on the model. This type of error is more subtle than the first and requires modifying or replacing the model.

All figures and tables in this chapter appear on the companion DVD. ON DVD

The Dartmouth/Thayer approach to solving engineering problems, proposed by Dartmouth College's Thayer School of Engineering, is another perspective on problem solving that requires students to work in teams and develop critical thinking skills while using written and oral communication skills. This problem-solving technique consists of 10 steps: state the problem, redefine the problem, identify constraints and general specifications, identify a first set of alternative solutions, select an alternative, redefine the problem using the selected alternative, refine and add specifications, brainstorm to generate other possible alternatives, develop a new set of specifications, and select the most viable alternative.[1]

A general approach to problem solving that is useful in this subject is an iterative approach that involves brainstorming, constructing models, analyzing, and testing until an acceptable solution is reached. It is important that all steps be done in order. These steps are:

1. State the problem.

Enter the creative problem-solving cycle with a statement of a problem that pertains to a specific topic. The problem may take the form of an open-ended question or an existing condition in need of change.

2. Redefine the problem.

The problem solvers should assume that the initial problem comes to them reflecting the bias of their customer and also his preconceived notion(s) of a desired solution. Therefore, the problem solvers must question the customer and redefine the problem in a "cleaner" form. The new problem statement is less ambiguous; it is open-ended and not constrained by the customer's preconceived ideas.

3. Identify constraints and set general specifications.

Once the problem solvers have a clear understanding of the customer and his needs, brainstorming will help generate a list of specifications that will establish the constraints the solution must meet. Some constraints (moral, legal, ethical) may apply to any problem, not just this one.

The first time through the cycle, the specifications will be general in nature. For optimal effectiveness, each should be defined, quantified, and justified.

4. Identify alternative solutions.

Brainstorm possible solutions to the problem using established specifications and constraints. Brainstorming is a free flow of ideas *without criticism,* since this would hinder tapping into the group's innate creativity. A long list assures the problem solvers that another solution can be found should a first or second choice fail.

5. Select the most viable alternative.

6. Redefine the problem.
The highest-ranking alternative is re-evaluated by redefining the problem as necessary.

7. Refine and add specifications.
Brainstorm to develop refined specifications for the alternative selected. In this second iteration, the specifications should be more specifically defined, quantified, and justified relative to the alternative solution that has been chosen.

8. Brainstorm alternatives.
Using the refined and more-focused specifications, additional appropriate alternatives should be generated.

9. Reiterate until the problem is solved.

10. Select the most viable alternative.

9.3 Important Discrete Distributions

As observed earlier, probability distributions arise in a variety of real-world situations, and many of these distributions fall into a small number of categories. We now examine several important discrete distributions.

9.3.1 The uniform discrete distribution

We have seen that a probability distribution, however represented (probability function, graph, cumulative density function, etc.), consists of an experiment, a sample space, a random variable, and the associated functions. A *uniform discrete distribution* arises when every one of the k values in the range of the discrete random variable has associated with it probability $\frac{1}{k}$. That is,

> Let X be a discrete random variable. X has a *uniform discrete distribution* if, for each of the k evenly spaced values in the range of X, $f(x) = P(X = x) = \frac{1}{k}$.

For example, let X be the number of dots on the upper face of a single die. X may take on the values 1, 2, 3, 4, 5, or 6, each with probability equal to $\frac{1}{6}$, and has the uniform discrete distribution shown in Figure 9.2.

$$f(x) = \begin{cases} \frac{1}{6}, x = 1,2,3,4,5,6 \\ 0, \text{otherwise} \end{cases}$$

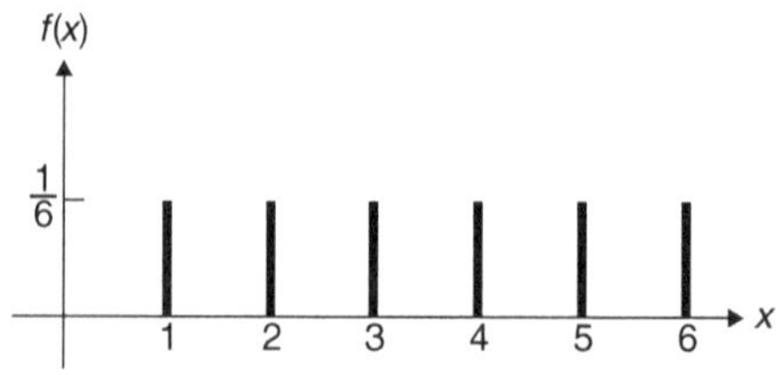

FIGURE 9.2 A uniform discrete distribution.

The graph of a uniform discrete distribution will always have an evenly spaced set of lines of equal height.

From the probability function, we construct the cumulative density function F and its graph, shown in Figure 9.3, recalling that

$$F(x) = \begin{cases} 0, & x < 1 \\ \frac{1}{6}, & 1 \le x < 2 \\ \frac{2}{6}, & 2 \le x < 3 \\ \frac{3}{6}, & 3 \le x < 4 \\ \frac{4}{6}, & 4 \le x < 5 \\ \frac{5}{6}, & 5 \le x < 6 \\ 1, & x \ge 6 \end{cases}$$

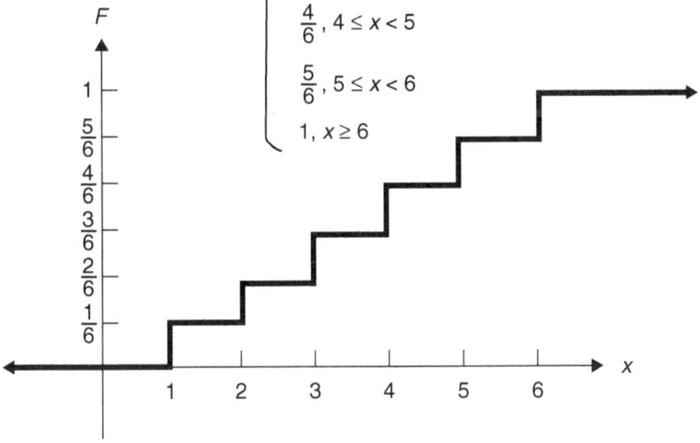

FIGURE 9.3 Cumulative distribution function for the discrete uniform distribution.

The mean of this example distribution is 2.917. This value is at the middle of the probability function graph and is horizontally its geometric center.

Using the simplified procedure for finding the variance:

$$\sigma = Var(X) = \sum_x x^2 f(x) - \mu^2 = (1^2 * \frac{1}{6}) + (2^2 * \frac{1}{6}) + (3^2 * \frac{1}{6}) + (4^2 * \frac{1}{6}) + (5^2 * \frac{1}{6}) + (6^2 * \frac{1}{6}) - 3.5^2$$

$$= \frac{91}{6} - 3.5^2 = \frac{105}{6} = 2.917,$$

And the standard deviation is:

$$\sigma = \sqrt{\sigma} = \sqrt{2{,}917} = 1.708.$$

9.3.2 The binomial distribution

The distribution we consider next arises in a wide variety of real-world situations, but rather than approach it directly, we first define a very simple experiment and its associated distribution.

> A *Bernoulli experiment*, named for the 17th-century mathematician Jacob Bernoulli, is one in which all the outcomes are in one of two mutually exclusive events, usually called "success" and "failure," with the probability of success p and the probability of failure $q = 1 - p$.

We perform a Bernoulli experiment when we roll a single die, with success occurring if six dots appear on the upper face. See Figure 9.4.

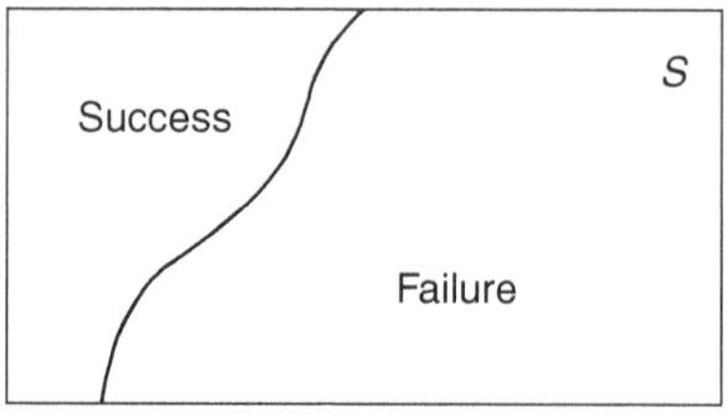

FIGURE 9.4 Sample space for a Bernoulli experiment.

For this simple experiment, we define an equally simple random variable. X is 1 if success occurs and 0 if failure occurs in the Bernoulli experiment (with the appearance of six dots equaling success). The probability function and graph associated with X are shown in Figure 9.5.

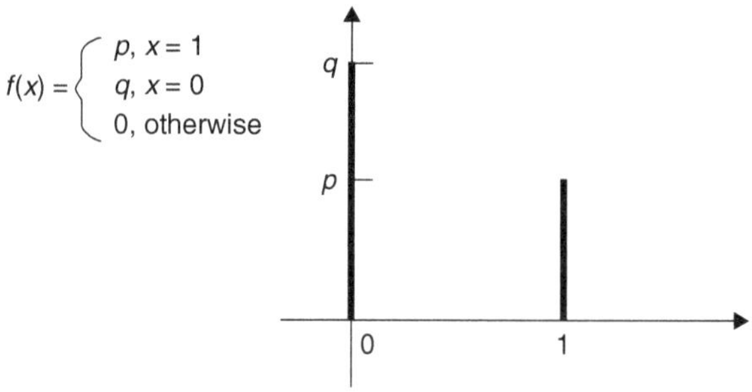

FIGURE 9.5 Bernoulli probability mass function.

The mean and variance of X are also easily found:

$$\mu = E(x) = \sum_x x f(x) = (1 * p) + (0 * q) = p$$

$$\sigma^2 = Var(X) = \sum_x x^2 f(x) - \mu^2 = (1^{2*} * p) + (0^2 * q) - p^2$$
$$= p - p^2 = p(1 - p) = pq.$$

If success is the occurrence of a six on a single roll of a die, and X is the number of successes on that single roll, then

$$P = \frac{1}{6},\ q = \frac{5}{6}$$

$$\mu = E(X) = p = \frac{1}{6}$$

$$\sigma^2 = Var(X) = pq = \frac{1}{6} * \frac{5}{6} = \frac{5}{36}, \text{ and}$$

$$\sigma = \sqrt{\frac{5}{36}} = 0.373.$$

Such a simple random variable is by itself of limited use, but we can put several of them together:

> A *sequence of Bernoulli trials* occurs when we perform a Bernoulli experiment *n independent* times, and the probability of success p is the same through all the trials.

The random variable associated with this larger experiment is the number of successes in n trials. If we roll a die 10 times, and success is the appearance of six dots, the random variable X is the number of times this occurs in the 10 trials. (Note that the successive rolls are independent, and the probability of success is always $\frac{1}{6}$.)

To find the probability function of X, first consider the probability of one particular sequence of, say, three successes (S) and 10–3 failures (F). Successes occur with probability $\frac{1}{6}$ and failures with probability $\frac{5}{6}$. The trials are independent, so the probability of this sequence of successes and failures is $\left(\frac{1}{6}\right)^3\left(\frac{5}{6}\right)^7$.

$$\begin{array}{ccccccccccccccccccc} F & & S & & F & & F & & F & & S & & S & & F & & F & & F \\ \frac{5}{6} & \times & \frac{1}{6} & \times & \frac{5}{6} & \times & \frac{5}{6} & \times & \frac{5}{6} & \times & \frac{1}{6} & \times & \frac{1}{6} & \times & \frac{5}{6} & \times & \frac{5}{6} & \times & \frac{5}{6} \end{array}$$

But there are $\binom{10}{3}$ sequences like this containing three successes and seven failures, so the probability of three successes in the 10 trials is

$$F(3) = P(X = 3) = \binom{10}{3}\left(\frac{1}{6}\right)^3\left(\frac{5}{6}\right)^7.$$

In general, if the random variable X is the number of successes in a sequence of n Bernoulli trials, the probability function of X is:

$$F(x) = P(X = x) = \binom{n}{x}^x p^n q^{n-x}, \quad 0 = 1,\ 2,\ 3,\ \ldots,\ n$$

0, otherwise.

Where p is the probability of success, and $q = 1 - p$ is the probability of failure. Such a random variable is said to have a *binomial distribution* and is identified as *b(n,p)*. The random variable of our example is $b(10, \frac{1}{6})$, and it has this probability function:

$$f(x) = P(X = x) = \binom{10}{x}^x \left(\frac{1}{6}\right)^n \left(\frac{5}{6}\right)^{n-x}, \quad x = 0,\ 1,\ 2,\ 3,\ \ldots,\ 10$$

0, otherwise.

Note that probabilities such as those shown in Figure 9.6 are given for a variety of binomial distributions in Table A-2.

x	$f(x)$
0	0.1615
1	0.3230
2	0.2907
3	0.1550
4	0.0543
5	0.0130
6	0.0022
7	0.0002
8	0.0000+
9	0.0000+
10	0.0000+

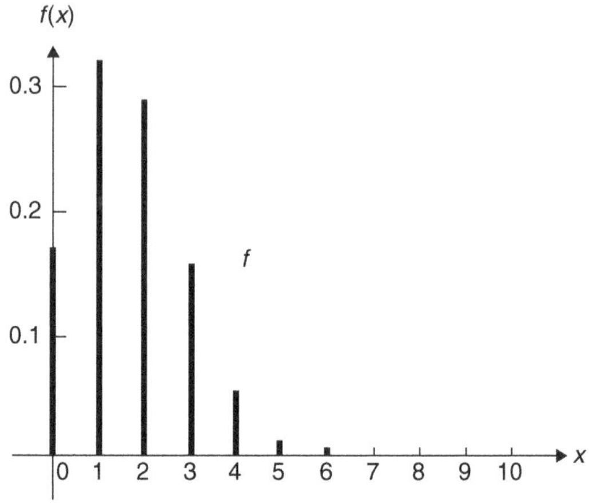

FIGURE 9.6 The probability function of the binomial distribution $b(10, \frac{1}{6})$.

The cumulative density function and its graph are shown in Figure 9.7.

x	$F(x) = P(X \leq x)$
0	0.1615
1	0.4845
2	0.7752
3	0.9302
4	0.9845
5	0.9975
6	0.9997
7	0.9999
8	1.0000
9	1.0000
10	1.0000

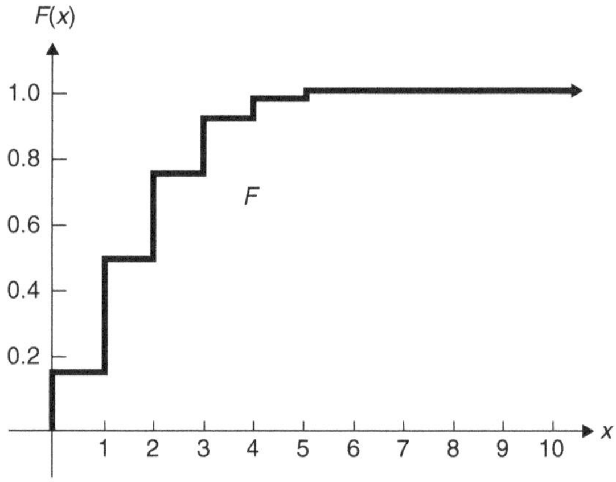

FIGURE 9.7 Cumulative density function and its graph.

To find the mean and variance of a binomial distribution, we need not find f(x) for all possible values of the random variable. Instead, we can develop some simple expressions that give us the values of these parameters for any binomial distribution.

Let X be the number of successes in a sequence of n Bernoulli trials, with probability of success p, and probability of failure $q = 1 - p$. For $i = 1, 2, \ldots, n$, let X be the number of successes on the ith trial; then X_i has mean p and variance pq, as previously described.

$$\text{But } X = X_1 + X_2 + \ldots + X_n \text{, so}$$

$$E(X) = E(X_1 + X_2 + \ldots + X_n)$$

$$= E(X_1) + E(X_2) + \ldots + E(X_n)$$

$$= p + p + \ldots + p$$

$$= np$$

The expected value of a sum of random variables is the sum of their expected values. Further, the Bernoulli experiments that comprise the sequence of trials are independent, so the variance of their sum is the sum of their individual variances:

$$Var(X) = Var(X_1 + X_2 + \ldots + X_n)$$

$$= Var(X_1) + Var(X_2) + \ldots + Var(X_n)$$

$$= \sigma_1^2 + \sigma_2^2 + \ldots + \sigma_n^2$$

$$= npq. \text{ Also, } \sigma = \sqrt{npq}.$$

In our example, X is $b(10, \frac{1}{6})$, so:

$$\mu = E(X) = 10 * \frac{1}{6} = \frac{10}{6} = \frac{5}{3} = 1.67,$$

$$\sigma^2 = Var(X) = 10 * \frac{1}{6} * \frac{5}{6} = \frac{50}{36} = \frac{25}{18} = 1.389, \text{ and}$$

$$\sigma = \sqrt{1{,}389} = 1.178.$$

The two parameters n and p determine which of the family of binomial distributions is being examined and, therefore, the shape of the distribution. Figures 9.8 and 9.9 illustrate the distributions $b(10, 0.5)$ and $b(10, 0.8)$.

With $p - 0.5$, the distribution is symmetrical about its mean, $np - 10 * 0.5 - 5$; when p is less than 0.5, the distribution is skewed to the left; and when p is greater than 0.5, the distribution is skewed to the left.

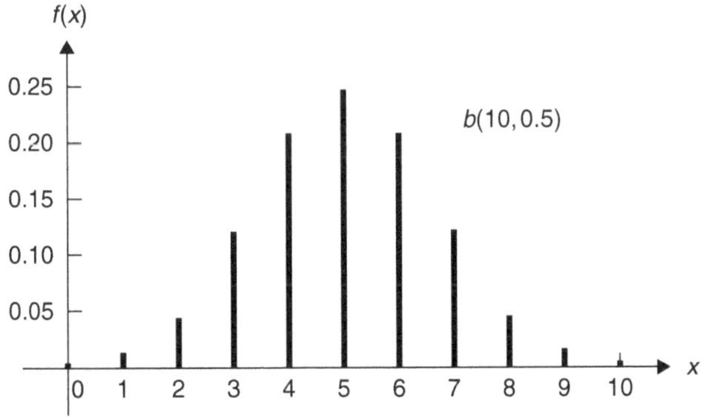

FIGURE 9.8 Distribution of $b(10, 0.5)$.

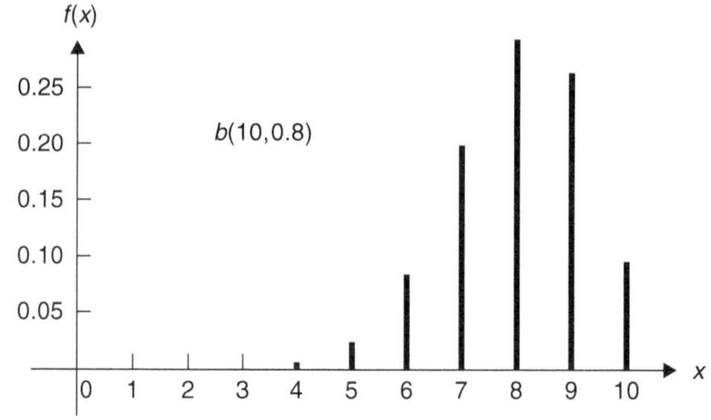

FIGURE 9.9 Distribution of $b(10, 0.8)$.

EXAMPLE 9.1

Consider this example of a binomial distribution. A salesman will make eight calls in a day. At each, he feels he has a 30% chance of making a sale, and whether he makes a sale on one particular call is independent of the others. If X is the number of sales the salesman will make, X has the binomial distribution $b(8, 0.3)$.

Solution:

The probability function of X is $f(x) = \binom{8}{3} 0.3^x 0.7^{8-x}$, and its values are given in Table 9.1 (as found in Table A-2).

ON DVD

TABLE 9.1 The binomial distribution b(8, 0.3).

X	**f(x) = P(X = x)**
0	0.058
1	0.198
2	0.296
3	0.254
4	0.136
5	0.047
6	0.001
7	0.000
8	0.000

The probability that the salesman will make four sales is f(4) = 0.136. The probability that he will make three or fewer sales is f(0) + f(1) + f(2) + f(3) = 0.058 + 0.198 + 0.296 + 0.254 = 0.806.

The expected number of sales is 8 * 0.3 = 2.4, with variance 8 * 0.3 * 0.7 = 1.68 and standard deviation $\sqrt{1.68} = 1.296$.

9.3.3 The hypergeometric distribution

Situations that give rise to the distribution known as *hypergeometric* often resemble those of binomial distributions but with an important difference. With the binomial distribution, the trials are *independent*, but in this new situation, the trials are *dependent* and in a specific way as shown in this example.

EXAMPLE 9.2

An employer has 18 employees, 10 of whom are competent, with the remaining 8 incompetent. She will assign 7 employees, randomly chosen, to a certain task. What is the probability that 4 of those chosen will be competent and the other 3 incompetent?

Solution:

The outcomes in this experiment are groups of 7 employees, and there are $\binom{18}{7}$ ways in which 7 employees can be chosen from the 18 (order of selection does not matter). We must count the number of outcomes, groups of 7 employees, containing 4 competent and 3 incompetent employees. The 4 competent employees may be chosen in $\binom{10}{4}$ ways and the incompetent employees in $\binom{8}{3}$ ways, so that there are $\binom{10}{4} * \binom{8}{3}$ outcomes in this event, and its probability is

$$\frac{\binom{10}{4} * \binom{8}{3}}{\binom{18}{7}} = \frac{210 * 56}{31824} = 0.3695 \cdot$$

In general, suppose that we are choosing a random group of n objects from a total of N objects, of which S are of a particular kind that we can call successes. The random variable X is the number of successes in the selected group. There are $\binom{N}{n}$ possible groups of n objects that can be chosen from N objects. To have exactly x successes in the chosen group requires that x objects be chosen from the S successes, and the remaining $(n - x)$ objects be chosen from the $(N - S)$ other objects; there are $\binom{S}{x}\binom{N-S}{n-x}$ ways for this to be done. We obtain this probability function for the random variable:

$$\mathrm{f(x)} = \mathrm{P(X = x)} = \frac{\binom{S}{x}\binom{N-S}{n-x}}{\binom{N}{n}}, x = 0, 1, 2, \ldots, \min(n, S),$$

0, otherwise.

A random variable whose distribution can be described with this function is called *hypergeometric*:

$$f(x) = P(X = x) = \frac{\binom{S}{x}\binom{N-S}{n-x}}{\binom{N}{n}}, x = 0, 1, 2, \ldots, \min(n, S),$$

$$= 0, \text{ otherwise.}$$

The selection of a handful of n objects from N objects is equivalent to selecting one object n times, if the selected objects are not replaced. This is the most important application of the hypergeometric distribution: sampling *without replacement*.

Finding the mean and variance of a hypergeometric distribution by using expected values is tedious. Applying intuition will, however, lead us to a correct result.

If S of N objects are of the kind labeled success, and we take a random sample of size n, then we would expect the proportion of successes in the sample to be the same as that in the original collection of objects (the population). If X is a hypergoemetric random variable, then

$$\mu = E(X) = n\frac{S}{N},$$

It can also be shown, using techniques more sophisticated than those available to us here, that

$$\sigma^2 = Var(X) = n\frac{S}{N}\left(1 - \frac{S}{N}\right)\left(\frac{N-n}{N-1}\right).$$

In our example, the expected number of competent people to be assigned to the task is

$$\mu = E(X) = 7 * \frac{10}{18} = 3.89, \text{with}$$

$$\sigma^2 = Var(X) = 7\frac{10}{18}\left(1 - \frac{10}{18}\right)\left(\frac{18-7}{18-1}\right) = 1.118, \text{ and}$$

$$\sigma = \sqrt{1{,}118} = 1{,}058.$$

Figure 9.10 shows the probability functions of hypergeometric distributions with $N = 10$ and $n = 5$ for three values of S, the number of successes in the population.

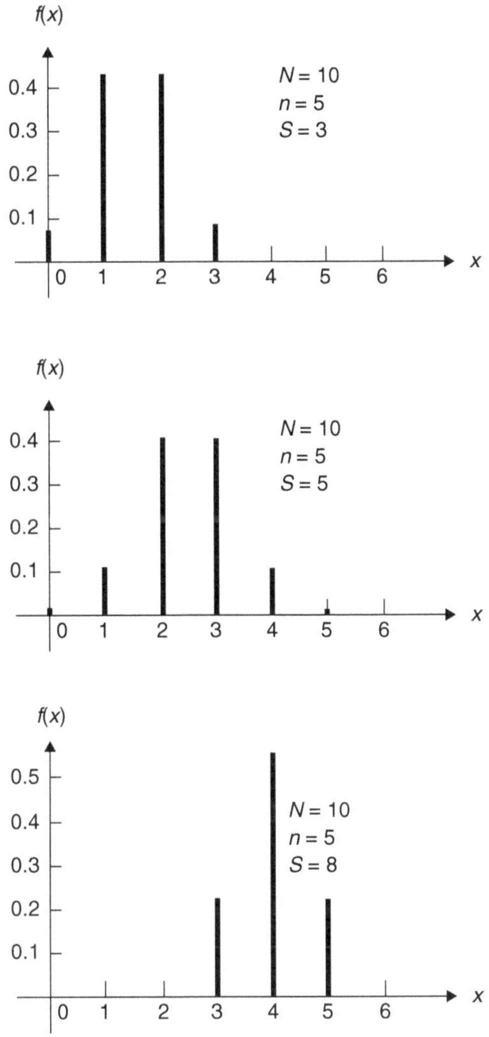

FIGURE 9.10 Probability functions of hypergeometric distributions.

Consider selecting n times from a group of N objects of which S are successes, but with *replacement*, returning each selected object to the original group before the next selection. In this case, the successive drawings are independent, and the probability of drawing a success remains $\frac{S}{N}$ throughout. This produces a sequence of n Bernoulli trials, and if X is the number of successes selected in the n drawings, X has the binomial distribution $b(n, \frac{S}{N})$, with mean $n\frac{S}{N}$ and variance $n\frac{S}{N}\left(1-\frac{S}{N}\right)$. This shows the relationship between sampling without replacement, which produces a hypergeometric distribution, and sampling with replacement, which gives rise to a binomial distribution. The means of the corresponding distributions are the same, but the variance of the hypergeometric distribution is less than the variance of the binomial distribution by a factor of $\frac{N-n}{N-1}$.

Notice, too, that as the size N of the original group increases, $\frac{N-n}{N-1}$ nears 1, and the hypergeometric distribution approaches its related binomial.

EXAMPLE 9.3

Hypergeometric distributions often occur in relation to quality control. Suppose that a manufacturer receives a certain component in lots of 10, and from each lot he randomly selects 3 components to be tested. If one or more of the tested components are defective, the lot is rejected. If a particular lot contains 2 defective components, what is the probability that it will be rejected?

Solution:

If X is the number of defective components in the sample of 3 that are tested, then X has a hypergeometric distribution with $N = 10$, $S = 2$, and $n = 3$. The lot will be rejected if the number of defective components is one or more, so

$$\begin{aligned} P(\text{Reject the lot}) &= P(X \geq 1) \\ &= 1 - P(X < 1) \\ &= 1 - P(X = 0) \\ &= 1 - \frac{\binom{2}{0}\binom{8}{3}}{\binom{10}{3}} \\ &= 1 - \frac{1 * 56}{120} \\ &= 1 - 0.467 = 0.533. \end{aligned}$$

Note that the expected number of defective components in the sample is

$$E(X) = \mu = 3 * \frac{2}{10} = 0.6, \text{ with variance}$$

$$\mathrm{Var}(X) = \sigma^2 = 3 * \frac{2}{10} * (1 - \frac{2}{10}) * (\frac{10-3}{10-1})$$

$$= 0.3733, \text{ and standard deviation}$$

$$\sigma = \sqrt{0.3733} = 0.611.$$

9.3.4 The poisson distribution

Consider the flow of customers at a fast-food restaurant, arriving "at random" and being served at one of four counter locations. This is an example of a "waiting-line" problem, in which customers wait in one or more queues to be served by one or more servers.

Waiting-line systems like this occur in such situations as:

- The assignment of operators to machines (one operator per machine). Here, the machines are customers, and the operators are servers.
- The execution of programs on a network. The programs are customers, and the network is the single server.
- The passage of motor vehicles through an intersection. Cars may form queues along the streets leading to the intersection.

In such situations, and many more like them, it is important to know how many customers might be expected to arrive, or how many service stations to set up so that waiting time is minimized without wasting resources. To answer these questions, and others related to waiting-line problems, we employ the *Poisson probability distribution*, which can be used to describe the arrivals of customers into queues.

The Poisson distribution applies only to situations called *Poisson processes*, and the occurrence of the event of interest (the submission of a job to a computer system, the arrival of a car at an intersection) in the process is called a *change*. We are interested in the number of changes in an interval of time.

A Poisson process is defined by three conditions. When you are applying a Poisson distribution to a real-world situation, it is wise to check experimentally that these conditions hold. They are:

- The numbers of changes in nonoverlapping intervals are independent.
- The probability of exactly one change in an interval is proportional to the length of the interval.
- In a small interval, the probability of more than one change is essentially 0.

Let α be the average number of changes in a one-unit interval of time, and let T be the length of an interval. Then the Poisson random variable X is the number of changes in the interval of length T, and its probability function is:

$$f(x) = P(X = x) = \frac{(aT)^x e^{-aT}}{x!}, \quad x = 0,\ 1,\ 2,\ 3,\ \ldots$$
$$= 0, \text{ otherwise}$$

The value of the parameter aT determines which Poisson distribution is being considered, and since this value is the average number of arrivals in time T, we would expect it to be the mean of X. Letting $\lambda = aT$, we see that this is the case:

$$\mu = E(X) = \sum_x x * f(x)$$
$$= \sum_{x=0}^{\infty} x \frac{\lambda^x e^{-\lambda}}{x!}$$
$$= \sum_{x=1}^{\infty} x * \frac{\lambda^x e^{-\lambda}}{x!}$$

$$= \sum_{x=1}^{\infty} \frac{\lambda^x e^{-\lambda}}{(x-1)!}$$

$$= \lambda \sum_{x=1}^{\infty} \frac{\lambda^{x-1} e^{-\lambda}}{(x-1)!}$$

$$= \lambda \sum_{x-1=0}^{\infty} \frac{\lambda^x e^{-\lambda}}{(x-1)!}$$

$$= \lambda * 1 = \lambda$$

Since f is a probability function,

$$\sum_{x-1=0}^{\infty} f(x-1) = \sum_{x-1=0}^{\infty} \frac{\lambda^{x-1} e^{-\lambda}}{(x-1)!}$$

a more surprising result (left as a special challenge for the reader) is that the variance of a Poisson distribution is also λ.

The graphs in Figure 9.11 show Poisson distributions with three different values for λ. When λ is small, the distribution is skewed to the right; as λ increases, it becomes more symmetrical.

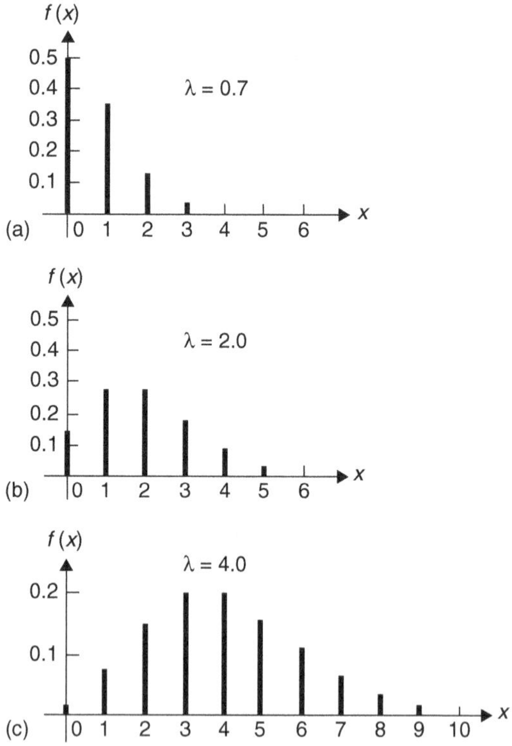

FIGURE 9.11 Poisson distributions with three different values for λ.

Suppose that a motel receives an average of one call per minute in the course of a working day and that the arrivals of these calls constitute a Poisson process. Note that this does not mean that the motel receives exactly one call every minute. What is the probability that in a given five-minute period, the motel receives six calls?

We must first observe that $a = 1$ and $T = 5$, so $\lambda = aT = 5$. Then the probability of six calls in five minutes is:

$$f(6) = P(X = 6) = \frac{5^6 e^{-5}}{6!} = \frac{5^6 * 0.006738}{6!} = 0.1462.$$

Note that Table A-3 provides values of $e^{-\lambda}$ for various λ, and Table A-2 gives the probabilities from selected Poisson distributions. ON DVD

Likewise, the probability of 8, 9, or 10 calls in 10 minutes is:

$$P(8 \le X \le 10) = f(8) + f(9) + f(10)$$

$$= \frac{10^8 e^{-10}}{8!} + \frac{10^9 e^{-10}}{9!} + \frac{10^{10} e^{-10}}{10!}$$

$$= 0.1126 + 0.1251 + 0.1251 = 0.3628.$$

9.3.5 Summary of discrete distributions

Table 9.2 summarizes the discrete distributions we have described.

TABLE 9.2 Summary of discrete distributions.

Type	Probability Function	$\mu = E(x)$	σ^2, σ	Distributional Shape
Uniform	$f(x) = 1/n$, where n is the number of values in the x	Depends on values of x	Depends on values of x	Rectangular
Binomial	$f(x) = \binom{n}{x} p^x q^{(n-x)}$, where N = number of trials, p = prob. of success and q = 1 - p	np	npq, $\sqrt{npq}$	if p = 1/2 symmetrical If p > 1/2 neg. skewed if p < 1/2 pos. skewed
Hyper-geometric	$f(x) = \frac{\binom{S}{x}\binom{N-S}{n-x}}{\binom{N}{n}}$ where n = size of sample, N = size of population, S = number of successes in Population	$n\frac{S}{N}$	$n\frac{S}{N}\left(1-\frac{S}{N}\right)\left(\frac{N-n}{N-1}\right)$, $\sigma = \sqrt{\sigma^2}$	similar to binomial with $p\frac{S}{N}$

(*Continued*)

TABLE 9.2 Continued.

Type	Probability Function	$\mu = E(x)$	σ^2, σ	Distributional Shape
Geometric	$f(x) = (1-p)^{(1-x)}\,p$	$\frac{1}{p}$	$\frac{(1-p^2)}{p}$	similar to binomial
Poisson	$f(x) = \frac{\lambda^x e^{-\lambda}}{x!}$, where $\lambda = \alpha(T)$ Average number of arrivals In time T	λ	$\lambda\sqrt{\lambda}$	small λ, pos. skewed as λ increases, more symmetrical

9.4 Exercises

1. At an automobile dealership, the probability that a salesperson will sell a car is 0.20 for each customer that walks in, with customer behaving independently of each other. If 20 customers enter the dealership on a given morning:

   a. What is the probability that 5 will buy cars?

   b. What is the probability that fewer than 4 will buy cars?

   c. What is the expected number of sales? The variance?

2. Shipments of raw materials arrive at a factory on time with probability 0.80. Assume that the shipments arrive independently.

   a. What is the probability that of eight shipments, fewer than six arrive on time?

   b. Is the assumption in the statement of the problem likely to be valid? Why or why not?

3. A group of 20 job applicants contains 12 college graduates. If 6 are chosen at random to be interviewed:

   a. What is the probability that 3 are college graduates?

   b. What is the probability that fewer than 3 are college graduates?

   c. What is the expected number of college graduates in the group? Its variance and standard deviation?

4. In a batch of 25 calculators, 3 are defective. If 5 are selected at random and tested:

   a. What is the probability that none of the defective calculators is found?

   b. What is the probability that all of the defective calculators are found?

   c. What is the probability that at least one defective calculator is found?

   d. What is the expected number of defective calculators found and the variance?

5. Customers arrive at a shoe store at an average rate of 0.3 per minute. In a 10-minute period:

a. What is the probability that exactly four customers arrive?

b. What is the probability that two or fewer customers arrive?

c. What is the probability that more than two customers arrive?

d. What is the expected number of customers? The variance and standard deviation?

6. In problem 5, what is the expected number of customers in five minutes? One hour?

7. Poisson distributions often occur in association with intervals of some quantity other than time: In a factory that makes coaxial cable, flaws occur with a Poisson distribution at an average rate of 1 per 1,000 feet of cable. In a 5,000-foot roll:

a. What is the expected number of flaws?

b. What is the probability of three or fewer flaws?

8. We draw a single card from a well-shuffled deck, and X is its rank.

a. Give the probability function of X and its graph.

b. Draw the of the cumulative density function of X.

c. Find E(X), Var(X), and σ.

d. Find $P(3 \leq X < 7)$.

Reference

1. Muller, C.B, "Dartmouth Project for Teaching Engineering Problem Solving," available online at *http://engineering.dartmouth.edu/teps/*, 1996–1997.

CHAPTER 10

CONTINUOUS PROBABILITY DISTRIBUTIONS

OVERVIEW AND LEARNING OBJECTIVES

In This Chapter

10.1 Introduction

The range of a discrete random variable is finite or countably infinite, allowing for a set of distinct values, each with positive probability defined in an interval or union of intervals.

But this approach for assigning probability values to a continuous random variable would violate the axiom of probability that requires the sum of all the probability values to equal one. In this chapter, we will provide an overview of continuous probability distributions that are commonly applied.

10.2 Introduction to Continuous Probability Distributions

The range of a continuous random variable is *uncountably infinite*, and as stated before, any assignment of probabilities that gives positive probability to each value in the range violates the axioms of probability since the total probability associated with the experiment would become infinite. We must find a different way to assign probabilities to events that involve continuous random variables. In particular, the probability associated with any single value must be 0.

We saw when using Venn diagrams that the *area* of a portion of the Venn diagram representing an event could be used to represent the *probability* of that event. We now take the analogy of probability and area a step further by defining a function under which *areas are equal to probabilities*:

> Let X be a continuous random variable. The function f(x) is the *probability density function* of X if the probability that X takes on a value in the interval [a, b] is equal to the area trapped, within the interval, between the horizontal axis and the f(x) curve. Using calculus, this is $P(a \le X \le b) = \int_a^b f(x)dx$.

The probability that the random variable X falls in the closed interval [a, b] is equal to the area under the graph of the probability function and above [a, b], as shown in Figure 10.1.

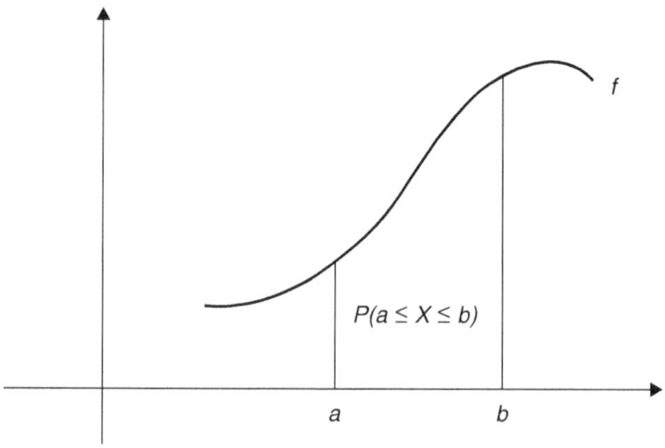

FIGURE 10.1 $P(a \le X \le b)$ for a continuous random variable X.

ON DVD

All figures and tables in this chapter appear on the companion DVD.

Consider the simple experiment of spinning a toy spinner with X the value at which it stops (see Figure 10.2). X is continuous, and its range is the interval [0, 1]. The probability function of f, as shown in the graph in Figure 10.3, is

$$f(x) = 1,\ 0 \le x \ge,$$
$$f(x) = 0,\ \text{all other } x$$

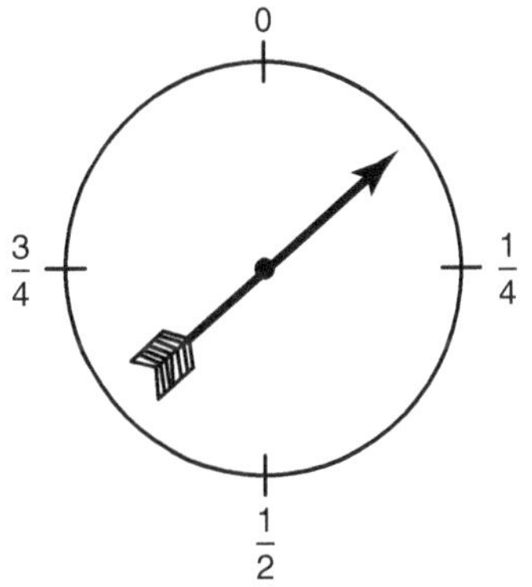

FIGURE 10.2 A toy spinner.

It is reasonable that $P(\frac{1}{4} \le X \le \frac{1}{2})$ is equal to the area of a rectangle, or base times height = $(\frac{1}{2} - \frac{1}{4}) * 1 = \frac{1}{4}$ (refer to Figure 10.3). This same result can be derived using calculus:

$$P(\frac{1}{4} \le X \le \frac{1}{2})\ \int_{\frac{1}{4}}^{\frac{1}{2}} f(x)dx = \int_{\frac{1}{4}}^{\frac{1}{2}} 1dx = x\Big|_{\frac{1}{4}}^{\frac{1}{2}} = \frac{1}{2} - \frac{1}{4}.$$

$$P(\frac{1}{4} \le X \le \frac{1}{2}).$$

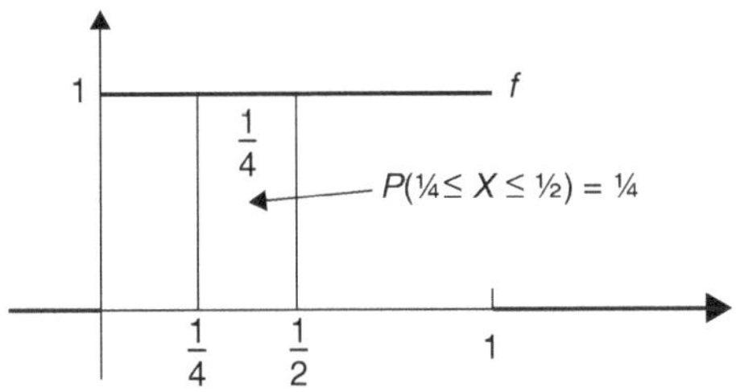

FIGURE 10.3 Continuous uniform distribution on [0,1].

Two observations should be made about a continuous probability function f:

- f(x) ≥ 0 for all x.

  If f were ever to dip below the x-axis, then there would be a tiny interval in the dip over which the integral of f, or area, would be negative. This would mean that the probability that X falls in that interval is negative, which cannot happen.

- $\int_{-\infty}^{\infty} f(x)dx = 1,$ or the total area under the graph and above the x-axis must be equal to 1.

  All the area under the graph of f, that is, all the probability associated with the random variable, must total 1, since this represents the probability that any outcome event occurs.

  We can see, too, that probabilities assigned by integrating f, finding areas, satisfy the axioms of probability:

- For every interval [a, b], the area trapped is greater than or equal to 0.

$$\mathrm{P}(a \le X \le b) = \int_a^b f(x)dx \ge 0, \quad \text{since} \quad \mathrm{f}(x) \ge 0.$$

- The total area is equal to 1.

$$\int_{-\infty}^{\infty} f(x)dx = 1, \quad \text{or} \quad \mathrm{P(S)} = 1.$$

- For disjoint intervals [a, b] and [c, d],

$$\mathrm{P}(a \le \mathrm{X} \le b \text{ or } c \le \mathrm{X} \le d) = \mathrm{P(a \le X \le b) + P(c \le X \le d)}.$$

$$\mathrm{P}(a \le \mathrm{X} \le b \text{ or } c \le \mathrm{X} \le d) = \int_a^b f(x)dx \ge 0 + \int_c^d f(x)dx \ge 0$$
$$= \mathrm{P(a \le X \le b) + P(c \le X \le d)}.$$

$$[\text{If } \mathrm{A} \cap \mathrm{B} = \phi, \quad \text{then} \quad \mathrm{P(A \cup B) = P(A) + P(B)}].$$

Finally, we can verify the earlier intuitive argument that for a continuous random variable, the probability associated with any single value must be 0, since there does not exist an area under one point. Using calculus we have:

$$\mathrm{P}(\mathrm{X} = c) = \mathrm{P}(\mathrm{c} \le X \le c) = \int_c^c f(x)dx = 0.$$

This means that including or excluding any single point, or a countable set of points, from the possible values of X that describe an event will not change the probability of that event.

$$\mathrm{P}(a \le X \le b) = \mathrm{P}(a < X \le b) = \mathrm{P}(a \le X < b) = \mathrm{P}(a < X < b) = \int_a^b f(x)dx.$$

We have seen that the cumulative density functions of discrete distributions are nondecreasing step functions because positive probabilities are associated with distinct individual values in the range of the random variable. In a continuous distribution, probability is not associated with particular values but is "spread out" over an interval; this changes the appearance of the cumulative density function.

Let X be a continuous random variable with probability function f. The *cumulative density function* F associated with X is defined to be the area under the curve in the interval $(-\infty, x_0)$, or:

$$F(x_0) = P(X \leq x_0) = \int_{-\infty}^{x_0} f(x)dx.$$

That is, the value of F at x_0 is the area under the graph of f and to the left of x_0.

Consider the following example: X is a continuous random variable with the probability function shown in Figure 10.4.

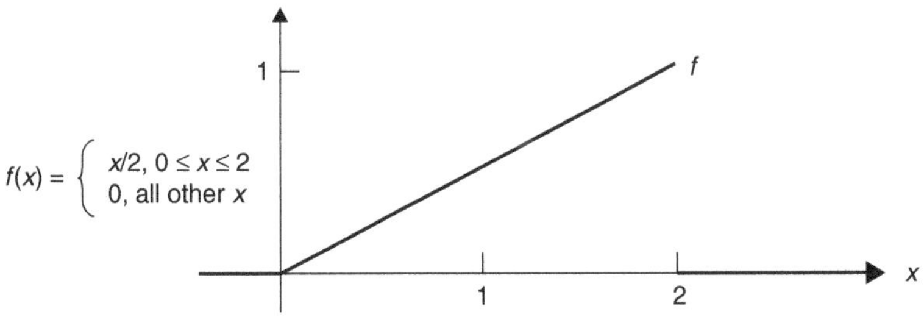

FIGURE 10.4 A continuous probability function.

The value of the cumulative density function F at 1 is equal to the area $\frac{1}{2} * (\frac{1}{2} * 1) = \frac{1}{4}$, or by using calculus we have:

$$F(1) = \int_{-\infty}^{1} f(x)dx = \int_{0}^{1} \frac{x}{2}dx = \left.\frac{x^2}{4}\right|_0^1 = \frac{1}{4} - \frac{0}{4} = \frac{1}{4}.$$

In general,

$$F(x_0) = \int_{-\infty}^{x_0} f(x)dx = \int_{0}^{x_0} \frac{x}{2}dx = (\frac{x^2}{4})_0^{x_0} = \frac{1}{4} - \frac{0}{4} = \frac{x_0}{4}, \quad \text{for } 0 \leq x_0 \leq 2,$$

and F has the graph as shown in Figure 10.5.

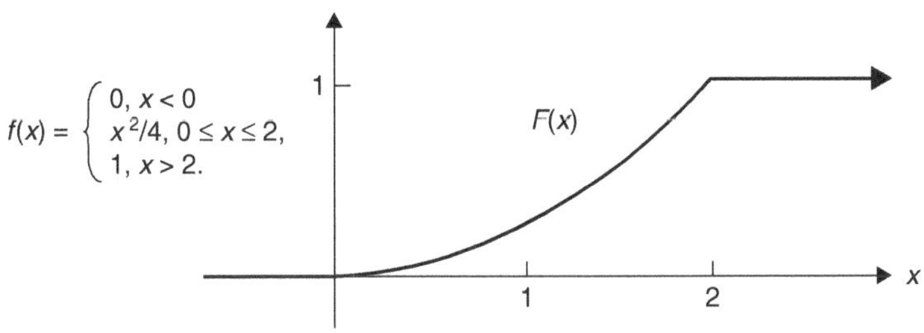

FIGURE 10.5 Cumulative density function of a continuous random variable.

Note that the cumulative density function of a continuous random variable is always nondecreasing and continuous.

10.3 Expected Value and Variance of Continuous Distributions

From the development of the definite integral in calculus, we know that the integral is the limit of the sum of an increasing number of ever-smaller values. We have also seen the correspondence between the integrals involved in the current discussion and summation expressions related to discrete distributions. It is not surprising that the definitions of the mean and the variance of a continuous distribution are nearly identical to their discrete counterparts. We integrate over all possible values of the random variable rather than taking a sum:

Let X be a continuous random variable with probability function f. The *expected value* or *mean* of X is defined to be

$$E(X) = \mu = \int_{-\infty}^{\infty} x f(x)dx$$

Let X be a continuous random variable with probability function f. The *variance* and *standard deviation* of X is defined to be

$$\text{Var}(X) = \sigma^2 = \int_{-\infty}^{\infty} (x-\mu)^2 f(x)dx, \text{ and}$$

$$\sigma = \sqrt{\sigma^2}.$$

As with discrete distributions, there is a shortcut formula, which is generally used to find the variance:

$$\text{Var}(X) = \sigma^2 = E(X^2) - \mu^2$$

$$= \int_{-\infty}^{\infty} x^2 f(x)dx - \mu^2.$$

In our last example, where

$$f(x) = \frac{x}{2}, 0 \le x \le 2,$$

$$f(x) = 0, \text{ elsewhere},$$

$$E(X) = \mu = \int_{-\infty}^{\infty} x f(x)dx = \int_0^2 x * \frac{x}{2}dx = \int_0^2 \frac{x^2}{2}dx$$

$$= \left(\frac{x^3}{6}\right)_0^2 = \frac{8}{6} - \frac{0}{6} = \frac{4}{3}.$$

$$\text{Var}(X) = \sigma^2 = \int_{-\infty}^{\infty} x^2 f(x)dx - \mu^2 = \int_0^2 x^2 \frac{x}{2}dx - \mu^2$$

$$= \int_0^2 \frac{x^3}{2}dx - \mu^2 = \left(\frac{x^4}{8}\right)_0^2 - \mu^2$$

$$= \left(\frac{16}{8} - \frac{0}{8}\right) - \left(\frac{4}{3}\right)^2 = \frac{2}{9} = 0.222, \text{ and}$$

$$\sigma = \sqrt{\sigma^2} = \sqrt{0.222} = 0.471$$

10.4 Particular Continuous Distributions

Certain continuous distributions occur in a wide variety of experimental situations. We will examine the continuous uniform and exponential distributions in this chapter. The normal distribution is a major part of the foundation of the field of statistics, and therefore the entire next chapter is devoted to this distribution.

10.4.1 The continuous uniform distribution

The range of the continuous uniform distribution is always an interval [a, b], and it is characterized by the fact that for any two intervals of equal width within [a, b], the probabilities that the random variable X will fall in each are equal. This is equivalent to saying that the areas under the graph of f, the probability function of X, above the two subintervals of equal width must always be equal. For this to be true, the graph of f must be a horizontal line above [a, b], as shown in Figure 10.6.

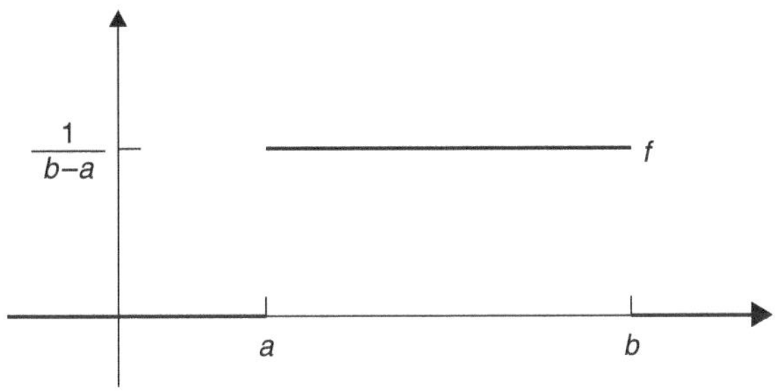

FIGURE 10.6 Probability function of a uniform distribution.

Let X be a continuous random variable with this probability function:

$$f(x) = \frac{1}{b-a}, \quad a \le x \le b$$

$$f(x) = 0, \text{ all other x.}$$

Then X has a *continuous uniform distribution.* It is clear from the graph of f why such a distribution is also sometimes called *rectangular.*

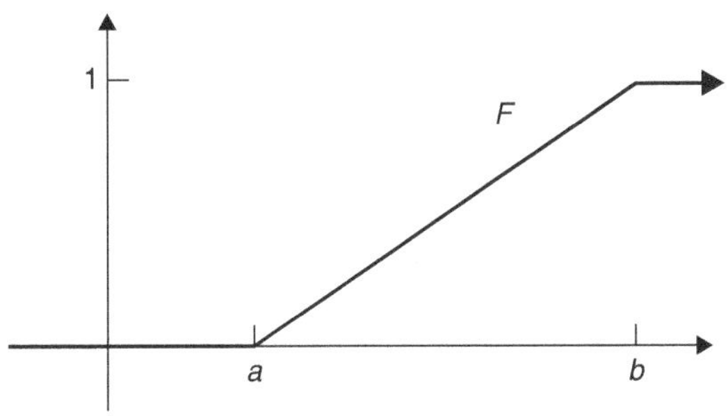

FIGURE 10.7 Cumulative density function F of the continuous uniform distribution.

It is not difficult to find the mean and variance of a uniform distribution as illustrated in Figure 10.7:

$$\mu = \frac{b+a}{2}, \text{ and}$$

$$\sigma^2 = \frac{(b-a)^2}{12}.$$

We have already seen a continuous uniform distribution in the toy spinner example in Figure 10.2. In that example, $a = 0$, $b = 1$, and the probability function is

$$\mathrm{F}(\mathrm{x}) = \frac{1}{1-0}, 0 \le x \le 1$$

$$\mathrm{F}(\mathrm{x}) = 0, \text{ for all other x.}$$

For this distribution,

$$\mu = \frac{1+0}{2} = \frac{1}{2} \quad \text{and} \quad \sigma^2 = \frac{(1-0)^2}{12} = \frac{1}{12}$$

10.4.2 The exponential distribution

ON DVD

Random numbers, as listed in Table A-28 in Appendix A, are generated by computer programs, and are intended to behave as though they come from uniform distributions. We can use these random numbers to generate numbers from other types of distributions by taking advantage of the fact that the cumulative density function F of any continuous distribution is monotonically increasing and continuous, and therefore has a well-defined inverse F^{-1}. To find a nonuniform random number x, we generate a uniform random number r, and let $x = \mathrm{F}^{-1}(r)$, as shown in Figure 10.8.

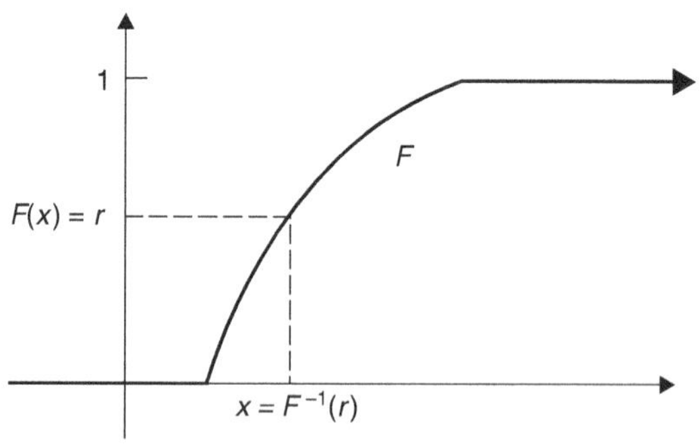

FIGURE 10.8 The exponential distribution.

Recall the Poisson process discussed in Chapter 9, "Discrete Probability Distributions," regarding the random arrivals of customers into a queue. Rather than count the number of arrivals in some interval of time T, let us now consider the length of time from some initial time t_0 until the next customer arrives, that is, the time until a change occurs in a Poisson process.

Let the random variable X be the time until a change occurs in a Poisson process. Then X has an *exponential* distribution, and its probability function f is:

$$f(t) = ae^{-at},\ t \geq 0$$
$$f(t) = 0,\ \text{otherwise}$$

Where t = time until the next change and α = average rate of changes per unit time. Note that since time intervals of negative length are impossible, t is always nonnegative.

We can find the mean of an exponential random variable by integration, according to the definition of the mean of a continuous distribution:

$$\mu = E(X) = \int_{-\infty}^{\infty} t * f(t)dt = \int_0^{\infty} t\left(ae^{-at}\right)dt = \int_0^{\infty} ate^{-at}\,dt.$$

Integrating by parts, let

$$u = at, \quad \text{so that} \quad du = a\,dt, \quad \text{and let}$$
$$dv = e^{-at}, \quad \text{so that} \quad v = \frac{-1}{a}e^{-at}.$$

Then

$$\mu = \left[(at)\left(\frac{-1}{a}e^{-at}\right)\right]_0^{\infty} - \int_0^{\infty}\left(-\frac{1}{a}e^{-at}\right)(adt)$$
$$\left[-te^{-at}\right]_0^{\infty} - \left[\frac{1}{a}e^{-at}\right]_0^{\infty}.$$

Using L'Hospital's rule, as applied here, if $\lim_{x\to\infty} f(x) = \infty$ and $\lim_{x\to\infty} g(x) = \infty$, then $\lim_{x\to\infty}\frac{f(x)}{g(x)} = \lim_{x\to\infty}\frac{f'(x)}{g'(x)}$.

$$= \left[\lim_{t\to\infty}\frac{-t}{e^{at}} - 0\right] - \left[\frac{1}{a}\lim_{t\to\infty}\frac{1}{e^{\partial t}} - \frac{1}{a}\right]$$
$$= \left[\lim_{t\to\infty}\frac{-1}{a\,e^{at}}\right] - \left[\frac{-1}{a}\right] = \frac{1}{a}.$$

This is reasonable; the average rate of change is a, and the expected time until a change is $1/a$.

Two important results are that the variance of an exponential distribution is $1/a^2$ and that the cumulative density function F is given by $F(t) = 1 - e^{-at},\ t \geq 0$.

The graphs in Figure 10.9 show the shapes of exponential distributions. Note that these curves always start at the point $(0, a)$ and are steeper for larger values of a.

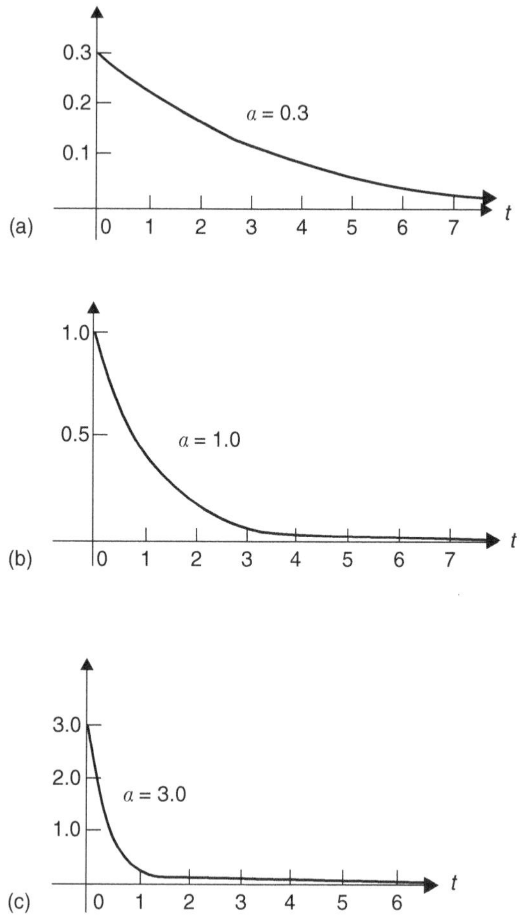

FIGURE 10.9 Shapes of exponential distribution.

This example illustrates the exponential model. Customers arrive at a bank at an average rate of 1.5 per minute, and their arrivals constitute a Poisson process. If one customer has just walked in, what is the probability that the next will follow in less than one minute?

Let X be the length of time until the next customer enters. Then X has an exponential distribution with $a = 1.5$. We seek P(X<1):

$$P(X < 1) = F(1) = 1 - e^{-1/5*1} = 1 - 0.2231 = 0.7769.$$

Similarly, the probability that the next customer will enter after a wait of two minutes or more is:

$$P(X \geq 2) = 1 - P(X < 2) = 1 - F(2).$$

$$= 1 - (1 - e^{-1.5*2}) = 0.0498.$$

The expected length of time between customers is $\frac{1}{a} = \frac{1}{1.5} = 0.67$ minutes, with variance $\frac{1}{a^2} = \frac{1}{1.5^2} = 0.44$ and standard deviation $\frac{1}{a} = 0.67$.

10.5 Exercises

1. Let $f(x)=\begin{cases} 2x, 0<x<1 \\ 0, all-other-x \end{cases}$.
   a. Verify that f is acceptable as the probability function of a continuous random variable, and draw its graph.
   b. Find $P(0 \leq x \leq 1/2)$.
   c. Find $P(1/2 < x < 2)$.
   d. State the cumulative density function of x and draw its graph.
   e. Find E(x).
   f. Find σ^2 and σ.
2. Let the random variable X be uniformly distributed on the interval [0, 3].
   a. Give the probability function of X and draw its graph.
   b. Find $P(1 \leq X \leq 2)$.
   c. Find $P(2.5 \leq X)$.
   d. Find E(X) and Var(X).
3. Oil deliveries at a refinery are uniformly distributed between 50,000 and 100,000 barrels per day.
   a. What is the probability that between 50,000 and 60,000 barrels will be delivered on a given day?
   b. If X is the amount of oil delivered on a given day, what is the probability function of X?
   c. What is the probability that more than 80,000 barrels will be delivered?
   d. What is the expected amount of a daily delivery? The variance and standard deviation?
4. Customers arrive at a shoe store at an average rate of 0.3 per minute. If the store opens precisely at 9:00 A.M.:
   a. What is the probability that the first customer enters within one minute?
   b. What is the expected length of time until the first customer enters? The variance?
5. Cars arrive at an intersection from the west at an average rate of four per minute.
   a. What is the expected length of time between two arrivals from the west?
   b. One car has just driven through the intersection from the west. What is the probability that another car will pass within 15 seconds? After at least a minute?
   c. Two cars have passed with a gap between them of only 5 seconds. What is the probability that another will pass within 15 seconds?

6. Z is the standard normal random variable, N(0, 1). Find the following probabilities:

a. $P(0 \le Z < 1.3)$

b. $P(Z \ge 2.15)$

c. $P(1.1 \le Z < 2.27)$

d. $P(-2.89 < Z \le 0)$

e. $P(Z < -1.77)$

f. $P(-2.14 < Z < -1.65)$

g. $P(Z \ge -1.00)$

h. $P(Z < 0.24)$

i. $P(-1.03 < Z \le 0.79)$

j. $P(Z > 3.0)$

CHAPTER 11

THE NORMAL DISTRIBUTION

OVERVIEW AND LEARNING OBJECTIVES

In This Chapter

11.1 Introduction

In this chapter, we present the normal probability distribution, a continuous probability distribution that is so commonly applied that it deserves its own chapter. This is an often used distribution in many fields of study because distributions of sample means approach a normal distribution, given a "large" sample size; the computations of probabilities is straightforward; and this distribution has enabled good decision making for many applications.

11.2 The Normal Distribution

Consider a person who is throwing darts and aiming at the bull's-eye. If the thrower is acquainted with the game, he will "on the average" hit the bull's-eye, and large deviations from it will occur less often than will small ones. Also, deviations above and below the bull's-eye will occur with approximately equal frequency.

These characteristics—values clustered symmetrically around a single central value and small deviations more common than large ones—can be observed in a great variety of phenomena. Measurements of men's heights are an example; very few men are more than a foot shorter or taller than the average height of 5′ 10″, but many are within two inches of this value. Other quantities that behave in this way are scores on standardized tests and dimensions of manufactured objects.

In the 18th century, Carl Fredrick Gauss derived a probability function that provided a model for many variables with the characteristics described above. This function, whose graph is called the *normal curve*, combines with the Central Limit Theorem (discussed in the next unit) to form the foundation of statistical inference. It is given in this definition:

Let X be a continuous random variable whose probability function f is

$$f(x) = \frac{1}{\sqrt{2\pi\sigma}} e^{\frac{-(x-\mu)^2}{2\sigma^2}}, -\infty \leq x \leq \infty.$$

Then X has a *normal* distribution with mean μ and variance σ^2, denoted $N(\mu,\sigma^2)$.

The probability function in Figure 11.1 describes what is traditionally known as the "bell-shaped curve."

ON DVD All figures and tables in this chapter appear on the companion DVD.

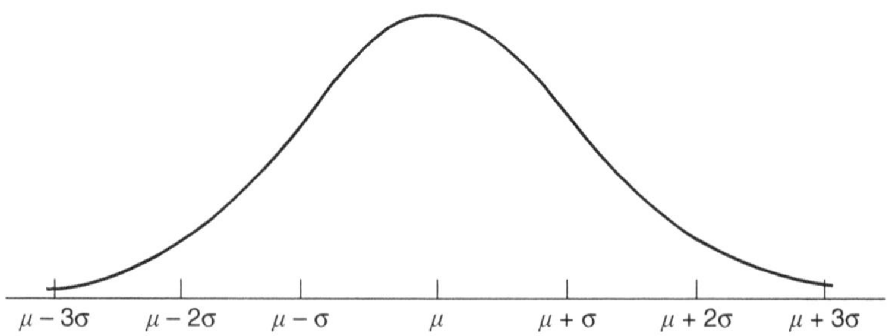

FIGURE 11.1 A normal distribution.

For a given value d, $f(\mu - d) = f(\mu + d)$; that is, the probability function is symmetrical around the mean μ. Also, the quantity expression $e^{\frac{-(x-\mu)^2}{2\sigma^2}}$ attains its maximum value of 1 when $x = \mu$, so the maximum value of f is $f(\mu) = \frac{1}{\sqrt{2\pi\sigma}}$. Finally,

$$\lim_{x \to -\infty} f(x) = \frac{1}{\sqrt{2\pi\sigma}} \lim_{x \to -\infty} e^{\frac{-(x-\mu)^2}{2\sigma^2}} = 0, \text{ and}$$

$$\lim_{x \to \infty} f(x) = \frac{1}{\sqrt{2\pi\sigma}} \lim_{x \to \infty} e^{\frac{-(x-\mu)^2}{2\sigma^2}} = 0.$$

The graph of f approaches the x-axis as a limit away from μ.

The shape of a normal distribution is determined by its standard deviation, while its location is determined by its mean, as shown in Figure 11.2.

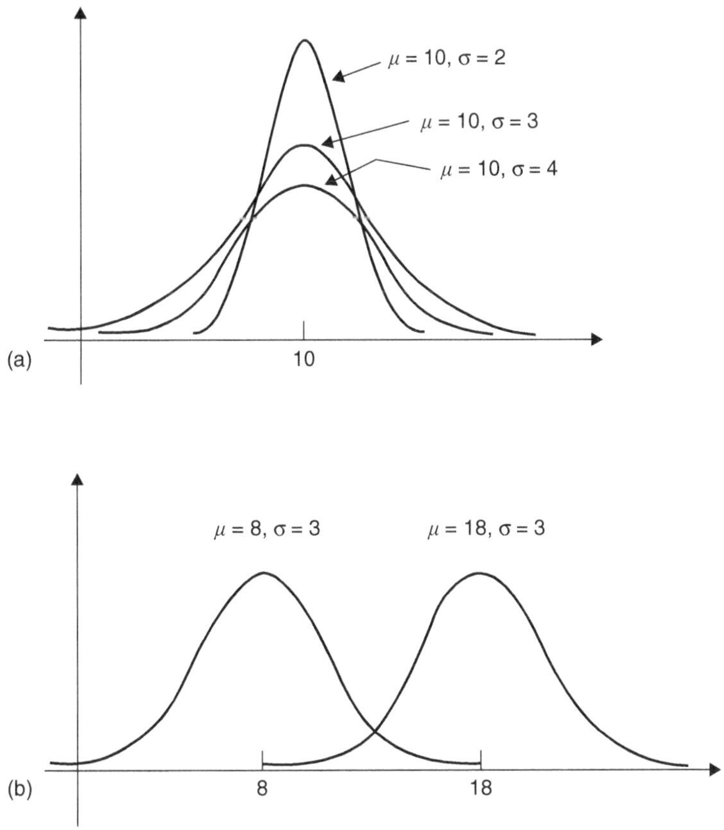

FIGURE 11.2 Shape of normal distributions for: (a) fixed μ and varying σ, and (b) fixed σ and varying μ.

The cumulative density function for a normal distribution is defined as:

$$F(X) = \int_{-\infty}^{\infty} f(x)dx = \int_{-\infty}^{\infty -} \frac{1}{\sqrt{2\pi\sigma}} e^{\frac{-(x-\mu)^2}{2\sigma^2}} dx.$$

Its graph, called an *S curve*, is shown in Figure 11.3.

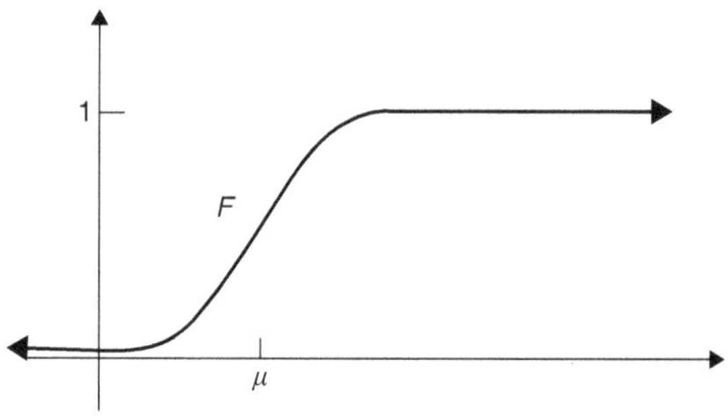

FIGURE 11.3 S-curve graph of the cumulative density function for a normal distribution.

There are an infinite number of normal distributions, each one characterized by its mean and standard deviation. This fact has great impact on finding probabilities involving normal random variables.

In our previous work with continuous random variables, we calculated probabilities by evaluating definite integrals, a process that involves finding the antiderivative of the probability function. We cannot do this with normal distributions, however, because there is no function whose derivative is e^{-x^2}. The normal probability function has no antiderivative.

However, there are numerical methods for approximating areas under a curve (the trapezoidal approximation, Simpson's rule, and others), and these methods can be used to build tables of probabilities involving normal distributions. Since there are an infinite number of normal distributions, we build such a table for only one, then relate all other normal distributions to it. The following definition describes this fundamental distribution.

> The *standard normal distribution*, whose random variable is generally indicated by Z, has mean 0, variance and standard deviation 1, and probability function:
>
> $$F(x) = \frac{1}{\sqrt{2\pi}} e^{\frac{-x^2}{2}}, \quad -\infty \leq x \leq \infty.$$

That is, Z is N(0,1).

The standard normal random variable can be transformed into another normal random variable with mean μ and standard deviation σ by multiplying by σ and adding μ. That is, if X is $N(\mu,\sigma^2)$,

$$X = \sigma Z + \mu.$$

Likewise, X can be transformed into Z by subtracting μ and dividing by σ.

$$(X - \mu)/\sigma = Z.$$

The pair of graphs in Figure 11.4 shows the relationship between a normal distribution with mean 10 and variance 9, and the standard normal distribution. Note that X = 3Z + 10 and (X – 10)/3 = Z.

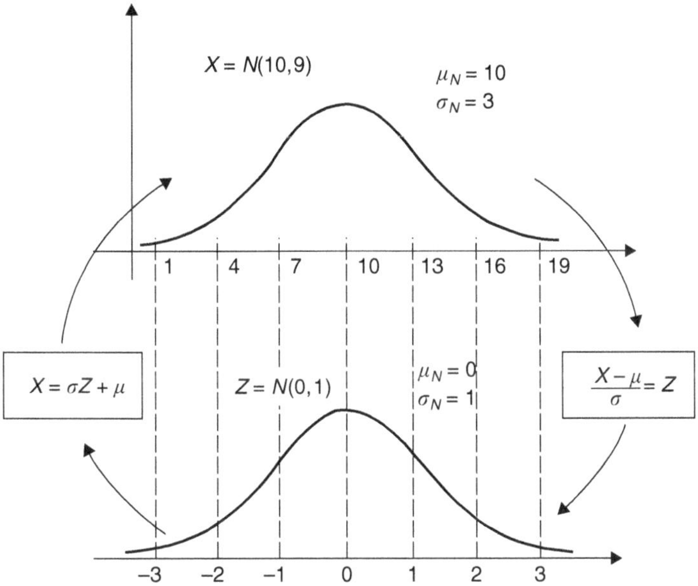

FIGURE 11.4 Relationship between a normal distribution, with mean 10 and variance 9, and the standard normal distribution.

A normal random variable X with mean μ and variance σ^2 is related to the standard normal random variable Z through the equation $(X - \mu)/\sigma = Z$. We will use this relationship to find probabilities involving normal random variables, but first we must become familiar with the distribution of Z. We do this through a series of examples.

EXAMPLE 11.1

Let Z be N(0, 1), the standard random variable. Find $P(0 \leq Z < 1)$ as depicted in Figure 11.5.

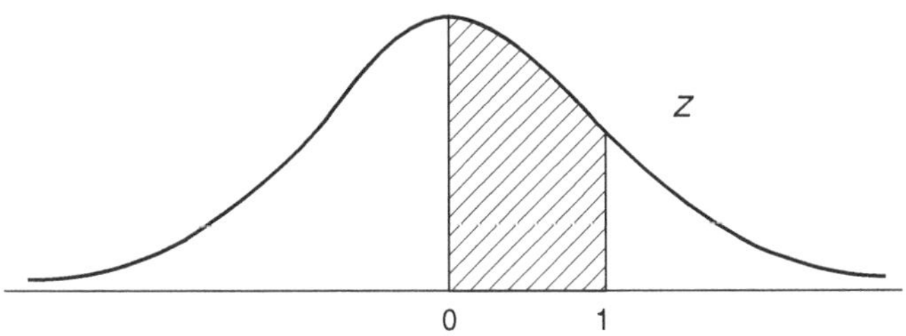

FIGURE 11.5 Area representation for $P(0 \leq z \leq 11)$.

Solution:

Table A-1 in Appendix A gives values of $P(0 \leq Z \leq z)$ for values of z from 0 to 3.9. From the table, $P(0 \leq Z \leq 1) = 0.3413$. Note that, since there is zero probability associated with any single value in a continuous distribution, $P(0 \leq Z < 1) = P(0 < Z \leq z) = P(0 \leq Z \leq z) = P(0 \leq Z \leq z)$.

EXAMPLE 11.2

Find $P(Z \geq 0)$ as depicted in Figure 11.6.

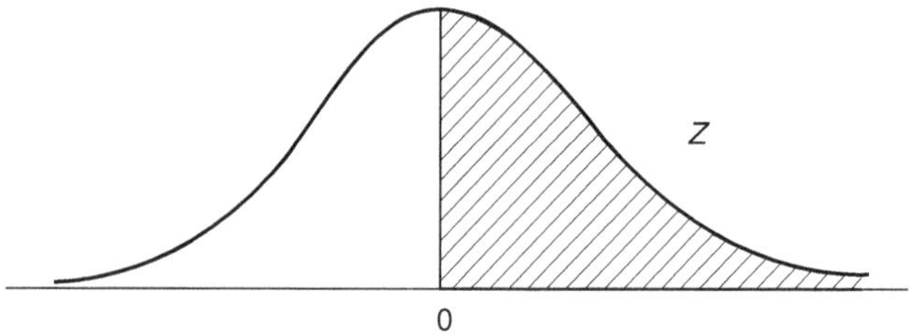

FIGURE 11.6 Area representation for $P(z \geq 0)$.

Solution:

We have seen that normal distributions are symmetric around their means. Therefore, Z is symmetric around 0. The total area under the curve is 1, so $P(Z \geq 0) = 0.5000$. Likewise, $P(Z \leq 0) = 0.5000$.

EXAMPLE 11.3

Find $P(Z > 1.5)$ as depicted in Figure 11.7.

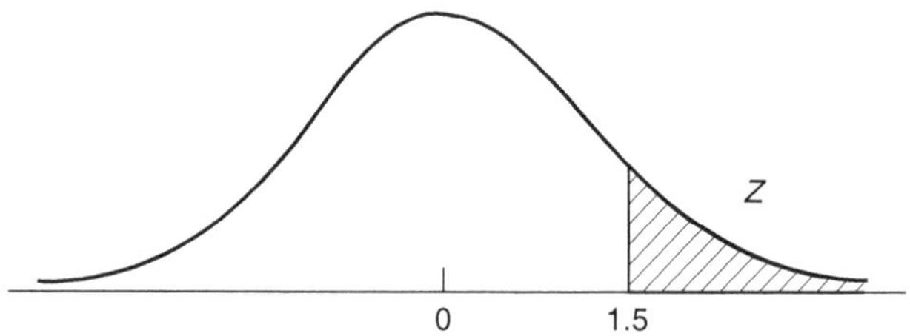

FIGURE 11.7 Area representation for $P(z \geq 1.5)$.

Solution:

$P(Z > 1.5)$ is not given in the table. However, $P(Z > 1.5) = P(Z > 0) - P(0 < Z < 1.5)$, and this latter value is 0.4332. Then

$$\begin{aligned} P(Z > 1.5) &= 0.5000 - P(0 < Z < 1.5) \\ &= 0.5000 - 0.4332 = 0.0688. \end{aligned}$$

EXAMPLE 11.4

Find P(1.25 ≤ Z ≤ 2.10) as depicted in Figure 11.8.

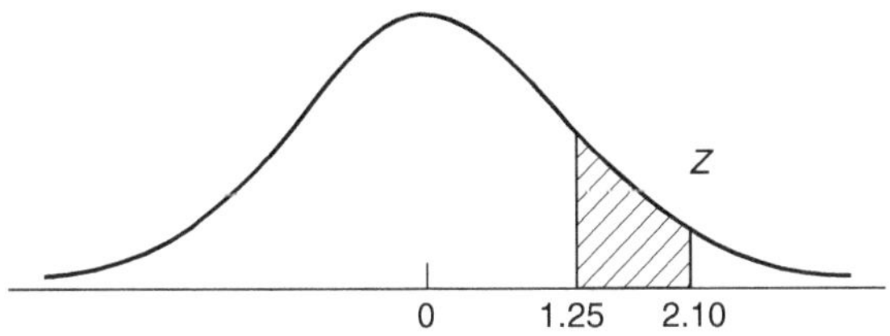

FIGURE 11.8 Area representation for P(1.25 ≤ z ≤ 2.10).

Solution:

To find this probability, we look up two values in Table A-1, P(0 ≤ Z < 1.25) and P(0 ≤ Z < 2.10). Then

$$\begin{aligned} P(1.25 \le Z \le 2.10) &= P(0 \le Z < 2.10) - P(0 \le Z < 1.25) \\ &= 0.4821 - 0.3944 = 0.0877. \end{aligned}$$

EXAMPLE 11.5

Find P(-1 ≤ Z ≤ 0) as depicted in Figure 11.9.

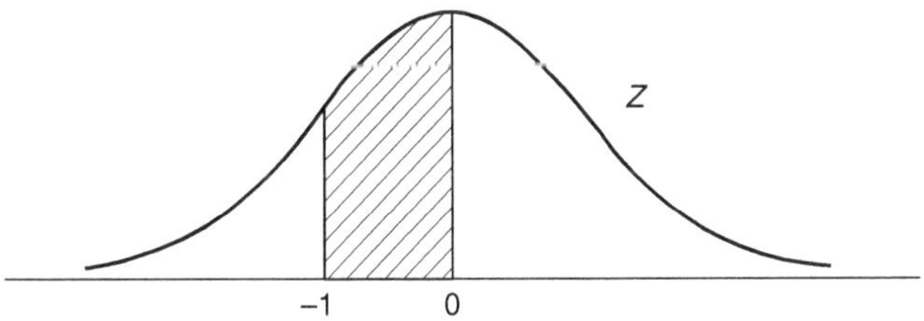

FIGURE 11.9 Area representation for P(-1 ≤ z ≤ 0).

Solution:

Because of the symmetry of Z around 0,

$$P(-1 \le Z \le 0) = P(0 \le Z \le 1) = 0.3413.$$

EXAMPLE 11.6

Find $P(Z < -1.5)$.

Solution:

$$\begin{aligned} P(Z < -1.5) &= P(Z \leq 0) - P(-1.5 < Z < 0) \\ &= 0.5000 - 0.4332 = 0.0688. \end{aligned}$$

EXAMPLE 11.7

Find $P(-2.10 \leq Z \leq -1.25)$.

Solution:

$$\begin{aligned} P(-2.10 \leq Z \leq -1.25) &= P(-2.10 \leq Z \leq 0) - P(-1.23 \leq Z \leq 0) \\ &= 0.4821 - 0.3944 = 0.0877. \end{aligned}$$

EXAMPLE 11.8

Find $P(Z \leq -1.2)$ as depicted in Figure 11.10.

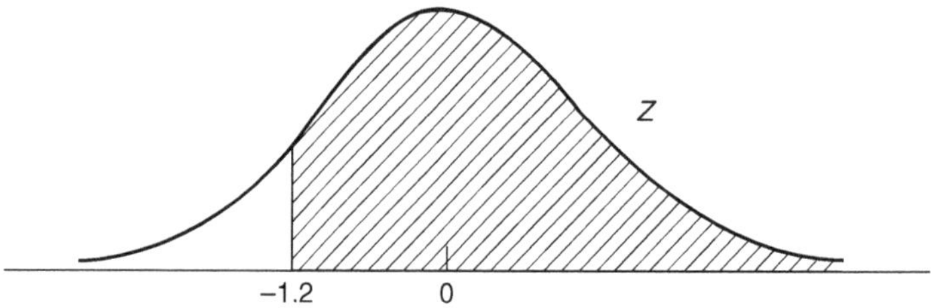

FIGURE 11.10 Area representation for $P(z \geq -1.2)$.

Solution:

Here, we must add two probabilities:

$$\begin{aligned} P(Z \geq -1.2) &= P(-1.2 \leq Z \leq 0) + P(Z > 0) \\ &= 0.3849 + 0.5000 = 0.8849. \end{aligned}$$

EXAMPLE 11.9

Find P(Z < 1.66) as depicted in Figure 11.11.

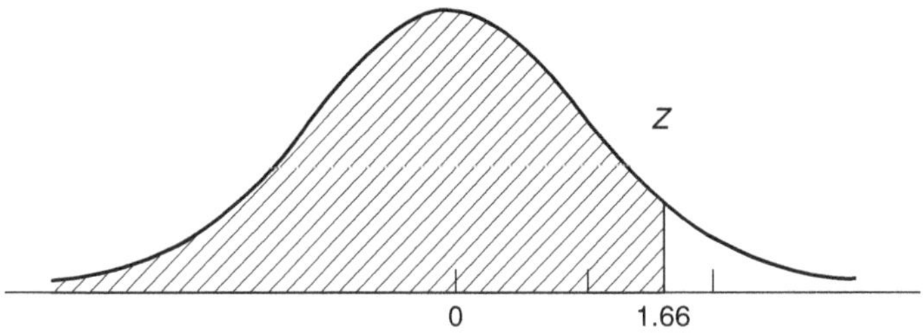

FIGURE 11.11 Area representation for $P(z \leq 1.66)$.

Solution:

This example is essentially the same as Example 8:

$$\begin{aligned} P(Z < 1.66) &= P(Z < 0) + P(0 \leq Z \leq 1.66) \\ &= 0.500 + 0.4575 = 0.9515. \end{aligned}$$

EXAMPLE 11.10

Find P(−0.57 < Z < 1.05) as depicted in Figure 11.12.

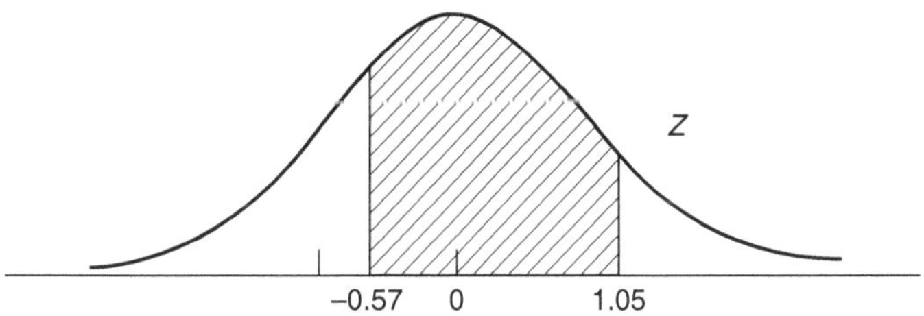

FIGURE 11.12 Area representation for $P(-0.57 \leq z \leq 1.05)$.

Solution:

Again, we add two probabilities:

$$\begin{aligned} P(-0.57 < Z < 1.05) &= P(-0.57 < Z \leq 0) + P(0 \leq Z < 1.05) \\ &= 0.2157 + 0.3531 = 0.5688 \end{aligned}$$

It should be clear from these examples that the graph of Z is a useful aid in finding probabilities involving the standard normal distribution.

Given a probability, we can also describe a corresponding event, as in the next example.

EXAMPLE 11.11

Find c so that $P(-c \le Z \le c) = 0.8132$ as depicted in Figure 11.13.

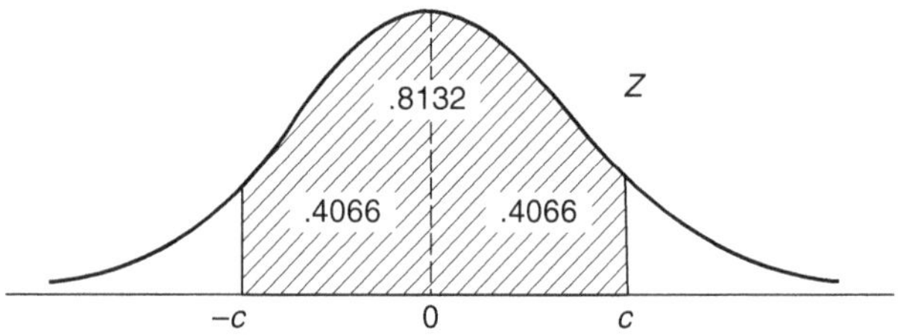

FIGURE 11.13 Area representation for $P(-c \le z \le c)$.

Solution:

DVD

We are looking for c to fit into this graph; if it does, then $P(0 \le Z \le c) = 0.4066$. From Table A-4, $c = 1.32$.

With this knowledge of the distribution Z, we can use the relationship $(X - \mu)/\sigma = Z$ to find probabilities for any normal random variable $X = N(\mu,\sigma^2)$.

Consider this next example.

EXAMPLE 11.12

A production process uses bearings whose lives in service are normally distributed with mean 740 hours and standard deviation 30 hours. What proportion of the bearings last more than 700 hours?

Solution:

This is the same as the probability that a randomly selected bearing will last more than 700 hours. That is, if X is the lifetime of a randomly selected bearing, X is N(740,700), and we seek $P(X > 700)$.

If $X > 700$, then $(X - 700) > (700 - 740)$, and

$$\frac{X-740}{30} > \frac{700-740}{30},$$

That is,

$$P(X>700) = P(\frac{X-740}{30} > \frac{700-740}{30}).$$

But 740 is the mean of X, and 30 its standard deviation. We know that $(X - \mu)/\sigma = Z$ for any normal random variable, so:

$$P(X > 700) = P(\frac{X-740}{30} > \frac{700-740}{30} = P(Z > -1.33).$$

and

$$P(Z > -1.33) = P(-1.33 < Z \le 0) + P(Z > 0)$$
$$= 0.4082 + 0.5000 = 0.9082$$

That is, $P(X > 700) = 0.9082$; approximately 91% of the bearings last 700 hours or more.

In general, if X is $N(\mu,\sigma^2)$, a normal random variable with mean μ and variance σ^2,

$$P(a \le X \le b) = P(\frac{a-\mu}{\sigma} < \frac{X-\mu}{\sigma} < \frac{b-\mu}{\sigma}) = P(\frac{a-\mu}{\sigma} < Z < \frac{b-\mu}{\sigma}).$$

Using the fact that $(X - \mu)/\sigma = Z$, we can reduce any problem involving a normal random variable to one involving the familiar Z.

EXAMPLE 11.13

Consider again the bearing example as depicted in Figure 11.14. If management wants to replace the bearings after no more than 2% of them have worn out, how long should they remain in service?

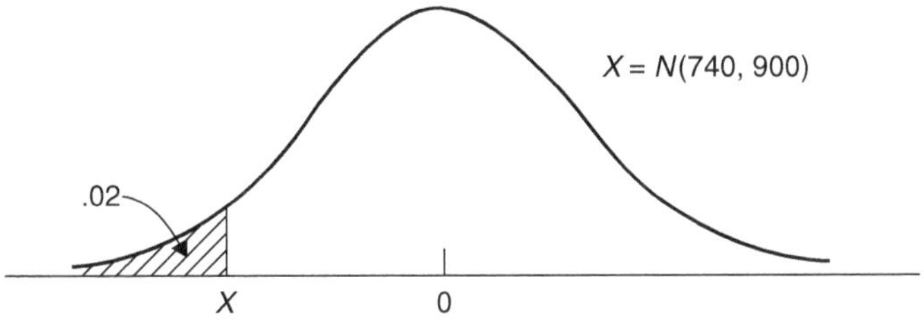

FIGURE 11.14 Area representation for $P(X < x) = 0.02$.

Solution:

Again, let X be the lifetime of a randomly chosen bearing. X is N(740,700), and we seek the value x for which $P(X < x) = 0.02$.

This kind of problem is also solved by changing it to one involving Z as depicted in Figure 11.15.

$$P(< x) = 0.02 \Rightarrow P(\frac{X-740}{30} < \frac{x-740}{30)}) = 0.02.$$

$$\Rightarrow P(Z < \frac{x-740}{30}) = 0.02$$

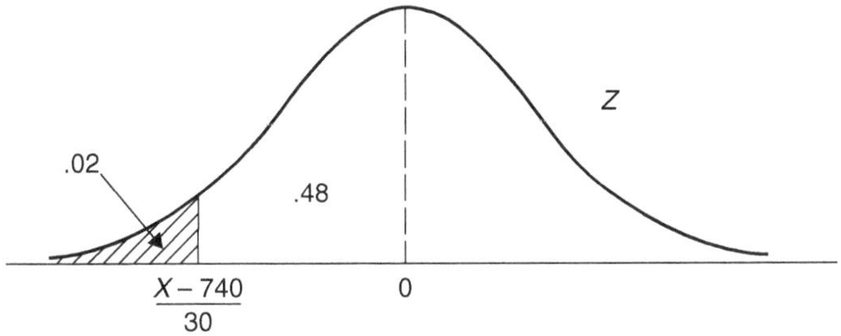

FIGURE 11.15 Area representation for $P(\frac{x-740}{20} \le z < 0)$.

EXAMPLE 11.13

Then

$$P\left(\frac{x-740}{30} < Z < 0\right) = 0.5 - 0.02 = 0.48$$

So from Table A-1, (s - 740)/30 = -2.06. Therefore, x = (30 * -2.06) + 740 = 678.2. If the bearings are replaced after 678.2 hours, only 2% of them will wear out in service.

11.3 Exercises

1. Z is the standard normal variable. In each of the following, find the value of z for which the statement is true:

a. $P(0 \le Z \le z) = 0.3925$

b. $P(Z > z) = 0.0281$

c. $P(Z \le z) = 0.8888$

d. $P(z \le Z < 0) = 0.1064$

e. $P(Z < z) = 0.1093$

f. $P(Z \le z) = 0.5239$

g. $P(-z \le Z \le z) = 0.6826$

h. $P(-z < Z < z) = 0.7372$

2. X has a normal distribution with mean 50 and standard deviation 5; that is, X is N(50,25). Find the following probabilities:

a. $P(50 < X < 57)$

b. $P(X > 57.5)$

c. $P(57.2 \le X \le 62)$

d. $P(42 < X \le 50)$

e. $P(X < 40)$

f. $P(X \le 51.7)$

g. $P(X \le 48.6)$

h. $P(41.7 < X < 49.3)$

i. $P(45 < X \le 52)$

j. $P(46.4 \le X \le 57.6)$

3. X is a normal random variable with mean 70.5 and standard deviation 7.6. Find x so that:

a. $(70.5 < X \le x) = 0.3686$

b. $P(X > x) = 0.3594$

c. $P(X \ge x) = 0.9082$

d. $P(x \le X \le 70.5) = 0.1628$

e. $P(X \le x) = 0.2776$

f. $P(X < x) = 0.7224$

g. $P(70.5 - x < X < 70.5 + x) = 0.6922$

4. During an interview process, a company's applicants are required to complete an aptitude test. If the times to complete the test are normally distributed, with mean 110 minutes and standard deviation 18 minutes, then answer the following:

   a. A person takes the test. What is the probability he finishes in less than 2 hours?
   b. What proportion of those who take the test require more than 90 minutes but less than 125?
   c. It is desired that 80% of those taking the test finish it. How long should be allowed?

5. Verify the empirical rule.

6. Bricks made at the Stonehenge Brickyard have weights that are normally distributed around a mean of 8.0 pounds, with standard deviation 0.25 pounds.

   a. We select one brick at random. What is the probability that its weight is greater than 7.9 pounds?
   b. What proportion of all the bricks have weights between 7.6 and 8.2 pounds?
   c. 90% of the bricks have weights greater than what value?

CHAPTER 12

DISTRIBUTIONAL APPROXIMATIONS

OVERVIEW AND LEARNING OBJECTIVES

In This Chapter

12.1 Introduction

Though the underlying concepts of quantities such as time and length are continuous, in practice we measure these with discrete approximations, such as tenths of a second or hundredths of an inch. In this chapter, we investigate using one distribution to approximate another.

12.2 Review of Discrete and Continuous Distributions

12.2.1 Summary of discrete distributions

In Chapter 9, we developed the discrete distributions shown in Table 12.1:

TABLE 12.1 Summary of discrete distributions.

Type	Probability Function	$\mu = E(x)$	σ^2, σ	Distributional Shape
Uniform	$f(x) = 1/n$, where n is the number of values in the x	Depends on values of x	Depends on values of x	Rectangular
Binomial	$f(x) = \binom{n}{x} p^x q^{(n-x)}$, where N = number of trials, p = prob. of success and q = 1 - p	np	npq, $\sqrt{npq}$	if p = 1/2 symmetrical If p > 1/2 neg. skewed if p < 1/2 pos. skewed
Hyper-geometric	$f(x) = \dfrac{\binom{S}{x}\binom{N-S}{n-x}}{\binom{N}{n}}$ where n = size of sample, N = size of population, S = number of successes in Population	$n\dfrac{S}{N}$	$n\dfrac{S}{N}\left(1-\dfrac{S}{N}\right)\left(\dfrac{N-n}{N-1}\right)$, $\sigma = \sqrt{\sigma^2}$	similar to binomial with $p\dfrac{S}{N}$
Geometric	$f(x) = (1-p)^{(1-x)}\, p$	$\dfrac{1}{p}$	$\dfrac{(1-p^2)}{p}$	similar to binomial
Poisson	$f(x) = \dfrac{\lambda^x e^{-\lambda}}{x!}$, where $\lambda = \alpha(T)$ Average number of arrivals In time T	λ	$\lambda\sqrt{\lambda}$	small λ, pos. skewed as λ increases, more symmetrical

ON DVD All figures and tables in this chapter appear on the companion DVD.

12.2.2 Summary of continuous distributions

Chapters 10 and 11 allow for the distributions shown in Figure 12.1:

TYPE	PROBABILITY FUNCTION	μ	σ^2, σ	DISTRIBUTION SHAPE
Uniform	$f(x) = \frac{1}{b-a}$, $a \le X \le b$	$\frac{a+b}{2}$	$\frac{(b-a)^2}{12}$, $\frac{(b-a)\sqrt{3}}{6}$	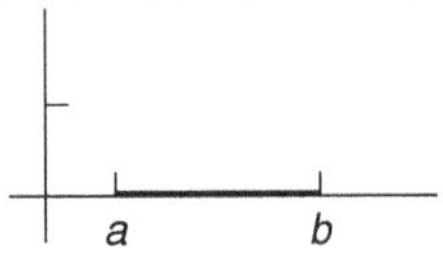
Exponential	$f(t) = \alpha e^{-\alpha t}$, $t \ge 0$, Where α is the average number of arrivals in a unit of time	$1/\alpha$	$1/\alpha^2$, $1/\alpha$	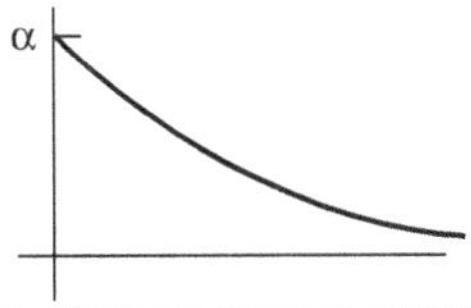
Normal	$f(x) = \frac{1}{\sqrt{2\pi}\,\sigma} \exp[-(x-\mu)^2/2\sigma^2]$, $-\infty < X < \infty$	μ	σ^2, σ	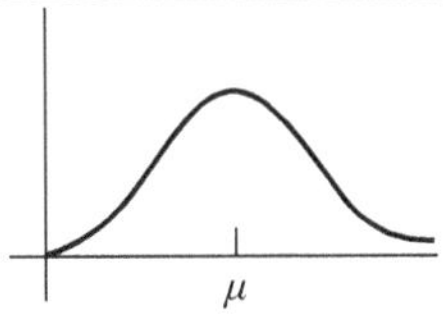

FIGURE 12.1 Continuous probability distributions.

12.3 Discrete Approximations of Discrete Distributions

We have already commented on similarities between the binomial and hypergeometric distributions. If we select a random sample of n items from a population of N items, of which S are of a particular type, the binomial distribution corresponds to sampling with replacement, and the hypergeometric distribution corresponds to sampling without replacement. Their means are identical, $\mu = np = n(S/N)$, while the variance of the hypergeometric distribution is less than that of the binomial:

$$\sigma^2_{binomial} = np(1-p)$$

$$\sigma^2_{hypergeometric} = n\left(\frac{S}{N}\right)\left(1-\frac{S}{N}\right)\left(\frac{N-n}{N-1}\right).$$

The difference is the factor $(N - n)/(N - 1)$, which is near 1 when N is large compared to n. In this situation, the hypergeometric distribution can be approximated with the binomial, as in the following example.

EXAMPLE 12.1

In a production lot of 200 integrated circuits, 30 are defective. If 10 circuits are randomly chosen to be tested, what is the probability that no more than 2 of those tested are defective?

Solution:

Note that N = 200, S = 30, and $n = 10$. Then

$$P = 30/200 = 0.15.$$

Using the binomial distribution $b(10, 0.15)$, we approximate the desired probability:

$$f(0) + f(1) + f(2) = \binom{10}{0} 0.15^0 0.85^{10} + \binom{10}{1} 0.15^1 0.85^9 + \binom{10}{2} 0.15^2 0.85^8$$
$$= 0.1969 + 0.3474 + 0.2759 = 0.8202$$

In general, if $p = S/N$ is near $\frac{1}{2}$, the heypergeometric and binomial distributions have the relationship shown in Figure 12.2.

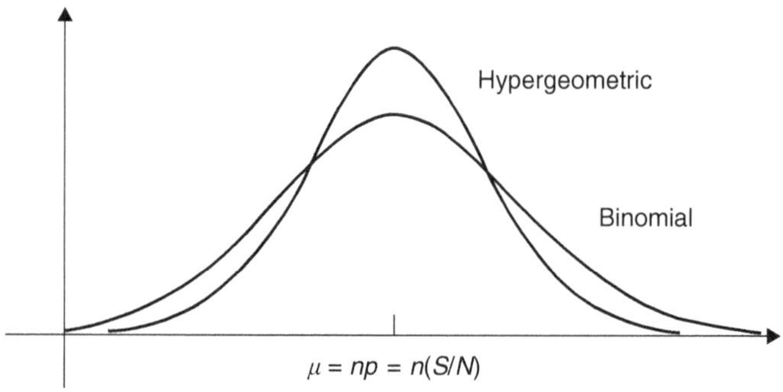

FIGURE 12.2 Comparison of hypergeometric and binomial distributions.

The Poisson distribution is also related to the binomial. When the probability of success p is small (that is, when $np \le 5$), the binomial may be approximated by the Poisson distribution whose mean is the same as that of the binomial. The Poisson distribution has $\lambda = np$, as shown in the next example.

EXAMPLE 12.2

The probability that a clock radio will require return to the factory for repairs is 0.04. If a department store has sold 50 of these clock radios, approximate the probability that more than one will be returned to the factory.

Solution:

First, P(more than one will be returned) = 1 - P(none or one will be returned). We approximate this latter probability using the Poisson distribution with $\lambda = np = 50 * 0.04 = 2$. (Since $np \leq 5$ w we may do this.)

$$f(0) + f(1) \cong \frac{2^0 e^{-2}}{0!} + \frac{2^1 e^{-2}}{1!} = 0.1353 + 0.2707 = 0.4060$$

The probability we seek is 1−0.4060 = 0.5940.

In general, the graphs of a binomial distribution and its corresponding Poisson distribution have the relationship shown in Figure 12.3.

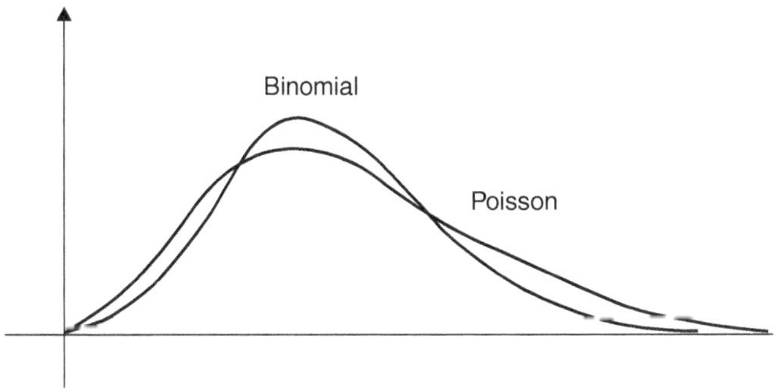

FIGURE 12.3 Comparison of binomial and Poisson distributions.

12.4 Continuous Approximations of Discrete Distributions

12.4.1 Normal approximation of a Poisson distribution

We have also seen that as λ increases, the Poisson distribution becomes less skewed and more bell-shaped, as shown in Figure 12.4.

This suggests that if λ is large enough, a normal distribution may be used to approximate the Poisson distribution, and this is the case. If $\lambda \geq 25$, then the Poisson distribution may be approximated by the normal distribution with the same mean and variance $N(\lambda, \lambda)$.

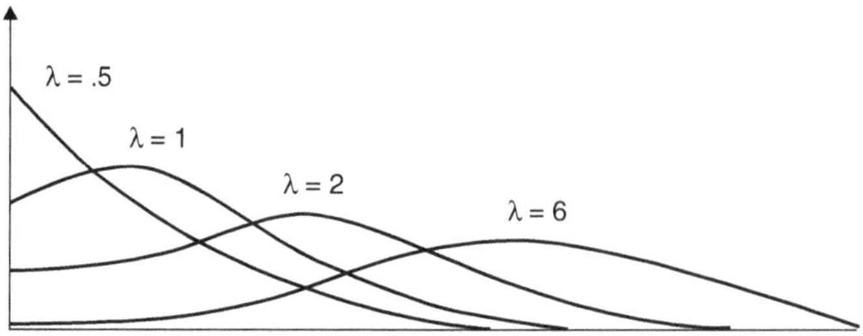

FIGURE 12.4 The Poisson distribution for several values of λ.

In a discrete distribution, there are positive probabilities associated with individual values in the range of the random variable, while in a continuous distribution there are not. When using a continuous distribution to approximate a discrete distribution, we include an interval of width 1 around each value of the random variable in the event whose probability we seek. That is, if continuous Y is used to approximate discrete X, $P(X=10) \cong P(9.5 \le Y \le 10.5)$. This adjustment is called the *continuity correction*, which is illustrated in Figure 12.5.

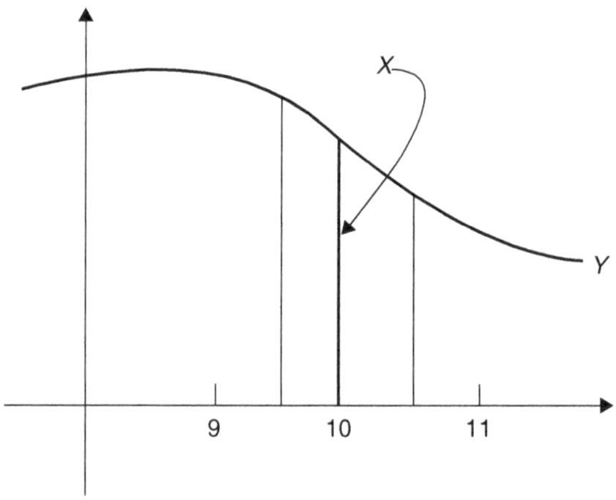

FIGURE 12.5 The continuity correction.

EXAMPLE 12.3

Now, let X be a Poisson random variable with $\lambda = 50$. We use the normal distribution $Y = N(50, 50)$ to approximate $P(52 \le X \le 60)$.

Solution:

$$
\begin{aligned}
P(52 \le X \le 60) &\cong P(51.5 \le X \le 60.5) \\
&= P\left(\frac{52.5-50}{\sqrt{50}} < \frac{Y-50}{\sqrt{50}} < \frac{60-50}{\sqrt{50}}\right) \\
&= P(0.21 < Z < 1.48) \\
&= 0.4306 \quad 0.0832 = 0.3474.
\end{aligned}
$$

12.4.2 Normal approximation of a binomial distribution

The binomial distribution is also bell-shaped when p is near $\frac{1}{2}$ or when n is large. When these conditions occur (when np and $n(1-p) \geq 5$), the binomial distribution $b(n, p)$ can be approximated with the normal distribution with mean np and variance npq, $N(np, np(1-p))$. Again, the continuity correction is used, as shown in the following example.

EXAMPLE 12.4

In a large city, 35% of the households have two incomes. If we randomly select 100 households, what is the probability that 40 or more will have two incomes?

Solution:

Let X be the number of sampled households with two incomes. Then X is $b(100, 0.35)$, with mean 35 and variance 22.75. We seek $P(X \geq 40)$, and we will approximate X with Y = N (35, 22.75).

$$P(X \geq 40) = P(Y > 39.5) = P\left(\frac{Y-35}{\sqrt{22.75}} > \frac{39.5-35}{\sqrt{22,75}}\right)$$

$$= P(Z > 0.94) = 0.5000 \quad 0.3264 = 0.1736.$$

12.5 Exercises

1. A pile of 60 tests contains 5 with scores of 100. If 7 tests are randomly selected, what is the probability that none or one of those selected have scores of 100? Solve this problem in two ways, using the hypergeometric distribution and the approximate binomial distribution.

2. From a group of 54 men and 46 women, a committee of 10 people is randomly assigned. Use both the hypergeometric distribution and the binomial approximation to find the probability that 5 members of the committee are women.

3. Two percent of the light bulbs produced by the Acme Light Company are defective. Use a Poisson distribution to approximate the probability that in a box of 100 bulbs, fewer than 4 are defective.

4. Having completed a special training program, the probability that a salesman will stay with the Ace Home Products Company is 96%. Use a Poisson distribution to approximate the probability that in a training group of 70 salesmen, less than 68 will stay with the company. (*Hint:* The probability that a salesman will leave is 4%.)

5. Cars arrive at a drive-up bank in a Poisson process with an average rate of 50 per hour. Use a normal distribution to approximate the probability that in an hour between 45 and 60 cars arrive.

6. Jobs are submitted at the input queue to a resolution center for the corporation in a Poisson process with an average of 15 per week. Use a normal distribution to approximate the probability that in a month, more than 70 jobs are submitted.

7. We flip a coin 50 times, and X is the number of heads. Use a normal distribution to approximate these probabilities:

   a. $P(20 \leq X \leq 30)$

   b. $P(20 < X < 30)$

   c. $P(X > 34)$

   d. $P(X < 27)$

8. The probability that a person will pass a particular standardized test is 0.65. If 90 people take this test, use a normal distribution to approximate the probability that between 50 and 60 (inclusive) pass.

UNIT 3

INTRODUCTION TO ESTIMATION

Several important security principles should be followed in an organizations IT facilities:

- *Default to access denial.* This makes users justify their need for access.
- *Non-secret design.* A system must be able to be described briefly in the open literature in order to sufficiently serve and be used with confidence.
- *User acceptability.*
- *Complete mediation.* Every access to every object must be asked for authority.
- *Least privilege.* Every user, programmer, networked computer, or other resource should use only the privileges necessary to complete the job.

Classify the questions in the CIO article, *8 Questions For Uncovering Information Security Vulnerabilities* by Andrew Jaquith, CSO at www.cio.com with the latter security principles.

Before the latter questions can be examined we must establish the sampling distributions for the commonly studied statistics. The Central Theorem, which establishes the sampling distribution for sampling means, provides the basis for considerable work in statistical analysis. Thus the researcher can obtain probability values for many observed sample means or sums.

You will find out in this unit that the Central Limit Theorem can be applied to both discrete and continuous random variables. With this machinery in hand you will posses the machinery necessary to answer the latter questions on information security matters.

CHAPTER 13

SAMPLING DISTRIBUTIONS

OVERVIEW AND LEARNING OBJECTIVES

In This Chapter

13.1 Introduction

We have considered descriptive statistics, methods by which large amounts of data are condensed and made more intelligent. This is a deductive process, reasoning from the whole to some part or characteristic of the whole. We now begin our examination of inductive reasoning, from a part to the whole, from a sample to the population from which it came and we begin our exploration of statistical inference.

The need for such processes is clear. Proportions of values in which we are interested might be too difficult, expensive, or time-consuming to obtain, or simply too large to be efficiently analyzed. Instead of investigating the entire population of values, we select a sample from it, analyze the sample, and from it infer characteristics of the population. By examining the incomes of 100 factory workers in Connecticut, for example we can estimate the income of all factory workers there.

In general, our process is this:

1. From a population of values, select a sample.
2. Compute one or more statistics of the sample.
3. Use these statistics to estimate or draw conclusions about parameters of the population.

In order to perform the third step, the sample must be related to the population in a way in which we can reason from the sample to the population. This can be done with samples chosen so that no element of the population is more likely than any other to be selected.

> A *simple random sample* is chosen from a population when elements of the population have the same probability of being included in the sample.

That is, the elements of the sample are randomly selected.

The selection of a random sample is a random experiment, upon which we may define random variables. In particular, we can consider statistics of the sample (its standard deviation or mean) as random variables, and examine their distributions:

> The distribution of a random variable that is a statistic of a sample is called a *sampling distribution*.

In this chapter, we describe some sampling distributions.

13.2 An Example of a Sampling Distribution

The most widely applied statistic is the sample mean, $\overline{X}$. We examine the sampling distribution of $\overline{X}$ by generating many samples from a known population and by comparing the observed distribution of sample means with the population means.

For illustration, a population composed of real numbers from 60 to 110 was generated, resulting in the distribution shown in Figure 13.1. The mean of the population is 87.60, and the standard deviation is 14.63.

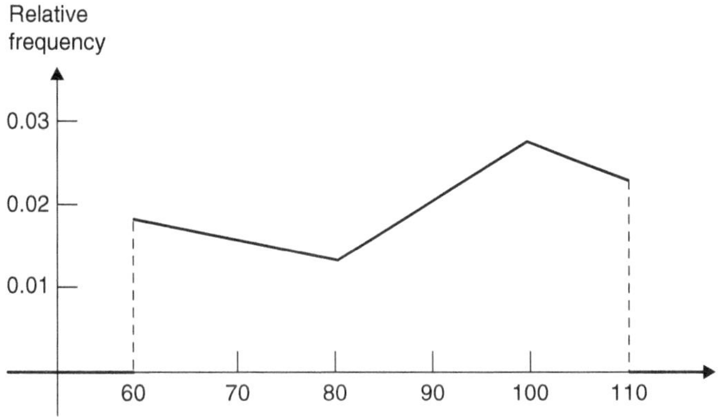

FIGURE 13.1 Population from which samples of size 50 were drawn.

All figures and tables in this chapter appear on the companion DVD.

This distribution was processed by selecting 200 random samples from this population of values and computing their sample means. A histogram of the sample means, which represents 200 sample means based on samples of size 50 from this given population, is shown in Figure 13.2.

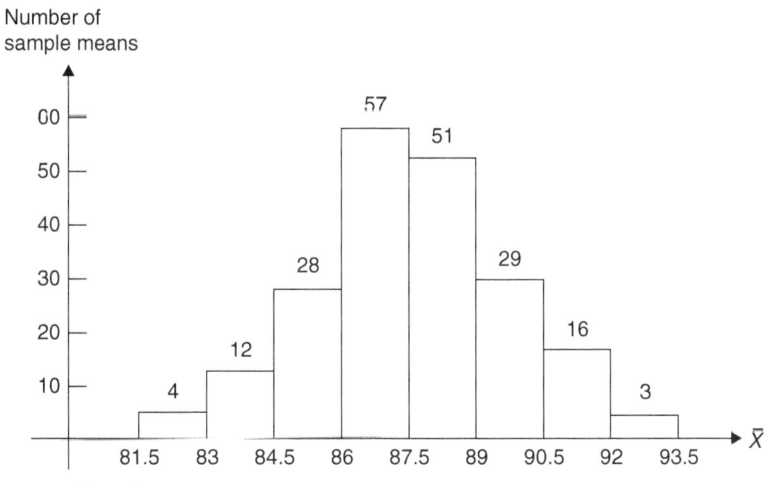

FIGURE 13.2 Histogram of sample means.

Three important observations can be made about the distribution of sample means:

- It appears normal, since it is symmetrical and bell-shaped.
- The mean of the distribution of sample means was found to be 87.5, very near the population mean of 87.6. This suggests that the expected value of $\overline{X}$, as a random variable, is near the population mean μ.

- The standard deviation of the sample means was found to be 2.17, which is much less than the population standard deviation of 14.63. That is, the distribution of sample means shows less variability than does the population.

We now consider the sampling distribution of $\overline{X}$ in general, and the theoretical foundations of the above observations.

13.3 The Sampling Distribution of $\overline{X}$

Suppose we have a population of values with mean μ, variance σ^2, and standard deviation σ. From this population, we select one element at random, and its value is the random variable X_i. It should be clear that the distribution of X_i is identical to that of the population.

Now select a random sample of n elements from the population. Let the random variables $X_1, X_2, \ldots, X_n$ be their values; the distributions of these random variables are identical to the distribution of X_i. All have mean μ, variance σ^2, and standard deviation σ.

The sample mean, $\overline{X}$, is the mean of the elements of the sample, so:

$$\overline{X} = \frac{X_1 + X_2 + \ldots + X_n}{n} = \frac{1}{n}\left(\sum_{i=1}^{n} X_i\right);$$

and the mean of the distribution of $\overline{X}$ is:

$$\mathrm{E}(\overline{X}) = \mathrm{E}\left(\frac{1}{n}\left(\sum_{i=1}^{n} X_i\right)\right) = \frac{1}{n}E\left(\sum_{i=1}^{n} X_i\right) = \frac{1}{n}\sum_{i=1}^{n} \mathrm{E}(X_i) = \frac{1}{n}\sum_{i=1}^{n} \mu = \frac{1}{n} * n\mu = \mu.$$

As predicted by our experience with the 200 samples of values, the expected value of the sample mean is equal to the population mean; the sample mean is "aimed at" the population mean.

To consider $\mathrm{Var}(\overline{X})$, first assume that the population size is large relative to the sample size. Then the elements of the sample will be independent, and we can apply rules for finding expected values to functions of the independent events. As a result, we know that the variance of a sum of independent random variables is the sum of their variances:

$$\mathrm{Var}(\overline{X}) = \mathrm{Var}\left(\frac{1}{n}\left(\sum_{i=1}^{n} X_i\right)\right) = \frac{1}{n^2} Var\left(\sum_{i=1}^{n} X_i\right) = \frac{1}{n^2}\sum_{i=1}^{n} Var(X_i)$$

$$= \frac{1}{n^2}\sum_{i=1}^{n} \sigma^2 = \frac{1}{n^2} * n\sigma^2 = \frac{\sigma^2}{n}.$$

Also, $\sigma_{\overline{X}} = \frac{\sigma}{\sqrt{n}}$. $\sigma_{\overline{X}}$ is often called the *standard error of the mean.*

If the population is small relative to the sample size, then the elements of the sample are not independent, and we must include a correction factor in the calculations of $\sigma_{\overline{X}}^2$ and $\sigma_{\overline{X}}$ for when $n \geq 5\%$ of N.

$$\sigma_{\overline{X}}^2 = \frac{\sigma^2}{n} * \frac{N-n}{N-1} \quad \text{and} \quad \sigma_{\overline{X}} = \frac{\sigma}{\sqrt{n}} * \left[\frac{N-n}{N-1}\right]^{\frac{1}{2}}.$$

Note that the relationship between the values of $\sigma_{\overline{X}}^2$ for independent and dependent sample elements is precisely the relationship between the variance of a binomial distribution and its corresponding hypergeometric distribution.

In our example, the standard error of the mean is

$$\sigma_{\overline{X}} = \frac{14.63}{\sqrt{50}} = 2.07.$$

This is very close to the observed standard deviation of the 200 sample means.

An important result in probability theory, though one whose proof is beyond the scope of this text, is that any linear combination of normal random variables is itself normally distributed. That is, if Y_i is $N(\mu_i, \sigma_i)$ for $i = 1, 2, \ldots, n$, and the a_i are constants, then $X = \sum_{i=1}^{n} a_i Y_i$ will have a normal distribution. The mean of X will be $\sum_{i=1}^{n} a_i \mu_i$, but the variance of X will not be $\sum_{i=1}^{n} a_i \sigma^2_i$, unless the Y_i are independent.

Since the mean of a sample is a linear combination of random variables (each of the coefficients is n^{-1}) a consequence of the above result is this: When sampling from a normally distributed population, the sampling distribution of $\overline{X}$ is also normal, with mean μ and variance $\frac{\sigma^2}{n}$.

For example, suppose that we take a random sample of 15 elements from a normally distributed population with mean 140 and variance 36. Then the sampling distribution of $\overline{X}$ is N(140, 36/15), and calculations like these may be performed:

$$P(\overline{X} > 141) = P\left(\frac{\overline{X} - 140}{6/\sqrt{15}} > \frac{141 - 140}{6/\sqrt{15}}\right) = P(Z > 0.65)$$
$$= 0.5000 - 0.2422 = 0.2578$$

13.4 The Central Limit Theorem

Populations need not be normally distributed, of course, and samples are taken from those that are not, so it is a pleasant surprise to find that $\overline{X}$ is approximately normal in a much wider range of situations. This remarkable result is the foundation of statistical inference.

If we take a sample of size $n \geq 30$ from a population with mean μ and standard deviation σ, then the sampling distribution of $\overline{X}$ is approximately normal with mean μ and standard deviation $\dfrac{\sigma}{\sqrt{n}}$.

That is, for a large enough sample, the sampling distribution of $\overline{X}$ is always approximately normal, regardless of the shape of the population distribution. With samples of size less than 30, a value dictated by experience and not theory, the formulas for $\mathrm{E}(\overline{X})$ and $\mathrm{Var}(\overline{X})$ hold, but the distribution of $\overline{X}$ is not near enough to normal to be useful.

The graphs in Figure 13.3 illustrate the central limit theorem. Note that as the sample sizes increase, the distributions of the sample means become closer to normal distributions.

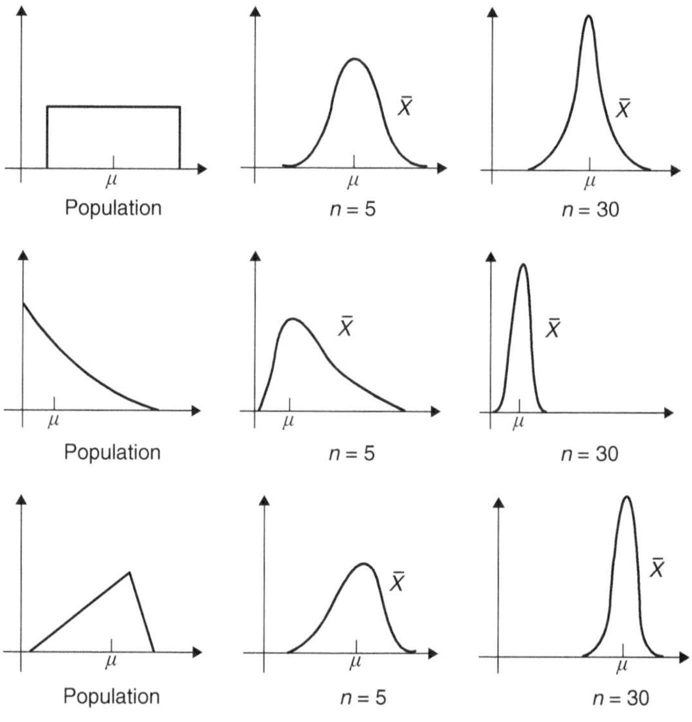

(c)

FIGURE 13.3 Graphs illustrating the central limit theorem.

These graphs also illustrate that as the size of the sample increases, $\sigma_{\overline{X}} = \dfrac{\sigma}{\sqrt{n}}$ decreases. The larger the sample size, the smaller the standard error, and the more closely packed around $\mathrm{E}(\overline{X}) = \mu$ is the distribution of $\overline{X}$. For a large sample, $\overline{X}$ is more likely to be near the population mean μ.

Given the parameters of a population, we can use the central limit theorem to find probabilities involving $\overline{X}$ as in this example.

EXAMPLE 13.1

Fluorescent bulbs manufactured by the All-Night Light Company have a mean lifetime 1700 hours, with standard deviation 300 hours. If 100 light bulbs are tested, what is the probability their mean lifetime falls between 1650 and 1725 hours?

Solution:

We are seeking P(1650 $<\overline{X}<$1725), and we know that $\overline{X}$ is approximately normal with mean 1700 and standard deviation $300\sqrt{100}$. Therefore,

$$\mathrm{P}(1650<\overline{X}<1725) \equiv P\left(\frac{1650-1700}{300/\sqrt{100}}<\frac{\overline{X}-1700}{300/\sqrt{100}}<\frac{1725-1700}{300/\sqrt{100}}\right)$$
$$=\mathrm{P}(-1.67<\mathrm{Z}<00.83)=0.4525+0.2967=0.7492.$$

EXAMPLE 13.2

Similarly, what is the probability that the mean lifetime of the sample is more than 1720 hours?

Solution:

$$\mathrm{P}(\overline{X}>1720)=\mathrm{P}(\frac{\overline{X}-1700}{300/\sqrt{100}}>\frac{1720-1700}{300/\sqrt{100}})$$
$$=\mathrm{P}(\mathrm{Z}>0.67)=0.5000-0.2486=0.2514.$$

In general, if $\overline{X}$ is the mean of a sample of size $n \geq 30$ from a population with mean μ and standard deviation σ,

$$\mathrm{P}(a<\overline{X}<b)=P\left(\frac{a-\mu}{\sigma/\sqrt{n}}<\mathrm{Z}<\frac{b-\mu}{\sigma/\sqrt{n}}\right).$$

EXAMPLE 13.3

Had the sample of 100 bulbs been taken from a production lot of 500 bulbs, we would include the correction factor $[(\mathrm{N}-n)/(\mathrm{N}-1)]^{1/2}$ in finding $\sigma_{\overline{X}}$:

Solution:

$$\sigma_{\overline{X}}=\frac{300}{\sqrt{100}}*[(500-100)/(500-1)]^{1/2} \equiv 28.86.$$

Then, for example,

$$\mathrm{P}(\overline{X}>1720)=\mathrm{P}(\frac{\overline{X}-1700}{28.86}>\frac{1720-1700}{28.86})$$
$$=\mathrm{P}(\mathrm{Z}>0.74)=).5000-0.2704=0.2296.$$

13.5 The Distribution of the Sample Median

We have seen that for a unimodal, symmetrical distribution, the mean and median are equal. When sampling from such a population, we might approximate the population mean with the sample median.

When a population is symmetrically distributed, the expected value of the sample median is the population mean. While the standard error of $\overline{X}$ is $\frac{\sigma}{\sqrt{n}}$, where σ is the population standard deviation, that of the median is larger, $1.253\frac{\sigma}{\sqrt{n}}$, so that the sample mean can be expected to be closer to the population mean than the sample median. The distribution of the sample median tends toward normality as the sample size increases, and when sampling from a normal population, the distribution of the sample median is itself normal, with mean equal to the population mean and standard deviation $1.253\frac{\sigma}{\sqrt{n}}$. Figure 13.4 compares the sampling distributions of the mean and medium.

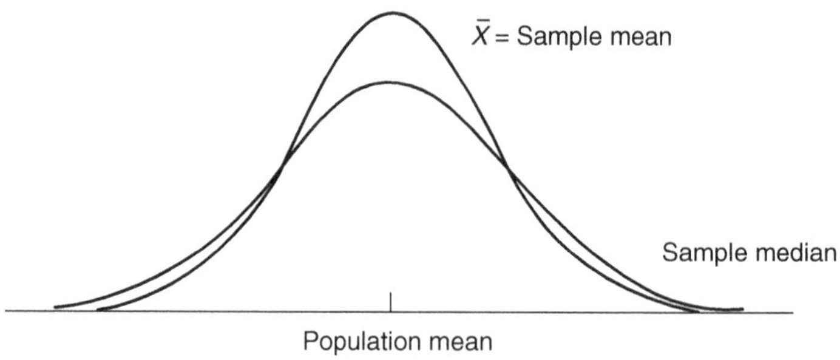

FIGURE 13.4 Sampling distributions of the mean and medium.

13.6 Sampling Distributions of Measures of Dispersion

Often the researcher will wish to investigate the variance or standard deviation of a population. In bottling soft drinks, for example, variations in the amounts in the bottles are of great importance. Here we consider some parameters of the distributions of the sample variance and sample standard deviation, though detailed examination of these distributions is postponed (the interested reader should study Chapter 39).

13.6.1 The expected value of the sample variance

While the definitions of the sample and population means are identical except for the symbols used, there is a significant difference in the calculations of variances. The population variance

$$\frac{1}{N}\sum_{i=1}^{N}(X_i-\mu)^2$$

is the mean of the squared deviations from the mean, but the sample variance

$$\frac{1}{n-1}\sum_{i=1}^{N}\left(X_i-\overline{X}\right)^2$$

is computed with a division by $n-1$, rather than n. When these definitions were made, the difference was justified by claiming that the sample variance was thereby made a better estimator of its corresponding population variance. We now have the mathematical tools to prove the claim. Because of the division by $n-1$ in the definition of the sample variance S^2, its expected value is the population variance σ^2:

$$\mathrm{E}\left(S^2\right)=\mathrm{E}\left(\frac{1}{n-1}\sum_{i=1}^{N}\left(X_i-\overline{X}\right)^2\right)$$

$$=\mathrm{E}\left(\frac{1}{n-1}\sum_{i=1}^{n}X_i^2-\frac{n}{n-1}\overline{X}^2\right)$$

$$=\frac{1}{n-1}\mathrm{E}\left(\sum_{i=1}^{n}X_i^2\right)-\frac{n}{n-1}\mathrm{E}\left(\overline{X}^2\right)$$

But $\mathrm{Var}(X_i)=\mathrm{E}(X_i^2)-(E(X_i))^2$, so that

$$\mathrm{E}(X_i^2)=\mathrm{var}(X_i)+(E(X_i))^2$$

$$=\sigma^2+\mu^2.$$

Then

$$\mathrm{E}\left(S^2\right)=\frac{1}{n-1}\sum_{i=1}^{n}\left(\sigma^2+\mu^2\right)-\frac{n}{n-1}\mathrm{E}\left(\overline{X}^2\right)$$

$$=\frac{n}{n-1}\left(\sigma^2+\mu^2\right)-\frac{n}{n-1}\mathrm{E}\left(\overline{X}^2\right).$$

But $\mathrm{Var}(\overline{X})=\mathrm{E}(\overline{X}^2)\ \ \mathrm{E}(\overline{X})^2$, so that

$$\mathrm{E}\left(S^2\right)=\frac{n}{n-1}\left(\sigma^2+\mu^2\right)-\frac{n}{n-1}\left(\mu^2-\frac{\sigma^2}{n}\right)$$

$$=\frac{n}{n-1}\sigma^2-\frac{n}{n-1}*\frac{\sigma^2}{n}$$

$$=\frac{n}{n-1}\sigma^2-\frac{1}{n-1}\sigma^2=\frac{n-1}{n-1}\sigma^2.$$

That is, $\mathrm{E}(S^2)=\sigma^2$.

Later, in Unit 5, we will consider sampling and S^2 in more detail, relating the distribution of S^2 to a positively skewed continuous distribution with the name "chi-square."

Intuitively, it might seem that the shape and parameters of the distribution of the sample variance would dictate the shape and parameters of the distribution of the sample standard deviation, but this is not entirely so. In particular, the expected value of the square root of a random variable is not necessarily the square root of the expected value. Though we have shown that $E(S^2) = \sigma^2$, we cannot necessarily say that $E(S) = \sigma$.

13.6.2 The sample range

The sample range provides an example of a sample statistic whose expected value is not, in general, equal to its corresponding population parameter, since the sample range will equal the population range only if the sample contains the two extreme values of the population. This is an unlikely situation, so the expected value of the sample range is always less than the population range.

13.6.3 The distribution of the sample proportion

Frequently, sampling is used to estimate the proportion of cases P in a population that have a particular characteristic. If we associate the value 1 with those cases that have the characteristic and the value 0 with those that do not, the mean of this population of 0s and 1s is the proportion P of the original population that has the characteristic of interest. If we take a sample of n cases from the population (the proportion $\hat{p}$ of the sample having the characteristic), the mean of the corresponding sample of 0s and 1s, is an estimate of the population proportion P.

The number of elements R of the sample having the characteristic has the binomial distribution $b(n,P)$, which has mean nP and variance $nP(1 - P)$. The proportion, $\hat{p}$, of the sample having the characteristic is R/n, so:

$$E(\hat{p}) = E\left(\frac{R}{n}\right) = \frac{1}{n}E(R) = \frac{1}{n}nP = P.$$

and

$$\mathrm{Var}(\hat{p}) = \mathrm{Var}\left(\frac{R}{n}\right) = \frac{1}{n^2}\mathrm{Var}(R) = \frac{1}{n^2}nP(1-P) = \frac{P(1-P)}{n}.$$

The distribution of the sample proportion has expected value P, the population proportion, and standard deviation $[P(1-P)/n]^{\frac{1}{2}} = \sigma_p$.

Also, we know by the central limit theorem that as n increases, the distribution of $\hat{p}$ becomes approximately normal. Therefore, for large samples, the distribution of the sample proportion is approximately normal with mean P and standard deviation $[P(1-P)/n]^{\frac{1}{2}}$.

EXAMPLE 13.4

Thirty-eight percent of all registered voters are Democrats. If we interview 300 randomly selected registered voters, what is the probability that between 36% and 40% of them are Democrats?

Solution:

We seek $P(0.36 < \hat{p} < 0.40)$, where $\hat{p}$ is approximately N(0.38, (0,38 * 0.62)/300):

$$P(0.36<\hat{p}<0.40)=P\left(\frac{0.36-0.38}{\left(\frac{0.38*0.62}{300}\right)^{1/2}}<\frac{\hat{p}-0.38}{\left(\frac{0.38*0.62}{300}\right)^{1/2}}<\frac{0.40-0.38}{\left(\frac{0.38*0.62}{300}\right)^{1/2}}\right)$$
$$=P(-0.71<Z<0.71)$$
$$=0.2611+0.2611=0.5222$$

Finally, if the sample size is large relative to the size of the population, we must adjust the standard error σ_p, as shown in the following example.

EXAMPLE 13.5

If a population of 500 elements has proportion 0.71, and we take a sample of 100 elements from it, find the standard deviation of the distribution of the sample proportion $\hat{p}$.

Solution:

The distribution of the sample proportion $\hat{p}$ is approximately normal with mean 0.71 and standard deviation:

$$\sigma_p=\left(\frac{0.71*0.29}{100}\right)^{1/2}*\left(\frac{500-100}{500-1}\right)^{1/2}.$$

13.7 Exercises

1. Using Table 13.1 as a reference, construct a spreadsheet of 700 rows where the first column contains a sequence number, the second column contains a random number from 1 to 75, and the third column contains the corresponding value of X7 from the American Cities database in Appendix B. ON DVD

TABLE 13.1 Random values of X7.

s	Random Number	X7
1	21	2.8
2	44	-1.9
3	40	0
4	42	-3.6
5	75	2.8
6	17	-5.1
7	48	16.3
8	40	0
9	51	-2.1
10	74	2.8
11	40	0
12	23	1.1
13	49	2.4

(Hints: Use the Fill/Series feature of the Edit menu to construct the first column. Use the LOOKUP(.) function to obtain the X7 values.)

ON DVD

2. By completing these steps and using the table in problem 1, construct the following table of 100 samples of size 7 of the values of X7 from the American Cities database in Appendix B:

TABLE 13.2 100 descriptive statistics samples of size for X7.

100 Samples of Size 7				
Num	Sample Mean	Median	Variance	Std Dev
1	2.86	2.40	32.87	5.73
2	5.86	6.40	76.60	8.75
3	1.40	2.80	108.40	10.41
4	1.84	-1.90	93.10	9.65
5	4.09	1.30	41.98	6.48
6	16.41	20.20	299.58	17.31
7	4.09	6.40	148.18	12.17
8	10.27	9.50	115.49	10.75
9	1.87	2.20	23.96	4.89
10	5.93	5.20	33.21	5.76
11	10.00	7.20	114.12	10.68
12	2.87	-1.80	82.83	9.10

a. Fill the first column using the Fill/Series feature. Store the address of the first value of X7 in problem 1 in a vacant cell, say P30. (Hint: Use a formula such as = ADDRESS(35 + (P30 – 1) * 7, 14).) Store the address of the seventh value of X7 in problem 1 in the adjacent row of the same column, say P31. (The objective is to find the mean of each group of seven X7 values starting with the first group of seven.)

b. Compute each mean by indirectly using the addresses stored in P30 and P31, through the use of the INDIRECT(.) function. (Hint: Use a formula such as = AVERAGE(INDIRECT(ADDRESS(35 + (P30 – 1) * 7, 14)): INDIRECT(ADDRESS(35 + (P31 – 1) * 7 – 1, 14))).)

c. Complete the table for the median, variance, and standard deviations for X7.

3. Using your table from problem 2, build histograms for the means, medians, variances, and standard deviations for the 100 samples of size 7 for X7. Compare the histograms and statistics of the population and the corresponding sample statistics. What observations can you make?

4. Select a variable from the American Cities database or some other datafile. Treating the values of this variable as a population, find its parameters. Then generate random samples of size 10, and find their statistics. Use these statistics to examine the sampling distributions of:

   a. The mean

   b. The median

   c. The variance

   d. The standard deviation

   Compare the observed values of parameters of these distributions with the values predicted by theory (in those cases where we have discussed them), and observe the shapes of the sampling distributions.

5. Redo problem 4 for sample sizes of 20 and 30. Discuss any observations that are related to distributional shape and values as a function of increasing sample size.

6. A normally distributed population of values has mean 56 and standard deviation 12. If $\overline{X}$ is the mean of a random sample of size 9 from this population, find:

   a. $P(54 < \overline{X} \leq 59)$

   b. $P(\overline{X} \leq 62.4)$

   c. $P(47 < \overline{X} < 53)$

   d. x for which $P(\overline{X} > x) = 0.0228$

7. We take a sample of size n from a population that has mean 100 and standard deviation 20. Find $P(98 < \overline{X} < 102)$ if:

   a. $n = 50$

   b. $n = 100$

   c. $n = 200$

8. We take a sample of size n from a population that has mean 5.75 and standard deviation 0.32. Find $P(5.70 < \overline{X} < 5.80)$ if:

a. $n = 40$

b. $n = 80$

c. $n = 150$

9. Salaries of white-collar workers in a large city have standard deviation $14,500. We take a random sample of 200 such workers.

a. Describe the distribution of the mean salary of the sample, and sketch its graph.

b. What is the probability that the mean salary of the sample is within $500 of the mean salary of all white-collar workers in the city?

c. How large a sample should be taken so that the probability that the sample mean is within $250 of the population mean is at least 0.90?

10. A large company has 140 accounts receivable, which have a mean value of $210 and a standard deviation of $75. If 35 accounts are randomly selected, what is the probability that their mean amount is:

a. Between $190 and $230?

b. Less than $202?

11. In problem 9, assuming that the population of white-collar incomes is normally distributed:

a. Describe the sampling distribution of the median, and sketch its graph.

b. What is the probability that the median of a sample of size 200 will fall within $500 of the mean salary of all white-collar workers?

c. How large a sample should be taken so that the probability that the sample median is within $250 of the population mean is at least 0.90?

d. Is the assumption of a normally distributed population likely to be valid? Why or why not?

12. You take a sample of size 50 from a large symmetrically distributed population, and compute the mean and medium of the sample. Which is likely to be closer to the population mean? Why?

CHAPTER 14

POINT ESTIMATION AND INTERVAL ESTIMATION

OVERVIEW AND LEARNING OBJECTIVES

In This Chapter

14.1 Introduction

Having examined the sampling distributions of several important statistics, we now apply the concept to *estimation*, the process of predicting population parameters by analyzing a random sample from the population. Two techniques for this analysis are point estimation and interval estimation.

A *point estimate* of a population parameter is a single sample statistic used to represent the true value of the parameter, and we endeavor to find the "best" point estimate for a given parameter. In a one-point estimation method, a population parameter is estimated simply with the corresponding sample statistic; the sample mean estimates the population mean, and so on. We will discuss several criteria for determining a "best" point estimator.

A point estimate of a population parameter may be precise or inaccurate; we have no way of knowing. We would like to have an indication of the accuracy of the estimate. This gives rise to confidence intervals, or *interval estimates* of parameters. The researcher defines sample statistics L_L and L_U for which

$$P(L_L < \text{true value of parameter} < L_U) = C$$

Where C is a value between 0 and 1, usually 0.90 or 0.95, when the sample is taken and L_L and L_U are the limits of a confidence interval for the parameter with confidence level C.

14.2 Point Estimation

Consider the following example, in which we estimate a population proportion P.

EXAMPLE 14.1

The manager of a hardware store has bought 200 light bulbs, and wishes to know what proportion of them will burn out prematurely. Testing them all is futile, so she decides to test a random sample of 20 of them. She will use the proportion $\hat{p}$ of the sample that is defective as an estimate of the defective proportion P in the population of 200 bulbs. If 4 of the sampled bulbs are defective, $\hat{p} = 4/20 = 0.20$, and she will estimate that 20% of the entire shipment is defective.

$\hat{p}$ is a point estimator of P, but is it a "good" point estimator? Might some other statistic of the sample, say $\hat{p} - .01$, be "better"? In evaluating a point estimator or comparing estimators, these five factors are often examined: *bias*, *efficiency*, *mean-square*, *consistency*, and *sufficiency*. We consider these factors throughout this section.

The sample statistic T is an *unbiased* estimator of the population parameter θ if for all possible values of θ, $E(T) = \theta$. If $E(T) \neq \theta$, then the *bias* of T as an estimator of θ is $B = E(T) - \theta$,

EXAMPLE 14.2

Consider a dart thrower. If, on the average, his throws cluster around the specified target, we would say he is an unbiased thrower. An unbiased estimator is "aimed at" its corresponding parameter.

Previously we demonstrated that $(E(\overline{X}) = \mu$; the sample mean is an unbiased estimator of the population mean. We also showed that $E(S^2) = \sigma^2$, that

$$S^2 = \frac{1}{n-1}\sum_{i=1}^{n}\left(X_i - \overline{X}\right)^2$$

is an unbiased estimator of the population variance. Note that the division simply by n in the sample variance produces a biased estimator of the population variance σ^2:

$$\text{Let } S_1^2 = \frac{1}{n}\sum_{i=1}^{n}\left(X_i - \overline{X}\right)^2$$

$$\text{Then } S_1^2 = \frac{n-1}{n} * \frac{1}{n-1}\sum_{i=1}^{n}\left(X_i - \overline{X}\right)^2$$

$$= \frac{n-1}{n}S^2.$$

$$\text{Therefore, } E(S_1^2) = E\left(\frac{n-1}{n}S^2\right)$$

$$= \frac{n-1}{n}E(S^2) = \frac{n-1}{n}\sigma^2.$$

S_1^2 is a biased estimator of σ^2, with a tendency to underestimate the population variance, while S^2 is an unbiased estimator of σ^2. This justifies the definition of sample variance.

But is our unbiased dart thrower necessarily the best contestant? Not when competing against someone who throws a tighter pattern. We would also prefer estimators with the property of *minimum variance*, or *efficiency*.

Let T_1 and T_2 be point estimators of the population parameter θ. We say that T_1 is more *efficient* than T_2 if $Var(T_1) < Var(T_2)$.

Given two unbiased estimators, the one with the smaller variance is more likely to be close to the estimated parameter.

Consider drawing a random sample of size 2 from a population that has mean μ and variance 1. The following three quantities are possible point estimators of μ:

1. $\overline{X} = \dfrac{X_1 + X_2}{2}$
2. X_1
3. $\dfrac{X_1 + 2X_2}{3}$

where X_1 and X_2 are the values in the sample.

You can show that each of these quantities is an unbiased estimator of μ, but consider their variances:

1. $\text{Var}(\overline{X})\dfrac{\sigma^2}{n} = \dfrac{1}{2}$
2. $\text{Var}(X_1) = \sigma^2 = 1$
3. $\text{Var}(\dfrac{X_1 + 2X_2}{3}) = \dfrac{1}{9}\text{Var}(X_1) + \dfrac{4}{9}\text{Var}(X_2)$
   $= \dfrac{1}{9} * 1 + \dfrac{4}{9} * 1 = \dfrac{5}{9}$

We prefer the estimator with minimum variance, so we use $\overline{X}$; of these three estimators of μ, it is the most efficient.

There are situations in which a more efficient estimator is biased, as when estimating the population proportion P. We have shown that $E(\hat{p}) = P$, the sample proportion $\hat{p} = X/n$ is an unbiased estimator of P. Consider, however, the estimator $\hat{p}_1 = \dfrac{(X+1)}{(n+2)}$, where X again is the number of elements in the sample of size n with the characteristic of interest.

$$E(\hat{p}_1) = E[\frac{(X+1)}{(n+2)}] = \frac{E(X+1)}{n+2} = \frac{E(X)+1}{n+2}$$

$$= \frac{nP+1}{n+2}$$

This quantity is equal to P only if P = ½, so in general, $\hat{p}_1$ is a biased estimator of P. However,

$$\text{Var}(\hat{p}) = \text{Var}\left(\frac{X}{n}\right) = \frac{1}{n^2}Var(X), \text{while}$$

$$\begin{aligned}\text{Var}(\hat{p}) &= \text{Var}\left(\frac{X+1}{n+2}\right) = \frac{1}{(n+2)^2} Var(X+1) \\ &= \frac{1}{(n+2)^2} Var(X) < \frac{1}{n^2} Var(X) = Var(\hat{p}),\end{aligned}$$

So $\hat{p}_1$ is a more efficient estimator of P than $\hat{p}$ is.

In any estimation process, we would like to minimize the error between the estimating statistic and the parameter. That error is measured in this way:

> T is an estimator of the parameter θ. Then T has *mean-square error (MSE)* defined as:
>
> $$\text{MSE(T)} = \text{E}[(\text{T} - \theta)^2]$$

The mean-square error measures the expected deviation of the estimator not from the center of its sampling distribution, which may not be equal to the parameter being estimated, but from the parameter itself. In fact, if the statistic T is an estimator of the parameter θ, it is always true that $\text{MSE(T)} = \text{Var(T)} + \text{B}^2$, where $\text{B} = \text{E(T)} - \theta$ is the bias of T as an estimator of θ. Both the variance and bias of the estimator contribute to its mean-square error.

In the previous example, in which $\hat{p} = \text{X/n}$ and $\hat{p}_1 = (\text{X} + 1)/(\text{n} + 2)$ are considered as estimators of the population proportion P, it can be shown that though $\hat{p}$ is unbiased and $\hat{p}_1$ is not, $\text{MSE}(\hat{p}_1) < \text{MSE}(\hat{p})$.

Not only should our dart enthusiast maintain a tight pattern and bunch his darts around the target, but we should require that he improve with practice. The corresponding characteristic in an estimator is called *consistency.*

> A sample statistic T is a consistent estimator of the parameter θ if, as the sample is increased, T becomes more likely to be near θ.

Consider $\overline{X}$ as an estimator of μ. $\overline{X}$ is unbiased, so its sampling distribution is centered on $\mu; \sigma_{\overline{X}} = \sigma/\sqrt{n}$, so as n increases, the standard deviation of the distribution of $\overline{X}$ decreases, and the distribution becomes more tightly packed around μ. As n increases, $\overline{X}$ becomes more likely to be near μ, so $\overline{X}$ is a consistent estimator of μ.

If we wanted to estimate the variability of a population, we could take a random sample of n values from the population and compute the range of the sample; that is, the difference between its two extreme values. But this statistic uses only two of the sample values; it ignores the information represented by the other values. To take advantage of all the information in the sample that we have probably gone to some trouble to collect, we should estimate the variability of the population with a statistic that is computed from all of the elements of the sample, such as the sample variance. No other point estimator contains more information from the sample, and the sample variance is then called *sufficient.*

> If a sample statistic contains all the information the sample provides about a parameter σ, then T is called a *sufficient* statistic for σ.

The usefulness of sufficiency derives from a theorem in statistics that states that any function of a sufficient statistic is itself sufficient, and this suggests the following strategy for finding a "best" estimator for a population parameter:

1. Consider the set of all possible estimators of the parameter (for example, μ, the population mean).
2. In this set, find a sufficient statistic (for example, $\sum_{i=1}^{n} X_i$).
3. Find a function of the sufficient statistic that is an unbiased estimator of the parameter ($\overline{X} = \frac{\sum_{i=1}^{n} X_i}{2}$).
4. Under certain conditions, this statistic will also be consistent and of minimum variance.

Note that since estimators for the most prominent population parameters are established ($\overline{X}$ for μ, S^2 for σ^2, etc.), we will not in fact employ this strategy.

14.3 Interval Estimation

Though the sample mean $\overline{X}$ is a "good" estimator of the population mean μ, $\overline{X}$ is not likely to be equal to μ, and we know nothing about the accuracy of such a point estimate. However, we can use our knowledge of the sampling distribution of $\overline{X}$ to construct an *interval* around the point estimate and to state our degree of certainty that the population mean μ is in that interval. This is the process of *interval estimation.*

Consider the construction of an interval estimate for the population mean. Suppose that we take a random sample of size $n \geq 30$ from a large population with mean μ and standard deviation σ. From the central limit theorem we know that the sampling distribution of $\overline{X}$ is approximately normal with mean μ and standard deviation $\frac{\sigma}{\sqrt{n}}$.

An immediate consequence of this is that

$$\frac{\overline{X} - \mu}{\sigma / \sqrt{n}} \cong Z,$$

where Z is the standard normal random variable whose distribution is N(0, 1).

ON DVD

Given a value C (like 0.90 or 0.95), we can find, using Table A-1 in Appendix A, a value z for which

$$P(= z < Z < z) = C.$$

Take a look at Figure 14.1.

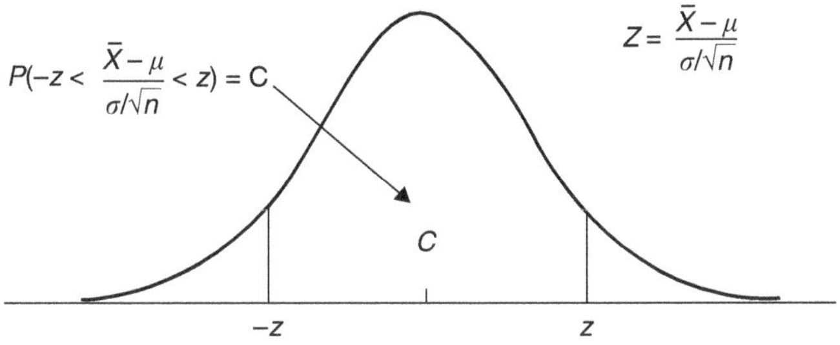

FIGURE 14.1 Construction of a confidence interval.

All figures and tables in this chapter appear on the companion DVD. ON DVD

For example, if C = 0.90, then

$$P(-z < Z < z) = 0.90$$
$$\Rightarrow P(0 < Z < z) = 0.4500$$
$$\Rightarrow z = 1.645$$

Algebraic manipulations produce the following sequence of equivalent statements:

$$P(-z < \frac{\bar{X}-\mu}{\sigma / \sqrt{n}} < z) = C$$

$$\Leftrightarrow P(-z\frac{\sigma}{\sqrt{n}} < \bar{X}-\mu < z\frac{\sigma}{\sqrt{n}}) = C$$

$$\Leftrightarrow P(-\bar{X}-z\frac{\sigma}{\sqrt{n}} < -\mu < -\bar{X}+z\frac{\sigma}{\sqrt{n}}) = C$$

$$\Leftrightarrow P(\bar{X}-z\frac{\sigma}{\sqrt{n}} < \mu < \bar{X}+z\frac{\sigma}{\sqrt{n}}) = C$$

From this we construct the estimate. Since $\bar{X}$ is a random variable, so are the two quantities $\bar{X} \pm z\left(\frac{\sigma}{\sqrt{n}}\right)$. The last equation above says that the probability that we will catch the population mean in the interval between these quantities is C. If we were to take a great number of random samples, each would produce a value of $\bar{X}$ at the center of one of these intervals, and the proportion of them that contain m would be approximately C. If we have one sample with its associated interval, we can think of that interval as having been randomly selected from all possible intervals associated with samples of size n, and we can be C confident that our particular interval does contain μ, as shown in Figure 14.2.

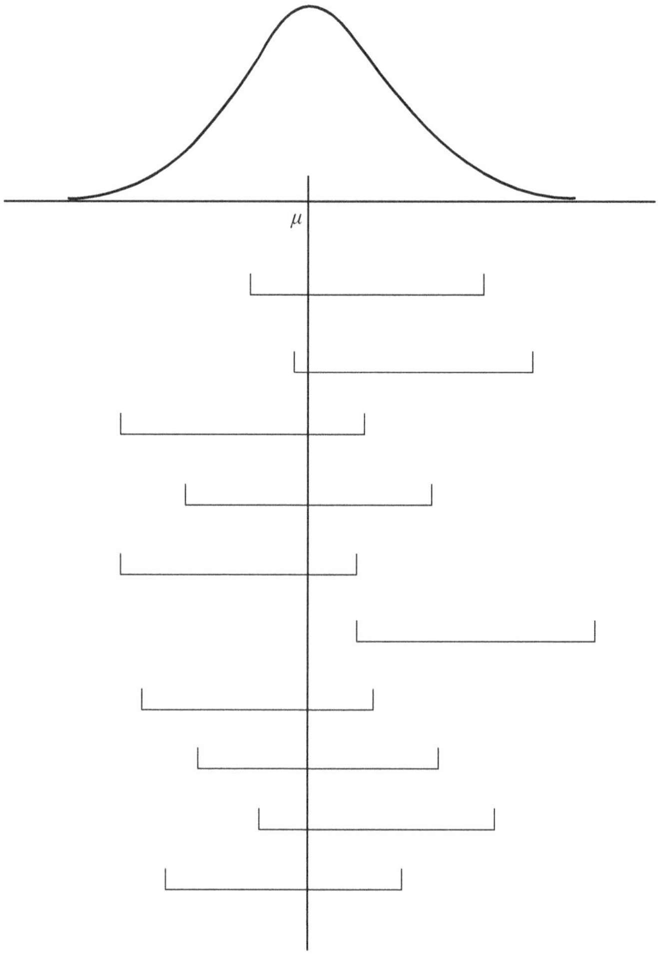

FIGURE 14.2 90% confidence intervals generated by repeated sampling.

This development gives us the following definitions:

> The *confidence level* or *degree of confidence* C associated with an interval estimate of μ is the proportion of all such interval estimates that contain μ. The interval $[\bar{X} - z\left(\frac{\sigma}{\sqrt{n}}\right), \bar{X} + z\left(\frac{\sigma}{\sqrt{n}}\right)]$, is a C $*$ 100% *confidence interval* for the population mean; the endpoints of this interval, $\bar{X} \pm z\left(\frac{\sigma}{\sqrt{n}}\right)$, are called *confidence limits*.

EXAMPLE 14.3

Suppose we take a random sample of size 100 from a population that has standard deviation 25, and we compute $\overline{X} = 78$. We construct a 90% confidence interval for the population mean.

Solution:

Corresponding to C = 0.90, we have seen that z = 1.645. The confidence limits are

$$78 \pm 1.645\left(\frac{25}{\sqrt{100}}\right) \cong 78 \pm 4.11 = 73.9, 82.1.$$

and [73.9, 82.1] is a 90% confidence interval for the population mean.

This means that we are 90% confident or 90% certain that μ is between 73.9 and 82.1. It does not mean that there is a 90% probability that the interval contains μ. The value of μ, though unknown to us, is fixed, and either is in the interval or is not, so the concept of probability does not apply.

When estimating the value of μ with a large sample ($n \geq 30$), we are not likely to know the value of the population standard deviation σ. When this happens, as it usually does, we can estimate σ with the sample standard deviation s; no other changes need to be made.

Given a desired confidence level and precision, how large a sample should be taken? Consider this example:

EXAMPLE 14.4

We plan to estimate the mean of a population of values whose standard deviation we know to be 32.6. How large a sample should we take to obtain a 95% confidence interval no more than 3 units wide?

Solution:

First, C = 95% ➔ $z = 1.96$

Then, for a confidence interval 3 units wide,

$$z\frac{\sigma}{\sqrt{n}} = 1.5 \rightarrow 1.96\frac{32.6}{\sqrt{n}} = 1.5.$$

Solving for n, we obtain

$$n = \frac{1.96 * 32.6}{1.5} = 42.60 \rightarrow n = 1814.5.$$

To meet the specified conditions, we should take a sample of at least 1,815 elements.

Our ability to construct confidence intervals for the population mean is based on knowledge of the distribution of the sample mean. Analogous processes allow us to generate confidence intervals for other parameters as long as the sampling distributions of their estimators are known, as in the following example:

EXAMPLE 14.5

Of a random sample of 300 Chicago voters, 84 support George Leroy Tyrebiter for mayor. We seek a 90% confidence interval for the true proportion of Chicago voters who support the illustrious Mr. Tyrebiter.

Solution:

We will find a confidence interval for the population proportion P, based on a sample proportion $\hat{p} = X/n$. We know that for a large sample, the distribution of $\hat{p}$ is approximately normal with mean P and standard deviation $\left[P(1-P)/n\right]^{1/2}$. Therefore,

$$\frac{\hat{p}-P}{\left[P(1-P/n\right]^{1/2}} = Z$$

And corresponding to the confidence level C we can find *z* so that

$$\mathrm{P}(-z < \frac{\hat{p}-P}{\left[P(1-P/n\right]^{1/2}} < \frac{\hat{p}-P}{\left[P(1-P/n\right]^{1/2}} < z) = \mathrm{C}.$$

Algebraically manipulating the expression within the parentheses, we obtain:

$$\mathrm{P}(\hat{p}-z\left[P(1-P)/n\right]^{1/2} < \mathrm{P} < +z\left[P(1-P)/n\right]^{1/2}) = \mathrm{C}$$

As in the development of confidence intervals for μ, the confidence limits are $\hat{p} \pm z\left[P(1-P)/n\right]^{1/2}$. We are estimating P with $\hat{p}$, however, so the confidence limits become

$$\hat{P} \pm z\left[\hat{p}(1-\hat{p})/n\right]^{1/2},$$

where *z* corresponds to the confidence level C.

In the example, C = 90%, so *z* = 1.645, *n* = 300, and $\hat{p} = 84/300 = 0.28. 90\%$ confidence limits for P are then

$$0.28 \pm 1.645\left(\frac{0.28 * 0.72}{300}\right)^{1/2} = 0.28 \pm 0.43$$
$$= 0.237, 0.323$$

We can be 90% certain that between 23.7% and 32.3% of Chicago's voters support George Leroy Tyrebiter.

14.4 Exercises

Interpretation Exercises

1. What is meant by statistical bias?
2. Under what conditions is the sample median an unbiased estimator of the population mean?
3. Of these sample statistics, which are sufficient: mean, standard deviation, range, maximum, skewness?
4. In general, and for practical purposes, what are the most desirable estimators of the population parameters μ, σ^2, σ, and P?
5. What is a confidence interval?
6. What is the importance of the central limit theorem in constructing confidence intervals for the population mean?
7. Consider the sample proportion $\hat{p}$ as an estimator of the population proportion P. Of the desirable characteristics that a point estimator might have, which are associated with $\hat{p}$?
8. You wish to estimate the mean educational level of the adults in a large city from statistics of a random sample of 500 adults. You may estimate the population mean with either the mean or median of the sample. Which do you choose, and why?

Computational Exercises

9. A random sample of size n from a large population has mean 58 and standard deviation 27. Find a 90% confidence interval for μ if:

   a. $n = 50$

   b. $n = 100$

   c. $n = 200$

   d. What general observation do these results suggest?

10. A random sample of 75 elements from a large population has mean $\overline{X} = 426$ and standard deviation $s = 91$. Find a confidence interval for the population mean μ if:

    a. C = 80%

    b. C = 90%

    c. C = 99%

    d. What general observations do these results suggest?

ON DVD

11. Consider the cities in the American Cities database, found in Appendix B, as a sample of all U.S. cities. The following is a printout from a standard statistical software system:

TABLE 14.1 Descriptive statistics for X3, unemployment.

VARIABLE X3		UNEMPLOYMENT				
MEAN	5.691	STD ERR	.197	STD DEV		1.647
VARIANCE	2.712	KURTOSIS	2.700	SKEWNESS		1.215
MINIMUM	3.000	MAXIMUM	12.000	SUM		398.400
C.V. PCT	28.936	.95 C.I.	5.299		TO	6.084
VALID CASES	70	MISSING CASES	4			

Use these statistics for the values of X3, unemployment, to find a 95% confidence interval for unemployment in all U.S. cities.

Several features of this exercise should be noted. First, because, four values are missing, the sample size is 70. Second, note that STD ERR = STD DEV/$\sqrt{n}$. Third, compare your computed results to the 95% confidence interval given above.

ON DVD

Excel Exercises

12. Using Excel, find confidence intervals for the mean change in construction activity for all U.S. cities, using the American Cities database as a sample, with confidence levels 90%, 95%, and 98%.

13. Based on the data in the American Cities database, find a 95% confidence interval for the proportion of U.S. cities in which construction activity had risen 10% or more from the previous year.

14. Use the American Cities database to find 90%, 95%, and 99% confidence intervals for the proportion of all U.S. cities in which unemployment is less than 4.0%.

15. Find a 95% confidence interval for the proportion of all U.S. cities in which department store sales increased by 10% or more, based on the American Cities database.

16. Find a 95% confidence interval for the proportion of all U.S. cities in which nonfarm employment increased by at least 5%, based on the American Cities database.

CHAPTER 15

INTRODUCTION TO HYPOTHESIS TESTING

OVERVIEW AND LEARNING OBJECTIVES

In This Chapter

15.1 Introduction

Hypothesis testing is the third estimation procedure presented, in addition to point estimation and interval estimation. In hypothesis testing, we pose two mutually exclusive statements, or *hypotheses*, about a population parameter, then develop a *decision rule*, which is a procedure based on one or more statistics of a random sample that will determine which of the two hypotheses we should accept as true.

We will use examples to illustrate the basic process for testing hypotheses. In hypothesis testing, two mutually exclusive hypotheses are given. The *null hypothesis* (designated H_0) states the *status quo*, the condition the researcher usually wishes to disprove, while the *alternative hypothesis* (H_a) is generally what the researcher wishes to show, that is, what will be true if conditions have changed.

EXAMPLE 15.1

Suppose that drug A has been used to treat a particular disease. A group of medical researchers develops drug B, which they hope will be more useful in combating the disease, but before it can be commonly used, the researchers must demonstrate that their new medicine is more effective than drug A. They will conduct an experiment to evaluate the two treatments, and will perform a *hypothesis test.*

Solution:

In the drug example, the hypotheses are these:

H_0: Drug A is at least as effective as drug B.

H_a: Drug B is more effective than drug A.

Hypothesis testing is a conservative process, based on the presumption that H_0 is true. The researcher hopes to reject H_0 and accept H_a on the basis of experimental (sample) evidence, but if H_0 is not decisively shown to be true, then the researcher fails to reject H_0, and no definite conclusion is reached.

What is needed is a procedure for deciding between the null and alternative hypotheses, a well-defined rule by which the decision, based on sample information, can be made. Consider another example:

EXAMPLE 15.2

A manufacturer of TV picture tubes has been producing a 19-inch tube that has expected life 3,100 hours, with standard deviation 450 hours. An engineer suggests a modification that she believes will increase the lifetimes of such tubes. To test her hypothesis, she tests a sample of 100 tubes built with the modification, and finds that the sample has mean lifetime 3,225 hours. Is this convincing evidence that the modification should be incorporated in all the company's picture tubes? Does it demonstrably extend expected tube life?

Solution:

The hypotheses being tested are:

H_0: The modification causes no improvement.

H_a: The modification extends tube life.

Rephrasing these in terms of a parameter, the population mean,

$$H_0: \mu \leq 3100$$

$$H_a: \mu > 3100$$

Where μ is the expected lifetime of picture tubes incorporating the modification. Recall that the expected life of tubes without the modification was 3,100 hours.

The engineer must decide whether to reject H_0 and accept H_a (conclude that the modification is effective) or to not reject H_0 (conclude that lifetimes of tubes incorporating the modification are no longer than those of the original design).

Hypothesis testing leads to four possible situations, as shown in Figure 15.1.

State of reality \ Decision	Do not reject the null hypothesis	Reject the null hypothesis
Null hypothesis true	Correct	Type I error
Null hypothesis false	Type II error	Correct

FIGURE 15.1 Outcomes of a hypothesis test.

EXAMPLE 15.2

All figures and tables in this chapter appear on the companion. ON DVD

Failing to reject H_0 when H_0 is true and rejecting H_0 when it is false are correct decisions. The other two possibilities represent errors. Rejecting the null hypothesis when it is, in fact, true is a *Type I error*, while failing to reject a false null hypothesis is a *Type II error*. In Example 2, a Type I error is committed if the engineer concludes that the modification is effective in prolonging tube life when it is not, and a Type II error is committed if the engineer fails to conclude that the modification is effective when, in fact, it is not.

The Type I error is generally more serious. In Example 2, it might result in spending time and money to incorporate an ineffective modification, so hypothesis tests are constructed to control the probability of a Type I error. This probability is indicated by α. The probability of a Type II error, discussed later in this chapter and in Chapter 16, "Single Large Sample Tests," is β.

In solving Example 2, we construct a rule by which we can decide if μ is greater than 3,100. We have seen the utility of estimating μ with the sample mean $\overline{X}$, so we will be convinced to reject the null hypothesis and accept the alternative if $\overline{X}$ is big enough, but how big is big enough? Since we also want to control α = P(Type I error), the probability that H_0 is rejected when it is true, consider the sampling distribution of $\overline{X}$ if H_0 is true, as shown in Figure 15.2.

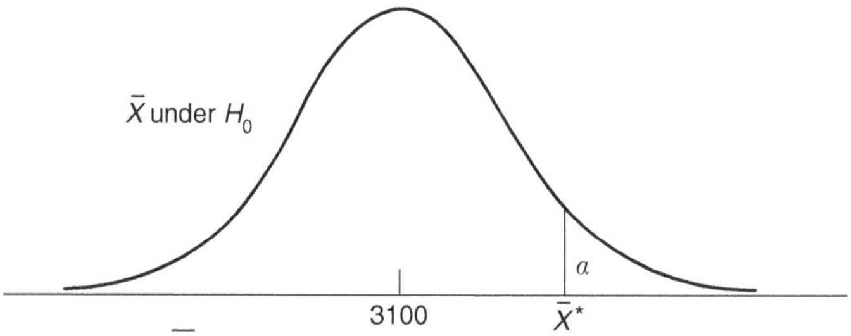

FIGURE 15.2 Sampling distribution of $\overline{X}$ given that H_0 is true.

If H_0 is true, then μ = 3100, and we know from the central limit theorem that the sampling distribution of $\overline{X}$ is approximately normal with mean 3,100 and standard deviation $450/\sqrt{100}$. Choose a value for α, say 5%, and cut off an area of size α in the upper tail of the graph of this distribution. Consider the value $\overline{X}^*$, that marks off its lower edge.

If H_0 is true, the probability that $\overline{X}$ will be greater than $\overline{X}^*$, is α, which we have chosen to be only 5%, so if the observed value of $\overline{X}$ is greater than $\overline{X}^*$, we can reject H_0 and accept H_a, concluding that the modification extends tube life, with only 5% probability of making a Type I error. We need only find the value of $\overline{X}^*$:

$$\alpha = \text{P(Type I error)} = \text{P(Reject } H_0 | H_0)$$

$$= P(\overline{X} > \overline{X}^* | H_0) = 5\% = 0.05$$

EXAMPLE 15.2

$$\rightarrow P\left(\frac{\overline{X}-3100}{450/\sqrt{100}} > \frac{\overline{X}^{*}-3100}{450/\sqrt{100}} \middle| \mu=3100\right)=0.05$$

$$\rightarrow P\left(Z > \frac{\overline{X}^{*}-3100}{450/\sqrt{100}}\right)=0.05$$

$$\rightarrow P\left(0 < Z < \frac{\overline{X}^{*}-3100}{450/\sqrt{100}}\right)=0.45$$

$$\rightarrow \frac{\overline{X}^{*}-3100}{450/\sqrt{100}}=1.645$$

$$\rightarrow \overline{X}^{*}=3100+1.645\frac{450}{\sqrt{100}}=3174.025.$$

This value completes construction of the decision rule:

> If the sample mean $\overline{X}$ is greater than $\overline{X}^{*} \cong 3174$, then we reject H_0 and accept H_a. If $\overline{X} < (\overline{X} \cong 3174)$, we cannot reject H_a.

The probability a is called the *significance level* of the test, and $\overline{X}^{*}$, is the *critical value*. Because our conclusion will be determined by the value of $\overline{X}$, the sample mean is called the *test statistic*. The interval above the critical value in this *upper-tail test* is called the *region of rejection,* since we reject H_0 when $\overline{X}$ falls in this region, while the interval below $\overline{X}^{*}$, is called the *region of acceptance,* or the area where we fail to reject.

In her examination of the 100 tubes made with the modification, the engineer found that $\overline{X}=3225$, which is greater than the critical value of 3174. According to the decision rule, she can reject H_0 at the 5% level of significance and conclude that the modification does extend the expected life of picture tubes.

In general, suppose we take a sample of size n>30 to perform an upper-tail test of the form

$$H_0: \mu \le \mu_0$$

$$H_0: \mu < \mu_0$$

We choose the significance level a, and must find the critical value $\overline{X}^{*}$, around which to build the decision rule. If H_0 is true, the distribution of the sample mean $\overline{X}$ is approximately normal with mean μ and standard deviation $\sigma/\sqrt{n}$, where σ is the population standard deviation and n is the sample size. Then

$$\alpha = P(\text{Type I error}) = P(\text{Reject } H_0|H_0) = P(\overline{X} > \overline{X}*)$$

$$P(\frac{\overline{X}-\mu_0}{\sigma/\sqrt{n}} > \frac{\overline{X}^*-\mu_0}{\sigma/\sqrt{n}}|H_0) = \alpha \rightarrow P(Z > \frac{\overline{X}^*-\mu_0}{\sigma/\sqrt{n}}) = \alpha$$

$$\rightarrow \frac{\overline{X}^*-\mu_0}{\sigma/\sqrt{n}} = z_\alpha.$$

Where Z_α, called the *critical normal deviate*, is chosen so that $P(Z > z_\alpha) = \alpha$. Then $\overline{X}^* = \mu_0 + z_0\left(\sigma/\sqrt{n}\right)$.

If $\overline{X} > \overline{X}^*$, we reject H_0 and accept H_q. If $\overline{X} < \overline{X}^*$, we fail to reject H_0, and no conclusion is reached, as shown in Figure 15.3.

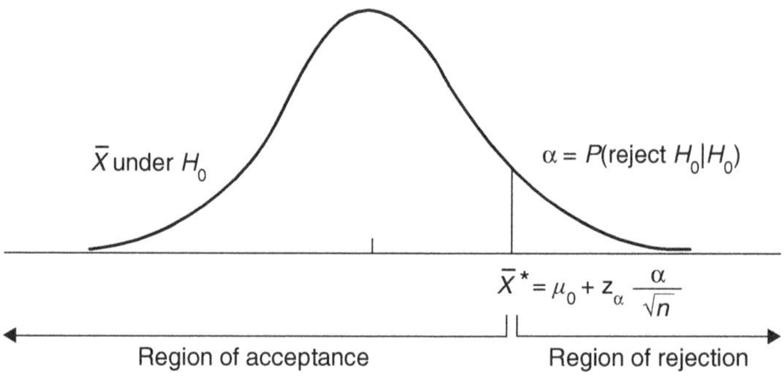

FIGURE 15.3 An upper-tail hypothesis test for μ.

If the population standard deviation is unknown, as it often is in sampling situations, the standard sample standard deviation is used instead of σ. Then $\overline{X}^* = \mu_0 + z_0\left(s/\sqrt{n}\right)$.

The decision rule for this kind of hypothesis test can be phrased in two different but equivalent ways. First, notice that if $\overline{X} > \overline{X}^* = \mu_0 + z_0\left(\sigma/\sqrt{n}\right)$, then $\frac{\overline{X}-\mu_0}{\sigma/\sqrt{n}} > z_\alpha$. That is, we can compare the value of the normal deviate $\frac{\overline{X}-\mu_0}{\sigma/\sqrt{n}}$ to the critical normal deviate, as shown in Figure 15.4.

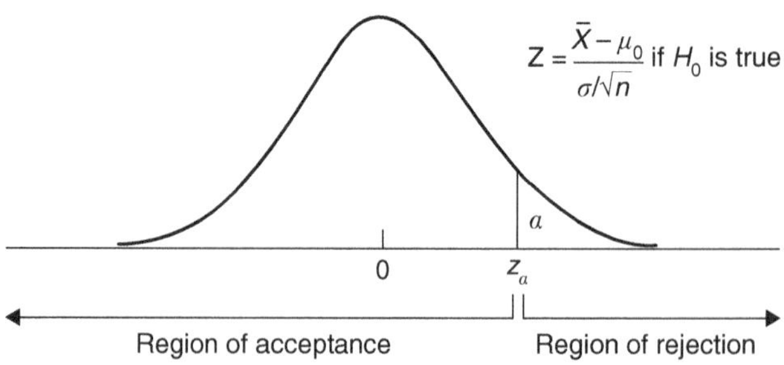

FIGURE 15.4 Regions of rejection and acceptance [fail to reject].

$$\text{If } \frac{\overline{X}-\mu_0}{\sigma/\sqrt{n}} > z_a, \text{ reject } H_0 \text{ and accept } H_a.$$

$$\text{If } \frac{\overline{X}-\mu_0}{\sigma/\sqrt{n}} < z_a, \text{ do not reject } H_0.$$

In Example 2, $a = 5\%$ so $z_a = 1.645$.

$$\frac{\overline{X}-\mu_0}{\sigma/\sqrt{n}} = \frac{3225-3100}{450/\sqrt{100}} = 2.78 > 1.645.$$

As before, we conclude that the engineer should reject the null hypothesis and conclude that the modification does increase expected tube life.

The second way to phrase the decision rule compares the area under the graph of the distribution of $\overline{X}$ when H_0 is true above the observed value of $\overline{X}$ to the significance level of a. If this area is less than a, then $\overline{X}$ itself must be above the critical value $\overline{X}^*$, and we reject H_0; if the area above the observed value of $\overline{X}$ is greater than a, then $\overline{X}$ must be below $\overline{X}^*$, and we cannot reject H_0, as shown in Figure 15.5. The area above $\overline{X}_0$, the observed value of $\overline{X}$, is $P(\overline{X} > \overline{X}_0 H_0)$.

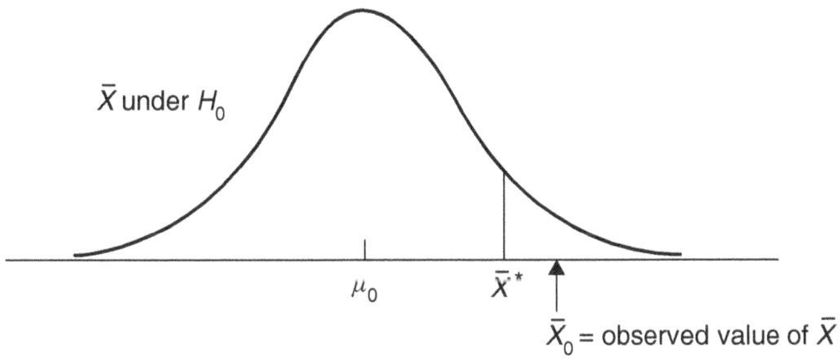

FIGURE 15.5 One tail probability.

Again looking at Example 2, $a = 5\% = 0.05$, and the observed value of $\overline{X}$ was $\overline{X}_0 = 3225$. Then

$$P(\overline{X} > 3225 | H_0) = P(\frac{\overline{X}-3100}{450/\sqrt{100}} > \frac{3225-3100}{450/\sqrt{100}} | H_0)$$

$$= P(Z > 22.78) = 0.5000 - 0.4973 = 0.0027.$$

$$0.0027 < 0.05 = a, \text{ so we reject } H_0.$$

The value $P(\overline{X} > \overline{X}_0 > | H_0)$ is sometimes called a *one-tail probability*. Note that this is the probability of a value of $\overline{X}$ at least as extreme as the observed value that will occur if H_0 is true.

Other forms of hypothesis tests are possible. We can perform *lower-tail* or *two-tail* tests with the population mean μ, as well as tests involving other population parameters. All statistical tests of hypotheses, however, will contain these elements:

- A formal statement of the null and alternative hypotheses, H_0 and H_a
- A test statistic and its sampling distribution
- A chosen level of significance, *a*
- A decision rule that defines the critical value(s) of the test statistic and the regions of acceptance and rejection
- A random sample from which to obtain the observed value of the test statistic

15.2 The Probability of a Type II Error

We have seen that the decision rule of a hypothesis test is developed to correspond to our choice of the significance level, the probability of a Type I error. We determine, and keep small, the probability of rejecting the null hypothesis when it is true. Suppose, however, that we fail to reject H_0. How much confidence can we have that we have not committed a Type II error? We must investigate β, the probability of failing to reject the null hypothesis when it is false.

Reconsider Example 2, in which the engineer hopes to demonstrate, using a sample of size 100, that a modification to her company's 19-inch picture tubes will increase their expected life beyond 3,100 hours. The population standard deviation is 450 hours and the hypotheses, to be tested at the 5% significance level, are these:

$$H_0: \mu \leq 3100$$

$$H_a: \mu > 3100$$

The critical value of this test is

$$\overline{X}^* = 3100 + 1.645 \frac{450}{\sqrt{100}} \cong 3174,$$

and the decision rule is:

If $\overline{X} > 3174$, reject H_0 and accept H_a;

If $\overline{X} < 3174$, do not reject H_0.

Suppose that the expected lifetime of tubes incorporating the modification is 3,200 hours. Then $\mu = 3200$, H_0 is false, and H_a is true. In this situation, the sampling distribution of $\overline{X}$ is approximately normal with mean 3200 and standard deviation $450/\sqrt{100}$, and the probability of a Type II error, or failing to conclude that H_0 is false even though H_a is true (since $\mu = 3200$), is:

$$\beta = \text{P}(\text{Type II error}) = \text{P}\overline{X} < 3174 | \text{H}_z \text{ with } \mu = 3200)$$

$$= P(\frac{\overline{X}-3200}{450/\sqrt{100}} < \frac{3174-3200}{450/\sqrt{100}} | H_z \text{ with } \mu = 3200)$$

$$= P(Z < -0.58) = 0.5000 - 0.2190 = 0.2810.$$

That is, if the true mean lifetime of tubes incorporating the new process is 3,200 hours, the probability that our test will nonetheless fail to reject H_0 is 28.10%. This probability is represented graphically in Figure 15.6 as the area under the curve of the distribution of $\overline{X}$ if $\mu = 3200$. Note that 3200 is to the left of $\overline{X}^* = 3174$.

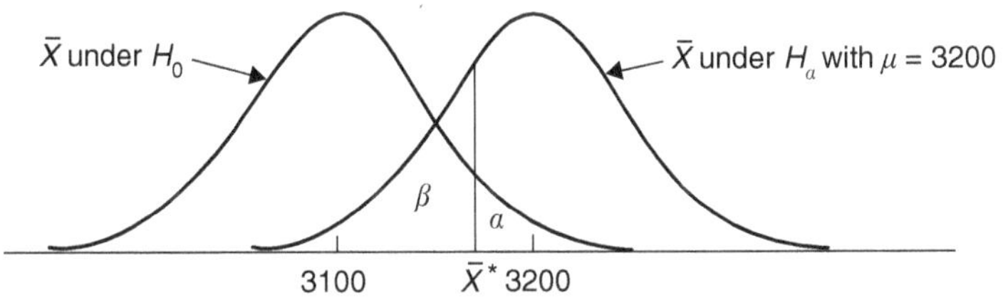

FIGURE 15.6 Comparison of H_0 distribution to an alternative distribution.

In general, for an upper-tail test of this kind,

$$\beta = P(\overline{X} < \overline{X}^* | H_a \text{ with } \mu = \mu_a) = P(Z < \frac{\overline{X}^* - \mu_a}{\sigma/\sqrt{n}}),$$

Where μ_a is a possible value of μ for which H_0 is false and H_a is true.

It is important to observe that the value of β depends on the true value of the population mean μ, which is also the center of the distribution of $\overline{X}$. If μ is, in fact 3,125,

$$\beta = P(\text{Type II error}) = P(\overline{X} < 3174 | H_z \text{ with } \mu = 3125)$$

$$= P(\frac{\overline{X}-3125}{450/\sqrt{100}} < \frac{3174-3125}{450/\sqrt{100}} H_z \text{ with } \mu = 3125)$$

$$= P(Z < -1.00) = 0.5000 \pm 0.3621 = 0.8621.$$

If the true mean is 3,125, the probability that we will mistakenly fail to reject H_0 is 86.21%, as shown in Figure 15.7.

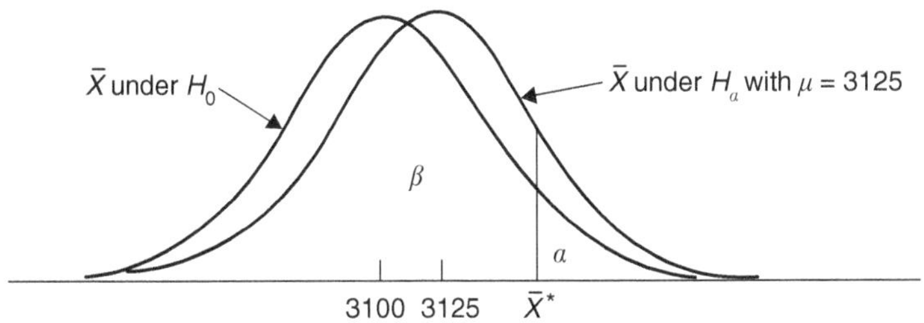

FIGURE 15.7 Failure to reject H_0.

There is a value of β corresponding to every possible value of μ for which H_0 is false. Since there are an infinite number of such values, it would be futile to attempt to calculate them all. We can, however, represent graphically the relationship between P(Accept H_0) and all possible values of μ.

If $\mu = 3100$, then H_0 is true, and P(Accept $H_0|H_0$) = 1 – P(Reject $H_0|H_0$) = 1 – a = 0.95. Calculations like these fill in the rest of Table 15.1, which shows the representative values of μ.

TABLE 15.1 Representative values of μ.

μ	P(Accept H_0)
3100	0.9500
3125	0.8621
3150	0.7019
3174	0.5000
3200	0.2810
3225	0.1292
3250	0.0455

Note that if $\mu = \overline{X}^*$, that is, if the population mean is equal to the critical value, the value of β is 0.5000.

From these values, we draw Figure 15.8 relating P(Accept H_0) to possible values of μ. It is called the *operating characteristic curve* or *OC curve* for the test, and at each possible value of μ except 3,100 (where H_0 is true), the height of the graph is the probability of a Type II error, β. The OC curve always begins at the point (μ_0, 1–a), where μ_0, is the null hypothesis value of μ, and passes through the point $(\overline{X}^*, 0.5)$.

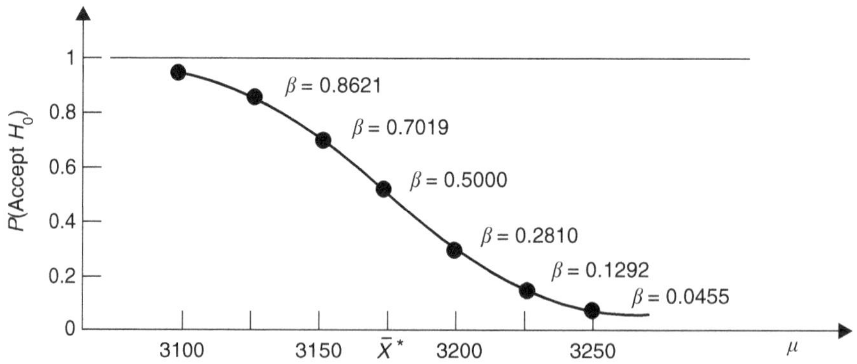

FIGURE 15.8 Operating characteristic curve.

Used as often as the operating characteristic curve is the *power curve,* which plots $1–\beta$ = P(Reject H_0) against possible values of μ, as shown in Figure 15.9.

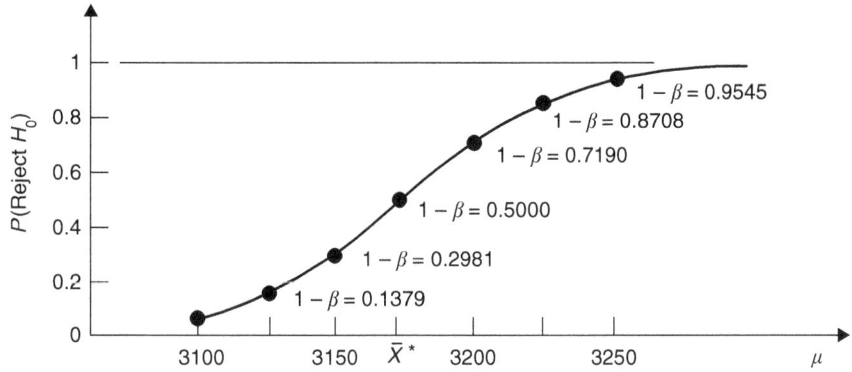

FIGURE 15.9 Power curve.

Note that the curve begins at the point (μ_0, a), also passes through $(\overline{X}^*, 0.5)$, and contains the same information as the operating characteristic curve.

The graphs in Figures 15.8 and 15.9 tell us the probabilities of accepting or rejecting the null hypothesis for all possible values of μ; they tell us how likely it is that our test will distinguish between μ_0 (31,00 in Example 2) and other possible values of μ.

These curves can also be used to illustrate how a test is affected by changes in significance level or sample size. For example, in an upper-tail test for μ of the type we have been considering, an increase in the significance level a will shift the starting point of the OC curve down and that of the power curve up, and will decrease the critical normal deviate Z_a so that $\overline{X}^* = \mu_0 + z_a \sigma/\sqrt{n}$ also decreases.

In Figure 15.10, the OC curve is moved down and the power curve up. That is, if the significance level is increased, the probability of a Type II error, β, is decreased. Conversely, decreasing a increases β. Again we encounter a trade-off, like that between certainty and precision in confidence intervals.

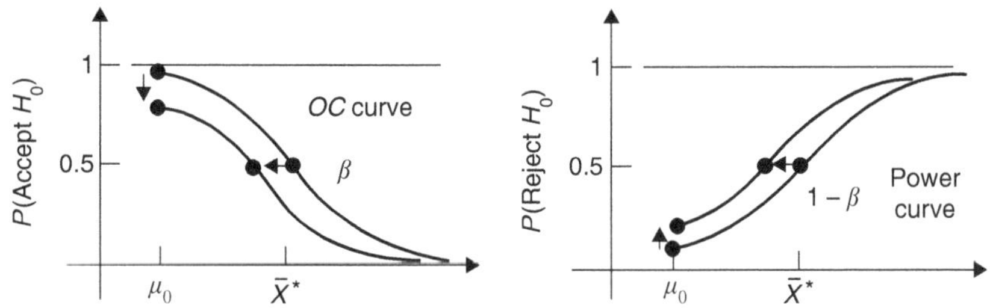

FIGURE 15.10 Effects on the OC and power curves of increasing a.

With confidence intervals, we found that we could improve certainty without degrading precision and vice versa by collecting more information; that is, by increasing the sample size n. In our upper-tail hypothesis test for μ, increasing n will decrease the standard deviation of $\overline{X}, \sigma/\sqrt{n}$, and will reduce the critical value $\overline{X}^* = \mu_0 + z_a \frac{\sigma}{\sqrt{n}}$. The starting points of the two curves, (μ_0, 1–a) and (μ_0, a), will not be affected, so the curves will become steeper, as shown in Figure 15.10. Except in a small region near μ_0, β will be reduced without degrading a; for most values of μ above μ_0, the test with larger n is more likely to be accurate.

In general, if the OC and power curves of one test are steeper than those of another, we say that the first test is more *powerful*, or better able to distinguish between μ_0 and other possible values of μ. We can increase the power of a test, its ability to distinguish between H_0 and H_a, by enlarging the sample.

15.3 Exercises

Interpretation Exercises

1. In Example 2, what is done if:

a. The null hypothesis is rejected?

b. The null hypothesis is not rejected?

c. A Type I error is committed?

d. A Type II error is committed?

ON DVD

2. Using the statistics of the variable X4 from the American Cities database given in Table 15.2, test the null hypothesis that nonfarm employment in U.S. cities rose no more than 2.2% against the alternative that the rise was greater than 2.2%.

a. Use $a = 5\%$.

b. Use $a = 10\%$.

c. Why is there a difference in the results of parts a and b? What generalization can be made?

TABLE 15.2 Descriptive statistics for nonfarm employment.

VARIABLE X4		CHANGE IN NONFARM EMPLOYMENT					
MEAN	2.699	STD ERR	.328	STD DEV	2.820		
VARIANCE	7.950	KURTOSIS	2.984	SKEWNESS	.937		
MINIMUM	– 4.400	MAXIMUM	13.900	SUM	199.700		
C.V. PCT	104.483	.95 C.I.	2.045	TO	3.352		
VALID CASES	74	MISSING CASES	0				

3. Using a sample of size 150, these hypotheses will be tested at the 5% level of significance:

$$H_0: \mu > 86$$

$$H_a: \mu > 86$$

a. If $\overline{X} = 89.3$ and $s = 19.6$, what conclusion is reached?

b. Consider the statistic $\left(\overline{X} - \mu_0\right) / \left(s/\sqrt{n}\right)$. Compare this value to the critical normal deviate x_a corresponding to $a = 5\%$. How can the decision rule be rephrased?

4. For $a = 1\%$, 2%, 5%, and 10%, find z so that $P(Z>z) = a$.

Excel Exercises

5. Using data from the American Cities database, test the null hypothesis that department store sales grew by no more than 8% against the alternative hypothesis that they grew by more than 8%. Test at the 5% significance level.

ON DVD

6. Using Excel along with data from the American Cities database, test the null hypothesis that the *proportion* of U.S. cities in which unemployment is less than 4% is no more than 10% against the alternative that the proportion is more than 10%. Use $a = 5\%$. (Hint: H_0: $P \leq 10\%$, H_a: $P > 10\%$. Find p^* for which $P(\hat{p} > \hat{p}^* | H_0) = a$.)

ON DVD

UNIT 4

HYPOTHESIS TESTING

We are now in a position to design methods to answer the questions in the CIO article *8 Questions For Uncovering Information Security Vulnerabilities* by Andrew Jaquith, available at *http://www.cio.com/article/109958/8_Questions_For_Uncovering_Information_Security_Vulnerabilities*. Additional security principles that enhance security through system design include:

- *Economy of mechanism.* Keep the design as simple and small as possible.
- *Separation of privilege.*
- *Least common mechanism.* Every shared mechanism represents a potential information path between users and therefore these should be minimized.

Study this article again and discuss the eight questions in light of the above security principles.

How do we decide to decide? How much evidence is enough? We will sample from a population in order to decide, or at least form an opinion on, something about the population. But a sample is only an example. One example does not necessarily prove a theory. On the other hand, a probabilistic statement about the range in which the characteristics of most systems would lie can be made. This unit will show you how to state a hypothesis clearly as involving two options and then select one option based on the results of a statistic computed from a random sample of data. Finally we will have a methodology for resolving the questions posed here.

Chapter 20 presents an alternative methodology to use called confidence intervals. While a hypothesis test is usually a yes-no decision, a confidence interval not only gives that answer but also provides information about the range of values for the parameter.

CHAPTER 16

SINGLE LARGE SAMPLE TESTS

OVERVIEW AND LEARNING OBJECTIVES

In This Chapter

16.1 Introduction

16.2 Sample Study

16.3 Lower-Tail Tests for the Population Mean

16.4 Two-Tail Tests for the Population Mean

16.5 Exercises

16.1 Introduction

In Chapter 15, we introduced hypothesis testing as a form of estimation in which we derive a rule that enables us to choose between two mutually exclusive statements about a population parameter, the null hypothesis H_0 and the alternative hypothesis H_a. We take a random sample from the population and compute a test statistic. The decision rule, chosen to control the probability of a Type I error, describes the values of the test statistic for which we reject H_0 and accept H_a, and the value for which we are unable to reject H_0. In this chapter, we extend our repertoire of techniques for performing single large sample hypothesis tests on the population mean.

16.2 Sample Study

The computer frees us from the drudgery of computations, and thus we can start "doing statistics" in a meaningful way almost immediately. We will pose questions, and you will be asked to make decisions and answer those questions. As various hypothesis tests are introduced in the chapters that follow, you will use them to make decisions and use Excel as a tool to perform the procedures.

ON DVD Most of our discussions and examples will be based on a fictitious database found in Table B-2, American Cities database – Version 2, which consists of economic and business data about 75 U.S. cities. Table 16.1 shows part of that database. We will use descriptive statistics to summarize the data, thereby making it more understandable, and with statistical inference we will draw interesting and useful conclusions about the economic climate that produced the data.

TABLE 16.1 Partial fictitious American Cities database - Version 2.

City Region	Change in Dept Store Sales	Unempl Rate	Change in Nonfarm Empl	Change in Income Factory Workers	Change in Factory Workers Income	Change in Construction Activity
X1	X2	X3	X4	X5	X6	X7
E	0.107	0.047	0.032	$43,432.00	0.124	0.005
E	0.088		0.011	$46,178.00	0.090	0.052
E	0.109	0.062	0.041	$44,132.00	0.057	0.215
E	0.127	0.065	0.022	$45,784.00	0.095	0.136
E	0.071	0.097	0.011	$48,362.00	0.025	0.214
E		0.064	0.049	$43,979.00	0.116	0.274
E	0.033	0.061	0.040	$43,002.00	0.087	0.118
E	0.119	0.063	0.002	$48,740.00	0.138	0.047
AVE(s)	**0.102**	**0.057**	**0.027**	**$45,073.70**	**0.085**	**0.055**
Count	70	70	74	73	73	74

ON DVD All figures and tables in this chapter appear on the companion DVD.

16.3 Lower-Tail Tests for the Population Mean

A *lower-tail* test for the population mean has this form:

$$H_0: \mu \geq \mu_0$$
$$H_a: \mu < \mu_0$$

In such a test we are attempting to determine if the population mean is less than some value μ_0; our test statistic is the sample mean, and we will be convinced that $\mu < \mu_0$ if $\overline{X}$ is small enough. The critical value $\overline{X}^*$ is below μ_0, and we will reject the null hypothesis and accept the alternative if $\overline{X} < \overline{X}^*$. The decision rule is derived as in the upper-tail test but with all the inequalities reversed, assuming a sample of size $n \geq 30$ and invoking the central limit theorem.

We select the significance level a, and must find the critical value $\overline{X}^*$ so that P(Type I error) = a:

$$\text{P(Type I error)} = \text{P(Reject } H_0 | H_0)$$
$$= P(\overline{X} < \overline{X}^* | H_0) = P\left(\frac{\overline{X} - \mu_0}{\sigma/\sqrt{n}} < \frac{\overline{X}^* - \mu_0}{\sigma/\sqrt{n}} | H_0\right)$$
$$= P\left(Z < \frac{\overline{X}^* - \mu_0}{\sigma/\sqrt{n}}\right) = a.$$

Therefore,

$$\frac{\overline{X}^* - \mu_0}{\sigma/\sqrt{n}} = -z_a, \quad \text{so} \quad \overline{X}^* = \mu_0 - z_a \frac{\sigma}{\sqrt{n}},$$

Where z_a corresponds to a as before, as shown in Figure 16.1. Note that a and $\overline{X}^*$ are in the lower tail of the distribution of $\overline{X}$ under H_0. The decision rule is this:

If $\overline{X}$, reject H_0 and accept H_a.

If $\overline{X} \geq \overline{X}^*$, do not reject H_0.

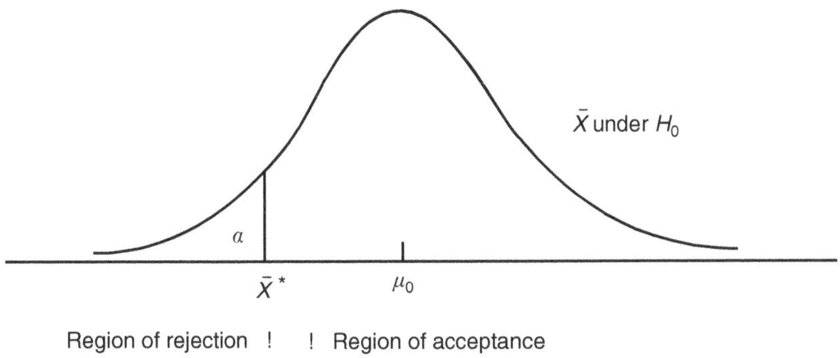

FIGURE 16.1 Lower-tail test for the population mean.

Again, we can compare the normal deviate $\frac{\overline{X}-\mu_0}{\sigma/\sqrt{n}}$ to the critical normal deviate $-z_a$, or the one-tail probability $P(\overline{X}<\overline{X}^*|H_0)$ to the significance level a, and if σ is unknown, we estimate it with s.

EXAMPLE 16.1

Using the data in the American Cities database, we wish to determine if average unemployment is below 5.8% in these cities, and we wish to limit the risk of making such a claim falsely to 2%.

Solution:

If μ is the mean unemployment in these cities, then we are testing these hypotheses at the 2% significance level:

$$H_0: \mu \geq 5.8$$
$$H_a: \mu < 5.8$$

The statistics of the sample of values of X3, unemployment, are provided in Table 16.2:

TABLE 16.2 Unemployment, X3, descriptive statistics.

Statistics	
Mean	5.708
Standard Error	0.196
Median	5.500
Mode	5.300
Standard Deviation	1.647
Sample Variance	2.710
Kurtosis	2.633
Skewness	1.184
Range	9.000
Minimum	3.000
Maximum	12.000
Sum	399.600
Count	70
Missing Cases	4
Confidence Level (95.0%)	0.392

EXAMPLE 16.1

Using the sample standard deviation to estimate σ, $z_\alpha = 2.05$ corresponding to $\alpha = 2\%$, and noting that the sample size is 70,

$$\overline{X}^* = 5.8 - 2.05\frac{1.647}{\sqrt{70}} = 5.396.$$

We will reject H_0 if $\overline{X}$ is less than $\overline{X}^* = 5.396$. The observed value of $\overline{X}$ is 5.708 > 5.396, so we are unable to reject H_0; we cannot conclude that average unemployment in the cities in the database is below 5.8%.

Note that equivalently, the z statistic $\frac{\overline{X}^* - \mu_0}{\sigma/\sqrt{n}} = \frac{5.708-5.8}{1.647/\sqrt{70}} = -0.47$ = is not less than the critical normal deviate −2.05, and the one-tail probability $P(\overline{X} < 5.704|H_0) = P(Z<-0.47) =$ 0,500 - 0.1808 = .319 is not less than $\alpha = 0.02$. We cannot reject H_0.

β, the probability of a Type II error, is found as in an upper-tail test; inequalities are again reversed, as shown in this calculation of β for the above example if $\mu = 5.65$:

$$\begin{aligned}\beta &= P(\text{Type II error}|H_a, \text{ with } \mu = 5.65)\\ &= P(\overline{X} > \overline{X}^*|\text{Ha, with } \mu = 5.65)\\ &= P\left(\frac{\overline{X}-5.65}{1.647/\sqrt{70}} > \frac{5.708-5.65}{1.647/\sqrt{70}}\,|\,H_a, \text{ with } \mu = 5.65\right)\\ &= P(Z > 0.295) = 0.5000 - 0.115 = 0.385.\end{aligned}$$

The operating characteristic and power curves of a lower-tail test are the mirror images of those for an upper-tail test. The OC curve begins at $(\mu_0, 1-\alpha)$, and the power curve begins at (μ_0, α); both pass through the point $(\overline{X}^*, 0.5)$, as shown in Figure 16.2.

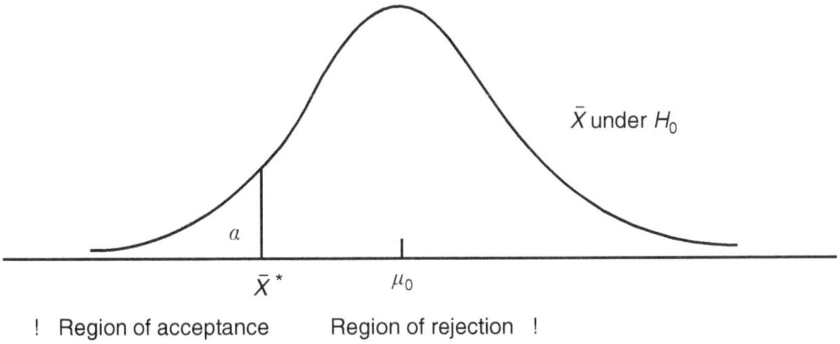

FIGURE 16.2 Operating characteristic and power curves of a lower-tail test for μ.

16.4 Two-Tail Tests for the Population Mean

Often we are concerned only that the population mean is different from some particular value and not with the direction of that difference. Thus we have *two-tail tests*:

$$H_0: \mu = \mu_0$$
$$H_a: \mu \neq \mu_0$$

In such cases, we will be convinced that H_0 is false if the test statistic $\overline{X}$ is far enough away from μ_0. There are two critical values, $\overline{X}_1^*$ and $\overline{X}_2^*$, again chosen to control α, and generally equidistant from μ. Because of this symmetry,

$$P(\overline{X} < \overline{X}_1^* | H_0) = \frac{\alpha}{2} \quad \text{and} \quad P(\overline{X} > \overline{X}_2^* | H_0) = \frac{\alpha}{2}$$

Derivations identical to those for the two one-tail tests give us formulas for the critical values:

$$\overline{X}_1^* = \mu_0 - z_{\alpha/2}\frac{\sigma}{\sqrt{n}} \quad \text{and} \quad \overline{X}_2^* = \mu_0 + z_{\alpha/2}\frac{\sigma}{\sqrt{n}}.$$

We reject H_0 if $\overline{X}$ is less than the lower critical value or greater than the upper critical value. If $\overline{X}$ falls between the critical values, we cannot reject H_0.

Again, we may compare either $\left(\frac{\overline{X} - \mu_0}{\sigma/\sqrt{n}}\right.$ with $\left.\pm z_{\alpha/2}\right)$ or (α with the *two-tail probability*) that a value of $\overline{X}$ is at least as far away from μ_0 as that observed would occur; see Figure 16.3.

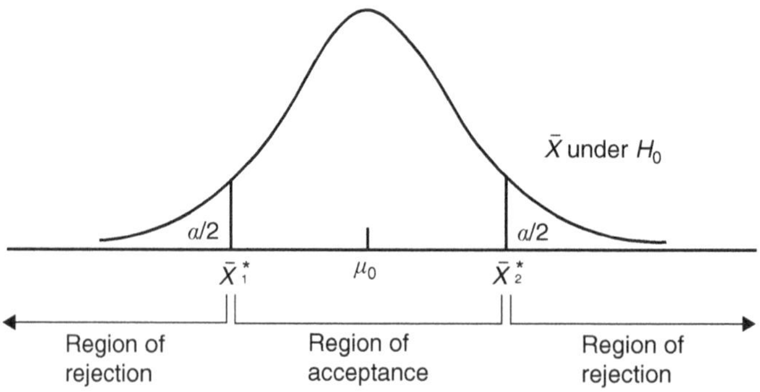

FIGURE 16.3 Two-tail test for the population mean.

Calculations of β and the forms of the OC and power curves are left as exercises for the reader.

EXAMPLE 16.2

A process makes machined parts to a mean diameter of 25.70 cm, with standard deviation 0.01 cm. Periodically, a sample of 50 parts is measured to see if the process requires adjustment. Management wishes to perform this task unnecessarily only 5% of the time. If one such sample has mean diameter 25.704 cm, should the process be stopped for adjustment?

Solution:

The hypotheses are these:

$$H_0: \mu = 2570$$
$$H_a: \mu \neq 2570$$

We are told that management wishes to limit the probability of a Type I error to 5%, so $z_{\alpha/2} =$ 1.96, and the critical values are:

$$\mu_0 \pm z_{d2} \frac{\sigma}{\sqrt{n}} = 25.70 \pm 1.96 \frac{0.01}{\sqrt{50}} = (25.6932, 25.7028).$$

The observed value of $\overline{X}$, 25.704, does not fall between the critical values, so management rejects H_0 and adjusts the manufacturing process.

EXCEL

SUMMARIZING DATA WITH DESCRIPTIVE STATISTICS IN EXCEL

Step 1. To create a set of descriptive statistics for a data sheet, choose **Tools** from the menu bar.

- Click **Data Analysis**.
- Select **Descriptive Statistics**.

Step 2. In the Descriptive Statistics dialog box:

- Select the input range by clicking and dragging the appropriate cells on the data sheet.
- If each data set is listed in a different column, then select **Columns**.
- Check the **Labels in First Row** option if the first row of the data range contains labels and not data.
- Make a selection for the output range (on the same or a different sheet).
- Select **Summary Statistics** to ensure that you get the most commonly used descriptive statistical measures.

Step 3. Click **OK**.

Step 4. Utilize the normative features of Excel to edit the output table.

16.5 Exercises

Interpretation Exercises

1. We will take a sample of size 80 from a population whose standard deviation we know to be 56 and test these hypotheses:

$$H_0: \mu \leq 300$$
$$H_a: \mu > 300$$

   a. At the 5% significance level, find the critical value and state the decision rule.

   b. For these possible values of μ, find the probability of a Type II error: 305, 300, 315, 320.

   c. Use the values found in part b to sketch the operating characteristic and power curves of the test.

   d. Find the critical value corresponding to $\alpha = 1\%$, restate the decision rule, and find β if $\mu = 315$. What generalizations does this suggest about α and β?

   e. In the original test, $\overline{X} = 312$. What conclusion is reached? What type of error might have been made?

   f. Perform the test in e again by comparing the z statistic to the critical normal deviate, then by comparing the probability that $\overline{X}$ would be at least as large as its observed value with α.

2. Show that for any upper-tail test of the population mean, the value of β if $\mu = \overline{X}^*$ is 0.5000.

3. What factors influence the power of a test of hypotheses?

4. Use the following data from the American Cities database to test the claim that construction in U.S. cities was up by more than 5% at the 5% significance level:

TABLE 16.3 Descriptive statistics for change in construction activity.

VARIABLE X7		CHANGE IN CONSTRUCTION ACTIVITY			
MEAN	5.523	STD ERR	1.081	STD DEV	9.301
VARIANCE	86.517	KURTOSIS	1.947	SKEWNESS	.924
MINIMUM	– 15.400	MAXIMUM	40.600	SUM	408.700
C.V. PCT	168.414	.95 C.I.	3.368	TO	7.678
VALID CASES	74	MISSING CASES	0		

a. Formally state the hypotheses and determine the decision rule.

b. If construction activity has in fact increased by 5.5%, find the probability that we will fail to reject H_0. Illustrate by drawing the distribution of $\overline{X}$ if $\mu = 5.5\%$ and indicate the area β.

c. Sketch the OC and power curves for this test.

d. Use the decision rule to come to a conclusion. What type of error might have been made?

5. Show that for any lower-tail test of hypotheses for μ, β is 0.50000 if $\mu = \overline{X}^*$.

6. Find $z_{\alpha/2}$ for $\alpha = 1\%$, 2%, 5%, and 10%.

Excel Exercises

7. Using the American Cities database, test the claim that the change in factory worker income was less than 10% at the 5% significance level.

a. Formally state the hypotheses and develop a decision rule.

b. Find β if the true mean change in factory worker income is 9.0%.

c. Sketch the OC and power curves of this test.

d. If the sample size were increased, how would the curves of Part C change?

e. Use the decision rule to reach a conclusion.

f. Repeat the test by comparing the normal deviate to the critical normal deviate, and by comparing $P(\overline{X} < 8.673 | H_0)$ to α.

CHAPTER 17

SINGLE SMALL SAMPLE TESTS

OVERVIEW AND LEARNING OBJECTIVES

In This Chapter

17.1 Introduction

In our work so far we have taken large samples and have found in the central limit theorem (CLT) the description of the sampling distribution of $\overline{X}$ and the equivalent statement that $(\overline{X}-\mu)/(\sigma/\sqrt{n})=Z$. Samples are not always of a size to allow us to invoke the CLT, however, though we often want to estimate μ with $\overline{X}$ when n is less than 30. In this chapter, we will extend our repertoire of techniques for performing hypothesis testing on the population mean when dealing with small samples ($n<30$), and for performing hypothesis testing with the population variance.

17.2 Small Sample Tests for the Population Mean

When we are dealing with small samples, we apply a result published in 1908 by the chemist William S. Gossett, who had studied the distribution of the quantity $t=\left(\overline{X}-\mu\right)/\left(s/\sqrt{n}\right)$ for many values of n. Gossett's employer, a brewery in Dublin, Ireland, did not wish him to publish, so this result first appeared under the pen name "Student."

> If we take a sample of size n from a normally distributed population with mean μ, the quantity $t=\left(\overline{X}-\mu\right)/\left(s/\sqrt{n}\right)$ has a *t distribution*, also called student's *t distribution*, with n–1 degrees of freedom.

The mean of every t distribution is 0, around which it is symmetrical and bell-shaped, though flatter than the standard normal distribution. The "number of degrees of freedom" is the parameter that determines which t distribution is being considered, and refers to the fact that for a given value of $\overline{X}$, n–1 of the sample elements are arbitrary; the last is determined by the others and $\overline{X}$. Mathematically, the t random variable is the standard normal random variable Z divided by the square root of a chi-square random variable (the chi-square distribution will be discussed shortly) over its own degrees of freedom.

As the number of degrees of freedom increases (as the sample size increases), the t distribution approaches that of Z. For 30 or more degrees of freedom, they are approximately equal, so we generally apply the t distribution to situations involving small samples, of size less than 30; see Figure 17.1.

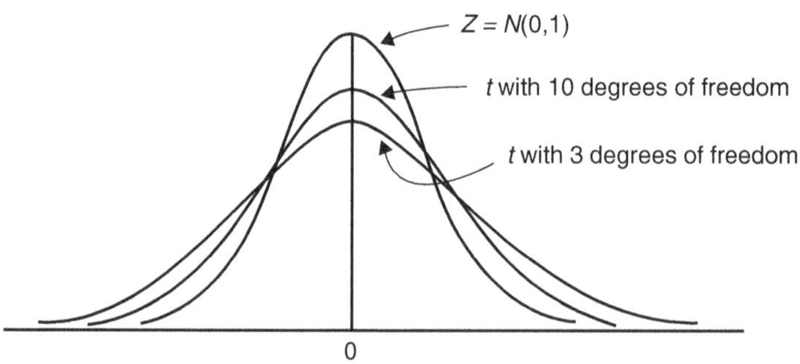

FIGURE 17.1 *t* distribution and the standard normal distribution.

All figures and tables in this chapter appear on the companion DVD. ON DVD

Though Gossett's result assumed that the population from which the sample was taken was normally distributed, further research has shown that this condition can be dropped. The t statistic is robust; it produces accurate results even without the assumption of normality, and we may use the t statistic without too much concern for the distributional shape of the underlying population.

Suppose that we wish to test these hypotheses at the 10% significance level, and we plan to take a sample of size 15:

$$H_0: \mu = 52$$

$$H_a: \mu \neq 52$$

If H_0 is true, the sampling distribution of the t statistic $(\overline{X}-52)/(s/\sqrt{n})$ is a t distribution with 15 – 1 = 14 degrees of freedom. We will be convinced to reject H_0 if t is less than $t_{a/2}$ or greater than $t_{a/2}$. From Table A-3, Critical Values of the t Distribution, in Appendix A, corresponding to a = 10% and 14 degrees of freedom, $t_{a/2}=1.761$, as shown in Figure 17.2. ON DVD

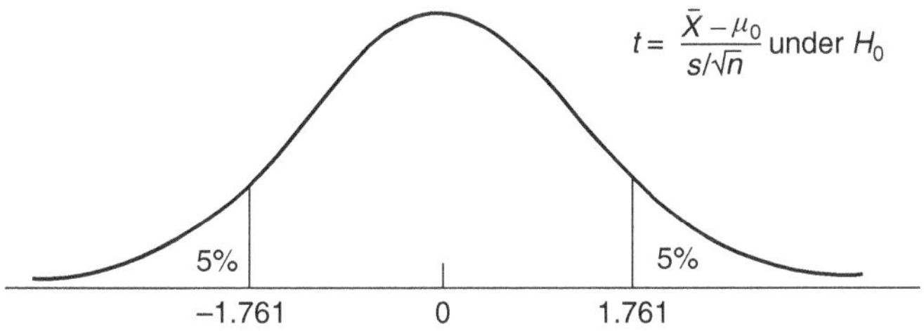

FIGURE 17.2 Critical values of the t statistic.

We take the sample and find that $\overline{X} = 47$ and $s = 12$. Then $t = (47-52)/(12/\sqrt{15}) = 1.614$, which is between the critical deviates, and we are not able to reject H_0.

Note that upper-tail and lower-tail tests may be performed in the obvious way using the t statistic, and that it is tests of this type that prove extremely useful in business decision making.

17.3 Hypothesis Tests with the Population Variance

Not all inferences concern the population mean; we are also interested in the value of the population variance. To investigate σ^2, we apply the following:

> If we take a sample of size n from a normally distributed population with variance σ^2, then the quantity $X^2 = (n-1)\, S^2/\sigma^2$ has a *chi-square distribution* with n–1 degrees of freedom.

The value $X^2 = (n-1)\, S^2/\sigma^2$ is called the *chi-square statistic*, and its distribution, like that of the t statistic, depends on the sample size; the number of degrees of freedom determines which chi-square distribution is being considered. A chi-square distribution with r degrees of freedom has mean r and variance $2r$. A chi-square random variable cannot take on negative values, and its distribution is skewed to the right, becoming more symmetrical and bell-shaped as the number of degrees of freedom increases, as shown in Figure 17.3.

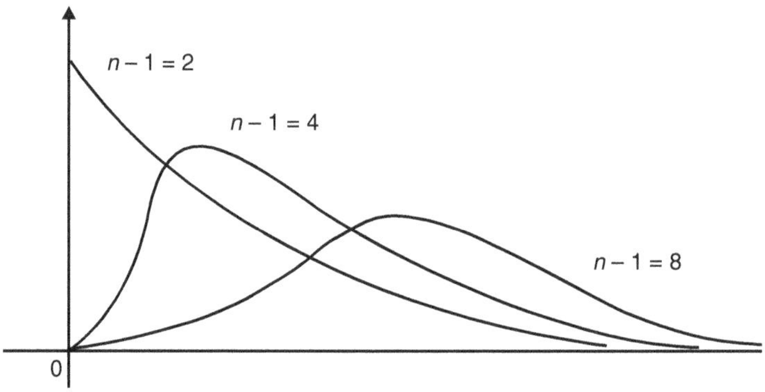

FIGURE 17.3 Chi-square distributions associated with various sample sizes.

Using the chi-square distribution, we can construct confidence intervals and perform hypothesis tests for the population variance, and because the chi-square statistic, like the t statistic, is robust, we are not restricted to normally distributed populations. Consider finding a C*100% confidence interval for σ^2 based on S^2, the variance of a sample of size n.

ON DVD

From Table A-7, Critical Values of the Chi-Square Distribution, found in Appendix A, we can find values $\chi^2_{1-\alpha/2}$ and $\chi^2_{\alpha/2}$ corresponding to n–1 degrees of freedom so that

$$P(\chi^2_{1-\alpha/2} < (n-1)S^2/\sigma^2 < \chi^2_{\alpha/2}) = C$$

as illustrated in Figure 17.4.

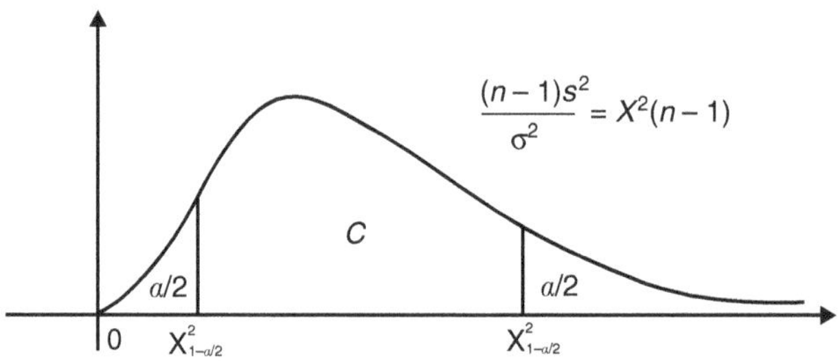

FIGURE 17.4 Construction for a two tail test for the population mean.

Algebraic manipulations transform the above statement into this:

$$P\left(\frac{(n-1)S^2}{\chi^2_{\alpha/2}} < \sigma^2 < \frac{(n-1)S^2}{\chi^2_{1-\alpha/2}}\right) = C.$$

Of all such intervals computed from observed values of σ^2, the proportion C contains σ^2. If we have found one such interval, we can be C certain that it contains σ^2, so a C*100% confidence interval for the population variance is

$$\left[\frac{(n-1)S^2}{\chi^2_{\alpha/2}}, \frac{(n-1)S^2}{\chi^2_{1-\alpha/2}}\right].$$

Consider finding a 90% confidence interval for the variance of unemployment in the American Cities database, based on the sample of 25 nonmissing values of eastern cities, shown in Table 17.1. From Excel, we have the following descriptive statistics:

ON DVD

TABLE 17.1 Descriptive statistics for the eastern cities.

City Region	Unempl Rate			
X1	X3			
E	0.047			
E			*Statistics*	
E	0.062			
E	0.065		Mean	5.87
E	0.097		Standard Error	.31
E	0.064		Median	5.8
E	0.061		Mode	6.5
E	0.063		Standard Deviation	1.57
E	0.039		Sample Variance	2.47
E	0.044		Kurtosis	42.354
E	0.048		Skewness	54.66
E	0.051		Range	6.6
E	0.035		Minimum	3.1
E	0.048		Maximum	9.7
E	0.058		Sum	146.9
E	0.088		Count	25
E	0.053			
E	0.066		***Note: Column 2 re-expressed into %s***	
E	0.071			
E	0.054			
E	0.081			
E	0.031			
E	0.065			
E	0.055			
E	0.049			
E				
E	0.074			

Corresponding to C = 0.90 and 25 – 1 = 24 degrees of freedom, the appropriate values from Table A-7 are $\chi^2_{0.95} = 14.611$ and $\chi^2_{0.05} = 37.652$. The variance S^2 of the sample of 25 values of X3 is 2.47, so the 90% confidence interval is:

$$[\frac{(25-1)2.47^2}{36.415}, \frac{(25--1)2.47^2}{13.848}] = [1.627, 4.281]$$

Note that the corresponding confidence interval for σ is:

$$\left[\sqrt{1.627}, \sqrt{4.281}\right] = [1.276,\ 2.069]$$

To test hypotheses concerning σ^2, we use the above result to select critical values. Consider a two-tailed test of the following form with significance level α.

$$H_0: \sigma^2 = \sigma_0^2$$
$$H_a: \sigma^2 \neq \sigma_0^2$$

We will reject H_0 if S^2 is far enough from σ_0^2, which will cause the value of $X^2 = (n-1)\, S^2 / \sigma^2$ to be extreme. We select critical values for the χ^2 statistic by chopping off an area of $\alpha/2$ from each tail, as shown in Figure 17.5.

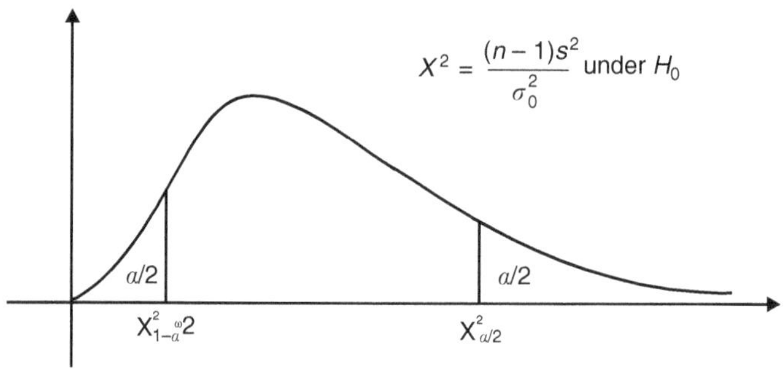

FIGURE 17.5 Two-tail test of the population variance.

The critical values are $\chi^2_{1-\alpha/2}$ and $\chi^2_{\alpha/2}$; we reject H_0 if the observed value of $X^2 = (n-1)\ S^2/\sigma^2$ is not between them.

For example, we test at the 2% significance level the null hypothesis that the variance in unemployment in eastern cities is 2.0 against the alternative that it is not, using the 27 cities given in Table 17.1.

$$H_0: \sigma^2 = 2.0$$
$$H_a: \sigma^2 \neq 2.0$$

Corresponding to $\alpha = 2\%$ and $25 - 1 = 24$ degrees of freedom, $\chi^2_{0.99} = 10.856$ and $\chi^2_{0.01} = 42.980$. The observed value of the chi-square statistic is

$$\chi^2 = \frac{(25-1)1.57^2}{2.0} = 29.58.$$

This value is between the critical values, so we cannot reject H_0; we cannot conclude that this variance is not 2.0.

For upper-tail tests, the critical value is simply χ^2_a, and for lower-tail tests, it is χ^2_{1-a}, with the appropriate changes in the decision rule.

17.4 Exercises

Interpretation Exercises

1. A sample of size 20 is taken from a large population to test the following hypotheses at the 5% significance level.

$$H_0: \mu = 50$$

$$H_a: \mu \neq 50$$

   a. State the decision rule for this test.

   b. The sample has mean 51.6 and standard deviation 9.1. What conclusion is reached?

2. The t distribution can also be used to compute confidence intervals for m based on small samples. As such, develop formulas for the endpoints of the interval shown in Figure 17.6. Given C, we can find t corresponding to n–1 degrees of freedom so that

$$P(-t < \frac{\overline{X} - \mu}{s/\sqrt{n}} < t) = C$$

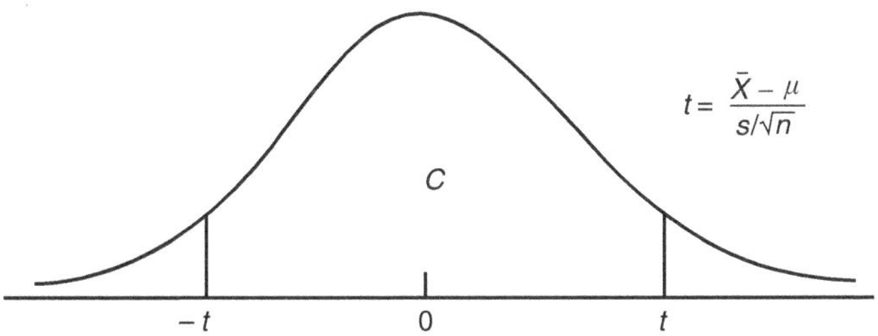

FIGURE 17.6 End Points of the N(0,1) distribution.

3. Show that if

$$P(\chi^2_{1-a} < \frac{(n-1)s^2}{\sigma^2} < \chi^2_{a/2}) = C$$

then

$$P(\frac{(n-1)s^2}{\chi^2_{a/2}} < \sigma^2 < \frac{(n-1)s^2}{\chi^2_{1-a/2}}) = C.$$

4. A process fills soft-drink bottles with an average of 16 oz. of soda, with standard deviation no more than 0.2 oz. If the standard deviation exceeds 0.2, the process must be adjusted; since this costs time and money, management hopes to adjust the process unnecessarily no more than 5% of the time. Periodically, a sample of 25 bottles is taken to see if adjustment is required. If one such sample has standard deviation 0.27 oz, should the process be adjusted?

5. Redo problem 4 using samples of 10 bottles.

Excel Exercises

ON DVD

6. Using the American Cities database in Table B-2, find a 90% confidence interval for the mean income of factory workers in all western U.S. cities.

7. We wish to test the null hypothesis that the average income of factory workers in all eastern U.S. cities is no more than $44,000 against the alternative that it is greater than $44,000, at the 5% significance level.

   a. State formally the hypotheses and the decision rule.

   b. Based on the statistics in problem 2 , draw a conclusion.

   c. Find a 90% confidence interval for mean income factory workers in all eastern cities.

8. Using the American Cities database, find a 90% confidence interval for the variance of the change in nonfarm employments in cities in the:

   a. East

   b. West

9. Test the null hypothesis that the variance of the change in nonfarm employment in the eastern cities is 4 against the alternative that it is not at the 10% significance level.

CHAPTER 18

INDEPENDENT SAMPLE TESTS

OVERVIEW AND LEARNING OBJECTIVES

In This Chapter

18.1 Introduction

18.2 Tests with Two Population Variances

18.3 Independent Sample Test for the Difference of Two Means

18.4 Exercises

18.1 Introduction

In most experiments where samples are taken to estimate population parameters, the researcher is not so much concerned with differences of parameters from particular predetermined values as with differences among corresponding parameters of several populations. For example, the marketing strategist who has to choose between two packages for a new product wants to know which package is more effective in promoting sales. In evaluating a new treatment for a disease, the question is not whether the new treatment shortens expected recovery time to less than 14 days, but whether the new treatment is more effective than the old. In each case, two samples will be taken and compared.

In this chapter, we explore techniques for comparisons of two populations, beginning with hypothesis tests for the quality of two population variances.

18.2 Tests with Two Population Variances

A company that packages liquid products such as lubricants and solvents plans to install a new automatic filling machine. The company's representatives have narrowed down the choice to two competing devices. An important factor in the decision is whether one device is consistently more accurate than the other, so they visit the two factories where the equipment is made to witness demonstrations. They wish to find out if the variances of the two processes are significantly different.

The presumption is that the accuracies of the two processes are about the same, so these hypotheses are being tested:

$$H_0: \sigma_A^2 = \sigma_B^2$$

$$H_a: \sigma_A^2 \neq \sigma_B^2$$

We now consider this problem in general. Given two populations, A and B, we wish to test for a significant difference in their variances. Are the values of one more widely dispersed than the values of the other? To do this, we take *independent samples* of sizes n_A and n_B from the populations and examine the value of the F statistic, $F = s_A^2 / s_B^2$.

If H_0 is true and the two population variances are equal, then we would expect the value of $F = s_A^2 / s_B^2$ to be near 1; a value of F far from 1 would be evidence that $\sigma_A^2 \neq \sigma_B^2$. To find how far F must be from 1 to enable us to reject the null hypothesis, we must know the sampling distribution of the F statistic when H_0 is true. This is provided by the following result, from Sir Ronald Fisher:

> If independent samples of sizes n_A and n_B are taken from two normally distributed populations with common variance σ^2, the sampling distribution of the F statistic, $F = s_A^2 / s_B^2$, is an *F distribution* with $(n_A - 1)$ and $(n_B - 1)$ degrees of freedom.

Mathematically, the F random variable is the ratio of two chi-square random variables each divided by its own number of degrees of freedom:

$$F = \frac{\chi_A^2 / df_A}{\chi_B^2 / df_B}.$$

The F distribution has a number of degrees of freedom associated with its numerator (with s_A^2 in the F statistic) and another with its denominator (with s_B^2). These two parameters determine and describe the F distribution.

The F random variable is the ratio of two values that are always non-negative, so it never takes on a value less than 0. Like the chi-square distribution, the F distribution is skewed to the right, and it becomes more symmetrical and bell-shaped as both numbers of degrees of freedom increase, as shown in Figure 18.1.

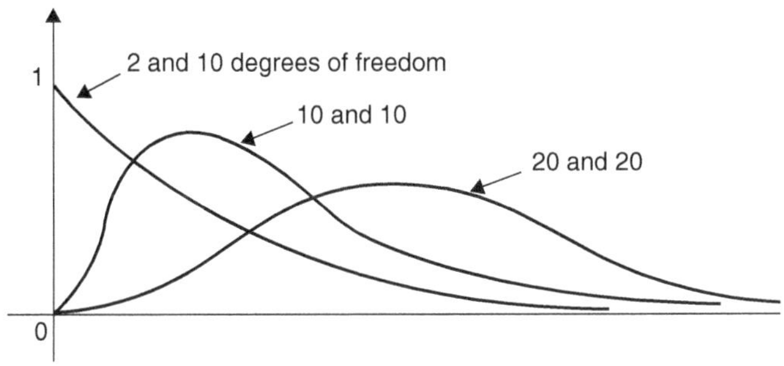

FIGURE 18.1 F distribution.

All figures and tables in this chapter appear on the companion DVD.

Given this information about the sampling distribution of the F statistic if H_0 is true, we can find critical values for our test. Corresponding to our chosen significance level a, and to $(n_A - 1)$ and $(n_B - 1)$ degrees of freedom, Table A-4, Critical Values of the F Distribution, in Appendix A provides us with the value $F_{a/2}$. If H_0 is true, the probability that $F = s_A^2/s_B^2$ will be greater than $F_{a/2}$ is $a/2$, so if this occurs, we will be convinced to reject H_0; see Figure 18.2.

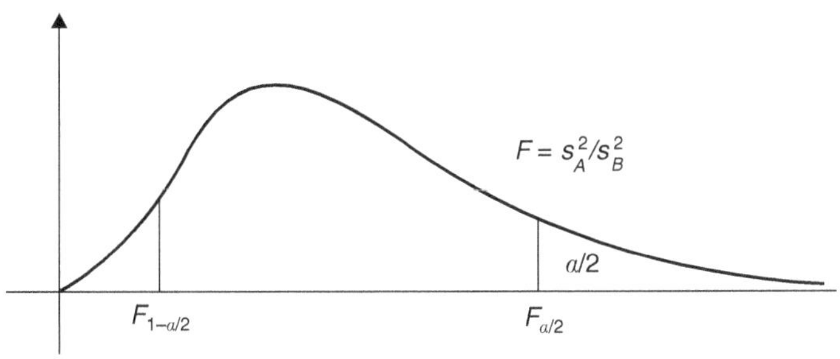

FIGURE 18.2 Critical values of the F distribution.

We will also be convinced to reject H_0 if F is less than $F_{1-a/2}$, but Table A-4 does not provide this value. However, while $F = s_A^2/s_B^2$ has an F distribution with (n_A-1) degrees of freedom in the numerator and $(n_B - 1)$ degrees of freedom in the denominator, the statistic $F' = s_B^2/s_A^2$

also has an F distribution, but with the numbers of degrees of freedom interchanged. For F′ we can also find an upper critical value $F'_{\alpha/2}$ such that we will reject H_0 if $F' > F'_{\alpha/2}$. But:

$$F' = s_B^2/s_A^2 > F'_{\alpha/2} \Leftrightarrow F = s_A^2/s_B^2 < \frac{1}{F'_{\alpha/2}}.$$

That is, the lower critical value of F is the reciprocal of the upper critical value corresponding to interchanged numbers of degrees of freedom.

Suppose that the test in our example was performed at the 10% significance level, and that representatives of the company weighed a sample of 21 drums filled by machine A and an independent sample of 25 drums filled by machine B, finding that $s_A^2 = 4.3lb^2$ and $s_B^2 = 2.2lb^2$.

Corresponding to 21–1=20 degrees of freedom in the numerator and 25 – 1=24 degrees of freedom in the denominator, $F_{\alpha/2} = F_{0.05} = 2.03$, as shown in Figure 18.3. Interchanging the numbers of degrees of freedom, $F'_{\alpha/2} = F'_{0.05} = 2.08$, so

$$F_{1-\alpha/2} = F_{0.95} = \frac{1}{2.08} \cong 0.481.$$

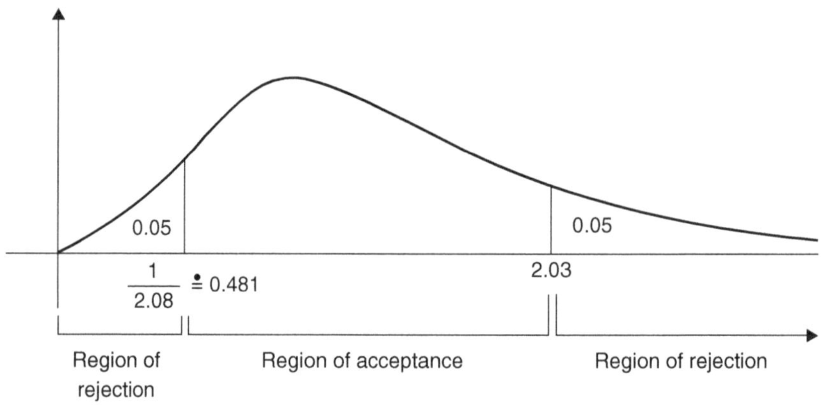

FIGURE 18.3 F with 20 and 24 degrees of freedom.

The null hypothesis will be rejected if $F < 0.481$ or $F > 2.03$. The observed value is $F = s_A^2/s_B^2 = 4.3/2.2 = 1.95$, which is between the critical values, so H_0 cannot be rejected. The company representatives cannot conclude at the 10% significance level that there is a significant difference in the accuracy of the two machines.

Upper-tail tests of the form

$$H_0: \sigma_A^2 \le \sigma_B^2$$

$$H_a: \sigma_A^2 > \sigma_B^2$$

can be performed by finding the one critical value F_α, and any lower-tail test can be transformed into an upper-tail test by placing the suspected larger variance in the numerator of the F statistic.

Unlike the *t* and chi-square tests discussed so far, the F-test for equality of variances is very sensitive to the assumption that both populations are normally distributed, particularly when the samples are small. The F-test should not be used with small samples unless the populations can be shown to be normally distributed. The chi-square goodness-of-fit test described in Unit 5, "Applications of Chi-Square Statistics," can be used for such a demonstration.

18.3 Independent Sample Test for the Difference of Two Means

Most often when we compare two populations, we wish to demonstrate that their respective means are either equal or unequal; that is, we test these hypotheses:

$$H_0: \mu_A = \mu_B$$

$$H_a: \mu_A \neq \mu_B$$

or equivalently,

$$H_0: \mu_A - \mu_B = 0$$

$$H_a: \mu_A - \mu_B \neq 0$$

This is often done by taking independent samples from the populations. Looking at the second of the two statements of hypotheses for such a test, it seems reasonable to use a test statistic related to $d = \overline{X}_A - \overline{X}_B$, where $\overline{X}_A$ and $\overline{X}_B$ are the means of the respective samples. The expected value of d is $\mu_A - \mu_B$, and we also have this important result:

> When independent samples of sizes n_A and n_B are taken from two normally distributed populations A and B *with common variance* σ^2, the sampling distribution of the statistic $t = [d - (\mu_A - \mu_B,) / ^{s_d}]$ is a t distribution with $n_A + n_B - 2$ degrees of freedom, where s_d is the standard deviation of the sampling distribution of d.

To describe the distribution of t, we must find s_d. Since the two samples, and therefore X_A and X_B, are independent, and both populations have variance σ^2,

$$\mathrm{Var}(d) = \mathrm{Var}(\overline{X}_A - \overline{X}_B,) = \mathrm{Var}(\overline{X}_A) + \mathrm{Var}(\overline{X}_B) \frac{\sigma^2}{n_A} + \frac{\sigma^2}{n_B}.$$

Then

$$\sigma_d = \left(\frac{\sigma^2}{n_A} + \frac{\sigma^2}{n_B} \right)^{1/2} = \sigma \left(\frac{1}{n_A} + \frac{1}{n_B} \right)^{1/2}.$$

s_A^2 and s_B^2 are both unbiased estimators of σ^2, so we can *pool* them to provide a better estimator of the common population variance:

$$\frac{(n_A - 1)s_A^2 + (n_B - 1)s_B^2}{n_A + n_B - 2} \text{ is an unbiased estimator of } \sigma^2.$$

Then

$$s_d = \left[\frac{(n_A - 1)s_A^2 + (n_B - 1)s_B^2}{n_A + n_B - 2}\right]^{1/2} \left(\frac{1}{n_A} + \frac{1}{n_B}\right)^{1/2}.$$

Further, if H_0 is true, then $\mu_A = \mu_B$, and our test statistic is simply

$$T = \frac{d - (\mu_A - \mu_B)}{s_d} = \frac{d}{s_d}.$$

Once again, corresponding to our choice of significance level a and to $n_A + n_B - 2$ degrees of freedom, we find $t_{a/2}$ so that, as shown in Figure 18.4,

$$P(t > t_{a/2}) = a/2 \text{ and } P(t < -t_{a/2}) = a/2.$$

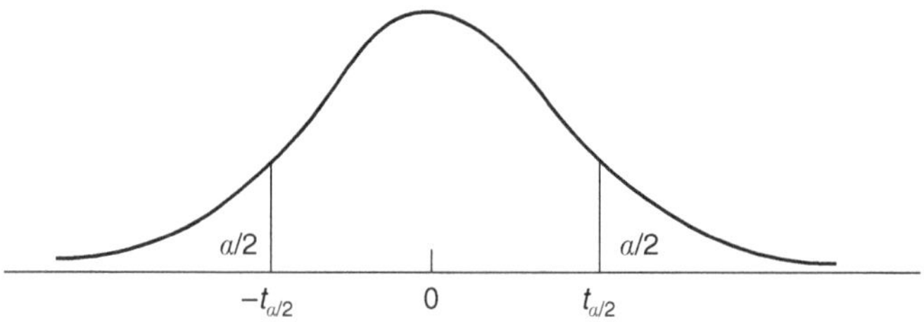

FIGURE 18.4 $t = d/s_d$ under H_0.

The critical values are $\pm t_{a/2}$. We will reject the null hypothesis and conclude that the population means are different if the t statistic is not between them. Keep in mind that this development is valid only if the population variances are equal.

ON DVD

Suppose that we want to determine if mean change in nonfarm employment in Eastern cities is different from that in the Central region, based on data in the American Cities database. Our hypotheses, which we will test at the 5% significance level, are:

$$H_0 : \mu_E = \mu_C$$
$$H_a : \mu_E \leq \mu_C$$

Using the Data Analysis feature of the Tools menu in Excel, the Descriptive Statistics selection generates the data provided in Tables 18.1 to 18.3 (after some editing efforts such as converting decimals into percentiles, etc.):

TABLE 18.1 Edited version of the American cities database.

City Region	Change in Dept Store Sales	Unempl Rate	Change in Nonfarm Empl	Change in Income Factory Workers	Change in Factory Workers Income	Change in Construction Activity
X1	X2	X3	X4	X5	X6	X7
E	10.700	4.700	3.200	$43,432.00	12.400	0.500
E	8.800	0.000	1.100	$46,178.00	9.000	5.200

(Continued)

TABLE 18.1 Continued.

City Region	Change in Dept Store Sales	Unempl Rate	Change in Nonfarm Empl	Change in Income Factory Workers	Change in Factory Workers Income	Change in Construction Activity
X1	X2	X3	X4	X5	X6	X7
E	10.900	6.200	4.100	$44,132.00	5.700	21.500
E	12.700	6.500	2.200	$45,784.00	9.500	13.600
E	7.100	9.700	1.100	$48,362.00	2.500	21.400
E	0.000	6.400	4.900	$43,979.00	11.600	27.400
E	3.300	6.100	4.000	$43,002.00	8.700	11.800
E	11.900	6.300	0.200	$48,740.00	13.800	4.700
E	13.300	3.900	1.500	$40,905.00	7.600	4.200
E	12.800	4.400	1.800	$41,299.00	10.600	-6.100
E	13.100	4.800	10.300	$46,547.00	7.900	20.200
E	-7.300	5.100	1.500	$43,790.00	7.700	3.300
E	0.000	3.500	1.900	$40,337.00	8.900	3.400
E	11.000	4.800	5.100	$40,718.00	9.200	22.500
E	11.100	5.800	1.600	$45,456.00	7.600	7.600
E	7.900	8.800	1.200	$42,285.00	8.700	4.900

TABLE 18.2 Excel descriptive statistics output for change in nonfarm employment, all cities and Eastern cities.

Criterion Variable X4		**X4 Eastern Cities**	
All Cities in Database			
Mean	2.734	Mean	2.459
Standard Error	0.324	Standard Error	0.422
Median	2.150	Median	1.800
Mode	1.800	Mode	1.100
Standard Deviation	2.785	Standard Deviation	2.194
Sample Variance	7.757	Sample Variance	4.815
Kurtosis	3.178	Kurtosis	5.157
Skewness	0.973	Skewness	1.862
Range	18.300	Range	10.700
Minimum	-4.400	Minimum	-0.400
Maximum	13.900	Maximum	10.300
Sum	202.300	Sum	66.400
Count	74	Count	27
Confidence Level (95.0%)	0.645	Confidence Level (95.0%)	0.868

TABLE 18.3 Excel descriptive statistics output for change in nonfarm employment, Central and Western cities.

X4 Central Cities		**X4 Western Cities**	
Mean	1.823	Mean	4.776
Standard Error	0.458	Standard Error	0.769
Median	1.900	Median	5.000
Mode	1.900	Mode	5.400
Standard Deviation	2.506	Standard Deviation	3.172
Sample Variance	6.283	Sample Variance	10.064
Kurtosis	1.616	Kurtosis	4.185
Skewness	0.175	Skewness	0.971
Range	13.100	Range	14.800
Minimum	-4.400	Minimum	-0.900
Maximum	8.700	Maximum	13.900
Sum	54.700	Sum	81.200
Count	30	Count	17
Confidence Level (95.0%)	0.936	Confidence Level (95.0%)	1.631

From these tables, the researcher is able to construct Table 18.4:

TABLE 18.4 Breakdown of change in nonfarm employment, X4, by city region, X1.

Change In Nonfarm Employment				
City Region				
Code	**Mean**	**Std Dev**	**N**	**Value Label**
	2.734	2.785	74	
E	2.459	2.194	27	EASTERN
C	1.823	2.506	30	CENTRAL
W	4.776	3.172	17	WESTERN

To employ the method derived above, the variances of the two populations must be equal; that is the case here, as can be verified by an F-test (an exercise left to the interested reader).

ON DVD

From Table A-4b, Critical Values of the t Distribution, in Appendix A, corresponding to $a = 5\%$ and $27 + 30 - 2 = 55$ degrees of freedom, $t_{a/2} = t_{0.025} = 2.004$, so we will reject H_0 and conclude that the two levels of unemployment are unequal if the value of the t statistic is less than –2.004 or greater than 2.004.

To find the value of the t statistic, we must first find s_d, based on the pooled variance estimate of the common population variance σ^2.

$$s_d = \left[\frac{(27-1)2.194^2+(30-1)2.506^2}{27+30-2}\right]^{1/2}\left(\frac{1}{27}+\frac{1}{30}\right)^{1/2} = 0.6270.$$

Then

$$T = \frac{d}{s_d} = \frac{\overline{X}_E - \overline{X}_C}{s_d} = \frac{2.459-1.823}{0.6270} = 1.1043.$$

t is between the critical values; it falls in the region of acceptance, so we do not reject H_0. We do not conclude that the two mean changes in nonfarm employment are different.

When the variances of the two populations are not equal, the sampling distribution of the t statistic, though symmetrical, is not a t distribution, and the technique developed above does not apply. The question of describing the distribution is called the *Fisher-Behrens problem*, and several statisticians (Lehman, Fisher, and Behrens, and others) have proposed solutions.

One such solution uses an approximation to the t statistic,

$$t = \frac{d-(\mu_A - \mu_B)}{(s_A^2/n_A + s_B^2/n_B)^{1/2}},$$

whose sampling distribution approximates a t distribution with this number of degrees of freedom:

$$\frac{\left[s_A^2/n_A + s_B^2/n_B\right]^2}{\left(s_A^2/n_A\right)^2/(n_A-1)+\left(s_B^2/n_B\right)^2/(n_B-1)}$$

This number is not usually an integer, but reasonable accuracy is obtained by rounding to the nearest integer.

As when the population variances are equal, we find the critical values $\pm t_{a/2}$ corresponding to a and the appropriate number of degrees of freedom, and compare the approximate t statistic to the critical values to draw a conclusion.

For example, suppose we wish to determine if the average change in factory worker income in the East is different from that in the West, at 5% significance level, based on the information in the American Cities database. That is, we wish to test these hypotheses:

$$H_0: \mu_E = \mu_W$$

$$H_a: \mu_E \neq \mu_W$$

The statistics of the values of variable X6 from the American Cities database, shown in Tables 18.5 through 18.7 are generated using Excel, via the Descriptive Statistics selection from Data Analysis under the Tools menu:

TABLE 18.5 Descriptive statistics output for change in factory workers income, X6, broken down by city region, X1, All cities and Eastern cities.

X6 All Cities		**X6 Eastern Cities**	
Mean	8.434	Mean	9.015
Standard Error	0.636	Standard Error	0.702
Median	8.650	Median	8.700
Mode	7.600	Mode	7.600
Standard Deviation	5.468	Standard Deviation	3.648
Sample Variance	29.903	Sample Variance	13.305
Kurtosis	3.927	Kurtosis	1.877
Skewness	-0.060	Skewness	0.836
Range	39.200	Range	17.400
Minimum	-10.900	Minimum	2.500
Maximum	28.300	Maximum	19.900
Sum	624.100	Sum	243.400
Count	74	Count	27
Confidence Level (95.0%)	1.267	Confidence Level (95.0%)	1.443

TABLE 18.6 Descriptive statistics output for change in factory workers income, X6, broken down by city region, X1, Central and Western cities.

X6 Central Cities		**X6 Western Cities**	
Mean	8.423	Mean	7.529
Standard Error	1.128	Standard Error	1.612
Median	8.250	Median	8.200
Mode	9.600	Mode	10.100
Standard Deviation	6.178	Standard Deviation	6.648
Sample Variance	38.166	Sample Variance	44.196
Kurtosis	3.502	Kurtosis	3.191
Skewness	0.683	Skewness	-1.128
Range	34.000	Range	30.600
Minimum	-5.700	Minimum	-10.900
Maximum	28.300	Maximum	19.700
Sum	252.700	Sum	128.000
Count	30	Count	17
Confidence Level (95.0%)	2.307	Confidence Level (95.0%)	3.418

TABLE 18.7 Summary statistics for change in factory workers income, X6, Broken down by city region, X1.

Change in Factory Income				
City Region				
Code	**Mean**	**Std Dev**	**N**	**Value Label**
	8.434	5.468	74	
E	9.015	3.648	27	EASTERN
C	8.423	6.178	30	CENTRAL
W	7.529	6.648	17	WESTERN

Critical values of the F statistic with 26 numerator and 15 denominator degrees of freedom are approximately 2.30 and 0.474 at the 10% significance level. The value of the F statistic is $3.648^2/6.648^2 = 0.301$, less than the lower critical value, so we can conclude that the population values are unequal.

Returning to consideration of the means, the number of degrees of freedom of the approximate t statistic is

$$\frac{\left[3.648^2/27+6.648^2/17\right]^2}{\left(3.648^2/27\right)^2/(26)+\left(6.648^2/17\right)^2/(16)}=\frac{9.5644}{0.0093+0.4224}=22.1553\cong 22$$

At the 5% significance level, the critical values are ±0.074. The value of the approximate t statistic is

$$t=\frac{d}{(s_E^2/n_E+s_W^2/n_W)^{1/2}}=\frac{(9.015-7.529)^2}{(3.648^2/27+6.648^2/17)^{1/2}}=\frac{1.486}{1.7586}=0.8320.$$

This is between the critical values, so we do not reject the null hypothesis; we do not conclude that the change in factory income in the East is different from that in the West.

We know that as the number of degrees of freedom increases, the t distribution approaches the standard normal. Therefore, with large sample sizes, the critical normal deviate $z_{\alpha/2}$ can be used to approximate $t_{\alpha/2}$. This can be done when the variances are equal or unequal.

From the above discussion and examples, we can also see how to perform one-tail tests, and any of these tests, including those when population variances are equal, can be performed by comparing the area under the graph of the sampling distribution of the t statistic beyond the observed value to the significance level of the test, as was done in Chapter 17, "Single Small Sample Tests," with one-sample tests.

EXCEL T-TEST FOR INDEPENDENT SAMPLES

Suppose we wanted to determine whether or not the change in factory workers income, X6, was equal for Eastern and Western cities. This question can be resolved by generating the following output from Data Analysis under the Tools menu using both the **F-Test Two Sample for Variances and T-Test for Two Samples Assuming Equal Variances** and **T-Test for Two Samples Assuming Unequal Variances** "options."

Two hypothesis tests are performed. The first is an F-test for the equality of the two population variances. Using the F-Test Two Sample for Variances, we get the results shown in Table 18.8:

TABLE 18.8.

F-Test Two-Sample for Variances		
	X6	
	EASTERN	***WESTERN***
Mean	9.015	7.529
Variance	13.305	44.196
Observations	27	17
Df	26	16
F	0.301	
P(F <= f) one-tail	0.003	
F Critical one-tail	0.487	
F-Test Two-Sample for Variances		
	X6	
	WESTERN	***EASTERN***
Mean	7.529	9.015
Variance	44.196	13.305
Observations	17	27
Df	16	26
F	3.322	
P(F <= f) one-tail	0.003	
F Critical one-tail	2.052	

Suppose that we are testing at α = 10%. Then $F_{0.05}$ = 2.03 and $F'_{0.05}$ = 2.21, and therefore $F_{0.95} = \frac{1}{2.21} = 0.4525$. The null hypothesis will be rejected since F= 0.301 does not lie between the critical values of $F_{0.05}$ = 2.03 and $F_{0.05}$ = 0.425, and therefore we conclude that the two variances are unequal. (Also note that the value of the F statistic, calculated with the larger variance in the numerator, is 3.322, and the probability, if the two population variances are equal, of obtaining a ratio of variances more extreme than this is 0.003, the *two-tail probability*. This is less than a likely significance level for such a test, and by this argument we can conclude that the two variances are unequal.)

To perform the t-Test, therefore, we select **t-Test for Two Samples Assuming Unequal Variances,** which produces the output shown in Table 18.9:

TABLE 18.9 t-Test for X6 on Eastern versus Western Cities.

t-Test: Two-Sample Assuming Unequal Variances		
X6		
	EASTERN	***WESTERN***
Mean	9.015	7.529
Variance	13.305	44.196
Observations	27	17
Hypothesized Mean Difference	0	
Df	22	
t Stat	0.845	
P(T<=t) one-tail	0.204	
t Critical one-tail	1.717	
P(T<=t) two-tail	0.407	
t Critical two-tail	2.074	

Note that the one-tail probability, P(T< = t) two-tail, is 0.407. Since this latter probability is less than the significance level of the test (10%), we reject the null hypothesis.

Reconsider the test for inequality of mean change in nonfarm employment (X4) between Eastern and Central cities, at a 95% confidence level, as shown in Tables 18.10 and 18.11:

TABLE 18.10 F-test for equal variances for Eastern versus Central Cities.

F-Test Two-Sample for Variances		
X4		
	EASTERN	***CENTRAL***
Mean	2.459	1.823
Variance	4.815	6.283
Observations	27	30
Df	26	29
F	0.766	
P(F <= f) one-tail	0.248	
F Critical one-tail	0.524	

TABLE 18.11 t-Test for X4 on Eastern versus Western Cities.

F-Test Two-Sample for Variances		
	X4	
	CENTRAL	*EASTERN*
Mean	1.823	2.459
Variance	6.283	4.815
Observations	30	27
df	29	26
F	1.305	
P(F <= f) one-tail	0.248	
F Critical one-tail	1.907	

Then $F_{0.25}$ = 2.12 and $F'_{0.25}$ = 2.06, and therefore $F_{0.975} = \dfrac{1}{2.06} = 0.4854$. The null hypothesis will not be rejected since F= 0.766 does not lie between the critical values of $F_{0.25}$ = 2.12 and $F_{0.975}$ = 0.4854, and therefore we conclude that the two variances are possibly equal, as we are unable to reject this possibility.

Next, utilizing the **t Test for Two Samples Assuming Equal Variances** we obtain the statistics shown in Table 18.12:

TABLE 18.12 SPSS printout for X6 by X1 breakdown.

t-Test: Two-Sample Assuming Equal Variances		
	X4	
	EASTERN	*CENTRAL*
Mean	2.459	1.823
Variance	4.815	6.283
Observations	27	30
Pooled Variance	5.589	
Hypothesized Mean Difference	0	
Df	55	
t Stat	1.014	
P(T<=t) one-tail	0.158	
t Critical one-tail	1.673	
P(T<=t) two-tail	0.315	
t Critical two-tail	2.004	

Because the P(T<=t) two-tail probability of 0.315 is more than 0.025 = $\alpha/2$, we cannot conclude that the mean changes in nonfarm unemployment are different between Eastern and Central cities.

18.4 Exercises

Interpretation Exercises

1. Corresponding to 10 degrees of freedom in the numerator and 12 degrees of freedom in the denominator, for the F distribution, find $F_{0.01}$, $F_{0.05}$, $F_{0.95}$, and $F_{0.99}$.

2. A sample of size 12 from population A has mean 12.7 and variance 23.2; a sample of size 15 from population B has mean 16.1 and variance 39.2. Assuming that both populations are normally distributed, can it be concluded at the 10% significance level that their variances are unequal?

Excel Exercises

ON DVD

3. Based on these statistics generated from the American Cities database (see Table 18.13), can it be concluded that:

   a. The variance of change in factory workers' income is different in Eastern and Central cities at the 10% significance level?

   b. The variance of change in factory workers' income is greater in Western cities than in Eastern cities at the 5% significance level?

TABLE 18.13 SPSS printout for X4 by X2 breakdown.

CRITERION VARIABLE X6 — CHANGE IN FACTORY WORKERS INCOME
BROKEN DOWN BY X1 — CITY REGION

VARIABLE	CODE	MEAN	STD DEV	N VALUE LABEL
FOR ENTIRE POPULATION		8.673	5.551	73
X1	1.	9.015	3.648	27 EASTERN
X1	2.	8.423	6.178	30 CENTRAL
X1	3.	8.562	7.121	16 WESTERN

TOTAL CASES = 74
MISSING CASES = 1 OR 1.4 PCT.

4. Based on the data in the American Cities database, can it be concluded that:

   a. The variance of the change in construction activity (X7) in Eastern cities is different from that in Western cities at the 2% significance level?

   b. The variance of the change in construction activity in the Central region is less than that in the Western region at the 5% significance level?

5. In problems 3 and 4, what assumption was made about the population from which the samples were taken?

6. Run a t-Test on the values of X7 from the American Cities database for the Central and Western cities.

   a. At the 5% significance level, are the respective population variances equal?

   b. Are changes in construction activity in these two regions unequal at the 5% significance level?

7. Run a t-Test at the 5% significance level for the claim that the average income of factory workers is greater in Western cities than in Eastern cities. What conclusions do you reach, and why?

8. Run a t-Test at the 5% significance level for the claim that the mean change in department store sales is greater in the Central region than in the Eastern region. What conclusions do you reach?

9. Run a t-Test at the 2% significance level for the claim that the mean change in non-farm employment is less in Eastern cities than in Western cities. What conclusions do you reach?

Computational Exercises

10. A sample of size 21 from population A has mean 56 and standard deviation 12. A sample of size 26 from population B has mean 63 and standard deviation 14. Assuming that the variances of their respective populations are equal, test the claim that the population means are unequal at the 5% significance level.

ON DVD

11. Consider the data from the American Cities database shown in Table 18.14:

TABLE 18.14 SPSS printout for X5 for Eastern and Central Cities.

CRITERION VARIABLE X4 CHANGE IN NONFARM EMPLOYMENT
BROKEN DOWN BY X1 CITY REGION

VARIABLE	CODE	MEAN	STD DEV	N	VALUE LABEL
FOR ENTIRE POPULATION		2.699	2.820	74	
X1	1.	2.363	2.302	27	EASTERN
X1	2.	1.823	2.506	30	CENTRAL
X1	3.	4.776	3.172	17	WESTERN

TOTAL CASES = 74

a. Assuming that the variances of X4 are equal for the Central and Western regions, test the claim that their means are different at the 10% significance level.

b. Test the claim that the mean value of X4 is greater for the Western region than for the Central region at the 1% significance level.

ON DVD

12. Table 18.15 shows output of a t-Test run on the values of X5 for the Eastern (GROUP 1) and Central (GROUP 2) cities in the American Cities database.

TABLE 18.15 SPSS – independent t-Test.

VARIABLE	NUMBER OF CASES	MEAN	STANDARD DEVIATION	STANDARD ERROR
X5				
AVERAGE INCOME OF FACTORY WORKERS				
GROUP 1	27	14504.8519	2636.825	507.457
GROUP 2	30	15723.6667	2019.017	368.620

		POOLED VARIANCE ESTIMATE			SEPARATE VARIANCE ESTIMATE		
F VALUE	2-TAIL PROB.	T VALUE	DEGREES OF FREEDOM	2-TAIL PROB.	T VALUE	DEGREES OF FREEDOM	2-TAIL PROB.
1.71	.164	– 1.97	55	.054	– 1.94	48.56	.058

Describe the two hypothesis tests being performed here.

a. At the 10% significance level, are the variances of average income in the Eastern and Central regions unequal? What importance does this have for the test of means?

b. At the 5% significance level, are the two mean incomes different?

c. Can it be concluded at the 10% significance level that the mean income of factory workers is larger in Central cities than in Eastern cities?

CHAPTER 19

MATCHED-PAIR TESTS

OVERVIEW AND LEARNING OBJECTIVES

In This Chapter

19.1 Introduction

19.2 Matched-pair Tests for the Difference of Two Means

19.3 t-Tests for the Difference of Population Proportions

19.4 Exercises

19.1 Introduction

When testing hypotheses concerning the means of two populations, thus far we have used independent samples; the values in the first sample have no effect on those in the second. Suppose, however, that we wish to assess the effectiveness of a training program by comparing pretraining test scores for a single group of people. An individual who scored well before the training could also be expected to score well after; the two samples are *dependent*, and rather than examine the two samples separately, we use as our sample *matched-pairs* of values the pairs of scores for each individual in the experiment. In this chapter, we will consider matched-pair hypothesis tests for the difference of two means and t-Tests for the difference in population proportions.

19.2 Matched-pair Tests for the Difference of Two Means

Samples of matched-pairs occur in two situations. In the first, of which the experiment described above is an example, measurements are made on the elements of a single sample group before and after the application of an experimental treatment, or after each of two different treatments. In the second, members of two groups are paired, and different treatments applied to the two groups. In a study of the effects of smoking on the incidence of heart disease, individuals might be matched by such characteristics as age, sex, weight, occupation, diet, family medical history, and environment, but with one member of each pair a smoker. Differences between members of a pair in subsequent medical history can then be more reasonably ascribed to smoking, since other factors were made identical.

Consider a matched-pair of samples (X_{Ai}, X_{Bi}) from two populations A and B. Again, we test these hypotheses:

$$H_0: \mu_A = \mu_B$$

$$H_a: \mu_A \neq \mu_B$$

We do not, however, find $\overline{X}_A$ and $\overline{X}_B$. Instead, for each pair we calculate the *paired difference* $d_i = X_{Ai} - X_{Bi}$. These values are really our sample. We find the *mean-paired difference* $\bar{d}$ and the standard deviation s_d in the usual ways:

$$\bar{d} = \frac{1}{n}\sum_{i=1}^{n} d_i;$$

$$s_d = \left[\frac{1}{n-1}\sum_{i=1}^{n}\left(d_i - \bar{d}\right)^2\right]^{1/2}$$

These can be thought of as statistics of a sample from the population of all paired differences. The mean-paired difference $\bar{d}$ is an unbiased estimator of $\mu_A - \mu_B$, and the sampling distribution of the quantity

$$t = \frac{\bar{d} - (\mu_A - \mu_B)}{s_d / \sqrt{n}}$$

is a t distribution with n–1 degrees of freedom.

If H_0 is true, $\mu_A - \mu_B = 0$, and $t = \dfrac{\bar{d}}{s_d/\sqrt{n}}$. We will be convinced to reject the null hypothesis if t is less than $-t_{\alpha/2}$ or greater than $t_{\alpha/2}$, where α is the significance level of the test, as shown in Figure 19.1.

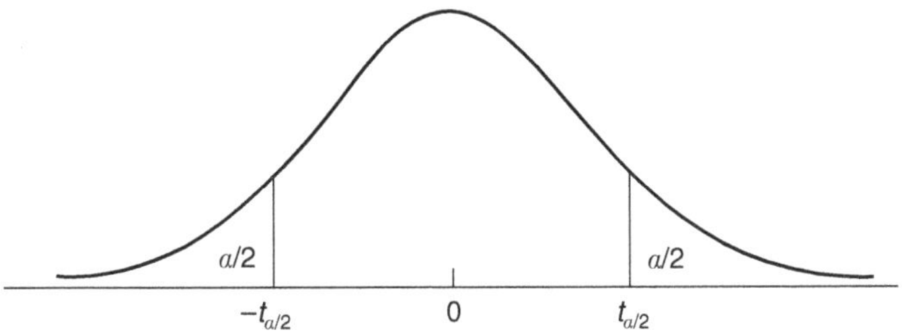

FIGURE 19.1 $t = \dfrac{\bar{d}}{s_d/\sqrt{n}}$ if H_0 is true.

All figures and tables in this chapter appear on the companion DVD. ON DVD

To draw another example from the American Cities database, consider testing to see if there is a difference in American cities between the mean change in nonfarm employment and the mean change in construction activity at $\alpha = 5\%$. The relevant variables are X4 and X7, and we use the pairs of values associated with each city in the database. Table 19.1 illustrates these computations. ON DVD

TABLE 19.1 Partial table including a column of *d* values.

City Region	Change in Nonfarm Employ	Change in Construction Activity	D
X1	X4	X7	
E	3.200	0.500	2.700
E	1.100	5.200	-4.100
E	4.100	21.500	-17.400
E	2.200	13.600	-11.400

(Continued)

TABLE 19.1 Continued.

City Region	Change in Nonfarm Employ	Change in Construction Activity	D
X1	X4	X7	
E	1.100	21.400	-20.300
E	4.900	27.400	-22.500
E	4.000	11.800	-7.800
E	0.200	4.700	-4.500
E	1.500	4.200	-2.700
E	1.800	-6.100	7.900
E	10.300	20.200	-9.900

For the first city, the pair is (3.2, 0.5), and $d = 3.2 - 0.5 = 2.7$.

Table 19.2 lists the descriptive statistics for the *d* column:

TABLE 19.2 Descriptive statistics for data in Table 16.1.

d statistics	
Mean	-2.753
Standard Error	0.936
Median	-1.8
Mode	2.9
Standard Deviation	8.055
Sample Variance	64.884
Kurtosis	0.613
Skewness	-0.670
Range	41.8
Minimum	-26.7
Maximum	15.1
Sum	-203.7
Count	74
Confidence Level (95.0%)	1.866

The mean paired difference $\bar{d}$ is –2.753 with standard deviation 8.055, and the critical values, corresponding to $a = 5\%$ and 74–1 = 73 degrees of freedom, are approximately ±1.99.

The value of the *t* statistic is $-2.753/(8.055/\sqrt{74} = -2.4546,)$ which is not between the critical values; we reject the null hypothesis and conclude that there is a difference between the mean change in nonfarm employment and the mean change in construction activity in U.S. cities.

Again, when the sample size is large, values from the standard normal distribution can be used to approximate the t values; here $z_{0.025} = 1.96$ is very close to the t deviate used and would have provided an acceptable test.

One-tail tests are performed by comparing to $-t_{\alpha}$ or to t_{α}.

EXCEL

EXCEL MATCHED-PAIR TEST

Step 1. Select the **Tools** menu.

- Select **Data Analysis**.

Step 2. Choose **t-Test Pair Samples for Means**.

- Choose the Variable 1 input range.
- Choose the Variable 2 input range.

Step 3. Click **OK**.

These steps, with additional editing, will generate the output shown in Table 19.3:

TABLE 19.3 Paired t-Test for nonfarm employment and construction activity means.

t-Test: Paired Two Sample for Means		
	X4	*X7*
Mean	2.734	5.486
Variance	7.757	86.802
Observations	74	74
Pearson Correlation	0.572	
Hypothesized Mean		
Difference	0	
Df	73	
	–	
t Stat	-2.940	
P(T<=t) one-tail	0.002	
t Critical one-tail	1.666	
P(T<=t) two-tail	0.004	
t Critical two-tail	1.993	

The P(T<=t) two-tail probability 0.004 is less than $\alpha = 0.05$ and we can reject H_0.

19.3 t-Tests for the Difference of Population Proportions

The t distribution can also be applied to testing the difference of two population proportions. For example, is the proportion of recalled automobiles produced by one manufacturer different from that of another? The hypotheses in such a situation are these:

$$H_0: P_A = P_B$$

$$H_a: P_A \neq P_B$$

Where P_A and P_B are the proportions of "successes" in two populations A and B. We take independent samples of sizes n_A and n_B, and find the sample proportions $\hat{P}_A$ and $\hat{P}_B$. We know that $\hat{P}_A$ and $\hat{P}_B$ are unbiased estimators of P_A and P_B, respectively, so $E(\hat{P}_A - \hat{P}_B) = P_A - P_B$. Since the samples are independent, so are $\hat{P}_A$ and $\hat{P}_B$, and the standard deviation of their difference is approximated by

$$\left[\frac{\hat{P}_A\left(1-\hat{P}_A\right)}{n_A}+\frac{\hat{P}_B\left(1-\hat{P}_B\right)}{n_B}\right]^{1/2}.$$

We test with the statistic

$$t=\frac{\hat{P}_A-\hat{P}_B}{\left[\frac{\hat{P}_A\left(1-\hat{P}_A\right)}{n_A}+\frac{\hat{P}_B\left(1-\hat{P}_B\right)}{n_B}\right]^{1/2}}$$

which has, when H_0 is true, a t distribution whose number of degrees of freedom is the maximum of $n_A - 1$ and $n_B - 1$.

Recall that when the elements of a sample that have the characteristic of interest are indicated by 1 and those that do not by 0, the sample mean $\overline{X}$ is also the sample proportion p. With such a coding, tests of two proportions may be performed with Excel's t-Test functions.

19.4 Exercises

Computational Exercises

1. A randomly chosen group of individuals is given a standardized reading test before and after taking a speed-reading course. Their scores are shown in Table 19.4:

TABLE 19.4 Before and after standardized reading data.

Individual	1	2	3	4	5	6	7	8	9	10
Before	86	83	86	70	66	90	70	85	77	86
After	82	79	91	83	68	86	81	90	85	94

a. Assuming that the population of score differences is normally distributed, do these results indicate, at the 55% significance level, that reading ability has been increased by the course?

b. If it is known that the variance of the population of score differences is $\sigma^2 = 48.4$, what conclusion is reached?

2. Two tire-tread materials are being compared. Pairs of tires of the two materials are mounted on test cars and their lifetimes measured. The results are shown in Table 19.5:

TABLE 19.5 Paired car tire data.

Pair	1	2	3	4	5	6	7	8	9	10	11	12
Life of Material *A*(1000s of miles)	42.6	43.1	41.9	45.2	43.6	41.0	42.6	43.7	42.2	46.3	42.4	41.8
Life of Material *B*(1000s of miles)	40.3	42.7	42.0	42.1	43.4	39.5	41.5	43.4	41.4	43.1	41.6	40.1

At the 5% significance level, can it be concluded that there is a difference in the mean lifetimes of the two materials?

3. Of 200 Republicans surveyed, 42 support Sam Rizzo for mayor of Dallas, while he has the support of 87 of 300 Democrats surveyed. At the 5% significance level, can it be concluded that Rizzo has more support among Democrats than among Republicans? (Note: For large numbers of degrees of freedom, $t_a \cong z_a$).

Interpretation Exercises

4. Tables 19.6 and 19.7 show the computer-generated output of a matched-pair t-Test using the values of X2 and X4 from the American Cities database:

TABLE 19.6 Change in mean nonfarm employment.

VARIABLE	NUMBER OF CASES	MEAN	STANDARD DEVIATION	STANDARD ERROR
X4 CHANGE IN NONFARM EMPLOYMENT				
	70	2.4971	2.538	.303
	70	10.1800	6.684	.799
X2 CHANGE IN DEPARTMENT STORE SALES				

TABLE 19.7 t-Test for department store sales means.

(DIFFERENCE) MEAN	STANDARD DEVIATION	STANDARD ERROR	CORR.	2-TAIL PROB.	T VALUE	DEGREES OF FREEDOM	2-TAIL PROB.
− 7.6829	6.767	.809	.157	.194	− 9.50	69	.000

a. Is a matched-pair t-Test appropriate here? Why or why not?

b. At the 5% significance level, is there a difference in the change in department store sales and the change in nonfarm employment in U.S. cities?

c. Can a stronger conclusion be made?

ON DVD

Excel Exercises

5. Test the claim, using the American Cities database, that there is a difference in the mean change in department store sales and the mean change in factory workers' income in U.S. cities. Use $\alpha = 5\%$.

6. Test the claim, using the American Cities database, that the mean nonfarm employment has risen more than factory workers' income in U.S. cities. Use $\alpha = 2\%$.

CHAPTER 20

HYPOTHESIS TESTING VERSUS CONFIDENCE INTERVALS

OVERVIEW AND LEARNING OBJECTIVES

In This Chapter

20.1 Introduction

Confidence intervals can be used as an alternative for hypothesis testing in business decision making. A hypothesis test is usually a yes-no decision. We either reject a hypothesis or fail to reject it. A confidence interval not only gives that answer, but it also provides information about the possible range of values for the parameter.

Confidence intervals are in general easier to understand and explain to decision makers than hypothesis tests. In this chapter, we will review hypothesis testing situations and resolve those situations with confidence intervals.

20.2 Choosing a Confidence Level

Often we have used confidence levels of 90% or 95% in our illustrations and examples. In reality, these confidence levels are not the only values used. The choice of the confidence level is based on the loss one would sustain if the parameter is outside the range and the gain one would have if the parameter is inside the range. If the loss is high compared to the gain, the confidence levels should be high. If the loss is negligible compared to the gain, a lower confidence level is fine.

In general, one must be cautious when interpreting confidence levels. When reviewing a research article with a 60% confidence level, do not make the assumption that the confidence is low and conclude that it is not worthy of making a decision based upon the underlying parameter. Even if the parameter is significant at the 99% confidence level, it is possible for decision makers to say that the confidence level is not high enough if the loss due to wrong decisions is large.

20.3 Hypothesis Testing versus Confidence Intervals

Rather than just a simple yes-no decision, as encountered with hypothesis tests, confidence intervals also provide the possible range of values for the parameter. A narrow confidence interval indicates that the parameter has been estimated with a high degree of precision. Wide confidence intervals indicate that a low precision is involved in the yes-no decision making. Decision makers often need to know the precision rather than or in addition to the simple yes-no answers in testing. When testing the null hypothesis that the means from two samples are the same and we have a [–50, 50] confidence interval for the difference in the population means, we can state there is no difference between the means since the interval includes zero. But think how much stronger the result would be if the confidence interval was [–1, 1].

Management is usually more receptive to confidence intervals since they are visual, making them easier to understand and communicate. Clearly it would be easier to comprehend that a parameter is in a specific range than to know that the probability of the parameter being one specific value is less than a specified percentage.

EXAMPLE 20.1

Suppose that the production time required to manufacture a product was obtained from two different manufacturers. The times submitted by manufacturer A are {5.36, 16.57, 0.62, 1.41, 0.64, 7.26}. The times submitted by manufacturer B are {19.12, 3.52, 3.38, 2.50, 3.60, 1.74}.

Solution (Method I):

For manufacturer A:

$$\overline{X}_A = 5.31$$

$$s_A^2 = 37.92$$

$$n_A = 6$$

For manufacturer B:

$$\overline{X}_B = 5.64$$

$$s_B^2 = 44.11$$

$$n_B = 6 \text{Then}$$

$$\overline{X}_A - \overline{X}_B = -0.33$$

$$S_d = 3.698$$

$$\text{degrees of freedom} = 11.921 \cong 12$$

$$90\% \text{ confidence interval for } d = [-6.92, 6.26]$$

The confidence interval includes 0. Therefore, at 90% confidence level we can conclude that the two manufacturers' average production times are not different.

Solution (Method II):

$$t_{5,0.05} = 2.015$$

$$90\% \text{ confidence interval for } d_A = 5.31 \pm (2.105)\sqrt{(37.92/6)}$$

$$= [0.24, 20.38]$$

$$90\% \text{ confidence interval for } d_B = 5.64 \pm (2.105)\sqrt{(44.11/6)}$$

$$= [0.18, 11.10]$$

The two confidence intervals overlap and the mean of one falls in the confidence interval for the other. Therefore, the two systems are not different at this level of significance.

20.5 One-sided Confidence Intervals

In two-sided confidence intervals, if the confidence level is $(1-a)*100\%$, there is a $\frac{a}{2}*100\%$ chance that the difference in a hypothesis test for the difference in population means will be more than the upper critical value. Similarly, there is a $\frac{a}{2}*100\%$ chance that the difference will be less than the lower critical value.

Managers, however, are often interested only in one-sided comparisons being made. When testing a hypothesis that the mean is greater than a specified value, the acceptance region is the confidence interval:

$$[\overline{X}-t_{1-a,\,n-1}\frac{s}{\sqrt{n}},\overline{X}]$$

The mirror image for the one-sided upper confidence interval for the population mean is given by:

$$[\overline{X},\overline{X}+t_{1-a,\,n-1}\frac{s}{\sqrt{n}}]$$

For large samples, z-values are used instead of t-values.

EXAMPLE 20.2

The time between home sales, in hours, for two salesmen at a local realty office were measured. The descriptive statistics for the salesman are shown in Table 20.1:

TABLE 20.1 Salesman descriptive statistics.

Salesman	Number	Mean	Std. Dev.
A	972	134.10	198.20
B	153	141.47	226.11

DVD All tables in this chapter appear on the companion DVD.

Management wants to see if Salesman A has faster turnaround times than Salesman B.

Solution:

We will use the procedure of unpaired observations. The mean difference is:

$$\overline{X}_A-\overline{X}_B = 124.10-141.47 = -17.37$$

EXAMPLE 20.2

$$S = \sqrt{\left(\frac{s_A^2}{n_A} + \frac{s_B^2}{n_B}\right)} = \sqrt{\frac{(198.20)^2}{972} + \frac{(226.11)^2}{153}} = 19.35$$

The degrees of freedom is:

$$df = \frac{\left(\frac{s_A^2}{n_A} + \frac{s_B^2}{n_B}\right)^2}{\frac{1}{n_A+1}\left(\frac{s_A^2}{n_A}\right)^2 + \frac{1}{n_B+1}\left(\frac{s_B^2}{n_B}\right)^2} - 2,$$

another solution to the Fisher-Behrens's problem:

$$df = \frac{\left((198.20)^2/972 + (226.11)^2/153\right)^2}{\frac{1}{972+1}\left((198.20)^2/972\right)^2 + \frac{1}{153+1}\left((226.11)^2/153\right)^2} - 2 = 191.05$$

Since the degrees of freedom are more than 30, we use the standard normal distribution instead of the *t* distribution. Thus, we use $z_{0.09}$ = 1.289 for a 90% one-sided confidence, which is

$$[-17.37, 17.37 + 1.28 * 19.35] = [-17.37, 7.402]$$

Since the confidence interval includes 0, we reject the hypothesis that Salesman A has faster turnaround times than Salesman B.

20.6 Confidence Intervals for Proportions

EXAMPLE 20.3

Suppose that 10 out of 1,000 pages printed on a laser printer are illegible. Management wants us to find a 90% confidence interval for the number of illegible pages produced by the printer.

Solution:

$$\hat{p} = 10/1000 = 0.01$$

EXAMPLE 20.3

Since the condition $\hat{p} \geq 10$ is satisfied, we can apply the standard normal distribution in constructing the confidence interval.

$$\hat{p} \pm z_{0.05}\sqrt{\frac{\hat{p}(1-\hat{p})}{n}}$$

$$= 0.01 \pm (1.645)\sqrt{\frac{0.01(0.99)}{1000}}$$

$$= 0.01 \pm 0.003(1.645)$$

$$90\% \text{ confidence interval} = [0.005, 0.015]$$

At 90% confidence we can state that 0.5% to 1.5% of the pages from the printer are illegible.

EXAMPLE 20.4

A pilot study found the laser printer to produce illegible print at the rate of 1 in 10,000. How many pages must be observed to get an accuracy of 1 in 1,000,000 at 95% confidence?

Solution:

We know that for a sample of size *n*, the $(1-a)*100\%$ confidence interval of the population mean is:

$$\overline{X} + z_{a/2}\frac{s}{\sqrt{n}}.$$

Then the desired accuracy of *r* percent implies that the confidence interval should be

$$[\overline{X}(1-r/100), \overline{X}(1+r100)]\cdot$$

Equating the desired interval with that obtained with *n* observations, we can determine the required sample size, *n*:

$$\overline{X} \pm z_{a/2}\frac{s}{\sqrt{n}} = \overline{X}\left(1+\frac{r}{100}\right)$$

$$z_{a/2}\frac{s}{\sqrt{n}} = \overline{X}\frac{r}{100}$$

$$n = \left(\frac{100 z_{a/2} s}{r\overline{X}}\right)^2$$

EXAMPLE 20.4

Then we have

$$\hat{p} = 1/10000 = 10^{-4},\ r = 10^{-6}.\ z_{0.25} = 1.96$$

$$n = (1.96)^2 \frac{10^{-4}\left(1-10^{-4}\right)}{\left(10^{-6}\right)^2} = 384{,}160{,}000$$

A total of 384,160,000 pages must be observed.

20.7 Exercises

Computational and Interpretational Exercises

1. The mean undergraduate grade point average (GPA) for students admitted to Super University was 3.57. Assume the estimate was based on a sample of 240 students admitted. Using past years' data, the population variance is known to be 0.08.

   a. At the 5% significance level, test the hypothesis that $\mu = 3.6$.

   b. Construct a 95% confidence interval for Part A.

   c. What conclusions can you reach when comparing your answers to Parts A and B?

2. The following are the prices per gallon for gasoline reported for a sample of gas stations along I-35 on a specific date:

   2.78 2.73 2.69 2.80 2.75 2.62 2.84 2.77 2.75 2.83

   a. What is the point estimate of the population mean price per gallon?

   b. What is the point estimate of the population standard deviation?

   c. Assuming the population is normally distributed, at the 5% significance level test the hypothesis that $\mu \le 2.80$.

   d. Assuming the population is normally distributed, at the 5% significance level test the hypothesis that $\mu \le 2.82$.

   e. Assuming the population is normally distributed, develop a 95% confidence interval estimate of the population mean price per gallon.

   f. What conclusions can you reach when reviewing your answers to c, d, and e?

3. A marketing firm conducted a pilot study to determine the rate of purchase potential for a particular product based on the participants before and after viewing a new

television commercial about the product. The purchase potential ratings were on a 0 to 10 scale, with 10 representing the highest rate of purchase potential. Table 20.2

TABLE 20.2 Potential Purchase Ratings.

Purchase Rating			Purchase Rating		
Individual	**After**	**Before**	**Individual**	**After**	**Before**
1	6	5	5	3	5
2	6	4	6	9	8
3	7	7	7	7	5
4	4	3	8	6	6

a. Run a hypothesis test where H_0 states that the mean rating after seeing the commercial is less than or equal to the mean rating before seeing the commercial, at the 5% significance level.

b. Redo the study in part a using confidence intervals.

4. A chemical process has produced, on the average, 700 tons of chemical per day. The daily yields for the past week are given in the following list. Does this data indicate that the average yield is less than 700 tons and hence that something is wrong? Use a 5% significance level.

Tons of Chemical
681
701
686
689
698

Test the hypothesis that the variance in tons of chemical produced daily is 30 against the alternative hypothesis that it is greater than 30 at the 1% significance level.

a. Solve this problem using hypothesis tests.

b. Solve this problem using confidence intervals.

c. State any conclusions that you can reach.

UNIT 5

APPLICATIONS OF CHI-SQUARE STATISTICS

Methods for simultaneously discovering exceptions and identifying incorrect data in large databases are referred to as anomaly detection methods. In *Chi Square Tests for Anomaly Detection in XML Documents Using Negative Association Rules*, by Permalatta and etc. found at *ccsenet.org/journal/index.php/cis/article/view/1852*, detected the detection anomalies in nested XML documents by independence between data.

This naturally leads the investigator to the application of the Chi Square test.

- What is a negative association rule?
- How does the approach of the authors differ from the conventional approach?

Why is the Chi-Square statistic so well known? It is the most commonly used test to determine if an observed dataset satisfies a specified distribution. The test is general and can be used for any distribution. It can be used for testing random numbers, as well as for testing random-variate generators Additionally, as in the header file, the Chi-Square Statistic can be used to determine whether or not two variates are related.

More importantly is that the chi-square distribution can be used to make inferences about a single population variance. Secondly, it is related to the F distribution, used to study two population variances. The F distribution is proportional to the ratio of two chi-square statistics.

CHAPTER 21

CHI-SQUARE TESTS OF MULTINOMIAL DATA

OVERVIEW AND LEARNING OBJECTIVES

In This Chapter

21.1 Introduction

Often a manager will suspect that orders come from several mutually exclusive categories, each with a specified percentage. The manager might attempt to obtain a sample of size n and determine the breakdown of the sample into the various categories. Using these sample percentages, the manager could then test the original hunch concerning the actual categorical percentages at a specified confidence level. In this chapter, we will study how to perform a hypothesis test using the *chi-square distribution* to solve such a problem, based upon a *multinomial distribution*.

21.2 Chi-Square Tests of Multinomial Data

We have seen that the binomial probability distribution $b(n, p)$ corresponds to selecting n subjects, with replacement, from a population of which the proportion p are "successes." The probability of drawing exactly r "successes" in the n independent trials is:

$$f(r)=\binom{n}{r}p^{r}(1-p)^{n-r}, \quad r=0,1,2,\ldots,n.$$

The *multinomial distribution* is an extension of the binomial: From a population containing proportion p_1 objects of type 1, p_2 objects of type 2, ..., p_k objects of type k, we select n objects at random and with replacement. The probability that we select n_1 objects of type 1, n_2 objects of type 2, ..., n_k objects of type k is:

$$\frac{n!}{n_1!n_2!\ldots n_k!}p_1^{n_1}p_2^{n_2}\ldots p_k^{n_k}$$

where $\sum_{i=1}^{k} p_k = 1$.

Calculations of probabilities for the multinomial distribution are tedious when n is large, but as binomial probability may be approximated with a normal, a multinomial may be approximated with a chi-square.

In a dataset, a *multinomial variable* is one whose values, not necessarily numerical, can be classified into three or more categories; that is, whose values are *nominal*. If we take a sample from a population of multinomial data (for example, ask a random group of voters to name their first choice for president), the probabilities with which various numbers of responses from the categories occur in the sample are described by the multinomial distribution, while the probabilities of various numbers of elements of type i occurring in the sample are given by the binomial distribution $b(n, p_i)$.

Given a population of multinomial values, we will want to test hypotheses concerning the proportions of various values (types of objects) in the population. That is, we wish to test whether a given set of values for the proportions accurately describes the population. Our hypotheses are:

$$\mathrm{H}_0: p_1 = p_{1_0};\; p_2 = p_{2_0};\; \ldots;\; p_k = p_{k_0}$$
$$\mathrm{H}_a: \text{At least one of the proportions is incorrect}$$

We take a random sample of size n from the population and compute a test statistic. Keep in mind that a test statistic should tend to indicate the truth or falsity of the hypotheses and should have a known distribution when H_0 is true and the postulated proportions hold.

If category i has population proportion p_{i_0}, the distribution of the number X_i of sample elements from that category is $b(n, p_{i_0})$, with expected value

$$E_i = E(X_i) = np_{i_0}.$$

If O_i is the number of elements from category i observed in the sample, then for large sample sizes the statistic

$$\chi^2 = \sum_{i=1}^{k} \frac{(O_i - E_i)^2}{E_i}$$

has an approximate chi-square distribution whose number of degrees of freedom is $k - 1$, one less than the number of categories. Note that the number of degrees of freedom of the chi-square distribution in this application does not depend on the sample size, as shown in Figure 21.1.

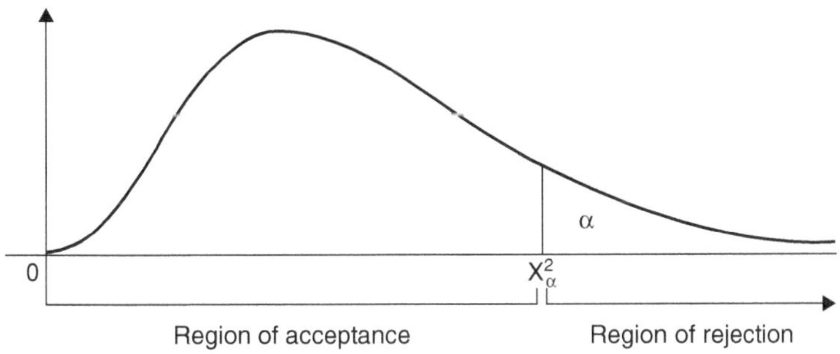

FIGURE 21.1 Chi-square with $k - 1$ degrees of freedom.

All figures and tables in this chapter appear on the companion DVD.

If the observed frequencies are near the expected frequencies supporting H_0, the value of the chi-square statistic will be near 0; if the O_i are far from the E_i, indicating rejection of H_0, the value of the statistic will be larger. Critical values of the chi-square statistic are found in the Table A-7, Critical Values of the Chi-Square Distribution, in Appendix A, and we reject H_0 when χ^2 is greater than the appropriate critical value. This is always an upper-tail test.

EXAMPLE 21.1

Suppose that W, X, Y, and Z were candidates for the CEO position in a local business. A survey was taken to determine which candidate had the greatest support among the employees. Additionally, suppose that intuition led management to test these hypotheses at the 1% significance level:

$$H_0: p_w = 0.45; p_X = 0.40; p_Y = 0.10; p_Z = 0.05$$

H_a: At least one of the above proportions is wrong

If a sample of 600 employees was surveyed, and 250 supported W, 240 supported X, 70 supported Y, and 40 supported Z, what conclusion can be reached by management?

Solution:

First, we construct a table of the observed and expected counts. For candidate W:

$$E_W = E(X_W) = np_{w_0} = 600 * 0.45 = 270.$$

Continuing in this manner, we can complete Table 21.1:

TABLE 21.1 Observed and expected counts.

	W	X	Y	Z
Observed	250	240	70	40
Expected	270	240	60	30

The value of the chi-square statistic is:

$$\chi^2 = \sum_{i=1}^{k} \frac{(O_i - E_i)^2}{E_i} = \frac{(250-270)^2}{270} + \frac{(240-240)^2}{240} + \frac{(70-60)^2}{60} + \frac{(40-30)^2}{30}$$
$$= 1.48 + 0 + 1.67 + 3.33 = 6.48.$$

Corresponding to $\alpha = 1\%$ and to $4 - 1 = 3$ degrees of freedom, the critical value is 11.345 and $\chi^2 = 6.48 < 11.345$, so we do not reject the null hypothesis. From the sample information, management cannot conclude that their intuitive guesses for the proportions were wrong.

EXCEL CHI-SQUARE TEST OF MULTINOMIAL DATA

ON DVD

Suppose it is claimed that in the American Cities database, one-third of the cities experienced increases of construction activity of 10% or less, one-third had increases of from 10% to 20%, and one-third had increases of more than 20%. At the 5% significance level, can we reject this claim using the variable X7 of the American Cities database?

Solution:

Step 1: Recode the variable X7 into the desired categories. The results of this action are illustrated in Table 21.2.

TABLE 21.2 Use of If and Max functions to recode X7.

Change in Construction Activity	**Category**
X7	
0.500	5
5.200	5
21.500	25
13.600	15
21.400	25
27.400	25
11.800	15
4.700	5
4.200	5
-6.100	5

(a)

Change in Construction Activity	**Category**
X7	
0.5	=IF(J3<=10,5,IF(J3<= 20,15,IF(J3<=MAX(J3:J76),25)))
5.2	=IF(J4<=10,5,IF(J4<= 20,15,IF(J4<=MAX(J4:J77),25)))
21.5	=IF(J5<=10,5,IF(J5<= 20,15,IF(J5<=MAX(J5:J78),25)))
13.6	=IF(J6<=10,5,IF(J6<= 20,15,IF(J6<=MAX(J6:J79),25)))
21.4	=IF(J7<=10,5,IF(J7<= 20,15,IF(J7<=MAX(J7:J80),25)))
27.4	=IF(J8<=10,5,IF(J8<= 20,15,IF(J8<=MAX(J8:J81),25)))
11.8	=IF(J9<=10,5,IF(J9<= 20,15,IF(J9<=MAX(J9:J82),25)))
4.7	=IF(J10<=10,5,IF(J10<= 20,15,IF(J10<=MAX(J10:J83),25)))
4.2	=IF(J11<=10,5,IF(J11<= 20,15,IF(J11<=MAX(J11:J84),25)))
-6.1	=IF(J12<=10,5,IF(J12<= 20,15,IF(J12<=MAX(J12:J85),25)))
20.2	=IF(J13<=10,5,IF(J13<= 20,15,IF(J13<=MAX(J13:J86),25)))

(b)

Step 2. Build the chi-square computational table shown in Table 21.3 in Excel:

TABLE 21.3 Chi-square computational data.

CATEGORY	5	15	25
OBSERVED	54	13	7
EXPECTED	24.67	24.67	24.67
CHI-SQUARE	53.054	D.F.	
		2	
DEVIATIONS	29.33	-11.67	-17.67
DEV. SQ.	860.44	136.11	312.11
DEV.SQ./EXP	34.883	5.518	12.653
SIGNIFICANCE			
0.000			

Table 21.3 was created using the functions shown in Table 21.4:

TABLE 21.4 Recoding of X7.

CATEGORY	5	15	25
OBSERVED	=COUNTIF ($K$3:$K$76,5)	=COUNTIF ($K$3:$K$76,15)	=COUNTIF ($K$3:$K$76,25)
EXPECTED	=74*0.33333333	=74*0.33333333	=74*0.33333333
CHI-SQUARE	=SUM(N16:P16)	D.F.	
		=COUNT(N16:P16) -1	
DEVIATIONS	=(N7-N8)	=O7-O8	=P7-P8
DEV. SQ.	=(N7-N8)*(N7-N8)	=(O7-O8)*(O7-O8)	=(P7-P8)*(P7-P8)
DEV.SQ./EXP	=N15/N8	=O15/O8	=P15/P8
SIGNIFICANCE			
=CHIDIST(N11,2)			

The significance of the chi-square statistic, whose number of degrees of freedom is one less than the total number of classes, is less than $\alpha = 5\%$, so we reject H_0. The suggested proportions are not correct.

21.3 Exercises

Computational and Interpretational Exercises

1. A large population of objects is one-half type As, one-third type Bs, and one-sixth type Cs. If a sample of 10 objects is randomly selected, what is the probability that it contains 6 type As, 3 type Bs, and 1 type C?

2. A machine is designed to mix nuts in the following proportions: 20% peanuts, 25% cashews, 15% pecans, 30% walnuts, and 10% acorns. In a bag of 200 nuts, there are 35 peanuts, 55 cashews, 25 pecans, 60 walnuts, and 25 acorns. At the 1% significance level, should it be concluded that the machine is not mixing nuts in the specified proportions?

3. The manager of a mail-order firm believes that 10% of the orders come from the West, 20% from the South, 40% from the East, 20% from the Midwest, and 10% from foreign countries. A sample of 125 orders has these destinations:

West	20
South	33
East	14
Midwest	48
Foreign	10
Total	125

   At the 5% significance level, is the manager correct about the proportions?

4. In your opinion, what proportions of your class would agree with each of the responses to this question:

   Check the one statement with which you most agree.

   ____________ On the whole, wars do the world some good.

   ____________ Peace and war are both essential to progress.

   ____________ War is hardly ever necessary in the modern world.

   ____________ There is no conceivable justification for war.

   Administer the question to the class, and test your proposed proportions at the 5% significance level.

Excel Exercises

5. It is claimed that the average income of factory workers is less than \$42,000 in 25% of American cities; between \$42,000 and \$46,000 in half; and greater than \$46,000 in the remaining 25%. Test this claim, using the American Cities database, at the 5% and 1% significance level.

6. Use the American Cities database to test the claim that the proportions of American cities having unemployment less than 4%, from 4% to 7%, from 7% to 10%, and above 10% are each one-fourth. Use the significance level of 5%.

CHAPTER 22

CHI-SQUARE TESTS OF INDEPENDENCE

OVERVIEW AND LEARNING OBJECTIVES

In This Chapter

22.1 Introduction

We are often concerned with finding whether two variables in a research study are connected in some manner; that is, whether a particular value of one increases the likelihood of a particular value of the other. If this is true, we say that the variables are *dependent*, and if the two variables do not interact, we say that they are *independent*, such as with events and random variables as discussed earlier. The two variables in a study can be thought of as random variables and are independent for all possible values of each, or $P(V_1 = v_1 \text{ and } V_2 = v_2) = P(V_1 = v_1) * P(V_2 = v_2)$. In this chapter, we will use the chi-square distribution to perform hypothesis tests of independence.

22.2 Chi-Square Tests of Independence

When the two variables being considered are qualitative, we can use a chi-square statistic to perform this test of their independence:

H_0: The variables are independent.

H_a: The variables are dependent.

In hoping to reject the null hypothesis, we seek to demonstrate that there is a relationship between the two variables.

ON DVD

Suppose that we wish to find if there is a relationship between city region and unemployment in American cities, based on the data in the American Cities database. In the database, the variable X3 is numeric, but we can obtain a qualitative measure of unemployment by grouping its values into two classes: 0 to 4.99% will be low employment, while values of 5% and above will be high. Tables 22.1a through 22.1d show this process.

TABLE 22.1a X3 variable and recoded X3 variable counts.

Unempl Rate	**Unemployment**	
X3	**Unempl**	**Count**
4.700	Low	29
0.000	High	45
6.200		
6.500		

TABLE 22.1b Excel functions for the recoding of X3.

Unempl Rate	**Unemployment**	
X3	**Unempl**	**Count**
4.7	Low	=FREQUENCY(A3:A76,{5})
0	High	=74–D3

(Continued)

TABLE 22.1b Continued.

Unempl Rate	Unemployment	
X3	**Unempl**	**Count**
6.2		
6.5		
9.7		
6.4		
6.1		

TABLE 22.1c Recoding of variable X3.

~	X3
X3	**Code**
4.700	low
0.000	low
6.200	high
6.500	high
9.700	high
6.400	high
6.100	high
6.300	high
3.900	low
4.400	low
4.800	low

TABLE 22.1d Excel functions to recode X IF(_,__,_) recoding for X3.3.

~	X3
X3	**Code**
4.7	=IF(A3<=4.99,“low”,“high”)
0	=IF(A4<=4.99,“low”,“high”)
6.2	=IF(A5<=4.99,“low”,“high”)
6.5	=IF(A6<=4.99,“low”,“high”)
9.7	=IF(A7<=4.99,“low”,“high”)
6.4	=IF(A8<=4.99,“low”,“high”)
6.1	=IF(A9<=4.99,“low”,“high”)
6.3	=IF(A10<=4.99,“low”,“high”)
3.9	=IF(A11<=4.99,“low”,“high”)
4.4	=IF(A12<=4.99,“low”,“high”)
4.8	=IF(A13<=4.99,“low”,“high”)

ON DVD

All figures and tables in this chapter appear on the companion DVD. We construct a *contingency table*, shown in Table 22.2, showing how many cases have each of the possible combinations of attributes of the two variables.

TABLE 22.2 Contingency table for unemployment (X3) broken down by city region (X1).

Count of X3	**X1**			
X3	C	E	W	**Grand Total**
HIGH	17	17	12	46
LOW	13	10	5	28
Grand Total	30	27	17	74

Table 22.2 was created by building a pivot table with the aid of Excel. (For instructions on constructing pivot tables, refer to Unit 1, "Descriptive Statistics.") Each city with no missing values of X3 will be counted in exactly one of the six cells of the table. Row and column totals are recorded in the margins, and the total number of cases is in the extreme lower right. The associated pivot table chart generated is shown in Figure 22.1.

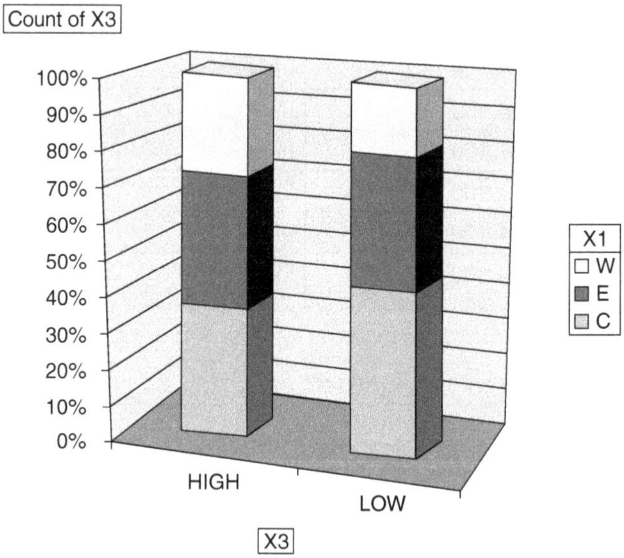

FIGURE 22.1 Pivot table chart for unemployment (X3) broken down by city region (X1).

Before we can perform the test of independence, we require some notation. In a contingency table, rows are numbered from top to bottom and columns from left to right, and cells are indicated by their row and column numbers, in that order. The low unemployment row is row 2, the central column is column 1, and the cell of central cities with low unemployment is cell 21. R_i is the total frequency of row i, C_j is the total frequency of column j, and the sample size, as always, is n.

Since we are testing the independence of two variables, we can rewrite the hypotheses in this way:

$$H_0: P_{ij} = P_i * P_j, \text{ for all } i \text{ and } j$$

$$H_a: P_{ij} \neq P_i * P_j, \text{ for some } i \text{ and } j$$

Where P_{ij} is the probability that a case falls in cell ij, P_i is the probability that a case falls in row i, and P_j is the probability that a case falls in column j. The null hypothesis in this form restates the independence of the variables in terms of independence of events.

Our test statistic is again calculated from a comparison of expected and observed frequencies, the frequencies of the cells in the contingency table. The observed frequencies O_{ij} have been recorded, and the expected frequencies E_{ij}, based on the assumption that the null hypothesis is true, are found from this formula:

$$\begin{aligned} E_{ij} &= \text{expected frequency of cell ij} \\ &= P_{ij} * n \\ &= P_i * P_j * n \\ &= \frac{R_i}{n} * \frac{C_j}{n} * n = \frac{R_i C_j}{n}. \end{aligned}$$

(Note that the sums of the expected frequencies in each row and column are equal to the marginal frequencies.)

For example, the expected frequency of cell 12 is

$$E_{12} = \frac{R_1 C_2}{n} = \frac{46 * 27}{74} = 16.78.$$

The test statistic is found by summing a quantity similar to that used in the multinomial test over all the cells in the contingency table:

$$\chi^2 = \sum_i \sum_j \frac{\left(O_{ij} - E_{ij}\right)^2}{E_{ij}}.$$

When the null hypothesis is true, this statistic has an approximate chi-square distribution with (R – 1)*(C – 1) degrees of freedom, where R is the number of rows in the contingency table and C is the number of columns. Again, the chi-square distribution is an approximation to a multinomial, and the sample size does not determine the number of degrees of freedom.

If the variables being tested are independent, the observed frequencies should be near those expected under independence, and χ^2 should be near 0. Observed frequencies far from E_{ij} tend to support H_a, and will cause a larger value of the chi-square statistic. The critical value can be found in Table A-7, Critical Values of the Chi-Square Distribution, in Appendix A, corresponding to the significance level a and to (R – 1)*(C – 1) degrees of freedom.

Using Excel, the computations and functions in Table 22.3 were used to run the hypothesis test:

TABLE 22.3 Excel formulas display.

	C	E	W	TOTAL
HIGH OBS	17	17	12	46
HIGH EXP	=($L$13*I17)/$L$17	=($L$13*J17)/$L$17	=($L$13*K17)/$L$17	
LOW OBS	13	10	5	28
LOW EXP	=($L$15*I17)/$L$17	=($L$15*J17)/$L$17	=($L$15*K17)/$L$17	
GRAND TOTAL	30	27	17	74

TABLE 22.4 EXCEL Spreadsheet generated by Table 22.3.

(O–E)	=I13–I14	=J13–J14	=K13–K14	=I15–I16	=J15–J16	=K15–K16
(O–E)*(O–E)	=I21*I21	=J21*J21	=K21*K21	=L21*L21	=M21*M21	=N21*N21
(O–E)*(O–E)/E	=I22/I14	=J22/J14	=K22/K14	=L22/I16	=M22/J16	=N22/K16
CHI-SQUARE	=SUM (I23:N23)					
SIGNIFICANCE	=CHIDIST (I25,2)					

The functions generated the table shown in Table 22.5:

TABLE 22.5 Chi-square analysis for unemployment – city region.

	C	E	W	GRAND TOTAL		
HIGH OBS	17	17	12	46		
HIGH EXP	18.649	16.784	10.568			
LOW OBS	13	10	5	28		
LOW EXP	11.351	10.216	6.4324			
GRAND TOTAL	30	27	17	74		
(O–E)	-1.649	0.216	1.432	1.649	-0.216	-1.432
(O–E)*(O–E)	2.718	0.047	2.052	2.718	0.047	2.052
(O–E)*(O–E)/E	0.146	0.003	0.194	0.239	0.005	0.319
CHI-SQUARE	0.906					
SIGNIFICANCE	0.636					

In this example, the value of the chi-square statistic is 0.906; if we choose 5% as the significance level of the test, the critical value corresponding to (2 – 1)*(3 – 1) = 2 degrees of freedom is 5.991. χ^2 is less than the critical value (also note the significance is larger than 0.05), so we cannot reject H_0. We cannot conclude from this test that there is a relationship between region and unemployment among American cities.

Note that tests of independence are always upper-tail tests.

22.3 Guidelines for Using the Chi-Square Test of Independence

To obtain meaningful results, these guidelines should be followed when using the chi-square test of independence:

1. For contingency tables with two rows and two columns, the average expected cell frequency should be at least 7.5.
2. For contingency tables with equal expected cell frequencies throughout, the expected cell frequencies should be at least 2 when $\alpha = 5\%$ and at least 4 when $\alpha = 1\%$.
3. For contingency tables with unequal expected cell frequencies, the average expected cell frequency should be at least 6 when $\alpha = 5\%$ and at least 10 when $\alpha = 1\%$.
4. The number of expected cell frequencies equal to 0 should be minimized, not by collapsing rows or columns but by taking sufficiently large samples.

Many mathematicians simply state that in any chi-square test of independence, every expected cell frequency should be at least 5.

In our example, the average expected cell frequency is 11.67, which justifies the use of the chi-square test by rule 3.

22.4 Exercises

Computational and Interpretational Exercises

1. In 1960, John F. Kennedy became the first Catholic president when he defeated Richard Nixon. A polling organization at the time wishes to know if there is a relationship between religious affiliation of voters and presidential preference; they survey 300 voters and obtain the results listed in Table 22.6:

   TABLE 22.6 Presidential Voting and Religious Preference.

	Voted for		
Religion	**Kennedy**	**Nixon**	**Totals**
Catholic	70	30	100
Protestant	70	80	150
Other or none	10	40	50
Totals	150	150	300

   At the 5% significance level, did the organization conclude that there was a relationship between a voter's religion and his or her presidential preference?

2. A paperback book has been bound in three differently colored covers. At the bookstore, 200 people buy copies of the book and provide the data for Table 22.7. At the 5% significance level, is there a relationship between the sex of the customer and the color of the book selected?

TABLE 22.7 Contingency table for sex of customer by color of book selected.

Color of Book Jacket			
	Blue	**Red**	**Yellow**
Men	43	30	14
Women	19	66	28

3. Administer this question to the members of your class, and record the response (yes or no) and the respondent's sex:

   Do you approve of the Patriot Act?

   At the 5% significance level, is there a relationship between support for the U.S. Patriot Act and a person's sex among members of your class?

4. Administer the following two questions to your class:

   a. Where did you buy your texts for this term?

   ____________Campus bookstore

   ____________Amazon.com

   ____________Another source

   b. What percentage of your total expenditures for supplies was spent at the campus bookstore?

   Use a chi-square test of independence to discover if there is a relationship between responses to the two questions at the 5% significance level.

5. Ask the following question of a random sample of students, teachers, and local community members unconnected with your college:

   Do you agree or disagree that varsity athletics are harmful to the purposes of a college or university?

   ____________Agree

   ____________Disagree

   ____________No opinion

   At the 5% significance level, is the response to this question independent of the respondents' position relative to your college?

ON DVD

6. Using the data in the American Cities database, test this claim at the 5% significance level: Change in construction activity in U.S. cities is dependent on region.

7. Using the data in the American Cities database, test this claim at the 5% significance level: In American cities, the average income of factory workers is dependent on region.

CHAPTER 23

CHI-SQUARE TESTS OF GOODNESS-OF-FIT AND MISSING DATA

OVERVIEW AND LEARNING OBJECTIVES

In This Chapter

23.1 Introduction

23.2 Chi-Square Tests of Goodness-of-Fit

23.3 Chi-Square Analysis of Missing Data

23.4 Exercises

23.1 Introduction

The chi-square test is used to test if a sample of data came from a population with a specific distribution.

An attractive feature of the chi-square goodness-of-fit test is that it can be applied to any univariate distribution for which you can calculate the cumulative distribution function. The chi-square goodness-of-fit test is applied to binned data (i.e., data put into classes). This is actually not a restriction since for non-binned data you can simply calculate a histogram or frequency table before generating the chi-square test. However, the value of the chi-square test statistic is dependent on how the data is binned. Note that the chi-square test requires a sufficient sample size in order for the chi-square approximation to be valid. In this chapter, we will conduct a chi-square test of goodness-of-fit and a hypothesis test on missing data.

23.2 Chi-Square Tests of Goodness-of-Fit

We can use a chi-square test based on the comparison of expected and observed frequencies to test whether a sample could have come from a population with a specified distribution. Such a test is called a *goodness-of-fit test* (or a *one-sample chi-square test*) and can be used to test the normality of a population involved in a t-test or to determine if the values produced by a random number generator are, in fact, selected from a uniform distribution. In general, these hypotheses are tested:

H_0: The sample is from a population having a specified distribution.
H_a: The population is from some other distribution.

ON DVD

Consider the values of X5 in the American Cities database. We will test at the 5% significance level the null hypothesis that the values of X5 were taken from a normally distributed population with mean \$45,000 and standard deviation \$2,300 against the alternative hypothesis that the population has some other distribution.

We begin by dividing the range of population values into cells and finding the number of sample values we should expect in each cell if the null hypothesis is true. If the population from which the values of X5 were taken is normally distributed with $\mu = 45{,}000$ and $\sigma = 2{,}300$, it seems reasonable to use cells of width 2000 from 40,000 to 50,000, with an unbounded cell at each end of this range. Note the maximum and minimum provided in Table 23.1.

TABLE 23.1 Descriptive statistics for X5.

X5	
Income of Avg. Factory Worker	
Mean	45073.699
Standard Error	271.830
Median	45226
Mode	45918

(Continued)

TABLE 23.1 Continued.

Standard Deviation	2322.514
Sample Variance	5394073
Kurtosis	-0.567
Skewness	-0.124
Range	10041
Minimum	40337
Maximum	50378
Sum	3290380
Count	73

All figures and tables in this chapter appear on the companion DVD. ON DVD

The probability that a single sample value would fall in the cell between 40,000 and 42,000 is

$$P(40{,}000 < X < 42{,}000) = P\left(\frac{40000-42000}{2300} < \frac{X-42000}{2300} < \frac{42000-42000}{2300}\right)$$
$$= P(-0.870 < Z < 0) = 0.3078.$$

We are taking a sample of size 73 (one value is missing), so the number of values that will fall in the cell from 40,000 to 42,000 has the binomial distribution $b(73, 0.3078)$ and expected value $73 * 0.3078 = 22{,}4694$.

In general, the expected frequency of cell i is:

$$E_i = n * P(\text{a randomly selected population value falls in cell } i).$$

Similar calculations provide us with the expected frequencies of the other cells, as shown in Figure 23.1.

$$P(X < 40{,}000) = P\left(\frac{X-42000}{2300} <= \frac{40000-42000}{2300}\right)$$
$$= P(Z < -0.870) = 0.5000 - 0.3078 = 0.1922$$

$$P(40{,}000 < X < 42{,}000) = P\left(\frac{40000-42000}{2300} < \frac{X-42000}{2300} < \frac{42000-42000}{2300}\right)$$
$$= P(-0.870 < Z < 0) = 0.3078$$

$$P(42{,}000 < X < 44{,}000) = P\left(\frac{42000-42000}{2300} < \frac{X-42000}{2300} < \frac{44000-42000}{2300}\right)$$
$$= P(0 < Z < 0.870) = 0.3078$$

$$P(44{,}000 < X < 46{,}000) = P\left(\frac{44000-42000}{2300} < \frac{X-42000}{2300} < \frac{46000-42000}{2300}\right)$$
$$= P(0.870 < Z < 1.739) = 0.4582 - 0.3078 = 0.1504$$

FIGURE 23.1 Calculations for the expected frequencies for X5—(*Continued*).

$$P(46{,}000 < X < 48{,}000) = P(\frac{46000 - 42000}{2300} < \frac{X - 42000}{2300} < \frac{48000 - 42000}{2300})$$

$$= P(1.739 < Z < 2.609) = 0.4964 – 0.4582 = 0.0382$$

$$P(48{,}000 < X < 50{,}000) = P(\frac{48000 - 42000}{2300} < \frac{X - 42000}{2300} < \frac{50000 - 42000}{2300})$$

$$= P(2.609 < Z < 3.478) = 0.0.5000 - 0.4964 = 0.0038$$

$$P(50{,}000 < X) = P(\frac{50000 - 42000}{2300} <= \frac{X - 42000}{2300})$$

$$= P(3.478 < Z) = 0.0000$$

FIGURE 23.1—Continued.

and by using Excel we obtain the observed frequencies shown in Table 23.2:

TABLE 23.2 Observed frequencies for X5.

Change in Income	Frequency
0–40000	0
40001–42000	8
42001–44000	15
44001–46000	24
46001–48000	19
48001–50000	6
50001–up	1
Total	73

Using the data in Tables 23.1 and 23.2, we can construct Table 23.3:

TABLE 23.3 Expected and observed frequencies in the goodness-of-fit test for X5.

Probability	0.1922	0.3078	0.3078	0.1504	0.0382	0.0038	0.0000
Expected Frequency	14.03	22.47	22.47	10.98	2.79	0.277	0.000
Observed Frequency	0	8	15	24	19	6	1

After recording the observed frequencies, we calculate the chi-square statistic in the usual way:

$$\chi^2 = \sum_{i=1}^{k} \frac{(O_i - E_i)^2}{E_i}$$

where E_i = expected frequency of cell i, O_i = observed frequency of cell i, and k = number of cells.

When the null hypothesis is true, the statistic has a chi-square sampling distribution with k – 1 degrees of freedom. If parameters of the population are being estimated from the sample, then the number of degrees of freedom is [k–1–(number of estimated parameters)], and when the number of degrees of freedom is 1, *Yates' correction* should be used in finding χ^2:

$$\chi^2 = \sum_{i=1}^{k} \frac{(O_i - E_i - 1/2)^2}{E_i}.$$

In our example, corresponding to 71 = 6 degrees of freedom and a = 5%, the critical value is $\chi^2_{0.05} = 12.952$. The value of the chi-square statistic is χ^2 = 236.688. See Tables 23.4a and 23.4b.

TABLE 23.4a Results of the χ^2 calculations.

Exp. Frequency (E)	14.03	22.47	22.47	20.98	2.79	0.277	0
Obs. Frequency (O)	0	8	15	24	19	6	1
(E-O)	14.03	14.47	7.47	-3.02	-16.21	-5.723	-1
(E-O)^2	196.84	209.38	55.80	9.12	262.76	32.75	1
(E-O)^2/E	14.03	9.32	2.48	0.43	94.18	118.24	
Chi-square	238.688						
Significance	0.000						

TABLE 23.4b Excel-generated table to calculate χ2.

Exp. Frequency (E)	14.03	22.47	22.47	20.98	2.79	0.277	0
Obs. Frequency (O)	0	8	15	24	19	6	1
(E-O)	=I7–I8	=J7–J8	=K7–K8	=L7–L8	=M7–M8	=N7–N8	=O7–O8
(E-O)^2	=I11*I11	=J11 *J11	=K11 *K11	=L11 *L11	=M11 *M11	=N11 *N11	=O11*O11
(E-O)^2/E	=I12/I7	=J12/J7	=K12/ K7	=L12/L7	=M12/ M7	=N12/N7	
Chi-square	=SUM (I13:O13)						
Significance	=CHIDIST (I15,6)						

Since the χ^2 is greater than the critical value, we reject the null hypothesis. We can conclude that the values of X5 did not come from the normal distribution described above. Also note that the significance of 0.000 is less than a = 0.05.

23.3 Chi-Square Analysis of Missing Data

In many data collection situations, particularly those involving human beings as the experimental units, missing values carry as much information as complete responses. If many people refuse to answer questions on a particular topic, this suggests that the topic is sensitive and perhaps important, and it may deserve further study.

Chi-square tests of independence can be used to analyze the missing values in a file in an exploratory way, by detecting significant dependence between the missing values of one variable and responses or nonresponses to another. If such a dependence is found, further qualitative (nonstatistical) analysis might explain the connection.

Datafiles can be broken down by variables, by cases, and by subfiles (subsets of the file). Exploratory chi-square analysis of missing data includes tests of independence for variable versus variable and for subfile versus variable. The contingency tables for such tests will have the form shown in Table 23.5.

TABLE 23.5 Tests of missing data.

Variable i		Variable j	
Response	**Nonresponse**	**Response**	**Nonresponse**
Response		S in Subfile	
V		U	
A		B	
R Nonresponse		F Not in Subfile	
I		I	
A		L	
B		E	
L			
E			

Such tests detect significant relationships between nonresponses to variable i and nonresponses to variable j, and between nonresponses to variable i and membership in sub file k. These relationships could then be explored in more detail.

23.4 Exercises

Computational and Interpretational Exercises

1. The values in Table 23.6 were randomly selected from a large population.

TABLE 23.6 Randomly selected values.

96	101	110	114	92
99	99	78	105	109
69	87	97	103	101
126	115	98	106	93
98	117	86	89	108
110	112	73	90	100

Test the hypothesis that the population is normally distributed with mean 100 and standard deviation 10. Use the 2% significance level.

2. Test the null hypothesis that the values in Table 23.7 were randomly selected from the uniform distribution U[0,10] against the alternative that they were selected from some other distribution at the 5% significance level.

TABLE 23.7 Uniform U(0,1) randomly selected values.

1.4	6.2	5.7	3.1	0.3	6.2	2.3
6.3	9.1	1.4	2.9	5.6	3.4	4.6
2.7	3.4	4.0	7.9	4.6	3.1	3.5
1.0	0.2	0.6	8.0	7.1	4.6	8.1
5.4	7.1	1.5	4.2	3.2	9.0	2.4
2.4	4.3	2.0	1.1	3.0	8.9	2.5
1.0	5.3	8.0	2.8	0.5	3.5	6.1
0.7	3.6	4.9	6.4	1.4	0.6	1.5

3. Table 23.8 lists the number of deaths in highway accidents per day in Fun Spot Resort Community.

TABLE 23.8 Number of highway deaths.

Number of Deaths	Number of Days
0	250
1	97
2	19
3	5
4	5

Using a chi-square goodness-of-fit test, test the hypothesis that the number of deaths per day possesses a Poisson distribution. Use $a = 5\%$. (Hint: Since you must estimate the rate of the Poisson distribution with the mean of the observed numbers of deaths, the number of degrees of freedom associated with the chi-square statistic is 5–1–1 = 3.)

4. Fun Spot Resort Community also found the statistics given in Table 23.9 for the length of time in days between highway deaths:

TABLE 23.9 Number of days between highway deaths.

Time in Days between Highway Deaths	Frequency
0	50
1	48

(Continued)

TABLE 23.9 Continued.

2	24
3	14
4	15
5	5
6	6
7	4
8	0
9	2
10	0
11	0
12	1

At the 5% significance level, test the hypotheses that these values came from an exponential distribution.

Excel Exercises

ON DVD

5. At the 5% significance level, test the hypothesis that the values of X3 in the American Cities database came from a normally distributed population with mean 5.5 and standard deviation 1.6.

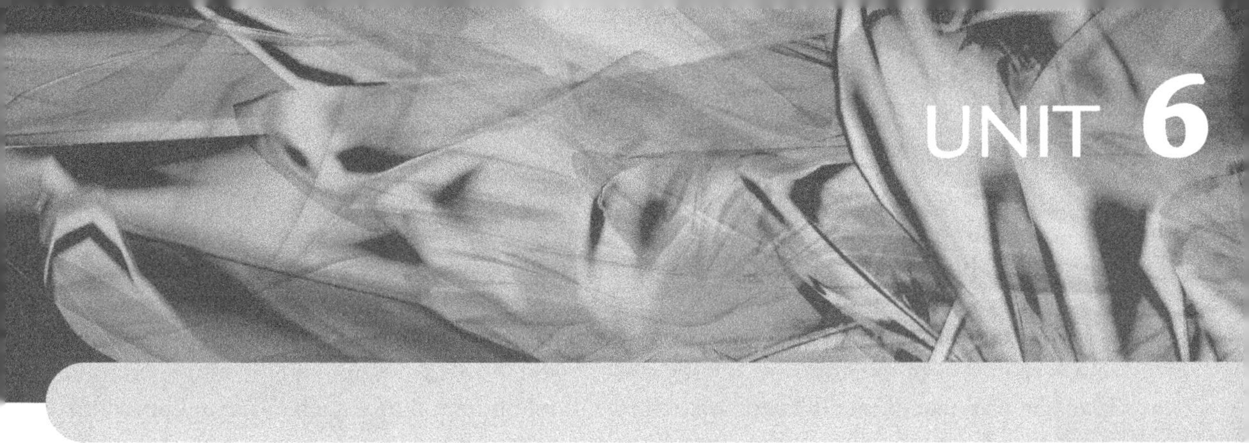

REGRESSION AND CORRELATION ANALYSIS

Regression and Curve Fitting

A CER is a mathematical expression, expressing cost as a function of one or more cost driving variables. In fact, the problem is that the CER complexity increases as one moves from linear to non-linear relationship, and from a single cost driver to more than one cost driver. But today, the necessary algorithms are well-known and widely available. The main methods to curve fitting are the multivariate regression techniques and cost improvement curve analysis. With the use of modern computer tools, it is possible to automatically investigate and test many combinations of CER's mathematical forms (linear, logarithm, exponential). One can also use more advanced black-box modeling techniques such as neural networks if the loss of the readability of the cost model is acceptable.

Test of the CERs Statistical Quality

Finally, we must verify the significance and accuracy of the cost estimation models through various statistical tests. The choice of the cost drivers can be statistically established and tested here. The CER can be simplified by selecting only the most statistically significant terms, for example by using stepwise regression techniques.

Selection of the Best CER

At this point, it is usual to obtain a great number of candidate models. In many cases, the statistical regression quality and estimation criteria are not sufficient. For their validation, the CERs must demonstrate reliability for an acceptable number of trials, and they should be representative of the database domain of application. It is also necessary to consider the technical coherence of the CER as defined by the experts in the application domain. At last, some input variables in the multiple regression do not have an important explanatory effect on the response. Then, it is a convenient simplification to keep only the statistically significant cost drivers in the model. The stepwise regression starts with all the terms in the model and removes the least significant terms until all the remaining terms are statistically significant. It is also possible to start with a subset of all the terms and then add significant terms or remove insignificant.

This illustration shows how linear statistical models for regression, analysis of variance, and experimental designs are widely used today in management. The next three units present a unified approach to the application of linear statistical models in regression, then in analysis of variance, and finally in experimental designs. The notion of a general linear statistical model which arises naturally in the context of regression models, is carried over to analysis of variance and experimental design. Application of the approach is readily understandable and brings out the relation of analysis of variance and experimental design to regression models.

CHAPTER 24

CORRELATION ANALYSIS

OVERVIEW AND LEARNING OBJECTIVES

In This Chapter

24.1 Introduction

In Unit 5, we saw that the chi-square test of independence can be used to detect a relationship between two variables. When such a test reveals a dependence, however, it tells us nothing about the nature of the relationship. Is it strong or weak? Do large values of one variable generally correspond to large or small values of the other? In this chapter, we will develop numerical measures of the *correlation* between the values of two variables.

24.2 Scatter Diagrams

We have seen how useful histograms can be in visualizing the distribution of the values of a variable. Similarly, we can create a graphical display of the values of two variables that illustrates the nature and degree of any relationship they may share. Such displays are called *scatter diagrams*, and are formed by plotting in the XY plane points whose coordinates are the paired values of the two variables. When one variable is thought to be dependent on the other, the values of the dependent variable are plotted along the vertical (Y) axis, while the values of the independent variable are plotted on the horizontal (X) axis.

For example, Table 24.1 gives values of floor area and natural gas used in January for 15 Midwestern homes heated with natural gas.

TABLE 24.1 Values of floor area and natural gas in January for 15 Midwestern homes.

FLOOR AREA (Thousands of Square Feet)	NATURAL GAS USED (Thousands of Cubic Feet)
1.4	31.6
1.7	40.3
2.1	52.0
1.9	41.5
2.4	50.7
1.8	43.8
1.7	44.0
1.9	48.2
1.3	30.5
1.2	30.7
2.0	55.2
1.9	47.6
2.2	57.1
1.4	36.2
1.6	34.5

All figures and tables in this chapter appear on the companion DVD.

We build the scatter diagram shown in Figure 24.1 by plotting the pairs of values given in Table 24.1. Since floor area can be expected to influence gas usage, the amount of gas is the dependent variable and is plotted on the vertical axis. The Chart Wizard in Excel was used to generate Figure 24.1.

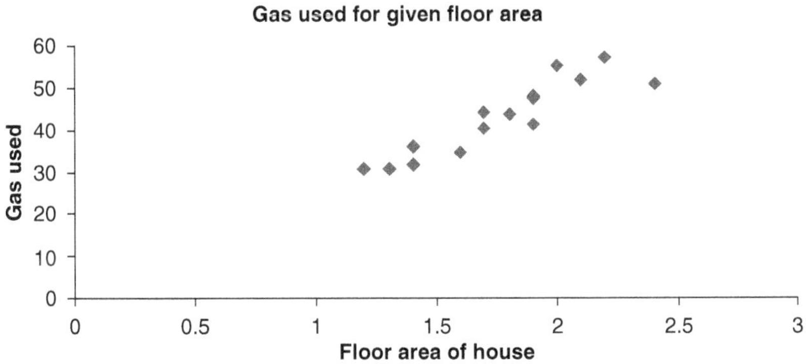

FIGURE 24.1 The pairs of values given in Table 24.1.

From Figure 24.1, we see that natural gas use is higher for larger houses and that the data points seem to cluster around a straight line. Such a relationship is said to be *positive* or *direct* and *linear*. Using the Chart Wizard allows the researcher to chart the most representative line for the data points, called the *trend line*, which is depicted in Figure 24.2.

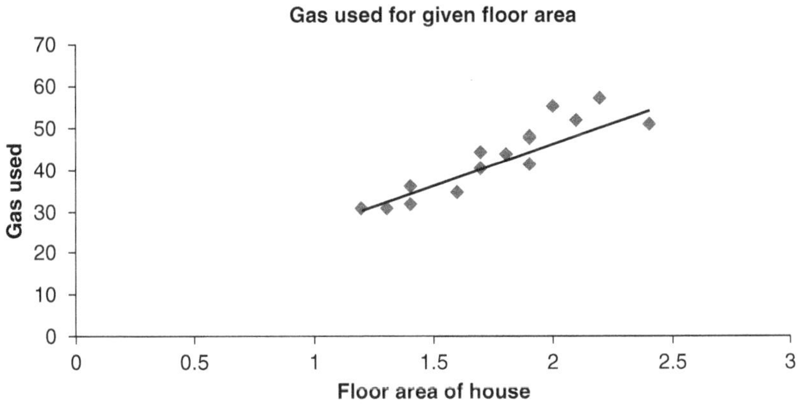

FIGURE 24.2 Trend line for gas used for specified floor areas.

Many kinds of relationships are possible, as Figure 24.3 illustrates. The strength of a relationship can also be inferred from a scatter gram. When the data points are closely bunched around a line or curve, the relationship between the two variables is strong; in a weak relationship, the points are scattered, as shown in Figure 24.4.

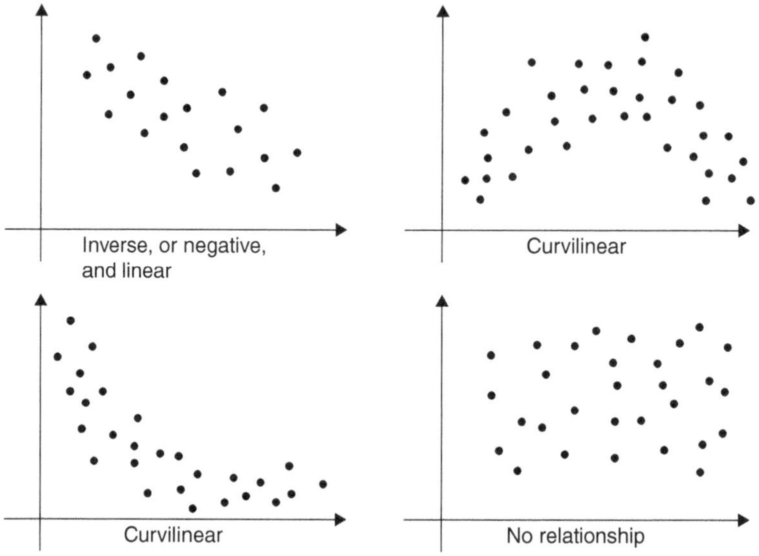

FIGURE 24.3 Relationships between two variables.

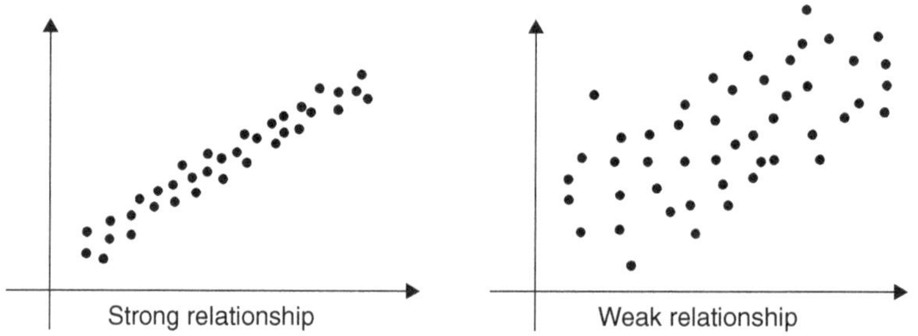

FIGURE 24.4 Diagrams showing the strength of relationships between two variables.

EXCEL

GENERATION OF SCATTER GRAMS IN EXCEL

Step 1. Enter data and functions/formulas into the spreadsheet.

Step 2. Click the **Chart Wizard** button on the standard toolbar.

- Choose **XY (Scatter)** in the Chart Type list.
- Choose **Scatter** from the Chart Subtype display.
- Click **Next**.

Step 3. Enter the Chart Source Data by clicking and dragging.

Step 4. Enter the **Title, Value(X)**, and **Value(Y)** labels.

- Select the **Legend** tab and choose the appropriate response.
- Click **Next**.

Step 5. Specify a location for the chart.

- Click **Finish**.

The following steps describe how to add a trend line:

Step 1. Position the mouse pointer over any data point in the scatter diagram and right-click to display a list of options.

Step 2. Choose **Add Trendline**.

Step 3. When the Add Trendline dialog box appears:

- Select the **Type** tab.
- Choose **Linear** from the **Trend/Regression** type display.
- Click **OK**.

24.3 The Pearson Correlation Coefficient

While a scatter diagram will illustrate the relationship (or lack of one) between two variables, we will seek *numerical* measures of correlation. The most widely used of these is applied to interval- or ratio-level data, like the house floor area and gas usage figures in the previous section.

Let us use the Descriptive Statistics function under the Data Analysis selection from the Tools menu to obtain statistical measures for the house area (X) and gas used (Y), as shown in Table 24.2.

TABLE 24.2 Descriptive statistics for gas used (Y) and house floor area (X).

Gas Used		Floor Area	
Mean	42.927	Mean	1.767
Standard Error	2.292	Standard Error	0.089
Median	43.8	Median	1.8
Mode	N/A	Mode	1.9
Standard Deviation	8.877	Standard Deviation	0.344
Sample Variance	78.802	Sample Variance	0.118
Kurtosis	−1.212	Kurtosis	-0.618
Skewness	0.005	Skewness	0.010
Range	26.6	Range	1.2
Minimum	30.5	Minimum	1.2
Maximum	57.1	Maximum	2.4
Sum	643.9	Sum	26.5
Count	15	Count	15

The means are $\overline{X} = 1.767$ and $\overline{Y} = 42.927$. Using these means, we create new pairs of values:

$$X_i' = \left(X_i - \overline{X}\right) \quad \text{and} \quad Y_i' = \left(Y_i - \overline{Y}\right).$$

Any relationship between the original X_i and Y_i will be preserved by the X_i' and Y_i', but the origin is at the point ($\overline{X}$, $\overline{Y}$), as shown in Figure 24.5.

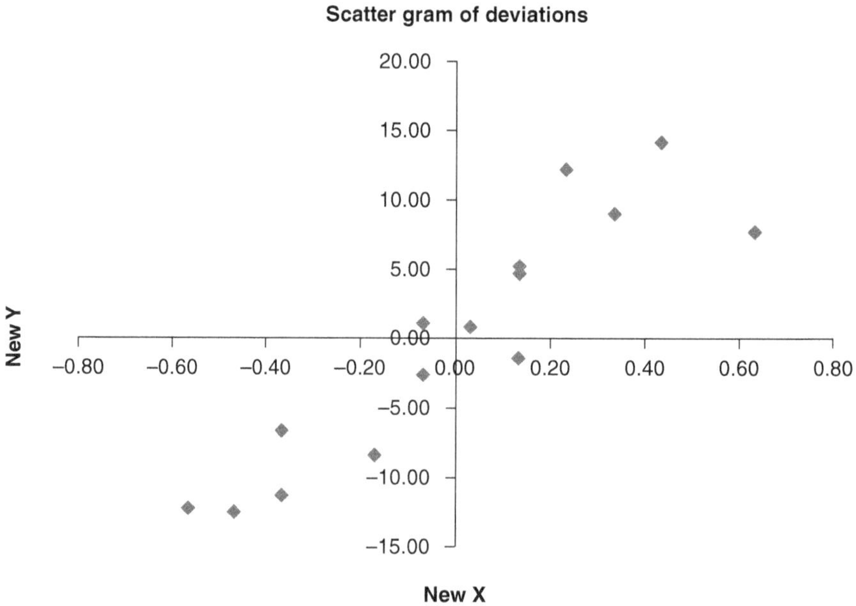

FIGURE 24.5 Transformation of the scatter diagram when $X_i' = \left(X_i - \overline{X}\right)$ and $Y_i' = \left(Y_i - \overline{Y}\right)$.

Most of the points of the original scatter diagram are now in the first and third quadrants relative to the new axes. If a generally negative relationship existed between the two variables, most of the points would be in the second and fourth quadrants; if no relationship existed, the points would be scattered in all four quadrants of the new diagram. From these distinctions we build our measure of correlation.

> The sum of the cross products of the deviations, SCPD, is commonly called the covariance of X and Y. The SCPD is a measure of the joint variation of the variables X and Y from their respective means $\overline{X}$ and $\overline{Y}$.

$$\text{SCPD} = \sum_{i=1}^{n} X_i' Y_i' = \sum_{i=1}^{n} \left(X_i - \overline{X}\right)\left(Y_i - \overline{Y}\right).$$

The cross product $X_i' Y_i'$ is positive for points in the first and third quadrants and negative for those in the second and fourth. When a positive relationship exists, this sum is positive; when negative, it is negative. When no relationship exists, the sum will be near 0. In our example,

$$\text{SCPD} = (1.4-1.767)(31.6-42.927) + \ldots + (1.6 - 1.767)(34.5 - 42.927)$$
$$= 39.013$$

This positive value corresponds to the positive relationship between X_i and Y_i.

While the SCPD is a statistic of the data pairs (X'_i, Y'_i) that indicates the direction of the relationship of the two variables, it does not tell us about their *degree* of correlation. Is the value found large or small? Further, the SCPD is not in the units of the original measurements; this prevents us from comparing such values for two different pairs of variables.

The next definition overcomes these obstacles:

The *sum of the cross products of the z-scores, the SCPZ,* is:

$$\text{SCPZ} = \sum_{i=1}^{n} \frac{X_i - \overline{X}}{s_X} * \frac{Y_i - \overline{Y}}{s_Y} = \frac{1}{s_X s_Y} * SCPD.$$

Dividing the SCPD by $s_X s_Y$ causes the resulting measure to be unit-less. Note also that SCPZ has definite maximum and minimum values.

If there is a perfect positive linear relationship between X_i and Y_i, then $Z_{X_i} = Z_{Y_i}$ for every data point, so that

$$\text{SCPZ} = \sum_{i=1}^{n} Z_{X_i} Z_{Y_i}$$

$$= \sum_{i=1}^{n} z_{X_i}^2$$

$$= \sum_{i=1}^{n} \frac{\left(X_i - \overline{X}\right)^2}{s_X^2}$$

$$= \frac{\sum_{i=1}^{n}\left(X_i - \overline{X}\right)^2}{\sum_{i=1}^{n}\left(X_i - \overline{X}\right)^2 / (n-1)} = (n-1).$$

Similarly, with a perfect negative linear relationship between the two variables, $Z_{X_i} = -Z_{Y_i}$ for every data point, and SCPZ $= -(n-1)$.

Thus, $-(n-1) \le \text{SCPZ} \le (n-1)$; in our example, the value of this statistic must fall between -14 and $+14$. From the previous Excel output, $s_X = 0.344$ and $s_Y = 8.877$. The sum of the cross products of the z-scores is:

$$\text{SCPZ} = \frac{SCPD}{S_X S_Y} = \frac{39.013}{0.344 * 8.877} = 12.776 \cdot$$

This indicates a strong relationship between the floor area of a house and the amount of gas used.

Still, this measure of correlation is influenced not only by the relationship between values in the data pairs, but also by the sample size. Division of the SCPZ by $(n-1)$ overcomes this deficiency and gives us this definition:

> Given n data pairs (X_i, Y_i), the *Pearson product-moment correlation coefficient*, named for Karl Pearson, who developed this measure in 1896 (or simply the *correlation coefficient*), R is defined to be:

$$R = \sum_{i=1}^{n} \frac{\left(X_i - \overline{X}\right)\left(Y_i - \overline{Y}\right)}{(n-1)S_X S_Y}.$$

Since $R = SCPZ/(n-1)$, it will always take on a value between -1 and $+1$; values near -1 indicate a strong negative relationship, while values near $+1$ indicate a strong positive relationship. If R is near 0, there is little correlation between the two variables.

In our example, $R = 12.776/14 = 0.913$, indicating a strong positive relationship between house area and gas use.

Note that the scatter diagram is helpful in checking that the three assumptions on which the calculation of the Pearson correlation coefficient is founded are satisfied. They are:

- A linear relationship exists between the two variables.
- The variables are continuously measured.
- For all values of X within the range of the study, the corresponding Y values have the same dispersion. This condition is called *homoscedasticity*.

In the graph on the left side of Figure 24.6, the distributions of Y values corresponding to each X value are equally dispersed, or *homoscedastic*. This condition is not satisfied by the data pairs depicted on the right side of the figure, where dispersion increases as X increases.

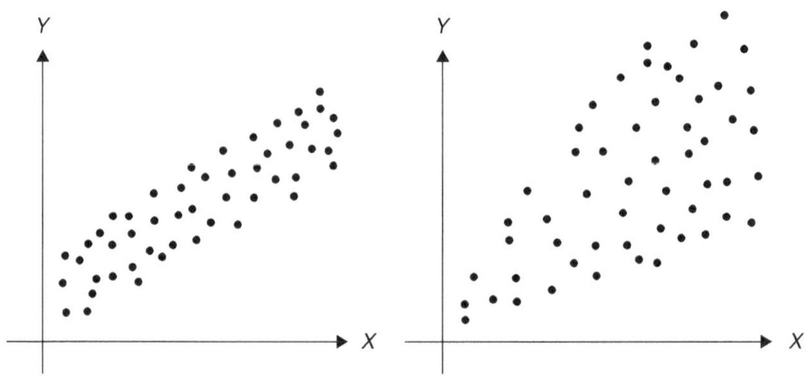

FIGURE 24.6 Homoscedastic versus not homoscedastic data.

24.4 Estimating the Population Correlation Coefficient

EXCEL

CORRELATION ANALYSIS IN EXCEL

Click the cell that you want to contain the correlation value.

Step 1. Click the f_x button on the address line.

Step 2. Select **Statistical** functions from the dialog box list.

Step 3. Select the **CORREL** function.

- Click and drag the first variable range.
- Click and drag the second variable range.

Step 4. Click **OK**.

The data pairs (X_i, Y_i) that we collect can be thought of as a sample from the population of all possible such data pairs. The computed correlation coefficient R is an estimate of the true population correlation coefficient ρ (rho), which indicates the true degree and kind of relationship (negative or positive) between the two variables.

The sampling distribution of R depends on the true value of ρ. When ρ is zero, the distribution of R is symmetrical, but as ρ departs from zero, the distribution of R becomes increasingly skewed, positively as ρ approaches –1, negatively as ρ nears +1. For sample sizes (numbers of data pairs) greater than 30, the distribution of R is approximately normal with mean ρ and standard deviation $\rho_R = \left(1-\rho^2\right)/(n-1)$, when ρ is near 0.

R.A. Fisher, however, identified a function of R whose sampling distribution is normal in general.[1] The result is this:

Let R be the Pearson correlation coefficient to be computed from n data pairs, with $n \geq 30$. The sampling distribution of the quantity

$$z_R = \ln_e\left[(1+R)/(1-R)\right]^{1/2}$$

is approximately normal with mean 0 and standard deviation $S_{Z_R} = 1/(n-3)^{1/2}$. For $n<30$, Fisher demonstrated that the statistic

$$t = R\left[(n-2)/\left(1-R^2\right)\right]^{1/2}$$

has a t distribution with $(n-2)$ degrees of freedom when $\rho = 0$.

With this knowledge of the sampling distribution of R, we can test hypotheses about the population correlation coefficient. In our example, suppose we wish to test the null hypothesis that the house area and January natural gas use are not related against the alternative that they are at the 5% significance level. Formally,

$$H_0: \rho = 0$$
$$H_a: \rho \neq 0$$

There are 15 cases in our example, so if H_0 is true, the quantity $t = R\left[(n-2)/(1-R^2)\right]^{1/2}$ has a t distribution with 15 – 2 = 13 degrees of freedom. Corresponding to $\alpha = 5\%$, the critical values are ±2.160, as shown in Figure 24.7, and we will reject the null hypothesis if the t statistic does not fall between them.

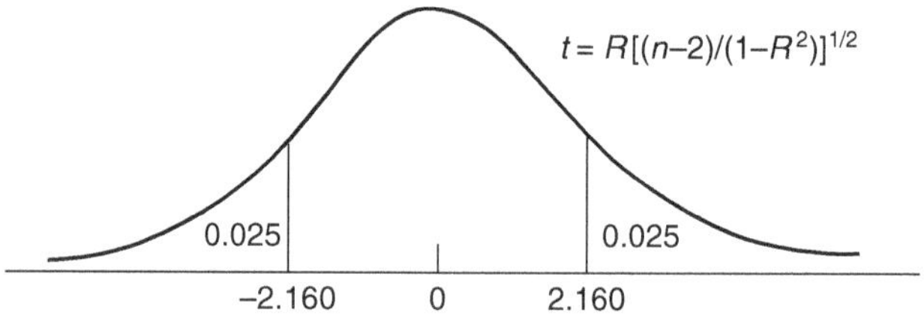

FIGURE 24.7 Critical values for the test of ρ.

By running the CORREL function in Excel, we find that R = 0.91255, so

$$t = R\left(\frac{n-2}{1-R^2}\right)^{1/2} = 0.91255\left(\frac{15-2}{1-0.91255^2}\right)^{1/2} = 8.045.$$

This value is outside the critical values, so we reject the null hypothesis and conclude that there is a significant correlation between house area and natural gas use.

Suppose that R had been 0.40. Then

$$t = R\left(\frac{n-2}{1-R^2}\right)^{1/2} = 0.40\left(\frac{15-2}{1-0.40^2}\right)^{1/2} = 1.56.$$

This value falls between the critical values, and we would not be able to conclude that a significant correlation exists.

The reader should be aware that the presence of a significant correlation between the values of two variables does not necessarily imply that a cause-and-effect mechanism links them. They might both be influenced by some third quantity, or their apparent connection might be a numerical coincidence. A significant correlation tells us only that some physical interaction might be present and that its possibility should be investigated.

24.5 Partial Correlation Coefficients

ON DVD

In many situations, an apparent correlation between two variables X and Y occurs when both are related to some third variable Z. In such cases, we would like to investigate the relationship between X and Y while *controlling for the effect* of Z. For example, in the

American Cities database, a small but significant correlation exists between X3 (unemployment) and X5 (average income of factory workers); both are related to X4 (change in non-farm employment). Could X4 be influencing both X3 and X5 to produce the appearance of a correlation? We measure the correlation of X3 and X5 while removing the effect of X4 on their relationship, as illustrated in Table 24.3.

TABLE 24.3 Intercorrelations among X3, X4, and X5.

Correlations	
X3 with X4	-0.178
X3 with X5	0.138
X4 with X5	-0.205

In general, using the values in Table 24.5, we calculate the *first-order partial correlation coefficient* of X and Y while controlling for Z, $R_{XY.Z}$, in this way:

$$R_{XY.Z} = \frac{R_{XY} - R_{XZ}R_{YZ}}{\left[\left(1-R_{XZ}^2\right)\left(1-R_{YZ}^2\right)\right]^{1/2}}.$$

The value of the first-order partial correlation coefficient identifies the part of the correlation of X and Y that is not an artifact of their relationship with Z.

In our example, computed from the three cases that have no missing values of X3, X4, or X5, we have:

$$R_{X3X5.X4} = \frac{R_{X3X5} - R_{X3X4}R_{X5X4}}{\left[\left(1-R_{X3X4}^2\right)\left(1-R_{X5X4}^2\right)\right]^{1/2}} = \frac{0.138-(-0.178)(-0.205)}{\left[\left(1-(-0.178)^2\right)\left(1-(-0.205)^2\right)\right]^{1/2}}$$

$$= \frac{0.1051}{[0.9611]^{1/2}} = 0.1091$$

We can conclude that some of the correlation between X3 and X5 can be associated with X4, but even when controlling for X4, there is a small positive correlation between X3 and X5.

Second-order partial correlation coefficients measure the interaction of two variables while controlling for the effects of two others; the partial correlation of X and Y while controlling for Z and W is:

$$R_{XY.ZW} = \frac{R_{XY.W} - R_{XZ.W}R_{YZ.W}}{\left[\left(1-R_{XZ.W}^2\right)\left(1-R_{YZ.W}^2\right)\right]^{1/2}}.$$

Second and higher-order partial correlation coefficients are difficult to interpret and are rarely used.

Zero-order partial correlation coefficients are the familiar Pearson correlations, when the effects of other variables are not controlled.

The significance of first-order partial correlation coefficients can be investigated with the following statistic, whose sampling distribution is a t distribution with $n - 3$ degrees of freedom, when the population partial correlation is 0:

$$t = \frac{R_{XY.Z}}{\left[\left(1 - R^2_{XZ.Z}\right)/(n-3)\right]^{1/2}}.$$

For example, we have found that $R^2_{X_3,X_5..X_5} = 0.1091$, so the value of the associated t statistic is

$$t = \frac{R^2_{X_3,X_5..X_5}}{\left[\left(1 - R^2_{X_3,X_5..X_5}\right)/(69-3)\right]^{1/2}} = \frac{0.1091}{\left[\left(1 - 0.1091^2/66\right)\right]^{1/2}} \cong 0.8917$$

Corresponding to 66 degrees of freedom (recall that $n = 69$), $t_{0.95} = 1.658$, so this positive correlation is not significant in a one-tail test at the 5% significance level.

24.6 Recommendations When Using Correlation Coefficients

When investigating relationships between variables with statistics such as those described in this chapter, the following points should be kept in mind:

- A large value of a correlation coefficient is not, by itself, meaningful; the significance of the coefficient is a function of sample size.
- The presence of a significant correlation between two variables does not necessarily imply a cause-and-effect relationship.
- The value of any correlation coefficient is meaningless if the assumptions on which it is based are not satisfied by the data being analyzed.

24.7 Exercises

Computational Exercises

1. Table 24.4 lists the curb weight in pounds, horsepower, and EPA gas mileage estimate for12 American automobiles:

TABLE 24.4 American automobile data.

Weight	Horsepower	Mpg
3205	125	14
2945	80	20

(Continued)

TABLE 24.4 Continued.

Weight	Horsepower	Mpg
4380	125	21
2040	74	29
3295	165	16
3520	195	16
2195	77	25
3205	195	16
1760	66	28
2515	130	21
3205	97	17
2480	93	28

a. Draw the scatter diagram associated with weight and mileage. What type of relationship, if any, is suggested?

b. Find the Pearson product-moment correlation coefficient R for weight and mileage.

c. At the 5% significance level, can we conclude from this data that a linear relationship exists between the weight of a car and its mileage? If so, describe that relationship.

2. For the data given in problem 1:

   a. Draw the scatter diagram associated with weight and horsepower. What type of relationship is suggested?

   b. Find the Pearson product-moment correlation coefficient R for weight and horsepower.

   c. At the 5% significance level, can we conclude from this data that a linear relationship exists between the weight and horsepower of an automobile?

3. For the data given in problem 1:

   a. Draw the scatter diagram associated with mileage and horsepower. What type of relationship is suggested?

   b. Find the Pearson product-moment correlation coefficient R for mileage and horsepower.

   c. At the 5% significance level, can we conclude from this data that a linear relationship exists between the mileage and horsepower of an automobile?

4. For the players on your school's varsity basketball team, obtain the number of points scored and the number of fouls committed in the last three games.

   a. Compute the Pearson correlation coefficient R for the two variables.

   b. Interpret the result of part a. Should the coach encourage more fouls? What explanation would you suggest for the value of R?

5. Collect the college and high school grade point averages (GPA) of the members of your class, making sure that everyone uses the same scale of measurement.

  a. Plot a scatter diagram of the value pairs. Which variable do you choose to plot on the vertical axis?

  b. Compute R for this data.

  c. At the 5% significance level, is there a relationship between high school and college GPAs? Interpret this result.

6. Repeat problem 5 comparing these two values for each member of your class: college GPA and entrance exam score (or SAT). If you had to predict success in college from high school GPA or entrance exam score, which would you use? Why?

7. For each of the following groups of data pairs, determine what type of relationship exists between X and Y, and try to find R:

a.

X	1	2	2	2	2	3	3	3	4	4	5
Y	2	2	2	2	2	2	2	2	2	2	2

b.

X	-1	-0.75	-0.5	-0.25	0	-0.5
Y	0	.60	.80	.95	1	.81

c.

X	0.25	0.50	1	0.75	0.75
Y	0.94	0.83	0.02	0.58	0.62

Excel Exercises

ON DVD

8. For the data in the American Cities database:

  a. Plot a scatter diagram of the values of X3 and X4 (unemployment and change in nonfarm employment).

  b. Find R for X3 and X4.

  c. At the 5% significance level, is there a relationship between X3 and X4? If so, describe and interpret the relationship.

ON DVD

9. Repeat problem 8 investigating X4 and X7.

10. Let X, Y, and Z be three variables in a data file. We might wish to investigate the relationships of both X and Y to Z. Do they both have the same correlation with Z? These hypotheses

$$H_0: \rho_{XY} = \rho_{YZ}$$

$$H_a: \rho_{XY} \neq \rho_{YZ}$$

can be tested using the test statistic

$$t = \frac{(R_{XY} - R_{YZ})\left[(n-3)(1+R_{XY})^{1/2}\right]}{\left[2\left(1-R_{XY}^2 - R_{YZ}^2 - R_{XY}^2 + 2R_{XY}R_{YZ}R_{XZ}\right)\right]^{1/2}}$$

whose sampling distribution, if $\rho_{XZ} = \rho_{YZ}$, is a t distribution with $n-3$ degrees of freedom. Use this result to test, at the 5% significance level, whether the correlation between X2 and X4 is equal to that between X4 and X7, using the American Cities database.

11. For the American Cities database, use zero-order partial correlations to calculate $R_{X2,X4.X7}$. What is the meaning of your answer? Use the statistic ON DVD

$$t = \frac{R_{X2,X4.X7}}{\left[(1-R_{X2,X4.X7})/(n-3)\right]^{1/2}}$$

to test these hypotheses:

$$H_0: \rho_{X2,X4.X7} = 0$$

$$H_a: \rho_{X2,X4.X7} \neq 0$$

Is the value of $R_{X2,X4.X7}$ significant at the 5% significance level?

12. Using the data given in problem 1, find the partial correlation of horsepower with mileage while controlling for weight. Is this value significant at $\alpha = 10\%$? What does it mean?

13. Consider the data given in Table 24.5 for a random group of 13 boys:

TABLE 24.5 Boys physical data.

Age (Years)	Height (Inches)	Weight (Pounds)
10	48	60
10	58	75
10	56	80
12	59	96
12	62	120
14	68	140
14	70	172
14	71	182
14	73	175
15	68	190
15	74	205
17	70	160
17	72	180

a. Why should a partial correlation coefficient be used to measure the linear relationship between height and weight?

b. Find $R_{Weight,Height.Age}$.

c. Is the value found in part b significant?

Interpretation Exercises

14. A sample of 50 data pairs (X_i, Y_i) produces a correlation coefficient of R = 0.4140. At the 5% significance level, can we conclude that $\rho > 0$?

15. Let X and Y be two variables. Show algebraically that:

a. $S^2_{X+Y} = S^2_X + S^2_Y + 2R_{XY}s_Xs_Y$,

b. $S^2_{X-Y} = S^2_X + S^2_Y - 2R_{XY}s_Xs_Y$.

(Hint: $S^2_{X+Y} = \frac{1}{n-1}\sum_{i=1}^{n}\left((X_i+Y_i)-(\bar{X}+\bar{Y})\right)^2 = \frac{1}{n-1}\sum_{i=1}^{n}\left((X_i+\bar{X})+(Y_i-\bar{Y})\right)^2$)

16. Let X and Y be two variables, and a, b, c, and d be constants, with a and c positive.

a. Show $R_{aX+b,cY+d} = R_{X,Y}$.

b. Interpret the result in part a.

ON DVD

17. For the American Cities database, use partial correlations to examine the relationship between X4 (change in nonfarm employment) and X7 (change in construction activity) while controlling for X5 (average factory workers' income). What interpretation might be placed on your results?

18. Consider the values in the American Cities database as a population rather than a sample; then $R_{X3,X5}$ becomes $\rho_{X3,X5}$.

ON DVD

a. Obtain four 10% random samples from the database, and for each, find $R_{X3,X5}$.

b. Compare the four sample values of $R_{X3,X5}$. Why are they different?

c. Would the mean of the four values of $R_{X3,X5}$ be a better estimate of $\rho_{X3,X5}$ than any of the individual values? Find that mean. If the sample sizes are different, compute the weighted mean.

d. The values of $R_{X3,X5}$ may also be averaged by finding the mean of the Fisher coefficients $Z_{R_{Xe,Xt}}$ and transforming the mean Z_R back into an estimate of $\rho_{X3,X5}$. Do this, again using a weighted mean if the sample sizes are different. Which value is nearer $\rho_{X3,X5}$, this value or the one found in part c? Which averaging technique appears to provide the better estimate of $\rho_{X3,X5}$?

(Hint: To obtain an estimate of $\rho_{X3,\, X5}$ from the mean of the Fisher coefficients, you must solve the equation:

$$z_a = \ln\left(\frac{1+R}{1-R}\right)^{1/2}$$

for R.)

Reference

1. Fisher, R.A, *Statistical Methods for Research Workers*, 10th ed. Oliver and Boyd, 1948.

CHAPTER 25

INTRODUCTION TO SIMPLE LINEAR REGRESSION

OVERVIEW AND LEARNING OBJECTIVES

In This Chapter

25.1 Introduction

In the previous chapter, we developed a measure for the degree of correspondence or relationship between two variables. These types of statistics provide information about the degree and, sometimes, the direction of the relationship. However, we can go beyond correlation analysis to predict, with known accuracy, the value of one of the variables when the value of the other is known. This prediction process is called *regression*.

The word "regression" comes from the first study that used the concept, which was published by Sir Francis Galton in 1885.[1] In the study, Galton examined the relationship between heights of parents and their children and found that the heights of children of unusually tall or short parents tended toward the average height of the population; that is, their heights regressed toward the average height. Regression has come to mean the prediction of the value of a variable from the values of one or more other variables.

Consider the simplest case, in which we have collected value pairs for two variables, X and Y, and we wish to predict the value of Y that might correspond to a given value of X.

We have seen that a scatter diagram graphically illustrates the relationship between two variables, and a linear relationship exists if the data points of the scatter diagram appear to cluster around a straight line. We know how to develop methods to identify the line that comes *closest* to all the data points, that *best* summarizes the apparent relationship between X and Y. This line is called the *regression line*, and we can predict the value of Y_0 corresponding to a given value X_0, as shown in Figure 25.1.

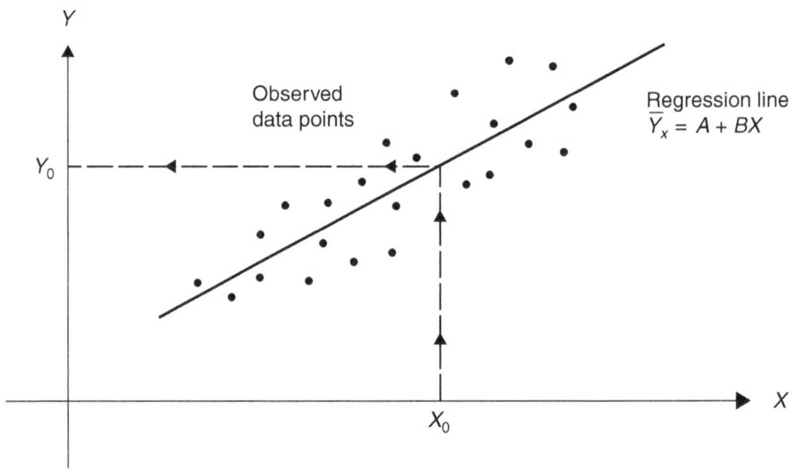

FIGURE 25.1 Regression line and its equation.

25.2 Sample Study

Returning to *Winning Database Configurations: An IBM Informix Database Survey* by Marty Lurie of the IBM Software Group, available at *www.ibm.com/developerworks/db2/zones/informix/library/techarticle/lurie/0201lurie.html*, we can improve on our analysis of the Informix database system by performing a regression analysis.

25.3 The Regression Line and Regression Equation

Using a straight-line approximation of the relationship between the amount of data and the number of CPUs gives a very good estimating tool for system sizing. This is called *regression analysis*.

Examining the graph, for XPS CPU versus data found in the article, reveals the actual CPUs, data points from the survey are plotted using a diamond. The "predicted values," shown as square symbols, are described by the following formula:

$$\text{XPS\_CPUs} = 0.01102 * \text{Data\_in\_GB} + 4.7$$

Continuing with the analysis, along the x-axis of the graph is the amount of data in gigabytes. The y-axis shows the number of CPUs. The largest value (labeled "Point 1") is an example of a surveyed customer using more CPUs than the model predicts. For 2.3 terabytes (2300 gigabytes), the model suggests using about 30 CPUs. The actual number of CPUs used by the customer is 40. Lurie noted that this particular site has very complex SQL, and they run many queries at the same time, so it makes sense for them to have more CPUs than predicted.

But how good is the model? The R^2 for this model is 64%. This means that if you use the above model to choose the number of CPUs, 64% of the differences in the values among XPS sites can be explained by the model.

What about the other 36% of differences between the sites? Different sites have different requirements for response time, and the complexity of queries varies greatly between sites.

Lurie then applied the same regression technique to the 7.x and 9.x servers. The model for making this comparison given in the article is overly optimistic.

The regression formula generated was:

$$\text{7.x/9.x\_CPUs} = 0.0106059 * \text{Data\_in\_GB} + 3.9$$

According to Lurie, this formula is misleading because it calls for fewer CPUs than XPS. The very large servers have skewed the regression analysis to the wrong conclusion. As noted in the lead article on regression, we develop an equation for the purpose of prediction. When we predict the value of a single dependent variable Y from one independent variable X, and the relationship between them is assumed to be linear, the equation is of the form

$$\overline{Y}_x = A + BX$$

where $\overline{Y}_X$ is the predicted value of Y corresponding to X, B is the slope of the regression line, and A is the y-intercept of the regression line. This is called *simple linear regression*, and is illustrated in Figure 25.1.

We may wish to incorporate more than one independent variable into our prediction of the dependent variable Y. When the relationship between each independent variable and Y is linear, this process is called *multiple linear regression*, and the regression equation is of the form

$$Y' = b_1X_1 + b_2X_2 + \ldots + b_kX_k + b_0,$$

where $b_1, b_2, \ldots, b_k$ are the coefficients of the independent variables $X_1, X_2, \ldots, X_k$, and b_0 is the constant term.

Regression analyses serve three purposes in statistical research: *description*, *statistical control*, and *prediction*. In some situations, previously collected data is examined to discover the relationship between some dependent quantity and one or more independent variables. In these cases, description is the goal.

In a study of the relationship between two variables, the researcher can use regression analysis to statistically control the influence of some other quantity. For example, in a study of two marketing strategies, surveys might be administered after each quarter. Regression equations can be found to relate the marketing strategy performance ratings to sales personnel, and these equations can be used to adjust the ratings, inflating those sales personnel with limited field experience and deflating those sales personnel with years of field experience. Remaining differences in performance ratings are then due to differences in the two marketing strategies.

Finally, a regression equation can be used to predict the value of a variable from the value of some other quantity, as in predicting sales averages from amount of advertisement funding.

25.4 Simple Linear Regression

We begin with the dependent variable related linearly to one independent variable. For example, suppose that over the past 12 quarters, a small business has spent the monthly amounts shown in Table 25.1 on advertising and has achieved the corresponding monthly sales.

TABLE 25.1 Monthly advertising costs and corresponding sales for a small business.

Quarter	Advertising (in \$10,000s)	Sales (in \$10,000s)
1	3.2	30
2	3.9	42
3	4.2	36
4	4.2	46
5	4.7	43
6	4.6	51
7	5.2	55

(Continued)

TABLE 25.1 Continued.

Quarter	Advertising (in $10,000s)	Sales (in $10,000s)
8	4.7	55
9	5.8	60
10	5.6	54
11	5.6	59
12	6.3	60

The scatter diagram in Figure 25.2 illustrates that a linear relationship appears to hold between the two variables; we now seek an equation with which to describe the relationship, the equation of a line *close* to all the data points.

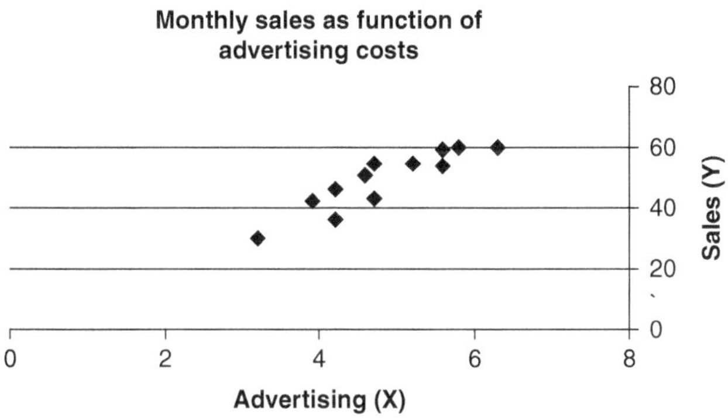

FIGURE 25.2 Scatter diagram showing a linear relationship between two variables.

In general, when the dependent variable is linearly related to one independent variable, the model that describes the relationship has this form:

$$Y_i = \beta_0 + \beta_1 X_i + \epsilon_i$$

Where Y_i is the *i*th observation of the dependent variable, X_i is the *i*th observation of the independent variable, β_0 and β_1 are the parameters of the model, and ε_i is the random error of Y_i.

We will estimate β_0 and β_1, and thereby arrive at an equation relating X and Y, which will allow us to predict Y given a value of X. In developing these two estimates, we will make the following four assumptions:

- X is not a random variable; that is, as the independent variable, X is thought of as influencing the value of Y, and is often under the control of the researcher.
- $E(\varepsilon_i) = 0$. The expected value of each error is 0. Therefore,

$$\begin{aligned} E(Y_i) &= E(\beta_0 + \beta_1 X_i + \varepsilon_i) \\ &= E(\beta_0) + E(\beta_1 X_i) + E(\varepsilon_i) \\ &= \beta_0 + \beta_1 X_i \end{aligned}$$

The random errors are independent; the magnitude of ε_i has no effect on ε_j when $i \neq j$.

- The random errors have common variance σ^2.

$$\begin{aligned} \text{Then} \quad \sigma^2(Y_i) &= \sigma^2(\beta_0 + \beta_1 X_i + \varepsilon_i) \\ &= \sigma^2(\varepsilon_i) \\ &= \sigma^2 \end{aligned}$$

That is, each Y_i comes from a probability distribution with mean $E(Y_i) = \beta_0 + \beta_1 X_i$ and variance σ^2, and the Y_i are independent of each other.

The line corresponding to the equation $Y = \beta_0 + \beta_1 X_i$ is sometimes called the *true regression line*, since it represents the true relationship between the independent and dependent variables.

Estimating β_0 and β_1 is the same as estimating the true regression line. It seems reasonable that the line that best matches the sample data points in the scatter diagram will also be the best estimate of the true regression line. Therefore, we seek A and B, estimates of β_0 and β_1, respectively, so that the line $\bar{Y}_i = A + BX_i$ matches as closely as possible the data points of the scatter diagram, as shown in Figure 25.3.

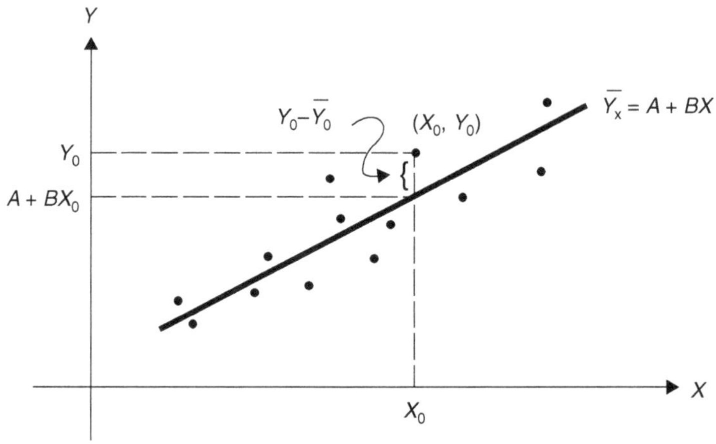

FIGURE 25.3 Matching the regression line to the data points.

We want to find A and B to minimize the errors between the Y_i and the corresponding values indicated by the regression line, $\bar{Y}_i = A + BX_i$. Some of these errors are positive and some negative; if we sum these errors they will tend to cancel out, so this sum is not a useful measure of the total error:

$$\sum_{i=1}^{n}(Y_i - \bar{Y}_i) = \sum_{i=1}^{n}(Y_i - (A + BX_i));$$

To provide a useful measure of the total error between the regression line and observed data points, we square each of the individual error terms indicated above, then sum over all the data points. This gives us the *error sum of squares*, ESS:

$$\text{ESS} = \sum_{i=1}^{n}\left(Y_i - \bar{Y}_i\right)^2 = \sum_{i=1}^{n}\left(Y_i - (A + BX_i)\right)^2 = \sum_{i=1}^{n}\left(Y_i - A - BX_i\right)^2$$

$$= \sum_{i=1}^{n}\left(Y_i^2 - 2AY_i - 2BX_iY_i + A^2 + 2ABX_i + B^2X_i^2\right).$$

We find the values of A and B that minimize the error sum of squares. This is the *method of least squares*, and the line found in this way, whose equation is $\bar{Y}_i = A + BX_i$, is the *least squares regression line*.

The formulas for A and B are:

$$A = \bar{Y} - B\bar{X}$$

$$B = \frac{n\left(\sum_{i=1}^{n} X_iY_i - \left(\sum_{i=1}^{n} X_i\right)\left(\sum_{i=1}^{n} Y_i\right)\right)}{n\sum_{i=1}^{n} X_i^2 - \left(\sum_{i=1}^{n} X_i\right)^2}$$

The derivation for these formulas is beyond the scope of this text, but the following discussion is for the interested student.

SIDE BAR

Derivation of A and B Formulas for Simple Linear Regression

In calculus, the location of a minimum or maximum value of a function of one variable is found by setting the derivative of the function equal to 0 and solving the resulting equation. Since the ESS is a function of both A and B, finding the location of its minimum value requires the simultaneous solution of the system of two equations obtained by setting equal to 0 the two partial derivatives of a function with respect to one variable. All other variables are treated as constants, and the derivative of a sum or difference is the sum or difference of the derivatives.

SIDE BAR

We take the partial derivative of ESS with respect to A:

$$\frac{\partial ESS}{\partial A} = \sum_{i=1}^{n}\left(0 - 2Y_i - 0 - 2A + 2BX_i + 0\right)$$
$$= \sum_{i=1}^{n}\left(-2Y_i + 2AA + 2BX_i\right)$$
$$= -2\sum_{i=1}^{n} Y_i + 2nA + 2B\sum_{i=1}^{n} X_i.$$

Next, the partial derivative of ESS with respect to B:

$$\frac{\partial ESS}{\partial B} = \sum_{i=1}^{n}\left(0 - 0 - 2X_iY_i + 0 + 2AX_i + 2BX_i^2\right)$$
$$= \sum_{i=1}^{n}\left(-2X_iY_i + 2AX_i + 2BX_i^2\right)$$
$$= -2\sum_{i=1}^{2} X_iY_i + 2A\sum_{i=1}^{n} X_i + 2B\sum_{i=1}^{n} X_i^2.$$

Setting the two partial derivatives equal to 0, we derive the *normal equations:*

$$\frac{\partial ESS}{\partial A} = 0$$

$$\Leftrightarrow -2\sum_{i=1}^{n} Y_i + 2nA + 2B\sum_{i=1}^{n} X_i = 0$$

$$\Leftrightarrow \sum_{i=1}^{n} Y_i = nA + B\sum_{i=1}^{n} X_i$$

$$\Leftrightarrow \mathrm{A} = \overline{Y} - B\overline{X}$$

and

$$\frac{\partial ESS}{\partial B} = 0$$

$$\Leftrightarrow -2\sum_{i=1}^{n} X_iY_i + 2A\sum_{i=1}^{n} X_i + 2B\sum_{i=1}^{n} X_i^2 = 0$$

SIDE BAR

$$\Leftrightarrow \sum_{i=1}^{n} X_i Y_i = A \sum_{i=1}^{n} X_i + B \sum_{i=1}^{n} X_i^2$$

$$\Leftrightarrow \sum_{i=1}^{n} X_i Y_i = \left(\overline{Y} - B\overline{X}\right) \sum_{i=1}^{n} X_i + B \sum_{i=1}^{n} X_i^2$$

$$\Leftrightarrow n\sum_{i=1}^{n} X_i Y_i = \left(\sum_{i=1}^{n} Y_i - B \sum_{i=1}^{n} X_i \right) \sum_{i=1}^{n} X_i + nB \sum_{i=1}^{n} X_i^2$$

$$\Leftrightarrow n\sum_{i=1}^{n} X_i Y_i = \left(\sum_{i=1}^{n} Y_i \right)\left(\sum_{i=1}^{n} X_i \right) - B\left(\sum_{i=1}^{n} X_i \right)^2 + nB \sum_{i=1}^{n} X_{i=1}^2$$

$$\Leftrightarrow n\sum_{i=1}^{n} X_i Y_i = \left(\sum_{i=1}^{n} Y_i \right)\left(\sum_{i=1}^{n} X_i \right) - B\left[\left(n\sum_{i=1}^{n} X_i^2 \right) - \left(\sum_{i=1}^{n} X_i \right)^2 \right]$$

$$\Leftrightarrow B = \frac{n\left(\sum_{i=1}^{n} X_i Y_i - \left(\sum_{i=1}^{n} X_i \right)\left(\sum_{i=1}^{n} Y_i \right) \right)}{n\sum_{i=1}^{n} X_i^2 - \left(\sum_{i=1}^{n} X_i \right)^2}.$$

Second-order partial derivatives can be used to show that the error sum of squares attains a minimum value with these expressions for A and B, and that the line whose equation is $\overline{Y}_i = A + BX_i$ therefore comes closest to all the data points, as measured by the ESS.

In our advertising and sales example, the following sums and means can be computed:

$$\sum_{i=1}^{12} X_i = 58.0 \qquad \sum_{i=1}^{n} X_i^2 = 289.160 \qquad \overline{X} = 4.033,$$

$$\sum_{i=1}^{n} Y_i = 591.0 \qquad \sum_{i=1}^{n} X_i Y_i = 2944.20 \qquad \overline{Y} = 49.250.$$

The regression coefficients are then:

$$B = \frac{12*2944.20 - 58.0*591.0}{12*9.936 - (58.0)^2} = 9.936$$

$$A = 49.250 - 9.936*4.833 = 1.229,$$

And the least squares regression equation is

$$\bar{Y}_i = 1.229 + 9.936X_i.$$

The regression equation can now be used to predict the value of $\bar{Y}_i$ that corresponds to a given value of X_i; that is, to estimate the expected value of Y for a given X. For example, if the business spends \$53,000 on advertising in a given quarter, then they can expect sales of \$538,900 in that quarter:

$$\bar{Y}_i = 1.229 + 9.936*5.3 = 53.89.$$

The computed regression equation is an estimate of the true regression equation, and the coefficients A and B are estimates of β_0 and β_1, respectively. By assuming that the random errors ε_i are normally distributed, the Y_i and therefore A and B are random variables, and we can describe the sampling distributions of A and B, which are statistics of the sample of data pairs. First, however, we consider a related quantity.

25.5 Exercises

Interpretation Exercises

1. From the general form of the normal equations for simple linear regression, obtain the formulas for the regression coefficients A and B.
2. Let $\bar{X}$ be the mean of the values $X_1, X_2, \ldots, X_n$.

   a. Show that $n\sum_{i=i}^{n}\left(X_i - \bar{X}\right)^2 = n\sum_{i=1}^{n}X_i^2 - \left(\sum_{i=1}^{n}X_i\right)^2$.

   b. Show that $\sum_{i=i}^{n}\left(X_i - \bar{X}\right)X_i = \sum_{i=1}^{n}\left(X_i - \bar{X}\right)^2$.
3. Show that the regression coefficient B can be written

$$B = \frac{\left[\sum_{i=1}^{n}\left(X_i - \bar{X}\right)Y_i\right]}{\left[\sum_{i=1}^{n}\left(X_i - \bar{X}\right)^2\right]}.$$

4. A correlation analysis is performed on the relationship between the percentage of defective parts produced (Y) and a quantitative measure of pressure on employees to speed production (X). A sample of four 15-minute time periods gave these values:

X = Pressure	−4	−1	1	4
Y = % Defective	2	3	5	10

a. Plot a scatter diagram of the relationship between X and Y.

b. Assuming the relationship between X and Y is linear, find the equation of least squares regression line, and plot that line on the scatter diagram.

c. How much of the variability of Y can be associated with the variability of X?

d. Suppose the correlation coefficient of X and Y is 0.41. What does this mean?

e. Is the correlation between X and Y significant at the 5% level? What does this mean?

Computational Exercises

5. Consider the data shown in Table 25.2:

TABLE 25.2 American automobile data(2).

Weight	Horsepower	MPG
3205	125	14
2945	80	20
4380	125	21
2040	74	29
3295	165	16
3520	195	16
2195	77	25
3205	195	16
1760	66	28
2515	130	21
3205	97	17
2480	93	28

a. Find the least squares regression equation relating mileage to automobile weight. Draw the line corresponding to this equation on the scatter diagram of mileage with weight.

b. If a car weighs 2,700 pounds, what mileage is predicted by the regression equation?

c. Find the standard error of estimate s_e. Of what is s_e an estimate?

6. Consider the data given in problem 5.

a. Find the least squares regression equation to predict mileage from horsepower. Draw its graph on the scatter diagram of mileage with horsepower.

b. If a car develops 145 horsepower, what mileage would we expect it to get?

c. Find the standard error of estimate s_e. Of what is s_e an estimate?

d. What does $R^2_{Horsepower, MPG}$ measure? Why is this important?

7. Table 25.3 gives the number of C# programs compiled and the number of times the programs needed to be recompiled at a college for each month of a 12-month period.

TABLE 25.3 Monthly C# recompiles.

Month	C# Programs	Number of Recompiles
1	800	34
2	500	0
3	1000	19
4	1600	24
5	1300	37
6	2100	32
7	1900	39
8	450	0
9	700	17
10	800	27
11	1200	41
12	750	22

a. By hand, find the least squares regression line for predicting the number of required recompiles from the number of C# programs.

b. Use Excel to repeat and verify the results of part a.

8. Suppose that in problem 6 we wish to predict the horsepower of a car from its mileage.

a. Using the formulas for A and B, find the equation of the regression line for this situation.

b. In the regression equation computed in problem 5, solve for horsepower in terms of mileage. Note that this is not the same equation as found in part a.

c. In general, to predict X from Y, which of these methods would you use? Why?

9. Compute the coefficient of determination R^2 for the data given in problem 7. What does this value tell us?

Excel Exercises

10. Redo problem 5 using Excel.

11. Redo problem 6 using Excel.

12. Answer the following questions about the variables X3 and X5 from the American Cities database. ON DVD

a. Plot the scatter diagram of X3 and X5 and find the value of R. What information do these convey?

b. Is there a significant relationship between these two variables? Explain.

c. Determine the least squares regression equation for predicting X3 from X5.

SIMPLE LINEAR REGRESSION PROJECT I

Companies whose products are based upon software development are concerned with the amount of code developed in a project as well as the development time needed. Clearly, the relationship between lines of code and function points, or a weighted average of the software's information domain and assessments of the software complexity, depends upon the programming language utilized and the quality of the software design.

An excellent article, complete with an associated data file, for developing a linear regression model that uses function points to predict development time is by Jack E. Manson and Brian R. Haggard in the *Journal of Statistics Education*, Vol. 15, Nov. 2 (2007) found at http:/www.amstat.org/publications/jse/v15n2/datasets.matson.html. This text can be freely shared among individuals.

After careful study of this article:

Part A:

You are to develop a linear regression model for estimating software development effort using historical function point data. In developing a useful regression model, a number of concerns must be addressed. For your first investigation you are to:

- Plot and interpret the distribution of function points.
- Plot and interpret the distribution of work hours.
- Plot and interpret work hours versus function points.
- Plot and interpret the natural logarithm of work hours versus the natural logarithm of function points.

Part B:

a. Review the plot of the dependent variable against the (various) predictor variables.

b. Hypothesize a model.

c. Check whether the statistical assumptions for the regression model are reasonably satisfied. If so, an appropriate model has been identified. If not, repeat steps (a) and (b).

Try out two models:

a. $E_{est} = b + a * FP$

b. $\ln(E_{est}) = b + a * \ln(FP)$

where E_{est} is the estimated development hours and FP is the size in function points.

Can you specify an upper and lower bound for the estimate in the form of a prediction interval for the point estimate of the efficiency of prediction for the chosen model? If so, complete this task and interpret the results.

SIMPLE LINEAR REGRESSION PROJECT II

Another application of linear regression in the *Journal of Statistics Education*, by Julie M. Hays, also complete with the associated data file is "Forecasting Computational Usage, Vol. II, No. I (2003) found at http://www.amstat.org/publications/jse/v11n1/datasets.hays.html. This text also can be freely shared among individuals.

A company is one of the largest volume specialty retailers of consumer electronics, personal computers, entertainment software, and appliances. In August of each year, the company purchases mainframe MIPS (Millions of Instructions Per Second, a measure of computing resources) in anticipation of the coming holiday season. Computing resources are needed to track and analyze retail information needed for billing, inventory, and sales. For planning and budgeting purposes, they also wish to forecast the number of MIPS needed the following year.

You are asked to forecast the MIPS needed for December 2000 and December 2001 using the company dataset. This dataset contains monthly data on computer usage (MIPS) and total number of stores from August 1996 to July 2000. Additionally, information on the planned number of stores through December 2001 is available.

After carefully studying the article:

a. Plot the computer usage versus time using SPSS and interpret the results.

b. Study the following concepts for forecasting models using MAD and CFE:

The evaluation of forecasting models is based on the desire to produce forecasts that are unbiased and accurate. The Mean Absolute Deviation (MAD) is one common measure of forecast accuracy.

1. Using the In-Class Data Table described and provided in the article:

a. Calculate forecasts using a three-period moving average and exponential smoothing with a smoothing constant of 0.2.

b. Calculate the MAD and CFE for both models. Determine the "best model."

c. Chart the actual and forecast MIPS by estimating MIPS for December 2000 and 2001 using a three-period moving average and exponential smoothing with a smoothing constant of 0.2.

d. Is there a "problem" with your forecasts? Defend your answer.

2. Go back and re-estimate your forecast using double exponential smoothing.

   a. How do the results compare to the forecasts from fitting a line to the data?

   b. How about the issue of seasonality? Try using an additive and multiplicative adjustment for seasonality using all the past data or only some of the past data. These two techniques can be found in Hanke's *Business Forecasting*[2] and Prins' article found at http://www.sbthik/princ/section5/princ52.htm, "6.5.2. What do we do when the data are 'non-normal'?"

References

1. Galton, Sir Francis, *Regression toward Mediocrity in Hereditary Stature, 1885.*
2. Hanke, J., and Wichern, R. (2008). *Business Forecasting*, Prentice Hall.

CHAPTER 26

SIMPLE LINEAR REGRESSION: HYPOTHESIS TESTING

OVERVIEW AND LEARNING OBJECTIVES

In This Chapter

26.1 Introduction

The computed regression equation is an estimate of the true regression equation, and the coefficients A and B are estimates of β_0 and β_1, respectively. By assuming that the random errors ε_i are normally distributed, the Y_i and therefore A and B are random variables, and we can describe the sampling distributions of A and B, which are statistics of the sample of data pairs. In this chapter, we will form a hypothesis test on the coefficients of the regression equation as well as a measure of the accuracy of the estimators for Y (the dependent variable) corresponding to a particular X (the independent variable) value.

26.2 The Standard Error of Estimate s_e

We assume that the random error terms ε_i, and therefore the Y_i have equal variances σ^2. We can estimate σ^2 by using the squared deviations of the Y_i, not from their mean $\bar{Y}$, but from the regression line. The deviations are $\left(Y_i - \bar{Y}_i\right) = \left(Y_i - \left(A + BX_i\right)\right)$, and the estimator of σ^2 is:

$$\frac{\sum_{i=1}^{n}\left(Y_i - \bar{Y}_i\right)^2}{(n-2)} = \frac{\sum_{i=1}^{n}\left(Y_i - A - BX_i\right)^2}{(n-2)} = \frac{ESS}{n-2}.$$

Note that the numerator is the error sum of squares that we minimized to find A and B, and that we divide by $(n - 2)$ in order that this quantity be an unbiased estimator of σ^2; A and B are being used to estimate β_0 and β_1.

More often used than this variance estimator is its square root s_e, the *standard error of estimate:*

$$s_e = \left[\frac{\sum_{i=1}^{n}\left(Y_i - \bar{Y}_i\right)^2}{(n-2)}\right]^{1/2} = \left[\frac{\sum_{i=1}^{n} Y_i^2 - A\sum_{i=1}^{n} Y_i - B\sum_{i=1}^{n} X_i Y_i}{(n-2)}\right]^{1/2}.$$

s_e estimates the standard deviation of the Y distribution corresponding to each possible X value, and is more commonly calculated by the second formula shown above. It is a measure of the dispersion of the data points around the regression line, shown in Figure 26.1.

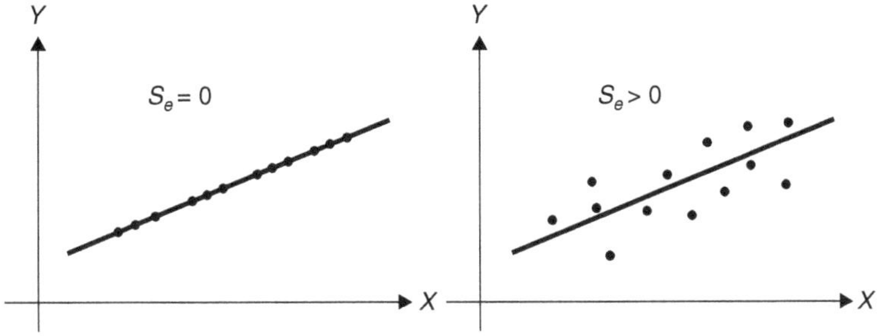

FIGURE 26.1 The standard error of estimate s_e.

All figures and tables in this chapter appear on the companion DVD. ON DVD

26.3 Sampling Distribution and Hypothesis Tests for β_0

26.3.1 The slope of the regression equation

We can *now describe the sampling distribution of* B, *the slope* of the regression line.

If the ε_i are normally distributed, then since β_0, β_1, and X_i are all constants, $\bar{Y}_i = \beta_0 + \beta_1 X_i + \varepsilon_i$ are also normally distributed. B can be written in this way:

$$B = \frac{n\left(\sum_{i=1}^{n} X_i Y_i - \left(\sum_{i=1}^{n} X_i\right)\left(\sum_{i=1}^{n} Y_i\right)\right)}{n\sum_{i=1}^{n} X_i^2 - \left(\sum_{i=1}^{n} X_i\right)^2}$$

and $A = \bar{Y} - B\bar{X}$

It can be shown that the sampling distribution of B is normal with mean β_1 and variance:

$$\sigma_B^2 = \frac{\sigma^2}{\sum_{i=1}^{n}\left(X_i - \bar{X}\right)^2}.$$

The interested reader can refer to the following sidebar for verification of this claim. Not only is B an unbiased estimator of β_1, but it can also be shown that as an estimator of β_1, B is sufficient and of minimum variance.

SIDE BAR

The Mean and Variance of B

If the ε_i are normally distributed, then since β_0, β_1, and X_i are all constants, $Y_i = \beta_0 + \beta_1 X_i + \varepsilon_i$ are also normally distributed. B can be written in this way:

$$B = \frac{\sum_{i=1}^{n}(X_i - \bar{X})Y_i}{\sum_{i=1}^{n}(X_i - \bar{X})^2}.$$

So if we let

$$k_i = \frac{(X_i - \bar{X})}{\sum_{i=1}^{n}(X_i - \bar{X})^2}, \quad \text{then} \quad B = \sum_{i=1}^{n} k_i Y_i.$$

But each k_i is a constant; B is a linear combination of independent normal random variables (Y_i), and must itself be normally distributed, as we pointed out earlier.

Further,

$$\begin{aligned} E(B) = E\left(\sum_{i=1}^{n} k_i Y_i\right) &= \sum_{i=1}^{n} k_i E(Y_i) = \sum_{i=1}^{n} k_i(\beta_0 + \beta_1 X_i) \\ &= \beta_0\left(\sum_{i=1}^{n} k_i\right) + \beta_1\left(\sum_{i=1}^{n} k_i X_i\right) \end{aligned}$$

But

$$\sum_{i=1}^{n} k_i = \left(\sum_{i=1}^{n} \frac{(X_i - \bar{X})}{\sum_{i=1}^{n}(X_i - \bar{X})^2}\right) = \frac{\sum_{i=1}^{n}(X_i - \bar{X})}{\sum_{i=1}^{n}(X_i - \bar{X})^2} = 0,$$

since $\sum_{i=1}^{n}(X_i - \bar{X}) = 0$ as previously discussed. And

$$\sum_{i=1}^{n} k_i X_i = \left(\sum_{i=1}^{n} \frac{(X_i - \bar{X}) X_i}{\sum_{i=1}^{n} (X_i - \bar{X})^2} \right)$$

$$= \frac{\sum_{i=1}^{n} (X_i - \bar{X}) X_i}{\sum_{i=1}^{n} (X_i - \bar{X})^2}$$

$$= \frac{\sum_{i=1}^{n} (X_i - \bar{X}) X_i}{\sum_{i=1}^{n} (X_i - \bar{X})^2}$$

But it can be easily shown that

$$\sum_{i=1}^{n} (X_i - \bar{X}) X_i = \sum_{i=1}^{n} (X_i - \bar{X})^2$$

Therefore

$$\sum_{i=1}^{n} k_i X_i = \frac{\sum_{i=1}^{n} (X_i - \bar{X})(X_i - \bar{X})}{\sum_{i=1}^{n} (X_i - \bar{X})^2} = 1,$$

Then $E(B) = \beta_0 * 0 + \beta_1 = \beta_1$.

Also,

$$\sigma^2 (B) = \sigma^2 \left(\sum_{i=1}^{n} k_i Y_i \right) = \sum_{i=1}^{n} k_i^2 \sigma^2 (Y_i) = \sum_{i=1}^{n} k_i^2 \sigma^2$$

$$= \sigma^2 \sum_{i=1}^{n} k_i^2 = \sigma^2 \sum_{i=1}^{n} \left[\frac{(X_i - \bar{X})}{\sum_{i=1}^{n} (X_i - \bar{X})^2} \right]^2,$$

$$= \sigma^2 \frac{\sum_{i=1}^{n} (X_i - \bar{X})^2}{\left[\sum_{i=1}^{n} (X_i - \bar{X})^2 \right]^2} = \frac{\sigma^2}{\sum_{i=1}^{n} (X_i - \bar{X})^2}$$

The statistic

$$z = \frac{B - \beta_{1_0}}{\sigma_B} \quad \text{or} \quad t = \frac{B - \beta_{1_0}}{\sigma_B} \ \text{(with } n-2 \text{ degrees of freedom)}$$

can be used, depending on the sample size, to test the hypothesis $H_0: \beta_1 = \beta_{1_0}$. Since

$$\sigma_B = \frac{\sigma}{\left[\sum_{i=1}^{n} (X_i - \bar{X})^2\right]^{1/2}}$$

And s_e is an estimate of σ, the statistic is usually calculated according to this formula:

$$z \text{ or } t = \frac{B - \beta_{1_0}}{s_e} \left(\sum_{i=1}^{n} X_i^2 - n\bar{X}^2 \right)^{1/2}.$$

When the slope of the regression line is 0, no relationship between the two variables is indicated. Thus, the hypothesis test most often performed with B is this:

$$H_0: \beta_1 = 0$$
$$H_1: \beta_1 \neq 0.$$

The null hypothesis says that there is no linear relationship between the variables; in attempting to accept the alternative, we hope to show that such a (numerical) relationship exists.

For example, to perform such a test with the advertising and sales data listed in Table 25.1 at the 5% significance level, the critical values corresponding to 12 – 2 = 10 degrees of freedom are $\pm t_{0.025} = \pm 2.228$. We have found that B = 9.936 and s_e = 4.4, so the value of the *t* statistic is

$$t = \frac{B - 0}{s_e} \left(\sum_{i=1}^{n} X_i^2 - n\bar{X}^2 \right)^{1/2} = \frac{9.936}{4.4} (289.16 - 12 * 4.833^2)^{1/2} = 6.72.$$

This value is beyond the critical values, so we can reject the null hypothesis and conclude, based on our sample data, that there is a linear relationship between advertising and sales for the business. Again, be aware that the existence of a significant *numerical* relationship does not necessarily imply a causal, physical relationship between the quantities represented. The existence of a causal relationship must be investigated through qualitative rather than quantitative methods.

26.4 Sampling Distribution and Hypothesis Tests for β_1

26.4.1 The Y-intercept of the regression equation

Of the two regression coefficients, β_1 is more important, since it indicates the expected change in Y corresponding to a unit change in X. Inferences about β_0, based on its estimator A, can also be useful; however, $\bar{X}$ and $\bar{Y}$ are constants, and $A = \bar{Y} - B\bar{X}$, so since the sampling distribution of B is normal (or a t distribution for small sample sizes), that of A is also. It can further be shown that

$$E(A) = \beta_0$$

$$\sigma^2(A) = \sigma^2 \left[\frac{1}{n} + \frac{\bar{X}^2}{\sum_{i=1}^{n} (X_i - \bar{X})^2} \right]$$

and that A is a sufficient, minimum variance estimator of β_0. The test statistic z or t (with $n-2$ degrees of freedom) for tests on β_0 is

$$\frac{A - \beta_0}{s_e \left(\frac{1}{n} + \frac{\bar{X}^2}{\sum_{i=1}^{n} X_i^2 - n\bar{X}^2} \right)^{1/2}}.$$

26.5 Hypothesis Test for the Conditional Mean of the Regression Equation

The fact that A and B are "good" estimators of β_0 and β_1 supports the claim that the least squares regression line is a good estimator of the true regression. Still, how accurate are estimates of Y corresponding to a particular value of X? To investigate this question, we examine the sampling distribution of $\bar{Y}_0$, the estimated expected value of Y corresponding to some value of X, X_0. (The true mean of the distribution of Y values corresponding to X_0 is the *conditional mean* $\mu_{X_0} = \beta_0 + \beta_1 X_0$. It is this quantity of which $\bar{Y}_0$, is an estimate.)

First,

$$\begin{aligned} E(\bar{Y}_0) &= E(A + BX_0) = E(A) + E(BX_0) \\ &= \beta_0 + X_0 E(B) = \beta_0 + \beta_1 X_0 = \mu_{X_0}. \end{aligned}$$

Second,

$$\begin{aligned}\sigma^2(\bar{Y}_0) &= \sigma^2(A+BX_0) = \sigma^2\left(\left(\bar{Y}-B\bar{X}\right)+BX_0\right)\\ &= \sigma^2\left(\bar{Y}+B\left(X_0-\bar{X}\right)\right)\\ &= \sigma^2\left(\bar{Y}\right)+\left(X_0-\bar{X}\right)^2\sigma^2(B),\end{aligned}$$

since $\bar{Y}$ and B are independent. But $\bar{Y}$ is simply the mean of the Y_i; if σ^2 is the common variance of the Y distribution, then $\sigma^2(\bar{Y}) = \sigma^2/n$, so that

$$\sigma^2(\bar{Y}_0) = \frac{\sigma^2}{n} + \frac{\left(\left(X_0-\bar{X}\right)^2\sigma^2\right)}{\sum_{i=1}^{n}\left(X_i-\bar{X}\right)^2},$$

since

$$\sigma^2(B) = \frac{\sigma^2}{\left[\sum_{i=1}^{n}\left(X_i-\bar{X}\right)^2\right]}.$$

$\sigma(\bar{Y}_0)$ is called the *Mean-Square Error (MSE)*, and is often written this way:

$$MSE = \left[\frac{\sigma^2}{n} + \frac{\left(\bar{X}-X_0\right)^2\sigma^2}{\sum_{i=1}^{n}X_i^2 - n\bar{X}^2}\right]^{1/2} = \sigma\left[\frac{1}{n} + \frac{\left(\bar{X}-X_0\right)^2}{\sum_{i=1}^{n}X_i^2 - n\bar{X}^2}\right]^{1/2}.$$

Again, σ can be estimated by the standard error of estimate s_e.

Finally, the sampling distributions for large n of both A and B are normal; so is the sampling distribution of $\bar{Y}_{X_0} = A + BX_0$. We can now construct confidence intervals for the conditional mean μ_{X_0} corresponding to X_0, and perform tests of the hypothesis $H_0\colon \mu_{X_0} = \mu_{X_0}$ using, depending on the sample size, the statistic

$$z = \frac{\bar{Y}_{X_0} - \mu_{X_0}}{MSE} \quad \text{or} \quad t = \frac{\bar{Y}_{X_0} - \mu_{X_0}}{MSE} \quad \text{(with } (n-2) \text{ degrees of freedom).}$$

Note that since the accuracy of estimates of conditional means depends on the MSE, we would like this quantity to be as small as possible, which can be accomplished by making n large.

26.6 The Coefficient of Determination

From the data pairs graphed on the scatter diagram, we can derive another measure of predictive accuracy of the regression line by considering the difference between each Y_i value and the mean of the Y_i, $\bar{Y}$.

As Figure 26.2 shows, the difference between the observed value Y_i and the mean $\bar{Y}$ of the Y_i values is the sum of two differences:

$$Y_i - \bar{Y} = (Y_i - (A + BX_i)) + ((A + BX_i) - \bar{Y}).$$

Y_i differs from $\bar{Y}$ because it is partly determined by its corresponding X value $((A + BX_i) - \bar{Y})$, and because of the variability of the Ys $(Y_i - (A + BX_i))$. Part of the variation of Y_i from $\bar{Y}$ can be explained by the regression line, and part cannot.

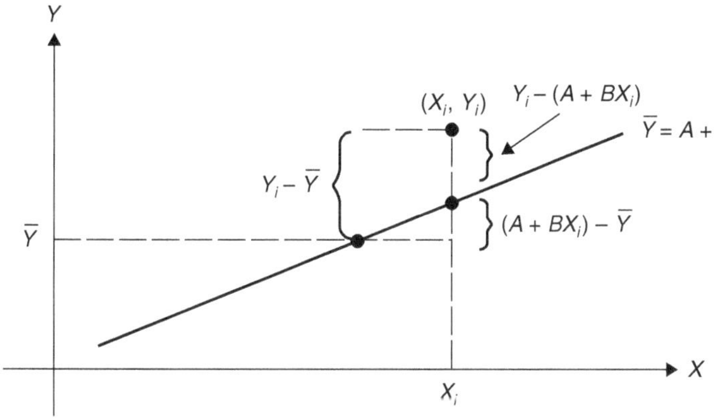

FIGURE 26.2 The components of the difference $Y_i - \bar{Y}$.

We can obtain an equation that considers the variations of all the Y values represented in the scatter diagram by squaring both sides of the above formula and summing over all the data points. Note the distinction between $\bar{Y}$, the mean of the Y values, and $\bar{Y}_i$, the estimated mean of the Y distribution corresponding to X_i. The result of this computation procedure is:

$$\sum_{i=1}^{n} (Y_i - \bar{Y})^2 = \sum_{i=1}^{n} (Y_i - \bar{Y}_i)^2 + \sum_{i=1}^{n} (\bar{Y}_i - \bar{Y})^2.$$

The interested reader can refer to the verification of this relationship in the following sidebar.

SIDE BAR

Sum of Squares Relationship

$$\begin{aligned}\sum_{i=1}^{n}(Y_i-\bar{Y})^2 &= \sum_{i=1}^{n}\left[\left((Y_i-\bar{Y}_i)+(\bar{Y}_i-\bar{Y})\right)\right]^2\\ &= \sum_{i=1}^{n}(Y_i-\bar{Y}_i)^2+2\sum_{i=1}^{n}\left(Y_i-\bar{Y}_i\right)\left(\bar{Y}_i-\bar{Y}\right)+\sum_{i=1}^{n}\left(\bar{Y}_i-\bar{Y}\right)^2\\ &= \sum\nolimits_{i=1}^{n}(Y_i-\bar{Y}_i)^2+2\sum\nolimits_{i=1}^{n}\bar{Y}_i(Y_i-\bar{Y}_i)\\ &\quad -2\bar{Y}\sum\nolimits_{i=1}^{n}Y_i(Y_i-\bar{Y}_i)+\sum\nolimits_{i=1}^{n}(\bar{Y}_i-\bar{Y})^2\end{aligned}$$

But

$$\sum_{i=1}^{n}\bar{Y}_i\left(Y_i-\bar{Y}_i\right)=0 \text{ and } \sum_{i=1}^{n}(Y_i-\bar{Y})=0, \text{ so}$$

$$\sum_{i=1}^{n}(Y_i-\bar{Y})^2=\sum_{i=1}^{n}(Y_i-\bar{Y}_i)^2+\sum_{i=1}^{n}(\bar{Y}_i-\bar{Y})^2$$

The sum on the left is called the *total sum of squares*, and is a measure of the total variability of the observed Y values about $\bar{Y}$. (Note that $\sum_{i=1}^{n}(Y_i-\bar{Y})^2=(n-1)Var(Y)$.) The sum on the right, the *regression sum of squares*, is a measure of the variability of the Y values that can be explained by the regression line, which can be thought of as being due to the influence of the X values. And the center sum, the *residual sum of squares*, measures the unexplained variation that is not due to the X values.

The Regression Sum of Squares (RSS) is related to the Pearson correlation coefficient (R) by:

$$\text{RSS}=\text{R}^2\sum_{i=1}^{n}(Y_i-\bar{Y})^2.$$

as shown in the following sidebar.

SIDE BAR

Relationship between RSS and R

$$
\begin{aligned}
\mathrm{RSS} &= \sum_{i=1}^{n}\left(\overline{Y}_i-\overline{Y}\right)^2=\sum_{i=1}^{n}\left(A+BX_i-\overline{Y}\right)^2\\
&=\sum_{i=1}^{n}\left(\overline{Y}-B\overline{X}+BX_i-\overline{Y}\right)^2\\
&=\sum_{i=1}^{n}\left(BX_i-B\overline{X}\right)^2\\
&=B^2\sum_{i=1}^{n}\left(X_i-\overline{X}\right)^2\\
&=\left[\frac{\sum_{i=1}^{n}\left(X_i-\overline{X}\right)\left(Y_i-\overline{Y}\right)}{\sum_{i=1}^{n}\left(X_i-\overline{X}\right)^2}\right]^2\sum_{i=1}^{n}\left(X_i-\overline{X}\right)^2\\
&=\frac{\left[\sum_{i=1}^{n}\left(X_i-\overline{X}\right)\left(Y_i-\overline{Y}\right)\right]^2}{\sum_{i=1}^{n}\left(X_i-\overline{X}\right)^2}
\end{aligned}
$$

But,

$$
R=\frac{\sum_{i=1}^{n}\left(X_i-\overline{X}\right)\left(Y_i-\overline{Y}\right)}{\left[\sum_{i=1}^{n}\left(X_i-\overline{X}\right)^2\right]^{1/2}\left[\sum_{i=1}^{n}\left(Y_i-\overline{Y}\right)^2\right]^{1/2}}
$$

Therefore,

$$
\mathrm{RSS}=\mathrm{R}^2\sum_{i=1}^{n}\left(Y_i-\overline{Y}\right)^2
$$

Substituting this quantity (the interested reader can refer to the next sidebar) into the original statement relating the three sums of squares and dividing the resulting expression by the total sum of squares derives the following formula:

$$R^2 = 1 - \frac{\sum_{i=1}^{n}\left(Y_i - \bar{Y}_i\right)^2}{\sum_{i=1}^{n}\left(Y_i - \bar{Y}\right)^2}.$$

SIDE BAR

Derivation of the Coefficient of Determination

$$\sum_{i=1}^{n}\left(Y_i - \bar{Y}\right)^2 = \sum_{i=1}^{n}\left(Y_i - \bar{Y}_i\right)^2 + R^2\sum_{i=1}^{n}\left(Y_i - \bar{Y}\right)^2$$

$$\Leftrightarrow 1 = \frac{\sum_{i=1}^{n}\left(Y_i - \bar{Y}_i\right)^2}{\sum_{i=1}^{n}\left(Y_i - \bar{Y}\right)^2} + R^2$$

$$\Leftrightarrow R^2 = 1 - \frac{\sum_{i=1}^{n}\left(Y_i - \bar{Y}_i\right)^2}{\sum_{i=1}^{n}\left(Y_i - \bar{Y}\right)^2}.$$

The quantity R^2, which is the square of the Pearson product-moment correlation coefficient R, is called the *coefficient of determination*. It is the proportion of the variability of the Y values that can be related to variation in the X values; R^2 measures the reduction in the sum of squares of deviations of the Y values from $\bar{Y}$, which can be achieved by considering their linear relationship to the X values. R^2 is often calculated using this formula:

$$R^2 = \frac{A\left(\sum_{i=1}^{n} Y_i\right) + B\left(\sum_{i=1}^{n} X_i Y_i\right) - n\bar{Y}^2}{\left(\sum_{i=1}^{n} Y_i^2\right) - n\bar{Y}^2}.$$

In our example, the coefficient of determination is 0.87123, indicating 87% of the variation of the Y values from their mean can be related to variation in the X values.

26.7 Observations about Linear Regression

Several observations can be made about linear regression:

- Valid predictions of $\overline{Y}_i$ are based on the assumption that future observations of X and Y will follow the pattern observed in the past.
- Do not attempt to predict Y from values of X beyond those examined when you are constructing the regression equation. There is no guarantee that the linearity observed in the study persists outside its bounds.
- In practice, the value of X from which Y is estimated is often itself an estimate. This introduces another element of uncertainty.
- In the z_X, z_Y plane, the regression equation may be written as $z_Y = Rz_X$.
- Since $B = (s_X/s_Y)R$, when B is 0, so must be R. Thus, when R is 0, the regression line is horizontal, and the significances of R, R^2, and B are always the same.

26.8 Exercises

Interpretation Exercises

1. Show that if two variables X and Y have a perfect positive linear relationship, then for any data point (X_i, Y_i),

$$\frac{X_i - \overline{X}}{s_X} = \frac{Y_i - \overline{Y}}{s_Y}.$$

(Hint: "Perfect positive linear relationship" means $Y_i = A + BX_i$, for all i.)

2. If $t = (B - \beta_{1_0})/\sigma_B$ and s_e is an estimate for σ, show that

$$t = \frac{B - B_{1_0}}{s_e}\left(\sum_{i=1}^{n} X_i^2 - n\overline{X}^2\right)^{1/2}.$$

3. In general, what is the meaning of the regression coefficient β_0? Does this always apply?

4. Using the fact that any linear combination of normal random variables is normal, argue that the sampling distribution of the least squares regression coefficient A is normal.

5. Show that

a. $B = \frac{s_Y}{s_X} R_{XIY}$, and

b. $Z_Y{}' = R_{XY} Z_X$, where R_{XY} is the Pearson correlation coefficient of the variables X and Y.

Computational Exercises

6. A correlation analysis is performed on the relationship between the percentage of defective parts produced (Y) and a quantitative measure of pressure on employees to speed production (X). A sample of four 15-minute time periods gave these values:

X = Pressure	-4	-1	1	4
Y = % Defective	2	3	5	10

a. If the pressure during a given time period is measured to be $x=2$, what is the expected percentage of defective parts? Find a 95% confidence interval for the expected percentage of defective parts.

b. Find a 95% prediction interval for the observed percentage of defective parts in a period in which pressure is measured to be $x=2$.

c. Find a 95% confidence interval for β_1, the slope of the true regression line.

TABLE 26.1 Automobile data.

Weight	Horsepower	MPG
3205	125	14
2945	80	20
4380	125	21
2040	74	29
3295	165	16
3520	195	16
2195	77	25
3205	195	16
1760	66	28
2515	130	21
3205	97	17
2480	93	28

7. Consider the data for cars shown in Table 26.1:

a. Find a 95% confidence interval for β_1, the slope of the true regression line. Based on this, can you reject the null hypothesis H_0: $\beta_1 = 0$ at the 5% significance level?

b. Find a 95% confidence interval for β_0, the Y-intercept of the true regression line. What meaning, if any, does the value of A have in this problem?

c. Corresponding to a weight of 2,700 pounds, find a 95% confidence interval for the expected mileage.

8. Consider the data given in problem

a. At the 1% significance level, can we conclude that there is a significant relationship between horsepower and mileage? Examine B to find out.

b. Find a 95% confidence interval for the value of β_0, the Y-intercept of the regression line. What meaning, if any, does A have in this situation?

c. If the horsepower of a car is increased by 25, what effect would you expect this to have on mileage?

d. For a car that develops 145 horsepower, find a 95% confidence interval for the corresponding expected mileage.

9. Table 26.2 gives the number of C# programs compiled and the number of times the programs needed to be recompiled at a college for each month of a 12-month period.

TABLE 26.2 C# Monthly recompiles data.

Month	C# Programs	Number of Recompiles
1	800	34
2	500	0
3	1000	19
4	1600	24
5	1300	37
6	2100	32
7	1900	39
8	450	0
9	700	17
10	800	27
11	1200	41
12	750	22

a. Test by hand the hypothesis H_0: $\beta_1 = 0$ at the 5% significance level.

b. Use Excel to verify the results in part c.

10. Consider the regression equation constructed for problem 9.

a. Find a 95% confidence interval for the expected number of recompiles corresponding to the submission of 1,000 C# programs.

b. The sampling distribution of a single observation of Y corresponding to a given X_0 is identical to that of $\overline{Y}_{X_0}$ except that it has standard deviation

$$\sigma\left[1+\frac{1}{n}+\frac{\left(X_0-\overline{X}\right)^2}{\sum_{i=1}^{n} X_i^{\,2}-n\overline{X}^2}\right]^{1/2}.$$

Use this information to find a 95% prediction interval for the value of a single future number of recompiles in a month in which 1,000 C# programs are submitted.

11. Consider the regression equation developed in problem 8. Find a 95% prediction interval for the mileage of a car that developed 145 horsepower. (Hint: See problem 10b.) What is the difference in meaning between this result and the result of Exercise 8d?

12. For the values given in problem 9:

a. Test H_0: $\rho = 0$ at the 5% significance level.

b. Test H_0: $\beta_1 = 0$ at the 5% significance level.

c. Show that when $\rho = \beta_1 = 0$ the statistics of these two tests are equal.

13. Table 26.3 gives the statistics produced from a statistical software package from a file of 28 data pairs(X_i, Y_i).

TABLE 26.3 Hypothetical paired data set (28 pairs).

CORRELATION (R)	—	.91745	R SQUARED	—	.84171
SIGNIFICANCE R	—	.00001	STD ERR OF EST	—	.96962
INTERCEPT (A)	—	.98862	STD ERR OF A	—	.49197
SIGNIFICANCE A	—	.02749	SLOPE (B)	—	.09611
STD ERR OF B	—	.00817	SIGNIFICANCE B	—	.00001
PLOTTED VALUES	—	28			
EXCLUDED VALUES	—	0			
MISSING VALUES	—	0			

a. Is there a significant linear relationship between X and Y here? Describe that relationship.

b. State the least squares regression equation. What is the expected value of Y corresponding to $X_0 = 28$?

c. Find a 95% confidence interval for β_1. Is there a way to do this without using the standard error of estimate value?

d. Find a 95% confidence interval for β_0.

14. Table 26.4 shows the statistics produced by a statistical software package from a file of 50 data pairs (X_i, Y_i).

TABLE 26.4 Hypothetical paired data set (50 pairs).

CORRELATION (R)	—	.89733	R SQUARED	—	.80520
SIGNIFICANCE R	—	.00001	STD ERR OF EST	—	8.08456
INTERCEPT (A)	—	– 10.59451	STD ERROR OF A	—	4.57615
SIGNIFICANCE A	—	.01246	SLOPE (B)	—	.78906
STD ERROR OF B	—	.05602	SIGNIFICANCE B	—	.00001

PLOTTED VALUES	—	50
EXCLUDED VALUES	—	0
MISSING VALUES	—	0

a. State the regression equation and find the value of Y expected to correspond to $X_0 = 95$.

b. Is the value of B significant? What does this tell us?

c. Find a 95% confidence interval for β_1.

d. Find a 95% confidence interval for β_1.

e. What is the meaning of R^2?

Excel Exercises

15. Redo problem 7 using Excel.

16. Redo problem 8 using Excel.

17. Answer the following questions about the variables X3 and X5 from the American Cities database.

a. Plot the scatter diagram of X3 and X5 and find the value of R. What information do these convey?

b. Is there a significant relationship between these two variables? Explain.

c. Determine the least squares regression equation for predicting X3 from X5.

d. What is the value of the coefficient of determination R^2, and what does it tell us?

e. Find a 95% confidence interval for β_1, the slope of the true regression line.

f. Why is the significance of B the same as the significance of R? What does it mean?

g. Find the expected value of X3 if X5 is $48,500.

CHAPTER 27

SIMPLE LINEAR REGRESSION: CASE STUDY

OVERVIEW AND LEARNING OBJECTIVES

In This Chapter

27.1 Introduction

27.2 The Statement for the Case Study

27.3 Case Study Analysis

27.4 Exercises

27.1 Introduction

In this chapter, we want to highlight mistakes commonly committed while using regression. Additionally, we would like to address the question of what constitutes a good model.

27.2 The Statement for the Case Study

Consider a small business with seven regional offices. Each office reports the total number of sales for the month on the last day of the month. Suppose that in August the regional offices data was {(14,2), (16,5), (27,7), (42,9), (39,110), (50,13), (83,20)}, where each data pair was (number of sales personnel, number of sales). The management would like to utilize this data to design and implement a linear model to predict the number of monthly sales as a function of the number of sales personnel.

27.3 Case Study Analysis

Given the data $n = 7$, $\sum xy = 3375$, $\sum x = 271$, $\sum x^2 = 13{,}885$, $\sum y = 66$, $\sum y^2 = 828$, $\bar{x} = 38.71$, and $\bar{y} = 9.43$. Therefore,

$$b_1 = \frac{\sum xy - n\bar{x}\bar{y}}{\sum x^2 - n(\bar{x})^2} = \frac{3375 - 7 * 38.71 * 9.43}{13{,}885 - 7 * (38.71)^2} = 0.2438$$

$$b_0 = \bar{y} - b_1\bar{x} = 9.43 - 0.2438 * 38.71 = -0.0083$$

The desired linear model is:

Number of monthly sales = – 0.0083 + 0.2438 (number of sales personnel)

A scatter plot of the data is shown in Figure 27.1. A straight line with intercept –0.0083 and slope 0.2438 is also shown in the figure. Notice that the line does give an estimate close to the observed values.

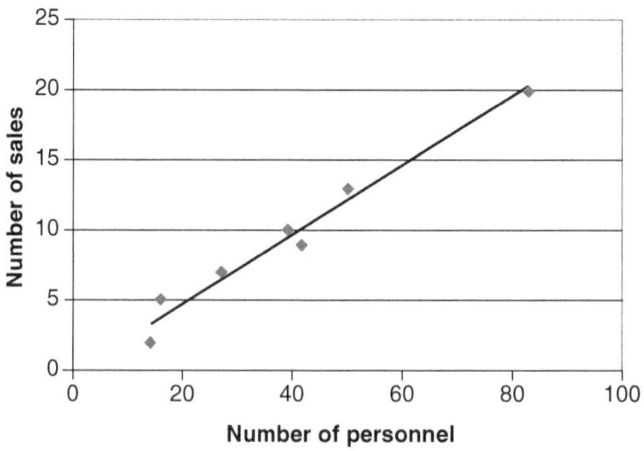

FIGURE 27.1 Scatter plot of number of monthly sales and number of available sales personnel.

ON DVD All figures and tables in this chapter appear on the companion DVD.

In Table 27.1, we have listed the number of monthly sales predicted by the model, the measured values, errors, and squared errors for each of the seven observations.

TABLE 27.1 Error computation for the number of monthly sales and number of available sales personnel.

Personnel	Sales	Estimate	Error	Error Sqr
X	Y	y′	y – y′	(y – y′)^2
14	2	3.4049	-1.4049	1.973744
16	5	3.8925	1.1075	1.226556
27	7	6.5743	0.4257	0.18122
42	9	10.2313	-1.2313	1.5161
39	10	9.4999	0.5001	0.2501
50	13	12.1817	0.8183	0.669615
83	20	20.2271	-0.2271	0.051574
	Totals	66	0	5.86891
	y′ = b0 + b1 * x			
	b0	-0.0083		
	b1	0.2438		

Note that the SSE is 5.869. This is the minimum possible SSE. Any other values of b_0 and b_1 would give a higher SSE.

$$SSE = \sum y^2 - b_0 \sum y - b_1 \sum xy = 828 - (-0.0083)(66) - (0.2438)(3375) = 5.869$$

Then

$$SST = SSY - SSO = \sum y^2 - n(\overline{y})^2 = 828 = 7 * (9.43)^2 = 205.71$$

and since the following table allows for the computation of the total sum of squares, SST, as illustrated in Table 27.2:

TABLE 27.2 Calculation of SST.

Sales		
Y	**y – y bar**	**y – y bar sqd**
2	-7.428571429	55.18367
5	-4.428571429	19.61224
7	-2.428571429	5.897959
9	-0.428571429	0.183673
10	0.571428571	0.326531
13	3.571428571	12.7551
20	10.57142857	111.7551
9.428571		205.7143
y bar	9.428571429	
SST	205.7143	

Having the values for SST and SSE allows one to find the value for the total sum of squares due to the regression, SSR.

$$\text{SSR} = \text{SST} - \text{SSE} = 205.71 - 5.87 = 199.84$$

$$R^2 = \frac{SSR}{SST} = \frac{199.84}{205.71} = 0.9715$$

Thus, the regression explains 97% of variation in the number of monthly sales.

The relationship between the sums of squares terms and the degrees of freedom is:

Sums of squares: $\text{SST} = \text{SSy} - \text{SSO} = \text{SSR} + \text{SSE}$

$$205.71 = 828 - 622.29 = 199.84 + 5.87$$

Degrees of freedom: $6 = 7 - 1 = 1 + 5$

The mean squared error is:

$$\text{MSE} = \frac{SSE}{df} = \frac{5.8}{5} = 1.17, \quad \text{where} \quad df = \text{degrees of freedom for errors}$$

The standard deviation of errors is:

$$S_e = \sqrt{MSE} = \sqrt{1.17} = 1.08$$

Then the standard deviations of b_0 and b_1 are:

$$S_{b_0} = S_e\left[\frac{1}{n} + \frac{x^{-2}}{\sum x^2 - n(\bar{x})^2}\right]^{1/2}$$

$$= 1.0834\left[\frac{1}{7} + \frac{(38.71)}{13{,}885 - 7 * 338.71 * 38.71}\right]^{1/2} = 0.8311$$

$$S_{b_1} = \frac{S_e}{\left[\sum x^2 - n(\bar{x})^2\right]^{1/2}}$$

$$= \frac{1.0834}{\left[13{,}885 - 7 * 38.71 * 38.71\right]^{1/2}} = 0.0187$$

From Table A-4, Critical Values of the t Distribution, in Appendix A, $t_{0.95}$ with five degrees of freedom is 2.015. Therefore, the 90% confidence interval for b_0 is:

$$-0.0083 \pm (0.2015)(0.8311) = -0.0083 \pm 1.6747 = (-1.6830, 1.6663)$$

Since the confidence interval includes 0, the hypothesis that this parameter is 0 cannot be rejected at the 0.10 significance level. Thus, b_0 is essentially 0.

Similarly, the 90% confidence interval for b_1 is:

$$0.2438 \pm (2.015)(0.0187) = 0.2438 \pm 0.0376 = (0.2061, 0.2814).$$

Since the confidence interval does not include 0, the slope b_1 is significantly different from 0 at this confidence level.

What would be the number of sales given that the regional office in question during the month had 100 sales personnel available?

Then

$$\text{Number of sales} = -0.0083 + 0.2438(100) = 24.3674$$

$$\text{Standard deviation of errors } s_e = 1.0834$$

The standard deviation of the predicted mean of a large number of observations is:

$$S_{-1} = S_e\left[\frac{1}{n} + \frac{(x_i - \bar{x})^2}{\sum x^2 - n(\bar{x})^2}\right]^{1/2} = 1.0834\left[\frac{1}{7} + \frac{(100 - 38.71)^2}{13{,}885 - 7(38.71)^2}\right]^{1/2} = 1.2159$$

From Table A-4, $t_{0.95}$ with five degrees of freedom is 2.015.

90% confidence interval for the predicted mean $= 24.3674 \pm (2.015)(1.2159)$

$$= 24.3674 \pm 2.4500$$

$$= (21.9174, 26.8174)$$

Therefore, we can say with 90% confidence that the mean monthly sales for a regional office with 100 sales personnel will be between 21.9 and 26.9. This prediction assumes that we will take a large number of observations for these types of regional sales offices and then take a mean.

A scatter plot of the data was shown in Figure 27.1. The plot does appear to satisfy a linear relationship. To check independence of errors, a plot of residuals as a function of the predicted number of monthly sales (y's versus errors (y – predicted y's)) is shown in Figure 27.2.

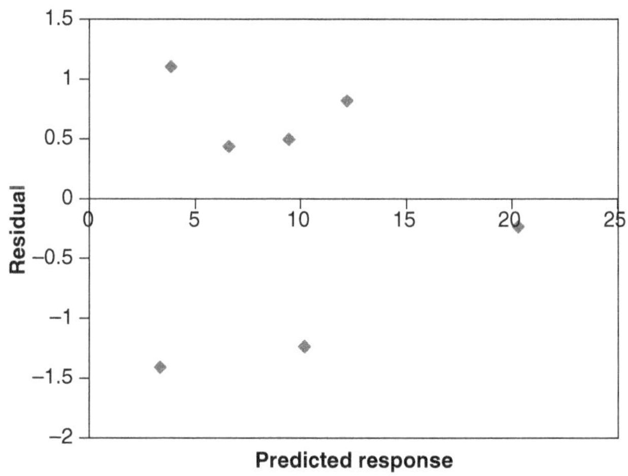

FIGURE 27.2 Graph of residual versus predicted response for the number of sales and number of sales personnel.

There does not seem to be any definite trend in the plot. A plot of errors as a function of observation numbers is shown in Figure 27.3. This graph also does not show any trends.

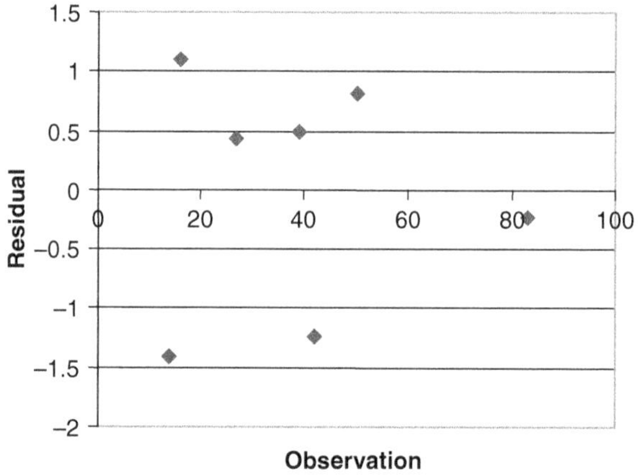

FIGURE 27.3 Residuals as a function of observation numbers for sales and sale personnel.

To check whether the normality assumption is valid, a normal quantile plot of the errors is shown in Figure 27.4. The graph is reasonably close to a straight line, leading us to believe that the normality assumption is approximately valid in this case.

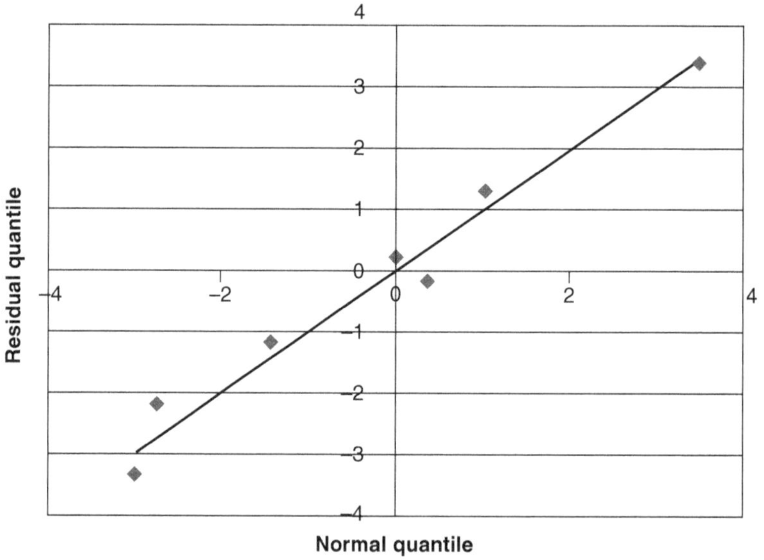

FIGURE 27.4 Normal quantile plot for the residuals of number of sales and number of sale personnel.

To check homoscedasticity, we notice that the errors do seem to have a larger spread toward the lower values of the predicted y's. It is difficult to make any judgments in this case due to a small number of observations. However, since the magnitude of errors is small relative to predictions, this is not a concern in this case.

27.4 Exercises

1. Browse the Internet and develop a spreadsheet containing the monthly returns for the S&P (Standard and Poor's) stock index and Dell. The beta of a stock is defined to be the slope of the least squares line used to predict monthly return for a stock from monthly return for the market.

   a. Estimate the beta of Dell.

   b. Interpret the meaning of Dell's beta.

   c. If you believe a recession is coming, would you rather invest in a high beta stock or a low beta stock?

   d. During a month in which the market goes up 5%, you are 95% sure that Dell's stock price will increase by between which range of values?

2. Browse the Internet and construct a spreadsheet containing the square footage and sales price for houses in a specific city.

   a. You are going to build a 500-square-foot addition to your home. By how much do you think your home value will increase?

   b. What percentage of the variation in home value is explained by variation in house size?

   c. A 3,000-square-foot house is selling for $500,000. Is this price out of line with typical real estate values in the city in your study? What might cause this discrepancy?

CHAPTER 28

INTRODUCTION TO MULTIPLE LINEAR REGRESSION

OVERVIEW AND LEARNING OBJECTIVES

In This Chapter

28.1 Introduction

Simple linear regression is restricted in several ways. Primarily, simple linear regression allows for only one predictor variable. Additionally, this predictor variable must be a quantitative variable. Finally, the response must be a linear function of the predictors. Using multiple linear regression, discussed in this chapter, we will describe techniques to overcome the restrictions of simple linear regression.

28.2 Sample Study

Let's continue with our study of *Winning Database Configurations: An IBM Informix Database Survey* by Marty Lurie of the IBM Software Group, available at *http://www.ibm.com/developerworks/db2/zones/informix/library/techarticle/lurie/0201lurie.html*, by answering the following questions:

- How is the data stored? What is actually stored the most? Why?
- Why would you expect OLTP databases to have significantly more tables than decision support databases?
- What was the big surprise in the physical design for Informix?
- What action was taken to avoid being I/O bound?
- What is meant by "fragment elimination"?
- Develop a multiple linear regression analysis for this study using the normalization, indexing, partitioning, and additional tuning parameters as the predictor variates and the number of CPUs as the predicted variable.

28.3 Multiple Linear Regression Model

In many situations, a dependent variable will be influenced by several independent variables rather than just one. For example, the mileage of a car will be influenced by its weight, engine displacement, horsepower, rolling resistance, axle ratio, and other factors.

When the relationships between the independent variables and the dependent variable are linear, we examine such a situation through *multiple linear regression analysis*, and we measure the collective relationship between the dependent variable and the independent variables with R^2, the square of the multiple correlation coefficient R. The multiple correlation coefficient is the Pearson correlation coefficient of the observed values of the dependent variable with those predicted by the regression model.

Assuming, as we did with the simple linear regression, that the error terms are normally distributed, the multiple linear regression model has the form:

$$Y_i = \sum_{j=1}^{p} \beta_j X_{ij} + \beta_0 + \varepsilon_i,$$

where X_{i1}, X_{i2}, … , X_ip are the values of the p independent variables (or *predictors*) for the ith case; Y_i is the observed value of the dependent variable; β_1, β_2, … , β_p are the parameters associated with the predictors; β_0 is the constant term; ε_i is a normally distributed error term with mean 0 and variance σ^2; and i is the number of the case. The estimate of Y_i based on the sample is:

$$E(Y_i) = \beta_0 + \sum_{j=1}^{p} \beta_j X_{ij}.$$

We wish to construct an estimate of the true model based upon sample data; that is, we wish to estimate the regression coefficients (sometimes called *weights*) β_1, β_2, … , β_p. The method used with simple linear regression to find a least squares solution can be extended to this larger situation, but such an approach creates enormously complex computations. A similar result can be achieved by using a matrix algebra notation that illustrates the concepts involved while avoiding tedious manipulations. It too provides a least squares solution.

We begin by defining the matrices:

$$Y = \begin{bmatrix} Y_1 \\ Y_2 \\ \cdot \\ \cdot \\ \cdot \\ Y_p \end{bmatrix}, \quad \beta = \begin{bmatrix} \beta_1 \\ \beta_2 \\ \cdot \\ \cdot \\ \cdot \\ \beta_p \end{bmatrix}, \quad \varepsilon = \begin{bmatrix} \varepsilon_1 \\ \varepsilon_2 \\ \cdot \\ \cdot \\ \cdot \\ \varepsilon_p \end{bmatrix}, \quad \text{and}$$

$$X = \begin{bmatrix} 1 & X_{1,1} & X_{1,2} & \cdot & \cdot & \cdot & X_{1,p} \\ 1 & X_{2,1} & X_{2,2} & & & & X_{x,p} \\ 1 & \cdot & \cdot & & & & \cdot \\ 1 & \cdot & \cdot & & & & \cdot \\ 1 & \cdot & \cdot & & & & \cdot \\ 1 & X_{n,1} & X_{n,2} & \cdot & \cdot & \cdot & X_{n,p} \end{bmatrix},$$

in which the various quantities are as defined previously. The general linear model can then be written as

$$Y = X\beta + \varepsilon,$$

where Y is the vector of dependent variable observations; X is the $n(p - 1)$ matrix of independent variable observations (n cases by p predictors and the constant term); β is the vector of beta coefficients; and ε is the vector of independent normally and identically distributed errors.

Then $E(Y) = X\beta$, since for each ε_i, $E(\varepsilon_i) = 0$. Now, let

$$b = \begin{bmatrix} b_0 \\ b_1 \\ \cdot \\ \cdot \\ \cdot \\ b_p \end{bmatrix} \text{ be the estimator vector for } \beta = \begin{bmatrix} \beta_0 \\ \beta_1 \\ \cdot \\ \cdot \\ \cdot \\ \beta_p \end{bmatrix}.$$

Then $Y' = E(Y) = S_b$. Using matrix algebra, we solve for b, multiplying on the left first by the transpose of X, then by $(X^{TR}X)^{-1}$. Recall that matrix multiplication is associative but not commutative.

$$Xb = Y' \Leftrightarrow (X^{TR}X)b = X^{TR}Y' \Leftrightarrow b = (X^{TR}X)^{-1}X^{TR}Y',$$

where X^{TR} is the transpose of X and $(X^{TR}X)^{-1}$ is the inverse of the $(p + 1)$ by $(p + 1)$ matrix $(X^{TR}X)^{-1}$.

As in simple linear regression, each element of the estimator vector b is an unbiased sufficient minimum variance that is a consistent estimator for the corresponding element of β.

Rather than solving the matrix equation developed above, some statistical software packages employ iterative techniques to find b. Such approaches are systematic, repetitive, easily programmed, and well suited to computer computation.

The Kelley-Salisbury technique is one of these methods. In it, successive approximations to β in the equation

$$R\beta_z = V$$

are found until a satisfactory solution is obtained. V is the vector of *validities*, or Pearson correlations between each of the predictor variables and the dependent variable, and R is the *intercorrelation matrix* of the predictors, the matrix of correlations of the independent variables with each other.

The vector β_z found by this method is not the same β discussed earlier; it is the vector of coefficients in the equation

$$z_Y = \sum_{j=1}^{p} \beta_{z_j} z_{X_{ij}}.$$

This equation is called the *normalized* solution of the multiple regression problem. We can use the facts that

$$z_{Y'} = \frac{Y' - \overline{Y'}}{s_{Y'}} \quad \text{and} \quad z_{X_i} = \frac{X_i - \overline{X_i}}{s_{X_i}}, \text{ for I} = 1, 2, \ldots, \text{p}$$

to find the vector b of weights in this form of the solution:

$$Y' = b_0 + \sum_{j=1}^{p} b_j X_{ij}.$$

Consider the following example, taken from a text by Walker and Lev and based on a sample of $n = 36$ cases.[1] Given the intercorrelation matrix for a dependent variable Y and the three predictors X_1, X_2, X_3, we seek the least squares multiple linear regression. The correlations of all the variables are given in Table 28.1.

TABLE 28.1 Intercorrelation.

	Y	X_1	X_2	X_3
Y	1.000	0.357	0.620	0.518
X_1	0.357	1.000	0.321	0.477
X_2	0.620	0.321	1.000	0.539
X_3	0.518	0.477	0.539	1.000

All figures and tables in this chapter appear on the companion DVD. ON DVD

The equation employed in the Kelley-Salisbury technique is this:

$R\beta_z = V$

$$\begin{bmatrix} 1.000 & 0.321 & 0.477 \\ 0.321 & 1.000 & 0.539 \\ 0.477 & 0.539 & 1.000 \end{bmatrix} x \begin{bmatrix} \beta_{z_1} \\ \beta_{z_2} \\ \beta_{z_3} \end{bmatrix} = \begin{bmatrix} 0.357 \\ 0.620 \\ 0.518 \end{bmatrix}$$

In general, the equation $\beta_z = R^{-1}V$ has a solution only if all the predictors are independent of one another. The iterative technique that is the heart of the Kelley-Salisbury method, however, will find a solution even when the predictors are statistically dependent.

We begin by choosing arbitrary values for the vector β_z of beta weights. We then perform one matrix multiplication, $R\beta_z = V_r$, where V_r is a vector of trial validities. The difference of largest magnitude between elements of V and elements of V_r is then added to the corresponding trial beta weight in the vector β_z, and then this process is repeated over again.

In our example, we set the weights equal to the validities. Our first multiplication is

$$R\beta_z = \begin{bmatrix} 1.000 & 0.321 & 0.477 \\ 0.321 & 1.000 & 0.539 \\ 0.477 & 0.539 & 1.000 \end{bmatrix} x \begin{bmatrix} 0.357 \\ 0.620 \\ 0.518 \end{bmatrix} = \begin{bmatrix} 0.803 \\ 1.014 \\ 1.022 \end{bmatrix} = V_r,$$

then

$$V - V_r = \begin{bmatrix} 0.357 \\ 0.620 \\ 0.518 \end{bmatrix} - \begin{bmatrix} 0.803 \\ 1.014 \\ 1.022 \end{bmatrix} = \begin{bmatrix} -0.446 \\ -0.394 \\ -0.504 \end{bmatrix}.$$

The difference of largest magnitude is $V_3 - V_{r_3} = -0.504$, so this amount is added to β_{z_3}. The vector of approximate weights is now

$$\beta_z = \begin{bmatrix} 0.357 \\ 0.620 \\ 0.014 \end{bmatrix}.$$

This process is repeated, adjusting one element of β_{z_3} at each iteration until V_r converges on V. In our example, this conversion occurs on the 18th iteration, as detailed in Table 28.2.

TABLE 28.2 The iterations of the Kelley-Salisbury method.

APPROXIMATIONS TO β_z			*VECTOR OF TRIAL VALIDITIES*		
β_1	β_2	β_3	V_1 0.357	V_2 0.620	V_3 0.518
0.357	0.620	0.518	0.803	1.014	1.022
0.357	0.620	0.014	0.563	0.742	0.518
0.151	0.620	0.014	0.357	0.676	0.420
0.151	0.620	0.112	0.403	0.729	0.518
0.151	0.511	0.112	0.368	0.620	0.459
0.151	0.511	0.171	0.397	0.652	0.518
0.111	0.511	0.171	0.357	0.639	0.499
0.111	0.492	0.171	0.350	0.620	0.489
0.111	0.492	0.200	0.364	0.635	0.518
0.111	0.477	0.200	0.360	0.620	0.510
0.111	0.477	0.208	0.363	0.625	0.518
0.105	0.477	0.208	0.357	0.623	0.515
0.105	0.474	0.208	0.356	0.620	0.514
0.105	0.474	0.212	0.358	0.622	0.518
0.105	0.474	0.212	0.358	0.620	0.516
0.105	0.472	0.214	0.359	0.621	0.518
0.103	0.422	0.214	0.3566	0.6204	0.5175
0.103	0.472	0.215	0.3571	0.6209	0.5185
0.103	0.471	0.215	0.3567	0.6199	0.5180

[a]The final values of the weights are $\beta_{z1} = 0.103$, $\beta_{z2} = 0.471$, and $\beta_{z3} = 0.215$.

It should be noted that the Artificial Neural Network (ANN) methodology uses a weighted sum of input variables to generate a set of output variables similar to the Kelley-Salisbury method. But what is the relationship between ANNs and multiple linear regression? The interested reader should study the following sidebar.

SIDE BAR

Artificial Neural Networks and Multiple Linear Regression Models

A neural network is essentially a way of simulating a human brain's activity. The brain is composed of billions of units referred to as neurons. Each neuron is connected to other neurons and communicates with them through electrochemical signals. These signals are captured by synapses, or end branches to the neuron's cell, called dendrites. Neurons sum up the inputs from the synapses, the inputs, and output a signal on a tail, to the neuron, referred to as an axon. Artificial neurons are simply an electronic simulation of biological neurons.

Each artificial neuron acts similar to the Kelley-Salisbury algorithm for a multiple linear regression equation. When input signals are triggered to the artificial neuron, then the sum of the weighted input values is submitted to the artificial neuron and an output signal is activated if it is greater than a threshold value.

In order to link many neurons together, one can start with an initial design of neurons which are organized into layers. These layers feed their output forward to the next layer until linking to the final output from the neural network. Input is sent to each neuron in the innermost layer, actually a hidden layer, and then each hidden layer's neuron output is sent to every neuron in the next layer. Such a neural network, referred to as a *feed forward network* can have any number of hidden layers.

Neural networks have to be trained. As with the Kelley-Salisbury algorithm for multiple linear regression, one way of training a neural network is with randomly assigned input weights. Then a series of inputs, representative of the final desired outputs, are fed into the artificial neural network. For each configuration of inputs one checks the output and adjust the weights accordingly until the network successively recognizes the desired result. The approach overview in this discussion is referred to as *supervised learning*. The data fed into the artificial neural network during training is the training set.

It should be noted that the Kelley-Salisbury algorithm for multiple linear regression employs only one output, whereas even the simplest neural network, a perception, can have one or more outputs. Each output has a separate bias and set of weights. Actually, a perception with a linear activation function is a linear regression model and a perception with a logistic function is a logistic regression model. In fact, to be more explicit, a perception with a threshold activation function is a linear discriminant function. With multiple outputs, the threshold perception is a *multiple discriminant function*.

Another major difference between artificial neural networks and multiple linear regression is the number of coefficients per input. Regression coefficients are determined by the method of least squared error, an optimization technique, whereas, neural nets employ the method of gradient descent, which allows for multiple coefficients per input.

SIDE BAR

In the method of gradient descent, the gradient vector of the error surface is calculated. Such vectors are directed along the steepest descent from the current point, therefore, movement in the current direction will decrease the error. A sequence of these movements will result in a local minimum decrease in error.

This discussion makes it apparent that researchers in information systems need to undertake a study of artificial neural networks. Multiple linear regression models are well equipped for linear modeling using optimization strategies. But neural networks are capable of modeling in nonlinear instances. The interested reader should view *Neural Networks: Tutorial Slides* by Andrew Moore found at http://www.autonlab.org/tutorials/neural.html and *Neural Networks and Statistical Models* by Warren Sarle, SAS Institute Inc. that was presented at the Proceedings of the Nineteenth Annual SAS Users Group International Conference in April, 1994.

In multiple linear regression, the coefficient of determination R^2 is an estimate of the proportion of the variability of the dependent variable that can be attributed to or associated with variation of the predictors, the independent variables. Thus, R^2 is a measure of the effectiveness of the multiple regression model.

R^2 can be found with a single matrix multiplication: $R^2 = \beta_z^{TR} V$. In our example,

$$R^2 = \begin{bmatrix} 0.103 \\ 0.471 \\ 0.215 \end{bmatrix}^{TR} x \begin{bmatrix} 0.357 \\ 0.620 \\ 0.518 \end{bmatrix} = \begin{bmatrix} 0.103 & 0.471 & 0.215 \end{bmatrix} x \begin{bmatrix} 0.357 \\ 0.620 \\ 0.518 \end{bmatrix} = 0.440.$$

That is, 44% of the observed variation of the dependent variable can be associated with the predictors X_1, X_2, and X_3.

28.4 The Relative Importance of Predictors

Having found the multiple linear regression equation, we would like to evaluate the relative importance of the independent variables; in other words, which predictors have a large influence on the dependent variable, and which are less important?

A straightforward and intuitive technique for such evaluations uses the multiple coefficient of determination R^2. To discover the importance of one or a group of predictors, the value of R^2 is calculated for the prediction model (called the *full model*), which includes all the predictor variables. Then a second model is constructed (the *reduced model*) from which the predictor or predictors of interest are *excluded*, and the value of R^2 for the reduced model is found.

The decrease in the value of R^2 from the full to the reduced model indicates the importance of the omitted predictors. If the drop in R^2 is large, the reduced model is not as effective as the full model in predicting the value of the dependent variable, and the omitted predictors must be important. If the drop in R^2 is small, the predictive ability of the model is not impaired, and the predictors being examined are unimportant.

A hypothesis test to determine if a drop in R^2 is significant can be performed with this F statistic:

$$F = \frac{\left(R_{FM}^2 - R_{RM}^2\right)/(df_{FM} - df_{RM})}{(1 - R_{FM}^2)/(n - df_{FM})}$$

where $R^2_{FM} = R^2$ for the full model, $R^2_{RM} = R^2$ for the reduced model, df_{FM} = the number of linearly independent predictors in the full model–1 (a group of predictors is linearly dependent if, when one is predicted by the others, the multiple coefficient of determination of the model is 1), df_{RM} = the number of linearly independent predictors in the reduced model–1, and n = the number of cases.

The number of degrees of freedom associated with the numerator of the F statistic is $(df_{FM} - df_{RM})$ and with the denominator is $(n - df_{RM})$.

If the coefficient of determination of the reduced model is significantly less than that of the full model, the quantity $R^2_{FM} - R^2_{RM}$ and, therefore, the F statistic itself, will be large; values of F near 0 occur when there is little difference between R^2_{FM} and R^2_{RM}. This is an *upper-tail* test of these hypotheses:

$$H_0: R_{FM}^2 = R_{RM}^2$$
$$H_a: R_{FM}^2 > R_{RM}^2.$$

Consider the reduced model created from the previous example by removing the predictor X_2. To compute R^2 for the reduced model, we perform another multiple linear regression. The intercorrelation matrix without X_2 is shown in Table 28.3:

TABLE 28.3 Setwise regression analysis.

	Y	X_1	X_3
Y	1.000	0.357	0.518
X_1	0.357	1.000	0.477
X_3	0.518	0.477	1.000

and the equation $R\beta_z = V$ becomes

$$\begin{bmatrix} 1.000 & 0.477 \\ 0.477 & 1.000 \end{bmatrix} x \begin{bmatrix} \beta_{z_1} \\ \beta_{z_3} \end{bmatrix} = \begin{bmatrix} 0.357 \\ 0.518 \end{bmatrix}.$$

Computation of the vector β_z is left as an exercise; the values are $\beta_{z_1} = 0.142$ and $\beta_{z_3} = 0.450$, and the coefficient of determination for the reduced model is

$$R^2_{RM} = \beta_z^{TR} V = [0.142 \quad 0.450] x \begin{bmatrix} 0.357 \\ 0.618 \end{bmatrix} = 0.284.$$

For the full model, $R^2 = 0.440$. To test the drop in R^2 from the full model to the reduced model at the 5% significance level, we compute the F statistic:

$$\begin{aligned} F &= \frac{\left(R^2_{FM} - R^2_{RM}\right)/(df_{FM} - df_{RM})}{\left(1 - R^2_{FM}\right)/(n - df_{FM})} = \frac{(0.440 - 0.284)/(3-2)}{(1-0.440)/(36-3)} \\ &= \frac{0.156/1}{0.560/33} = 9.47. \end{aligned}$$

Corresponding to (3–2) degrees of freedom in the numerator and (36–3) in the denominator, the critical value of the F statistic is $F_{0.05} \cong 4.17$. The F statistic exceeds the critical value, so the drop in the value of R^2 from the full model to the reduced model is significant at $\alpha = 5\%$; predictor X_2 is important in the regression model.

28.5 The Significance of R^2

We can also investigate the significance of the multiple coefficient of determination R^2 with an F test.

As with simple linear regression, the total sum of the squared deviations of the observed values of the dependent variable from the mean value of the dependent variable is:

$$SS_{total} = \sum_{i=1}^{n} \left(Y_i - \overline{Y}\right)^2,$$

and the regression sum of squares is:

$$SS_{Reg} = \sum_{i=1}^{n} \left[\left(b_0 + \sum_{j=1}^{p} b_j X_{ij}\right) - \overline{Y}\right]^2,$$

SS_{Reg} is a measure of the variability of the dependent variable that can be related to the predictors, and the residual sum of squares is:

$$SS_{Res} = \sum_{i=1}^{n} \left(Y_i - b_0 - \sum_{j=1}^{n} b_j \overline{X}_{ij}\right)^2$$

and is a measure of the remaining variability of the dependent variable. Associated with the SS_{Reg} is a number of degrees of freedom equal to the number of linearly independent predictors in the model minus 1, while the SS_{Res} has an associated number of degrees of freedom equal to the number of cases in the regression minus the number of linearly independent predictors.

Dividing each sum of squares by its respective number of degrees of freedom, we obtain an estimate of the variance of the dependent variable associated with the predictors (called the *mean square for regression*, MS_{Reg}) and an estimate of the variance of the dependent variable not associated with the predictors (the *mean square for residuals*, MS_{Res}). That is,

$$MS_{Reg} = SS_{Reg}/df_{Reg} \quad \text{and} \quad MS_{Res} = SS_{Res}/df_{Res}.$$

The upper-tail test of the hypotheses

$$H_0: \sigma^2_{Reg} = \sigma^2_{Res}$$
$$H_a: \sigma^2_{Reg} > \sigma^2_{Res}$$

is then carried out with the F statistic

$$F = \frac{MS_{Reg}}{MS_{Res}}.$$

This test is equivalent to testing H_0: $\rho^2 = 0$ against H_a: $\rho^2 > 0$, where ρ^2 is the population coefficient of determination.

When a multiple linear regression is performed using a statistical program software package, the printout will generally include the ANOVA (Analysis of Variance) table, similar to that shown in Table 28.4. Such a table assumes that all the predictors are linearly independent, and the significance of F corresponds to the test described above.

TABLE 28.4 ANOVA table.

ANOVA	DF	SUM OF SQUARES	MEAN SQUARE	F
Regression	$p - 1$	SS_{Res}	SS_{Res}/df_{Res}	MS_{Res}/MS_{Res}
Residual	$n - p$	SS_{Res}	SS_{Res}/df_{Res}	

28.6 Inferences about the Regression Coefficients

Such printouts often contain the variance-covariance matrix $s^2(b)$ of the variances and covariances of the sampling distributions of the b_i.

$$s^2(b) = \begin{bmatrix} s^2(b_0) & s(b_0, b_1) & . & . & . & s(b_0, b_p) \\ s(b_1, b_0) & s^2(b_1) & . & . & . & s(b_1 b_p) \\ . & & & & & \\ . & & & & & \\ . & & & & & \\ s(b_p, b_0) & s(b_p, b_0) & . & . & . & s^2(b_p) \end{bmatrix},$$

where $s^2(b_i)$ is the variance of the sampling distribution of b_i, and $s(b_i, b_j)$ is the covariance between b_i and b_j. The variance-covariance matrix can be found in this way:

$$s^2(b) = Y^{TR}Y - b^{TR}X^{TR}Y(X^{TR}X)^{-1}.$$

Since we also know that $E(b) = \beta$, we have here important information about the sampling distributions of the elements of b.

In particular, tests of the hypothesis H_0: $\beta_k = \beta_{k_0}$ can be performed using the statistic $t = (\beta_k - \beta_{k0})/s(b_k)$, which has associated with it $(n - p)$ degrees of freedom. Also $b_k \pm t_{a/2}s(b_k)$ are the endpoints of a confidence interval for β_k with confidence level $(1 - a) * 100\%$.

EXCEL

MULTIPLE LINEAR REGRESSION IN EXCEL

Suppose we are given the data shown in Table 28.5:

TABLE 28.5 Sports car data.

Car	Basic Price	Displacement cc	Horsepower	Curb Weight lb.	MPG
Maserti Merak	$31,000	2965	182	3185	14
Maserti Bore	$39,927	4931	315	3540	12
Maserti Khamsin	$43,587	4931	315	3800	12
Mazda GLC	$3,895	1415	65	1995	30
Mazda RX-7	$6,395	1146	100	2350	17
Mercedes-Benz 240D	$20,775	2746	142	3560	14
Mercedes-Benz 300TD	$25,000	2998	77	3780	23
Mercedes-Benz 280CE	$22,481	2998	77	3475	23
Mercedes-Benz 450SL	$34,760	4520	180	3795	12

(*Continued*)

TABLE 28.5 Continued.

Car	Basic Price	Displacement cc	Horsepower	Curb Weight lb.	MPG
MG Midget	$5,200	1493	50	1850	23
MGB	$6,550	1798	67	2415	16
Peugot 504	$7,922	1971	88	2905	22

We seek to predict MPG in terms of the predictors (displacement, horsepower, and curb weight). To run a multiple linear regression, we simply choose Regression from the Data Analysis list in the Tools menu. Table 28.6 shows the output.

TABLE 28.6 Regression analysis for sports car data.

Summary Output					
Regression Statistics					
Multiple R	0.77				
R^2	0.60				
Adjusted R^2	0.44				
Standard Error	4.38				
Observations	12				
ANOVA					
	df	**SS**	**MS**	**F**	**Sign F**
Regression	3	226.362	75.454	3.937	0.054
Residual	8	153.304	19.163		
Total	11	379.667			

From this output we see that 60% of the observed variation of the MPG can be associated with the predictors displacement, horsepower, and curb weight. The critical value for F at $\alpha = 5\%$ with 3 and 8 degrees of freedom is 8.84. Therefore, since our F value is 3.937, which is less than the critical value, we cannot conclude that the coefficient of correlation for this model is statistically significant, or that the true value of ρ^2 could actually be equal to 0. Also notice that the significance of F is 0.054, which is greater than, rather than less than, $\alpha = 5\%$.

The intercorrelation matrix can also be retrieved from the datafile by simply executing CORREL, under Data Analysis in the Tools menu, with the input range to include all the predictors and the dependent variables data values. For our datafile, this intercorrelation matrix is shown in Table 28.7:

TABLE 28.7 Sports car study inter-correlation matrix.

	Displ	**Horsepower**	**Curb Wt**	**MPG**
Displ	1			
Horsepower	0.86	1		
Curb Wt	0.84	0.60	1	
MPG	-0.65	-0.75	-0.55	1

From this matrix we see that all of the predictors appear to be highly correlated in the positive direction with each other, while all of the predictors are highly negatively correlated with the dependent variable MPG.

Using the Regression function one can create a wide variety of residual graphs. This topic will be covered in Chapter 30, "Multiple Linear Regression: Handling Violations of Restrictions."

Let's conclude this chapter, for the interested reader, with the question: How are artificial neural networks, ANNs, related to statistical methods?

28.7 Exercises

Interpretation Exercises

1. In a study of subscribers to two computing journals, a regression analysis was based on the data file to predict the value of the dependent variable TYPE (1 = major journal of the field, 0 = proceedings of national meeting) from the predictors ORGAN (professional organization affiliation), EDUC (degree held), INCOME (total household income), and INTEREST (research interests). The data file contained the values of these five variables for 225 individuals, and the partial chart in Figure 28.1 summarizes the results:

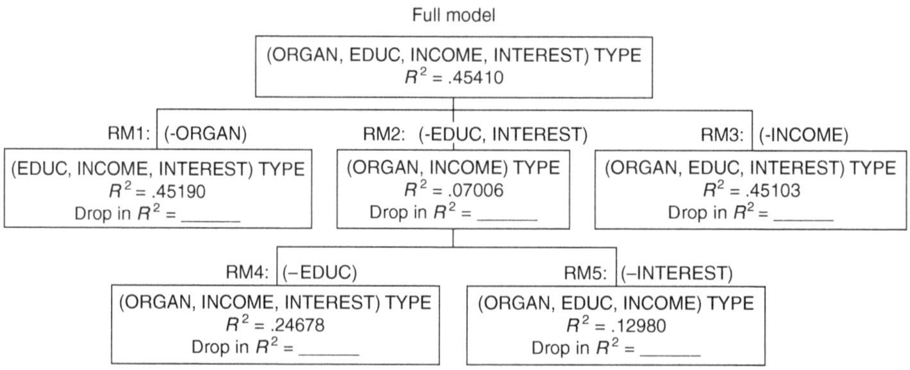

FIGURE 28.1 Regression analysis process.

The partial printout in Table 28.8 corresponds to the full regression model:

TABLE 28.8 Hypothetical data file.

MULT R	0.67387	ANOVA	DF	F
RSQUARE	0.45410	REGRESSION	2	23.09671
ADJ R SQ	0.45308	RESIDUAL	222	SIGN
ST ERR	0.79004			.002

a. Is the full model a "good" model from which to develop the regression equation? Why or why not?

b. Which predictors are significant in predicting TYPE? Which should be included in the final regression equation? Why?

2. Consider the data file in Table 28.9:

TABLE 28.9 Case based hypothetical inter-correlation matrix.

Case	X_1	X_2	Y
1	2	6	23
2	8	9	83
3	6	8	63
4	10	10	103

a. Find a least squares equation to predict Y from X_1 and X_2, and its associated value of R^2.

b. Show that the equations $Y = -7 + 9X_1 + 2X_2$ and $Y = -87 + X_1 + 18X_2$ are just as good as the one found in part a by computing their coefficients of determination.

c. Why are there several equally good regression equations? (Hint: Consider R_{X_1, X_2}.)

Computational Exercises

3. The intercorrelation matrix in Table 28.10 relates overhead cost in a factory (Y) to units produced (X_1), direct labor cost (X_2), weight of output (X_3), and research and development costs (X_4).

TABLE 28.10 Overhead costs study data.

	Y	X_1	X_2	X_3	X_4
Y	1.000	0.562	0.401	0.197	0.465
X_1	0.562	1.000	0.396	0.215	0.583
X_2	0.401	0.396	1.000	0.345	0.546
X_3	0.197	0.583	0.546	1.000	0.365
X_4	0.465	0.583	0.546	0.365	1.000

a. Write out the equation $R\beta_z = V$.

b. Find the normalized regression equation $z_Y = \sum_{i=1}^{4} \beta_i z_{X_i}$ by the iterative Kelley-Salisbury technique.

c. Find and interpret the value of R^2.

4. Consider the data in Table 28.11:

TABLE 28.11 American auto data(3).

Weight	Horsepower	MPG
3205	125	14
2945	80	20
4380	125	21
2040	74	29
3295	165	16
3520	195	16
2195	77	25
3205	195	16
1760	66	28
2515	130	21
3205	97	17
2480	93	28

We will predict mileage (Y) with the two predictors weight (X_1) and horsepower (X_2).

a. Find the mean and standard deviation of X_1, X_2, and Y.

b. Construct the intercorrelation matrix of X_1, X_2, and Y.

c. Write the vector equation $R\beta_z = V$.

d. Use the Kelley-Salisbury technique to find the coefficients β_{X1} and β_{X2} in the normalized regression equation $z_Y = \beta_{X_1} X_1 + \beta_{X_2} X_2$.

e. Using the normalized equation found in part d, predict the mileage of a 2,500-pound car that develops 105 horsepower. (Hint: In the equation, use

$$z_{X_1} = \frac{2500 - \overline{X}_1}{S_{X_1}} \quad \text{and} \quad z_{2_1} = \frac{105 - \overline{X}_2}{S_{X_2}},$$

then find Y so that $z_Y = \dfrac{(Y' - \overline{Y})}{S_Y}$.)

SIDE BAR

Neural Net Projects (for the interested reader)

1. Answer these questions:
   a. What is a neural network?
   b. How many kinds of NNs exist?
   c. How are NNs related to statistical methods?
   d. How are layers counted?
   e. What are cases and variables?
   f. What are the population, sample, training set, design set, validation set, and test set?
2. Create a neural network, in Excel, with one input layer (two nodes), one hidden layer (two nodes), and one output layer (one node). All units are sigmoids and all weights are initialized to small values (keep in mind that every unit will have a bias weight, whose input is hardwired to +1). Train the neural network using the back propagation algorithm on the XNOR function (this is the negation of XOR).

Does your network converge? If so, present the final weights neatly labeled on the diagram of the network.

Reference

1. Helen M. Walker and Joseph Lev, *Statistical Inference*, New York.

CHAPTER 29

MULTIPLE LINEAR REGRESSION: CASE STUDY

OVERVIEW AND LEARNING OBJECTIVES

In This Chapter

29.1 Introduction

Multiple linear regression allows the researcher to provide estimates of the effects of each variable in combination with other variables. We seek to develop a model for prediction, but the first step is a careful analysis to include only "good" predictor variables in the model. This chapter presents the analysis and prediction phases of the model building process.

29.2 Case Study: Prediction of Rental Car's Basic Price

The Easy Rider Car Rental Service desires to develop a predictive model for the basic price of a rental car in terms of a variety of the car's technical specifications. After extended discussions with the maintenance and service shop, the researchers develop a reasonable understanding of potential technical specifications that will be included in the case study. These discussions are followed by the review of existing literature on the topic. Finally, after all parties understand the topic and agree to pursue the development of such a model, the study is given the green light. The decision is that the dependent variable of basic price (Y) should potentially consider the following predictor variables: displacement in cc (X1), horsepower (X2), curb weight in lbs. (X3), and MPG (X4).

Solution:

Step 1: Descriptive statistical analysis of the potential variables.

ON DVD

The data collected for this study is listed in Table B-3, Auto File, in Appendix B and illustrated in Table 29.1 with the resultant output from the Descriptive Statistics function of the Data Analysis selection under the Tools menu in Excel.

TABLE 29.1 Descriptive statistics for the price study.

	Y	X1	X2	X3	X4
Mean	44880.856	2768.619	125.753	2773.907	20.222
Standard Error	1805.957	150.617	6.295	78.594	0.592
Median	37495	1998	110	2630	20
Mode	55000	1588	110	3205	22
Standard Deviation	17786.615	1483.406	62.002	774.057	5.827
Sample Variance	316363664.333	2200494.280	3844.251	599164.793	33.953
Kurtosis	11.238	0.680	1.480	2.044	-0.271
Skewness	3.015	1.308	1.358	0.661	0.327
Range	106801	5688	277	5070	25
Minimum	33199	1062	48	240	10
Maximum	140000	6750	325	5310	35
Sum	4353443	268556	12198	269069	1961.5

(*Continued*)

TABLE 29.1 Continued.

	Y	X1	X2	X3	X4
Count	12	12	12	12	12
Largest (1)	140000	6750	325	5310	35
Smallest (1)	33199	1062	48	240	10
Confidence Level (95.0%)	3584.796	298.973	12.496	156.007	1.174

All figures and tables in this chapter appear on the companion DVD. ON DVD

In this table:

- Y = Basic price
- X1 = Displacement in cc
- X2 = Horsepower
- X3 = Curb weight in lbs.
- X4 = MPG

We see that, on the average, the basic car price is $44,880.85 and that the basic price distribution is skewed. Note that the mean and median are unequal. This study included 12 observations. In fact, the standard deviation for the basic car prices is $17,786.62, indicating a wide dispersion of values. In a similar fashion, displacement also has a wide dispersion relative to a mean of 2768.62. Horsepower and curb weight, on the other hand, appear to possess a tendency toward a tight symmetrical distribution with a mean of approximately 125 and 2774, respectively. MPG also appears to have a tendency toward a symmetrical distribution with a mean of 22 and standard deviation of approximately 6. At the time the study was conducted, this sample appeared to adequately reflect the population.

Step 2: Strategy for conducting the analysis phase.

We would like to make our prediction model as economical as possible, identifying and including only the predictors considered important. This can be accomplished by first constructing the full model that relates all four predictors in estimating the basic price. Then a series of reduced models from which various predictors or groups of predictors are excluded will be implemented. If the drop in R^2 from the full model to a reduced model is significant, then the omitted predictors are making a contribution to the estimation of the basic price; if the drop in R^2 is not significant, then the omitted predictors are not important in predicting the value of Y.

Construction of reduced models can be implemented in Excel simply by executing the Regression function under the Data Analysis selection in the Tools menu. But how do you determine the groups of predictors to be considered? A reasonable approach is to first analyze the output for the cross-correlations, obtained with the Correlation function under Data Analysis, as illustrated in Table 29.2.

TABLE 29.2 Cross-correlations.

	Y	X1	X2	X3	X4
Y	1.00				
X1	0.41	1.00			
X2	0.53	0.59	1.00		
X3	0.54	0.72	0.66	1.00	
X4	−0.43	−0.56	−0.75	−0.68	1.00

Since X1, X2, and X3 have high positive correlations with respect to one another, it seems reasonable to consider them together when forming reduced models. Of this set of three, X1 and X3 possess the largest positive correlation with respect to one another; therefore, it also seems reasonable to consider them together in the formation of a distinct reduced model. Likewise, X4 is negatively correlated with all of the other predictors, suggesting that X4 only be considered for forming a reduced model.

Considering the intercorrelations and a commonsense look at which predictors might reasonably be linked, the hierarchical chart of reduced models is constructed, as shown in Figure 29.1. Note that each model is reduced from the full model, not from those above it.

Step 3: Perform the analysis phase.

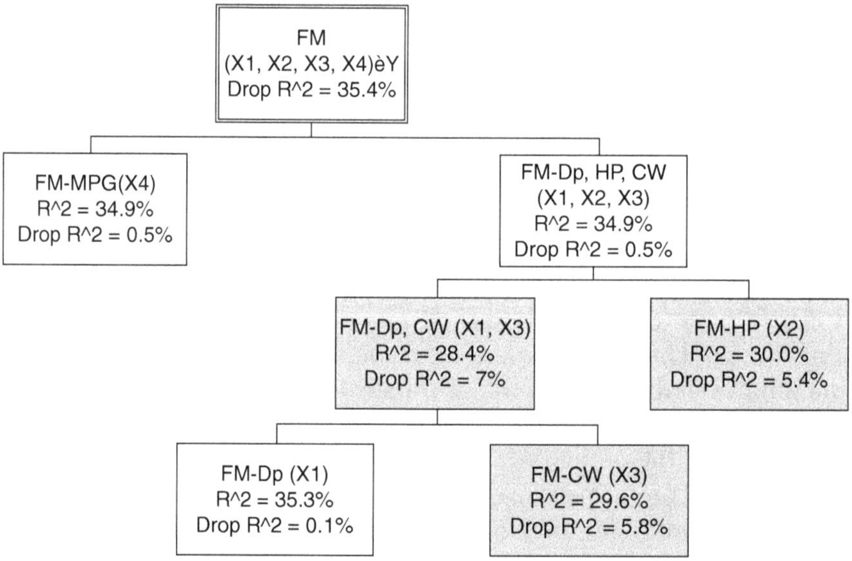

FIGURE 29.1 Hierarchical chart of reduced models.

Significant drops in R^2, indicating significant contributions to the dependent variable Y by the omitted predictors, are identified by the shaded reduced model boxes. The design of a hypothesis test for the drop in R^2 from the full model to a reduced model, especially at lower levels of the hierarchical chart, is beyond the scope of this text (for the interested reader, this test would require a sequential probability ratio test to be performed). An intuitive process will be employed instead. In this case, a drop of less than or equal to 5% (regardless of the level of the structure) will be considered significant. Only the three reduced models [FM–Dp,CW], [FM–HP], and [FM–CW] register significant drops in R^2 relative to the full model. The execution of the Excel Regression function for the [FM–Dp,SW] model, as depicted in Table 29.3, illustrates the regression run output generated for each reduced model.

TABLE 29.3 Output from an excel regression function execution.

Summary Output for FM–Dp,CW					
Regression Statistics					
Multiple R	0.533				
R^2	0.284				
Adjusted R^2	0.269				
Standard Error	15211.782				
Observations	97				
ANOVA					
	Df	**SS**	**MS**	**F**	**Significance F**
Regression	2	8619471743.676	4309735871.838	18.625	0.000
Residual	94	21751440032.303	231398298.216		
Total	96	30370911775.979			

The reduced models with significant drops in R^2 from the full model indicate that independent variables for displacement, curb weight, and horsepower should be included when predicating the basic price. Notice that MPG can safely be ignored when predicting basic price, since the drop in predictive efficiency when omitting MPG is only 0.5%.

Note that [FM–CW] has a significant drop in R^2 from the full model, while [FM–Dp] did not possess a significant drop, and therefore curb weight should be included in the future predictive model even though [FM–Dp,CW] possessed a significant drop. If neither [FM–Dp] nor [FM–W] had significant drops, but [FM–Dp,CW] did have a significant drop, then both curb weight and displacement would be included in the future predictive model. This latter case illustrates an interaction contribution to the prediction of the independent variable on the part of a group of predictor variables. Horsepower is to be included in the future predictive model based primarily on its own singular impact on the predictive efficiency.

At this point we have determined that only horsepower and curb weight are to be considered in predicting the basic price.

Step 4: Perform the predictive phase of the study.

We simply need to run the Regression function based upon the predictors (curb weight and horsepower) that our earlier work has shown are important. The results of the regression, as shown in Tables 29.4 and 29.5, provide the information for completing the predictive phase of the study.

TABLE 29.4 Predictive regression equation (CW,Dp)➔Price.

Regression Statistics					
Multiple R	0.590				
R^2	0.349				
Adjusted R^2	0.335				
Standard Error	14506.832				
Observations	97				
ANOVA					
	Df	SS	MS	F	Significance F
Regression	2	10588784066.701	5294392033.350	25.158	0.000
Residual	94	19782127709.279	210448167.120		
Total	96	30370911775.979			

TABLE 29.5 An ANOVA table for a regression analysis.

	Coefficients	Standard Error	t Stat	P-value	Lower 95%	Upper 95%	Lower 95.0%	Upper 95.0%
Intercept	12048.96	5542.96	2.17	0.03	1043.27	23054.64	1043.27	23054.64
X2	88.30	31.65	2.79	0.01	25.46	151.13	25.46	151.13
X3	7.83	2.53	3.09	0.00	2.80	12.87	2.80	12.87

The regression equation is

$$\text{Price} = 7.83(\text{curb weight}) + 88.30(\text{horsepower}) + 12{,}048.96$$

For example, if we look at the third observation, where curb weight is 2660 and horsepower is 111, then the predicted price is

$$\text{Price} = 7.83 * 2660 + 88.30 * 111 + 12048.96 = 42{,}677.10.$$

The actual basic price of the third observation is 40,495.00.

From the ANOVA table, we see that the value of the F statistic (0.000) is significant. We can conclude, then, that $\sigma^2_{\text{Reg}} > \sigma^2_{RES}$, or equivalently, that $\rho^2 > 0$. The sample estimate of ρ^2 is $R^2 = 0.349$, and the predictor equation is:

$$\text{Price} = 7.83(\text{curb weight}) + 88.30(\text{horsepower}) + 12{,}048.96.$$

with predictive efficiency 34.9%.

The coefficients can all be considered not equal to 0 at the 10% significance level, as illustrated by the P-values for the intercept and coefficients all being less than 0.05.

There are other methods used to determine which independent variables should be retained in a regression model. The most popular of these is *stepwise regression*, in which a sequence of regression equations is constructed. At each step, an independent variable is added to or deleted from the model.

Forward stepwise regression begins by including in the regression equation only the single independent variable that by itself produces the largest coefficient of determination R^2 as measured by an F test. In the second step, another predictor is added to the developing equation, one that with the predictor of the first step produces the largest value of R^2. This second step is then iterated, including one new predictor at each repetition until all are included in the final full model. The predictors entered first are the most important, and we can gauge the relative importance of predictors by watching R^2 increase at each step of the development.

The stepwise regression process also allows predictors to be removed from the equation. If subsequent inclusions have made earlier additions unimportant, this assumes that there is a single "best" set of predictors and we seek to identify them in the full model.

Stepwise regression can also be extended to *forward* and *backward set selection*, in which the researcher selects sets of predictors to be included and deleted at each step.

29.3 Forward, Stepwise, and Backward Selection

Most statistical packages use a variety of criteria for determining what subset of variables will be used as predictors in a regression model. Forward selection, backward elimination, and stepwise algorithms are commonly employed for the selection process. The selection criteria are usually based on the significance of the partial F statistic or on the increases of the multiple coefficient of determination when a variable is to be added to a candidate list of predictors. The potential predictor variables are added or deleted one at a time or they may be exchanged for one of the other predictor variables.

29.3.1 Forward selection

This is an automated, or software controlled, selection of predictor variables method. The method starts by completing simple regression analysis utilizing the single "best" predictor, or the predictor with the largest R^2 value. Next the predictor with the largest partial F value is chosen and a two-variable multiple linear regression analysis is performed. This process of adding predictors is continued until all variables have been added, or until the largest remaining partial F value is too small to consider as statistically significant. One of the problems with this approach is that it is possible to have several predictors that are nonsignificant at an early step in this process and then all become significant at a later step in the process. Note that once a predictor variable is added to the model, then that variable permanently remains in the model from that step forward.

29.3.2 Stepwise selection

In the stepwise selection algorithm any predictor variable having a partial F value not greater than or equal to a minimal significance level is dropped from the model and placed in the queue of candidates for forward selection and are, thus, eligible for re-inclusion in the model at a later stage in the process.

29.3.3 Backward elimination

A major problem with forward selection and stepwise selection is that a set of variates which jointly predict the criterion variable, yet individually have little to no prediction on the criterion variable, will cause trouble. Neither forward nor stepwise selection will discover the joint prediction for these predictors since both selection methods assume that the variable to be entered at any step is nonsignificant. As a result, any weaker variable cannot be significant.

Backward elimination starts with all predictor variables in the model, then variables are removed one at a time based upon providing the smallest partial F value. This procedure terminates when all partial F values are significant at a threshold value for staying in the model. Note that the backward elimination method will discover a set of variates which jointly predict the criterion variable yet individually have little or no predictive value for the criterion variable since their joint prediction value will be evident. Only the backward elimination algorithm allows for the discovery of such jointly strong predictor sets.

Good material on the selection methods are *Stepwise regression* from Wikipedia at *http://en.wikipedia.org/wiki/Stepwise_regression* and *Simplifying a Multiple Regression Equation* at *http://www.jerrydallai.com/LHSP/simplify.htm*.

29.3.4 Setwise selection

In this text, the recommendation is to determine the unique contribution of proper sets of the predictor variables to the prediction of the criterion. The contribution of a set of variables to prediction may be measured by the difference between squares of multiple correlation coefficients, one obtained for a regression model in which all predictors are used, called the full model (FM), and the other obtained for a regression equation in which a proper subset of variables under consideration have been deleted. This model is called the restricted model, (RM). The R^2 for the RM can never be larger than the R^2 for the FM. This difference between the two R^2s may be tested for statistical significance with the variance ratio test, or an F test.

The unique contribution may be interpreted as follows: if a variable is making a unique contribution, then knowledge of that variable furnishes information about the criterion. It is preferable to group predictors into logical sets, subsets, sub-subsets, and continue the process on individual variables. If two highly correlated predictor variables occur in a regression model, and one is deleted in order to determine its unique contribution, it will be found that there is little or no drop in the R^2 of the FM to the R^2 of the RM. This indicates that the deleted variable is not making a unique contribution and the presence of the other variable holds up the value of the value of the R^2. Finally, a tree structure of FM to RM tests can be devised to guide the testing or subjective analysis under the control of the researcher rather than under software system control.

29.4 Exercises

1. A statistician at Arcane College has been asked to analyze the hiring policies of the college's administrators. From 330 recent applications, a 10% random sample is selected, and for each sample applicant, the statistician constructs the profile in Figure 29.2:

APPLICANT PROFILE

Profile items

X1—Age
X2—Sex (1 = male, 2 = female)
X3—Degree held (1 = M.A., 2 = M.A. + 30 sem hr, 3 = Ph.D.)
X4—Cumulative GPA for graduate study
X5—Years of teaching and/or professional experience
X6—Number of publications
X7—Rank requested (1 = instructor, 2 = asst. prof., 3 = assoc. prof., 4 = prof.)

		Low		Average		High	
X8—Personal characteristics (enthusiastic, friendly, mature)	1	2	3	4	5	6	7
X9—Professional qualities and attitudes (knowledge, preparation)	1	2	3	4	5	6	7
X10—Interpersonal relationships (approachable, fair, honest)	1	2	3	4	5	6	7
X11—Technical skills of teaching (testing, grading, methods, organization)	1	2	3	4	5	6	7
X12—Communication skills (open, rapport, relevant)	1	2	3	4	5	6	7
X13—Recommendation of major professor	1	2	3	4	5	6	7
X14—Recommendation of previous employer	1	2	3	4	5	6	7
X15—Interview outcome	1	2	3	4	5	6	7

FIGURE 29.2 Applicant profile.

Then 10 judges (in this case, administrators responsible for hiring new personnel) are asked to sort the 33 applicants into seven categories as shown on the form in Figure 29.3:

JUDGMENTAL RATING SHEET

Rating sheet   ID number____________

1. Two highest profiles ____ ____
2. Four higher profiles ____ ____ ____ ____
3. Six high profiles ____ ____ ____ ____ ____ ____
4. Nine middle profiles ____ ____ ____ ____ ____ ____ ____ ____ ____
5. Six low profiles ____ ____ ____ ____ ____ ____
6. Four lower profiles ____ ____ ____ ____
7. Two lowest profiles ____ ____

Note: Within a group (e.g., the six high profiles) the order of preference is not noted.

FIGURE 29.3 Judgmental rating sheet.

Data from the two sheets—the applicants' profiles and ranks as assigned by each of the 10 judges—is coded and executed on a statistical software package to obtain the statistics given in Table 29.6. The variables J1 to J10 are the judges' ratings for each candidate.

TABLE 29.6 Applicant profile statistics.

VARIABLE	MEAN	STANDARD DEVIATION	STANDARD ERROR
J1	4.000	1.581	0.275
J2	4.000	1.581	0.275
J3	4.000	1.581	0.275
J4	4.000	1.581	0.275
J5	4.000	1.581	0.275
J6	4.000	1.581	0.275
J7	4.000	1.581	0.275
J8	4.000	1.581	0.275
J9	4.000	1.581	0.275
J10	4.000	1.581	0.275
X1	30.242	5.602	0.975
X2	1.576	0.502	0.087
X3	3.091	0.879	0.153
X4	3.582	0.248	0.043
X5	3.879	2.713	0.472
X6	2.000	2.179	0.379
X7	1.758	0.867	0.151
X8	5.636	1.055	0.184
X9	5.303	1.104	0.192
X10	5.455	1.063	0.185
X11	5.424	1.091	0.190
X12	5.273	0.944	0.164
X13	5.545	0.666	0.116
X14	5.364	0.653	0.114
X15	5.394	0.864	0.150

a. Interpret the data for profiles of the average applicant, and discuss your reaction to the judges' performance of their tasks

Restricting attention to the ratings given by judge 1, we construct a multiple regression equation relating the judge's rating as the dependent variable to the elements of the applicant's profiles as predictors. In addition, it would be desirable to make our prediction model as economical as possible, identifying and including the predictors that judge 1 considers important but excluding those that have no significant impact on the judge's ratings.

We construct a full model that relates the value of J1 to all 15 predictors, then a series of reduced models from which various predictors or groups of predictors are excluded. Again, if the drop in R^2 from the full model to a reduced model is significant, the omitted predictors are making a contribution to the value of J1; if the drop in R^2 is not significant, then the omitted predictors are not important in predicting the value of J1.

With the aid of a statistical software package, the intercorrelation matrix in Table 29.7 of all the variables involved in the regression was obtained:

TABLE 29.7 Applicant profile inter-correlational matrix.

CORRELATION COEFFICIENTS.

A VALUE OF 99.00000 IS PRINTED
IF A COEFFICIENT CANNOT BE COMPUTED.

	J1	X1	X2	X3	X4	X5
X1	-.21170					
X2	-.27566	.21558				
X3	-.33725	.52849	.09015			
X4	-.43798	.19368	.18949	.12093		
X5	-.27684	.61481	.02991	.22753	.26176	
X6	-.29926	.58874	.19998	.65245	.35710	.36468
X7	-.27352	.69446	.11533	.68578	.27485	.53177
X8	.22474	-.17494	.05364	.03675	-.23319	-.20144
X9	.17910	.15963	.07010	-.12593	.06238	.14835
X10	-.72489	.13307	.13840	.08814	.12262	.10636
X11	.39869	-.18106	-.46020	-.17187	-.15313	-.23557
X12	.02093	.14660	.25171	.15740	.23348	.07429
X13	-.08907	.18132	.24658	.01942	.21721	.10696
X14	-.21194	-.22997	.00867	-.05941	-.07905	-.25667
X15	.64063	-.26577	-.32326	-.41902	-.50254	-.16567

	X6	X7	X8	X9	X10	X11
X7	.82679					
X8	.01359	-.20180				
X9	-.20789	.01385	-.35860			
X10	.12136	.22492	-.23798	-.20094		
X11	-.19721	-.18525	-.10614	-.26595	-.33317	
X12	.28845	.27404	-.43040	.12811	-.12729	-.35855
X13	-.04308	.07382	-.28711	.06574	-.00803	-.19956
X14	-.06589	-.11543	.19795	-.07099	.38472	-.04789
X15	-.28218	-.28572	.19633	-.26027	-.20103	.28145

	X12	X13	X14
X13	.50153		
X14	-.31794	-.25495	
X15	-.21240	-.27665	-.09572

b. By analyzing the intercorrelation matrix for judge 1, what grouping of variables would you recommend for the reduced models? Why?

Figure 29.4 shows the hierarchical model of reduced models that was constructed for judge 1. Note that each model is reduced from the full model, not from those above it.

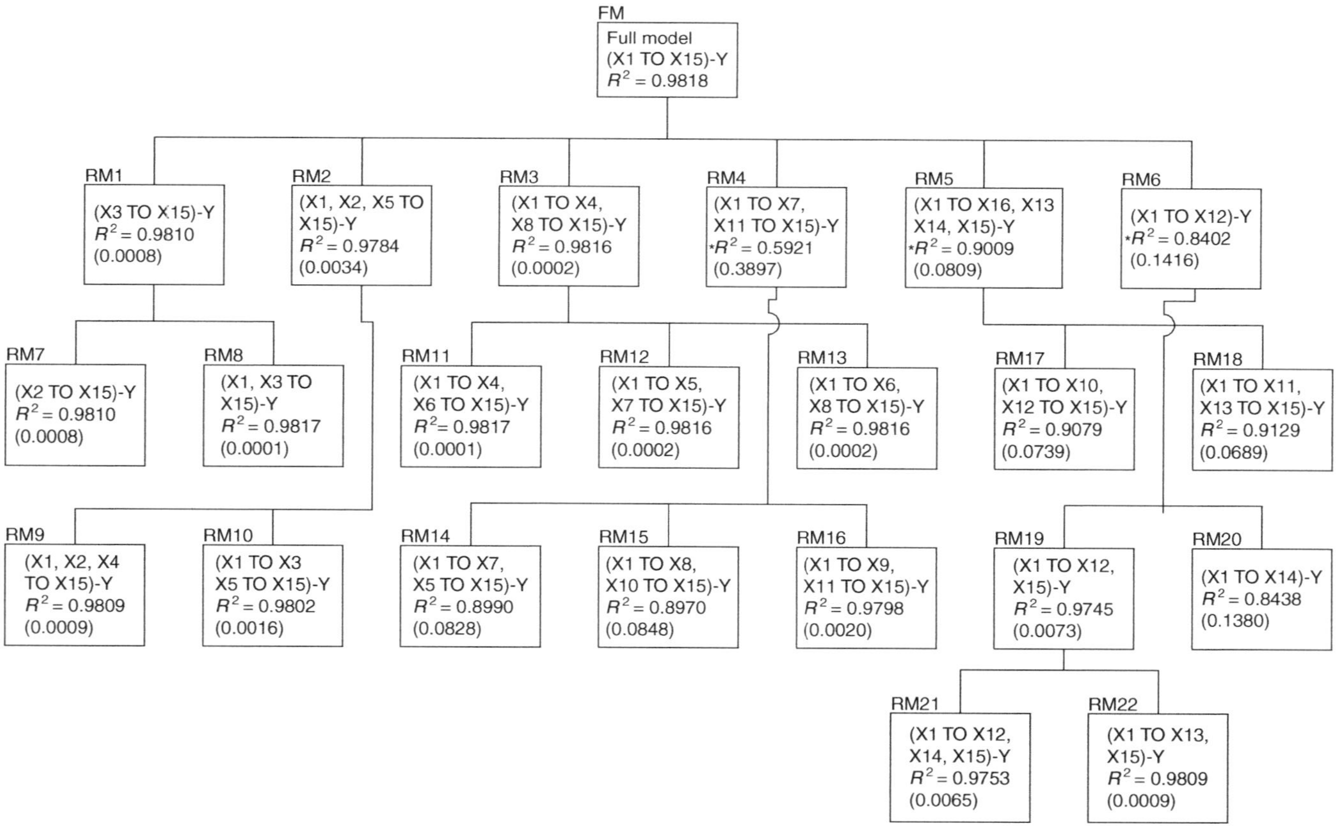

FIGURE 29.4 Judge one's judgmental policy analysis plan.

c. Complete the analysis phase of this regression study by interpreting the hierarchical chart of reduced models for J1.

d. Complete the prediction phase of this regression study by interpreting the computer output for J1 given in Table 29.8.

TABLE 29.8 Judge one's judgmental policy analysis.

DEP. VAR. . . J1 JUDGE 1

MEAN RESPONSE 4.00000 STD DEV 1.58114

FINAL STEP

			DF	SQUARES	MEAN SQ.	F
MULTIPLE R	.9888	ANOVA	DF	SQUARES	MEAN SQ.	F
R SQUARE	.9778	REGRESSION	8.	78.221	9.778	131.917
STD DEV	.2722	RESIDUAL	24.	1.779	.074	SIG. .000
ADJ R SQUARE	.9704	COEFF OF VARIABILITY		6.8PCT		

VARIABLE	B	S.E. B	F	SIG.	BETA	ELASTICITY
X8	1.175	.105	125.745	.000	.78405	1.65531
X11	1.133	.094	144.845	.000	.78157	1.53657
X14	−.060	.092	.428	.519	−.02477	−.08044
X15	1.179	.063	345.399	.000	.64438	1.59047
X13	.230	.087	6.988	.014	.09682	.31880
X9	1.097	.080	187.982	.000	.76597	1.45498
X12	1.060	.100	111.953	.000	.63319	1.39729
X10	.143	.099	2.089	.161	.09645	.19558
CONSTANT	−28.274	2.374	141.852	.000		

ALL VARIABLES ARE IN THE EQUATION.

COEFFICIENTS AND CONFIDENCE INTERVALS.

VARIABLE	B	95P	C.I.
X8	1.1747	.9585	1.3910
X11	1.1331	.9388	1.3274
X14	−.0600	−.2493	.1293
X15	1.1794	1.0485	1.3104
X13	.2300	.0504	.4095
X9	1.0975	.9323	1.2627
X12	1.0600	.8532	1.2668
X10	.1434	−.0614	.3483
CONSTANT	−28.2742	−33.1738	−23.3746

VARIANCE/COVARIANCE MATRIX OF THE UNNORMALIZED REGRESSION COEFFICIENTS.

	X8	X9	X10	X11	X12	X13
X8	.01097					
X9	.00658	.00641				
X10	.00851	.00608	.00985			
X11	.00797	.00567	.00752	.00886		
X12	.00790	.00499	.00693	.00708	.01004	
X13	.00069	.00057	.00022	.00043	−.00175	.00757
X14	−.00353	−.00210	−.00418	−.00247	−.00144	.00106
X15	.00055	.00117	.00110	.00024	.00081	.00121

	X14	X15
X14	.00841	
X15	.00049	.00403

2a. Have a group of at least 10 members of your class rate 20 famous paintings on the three scales in Figure 29.5:

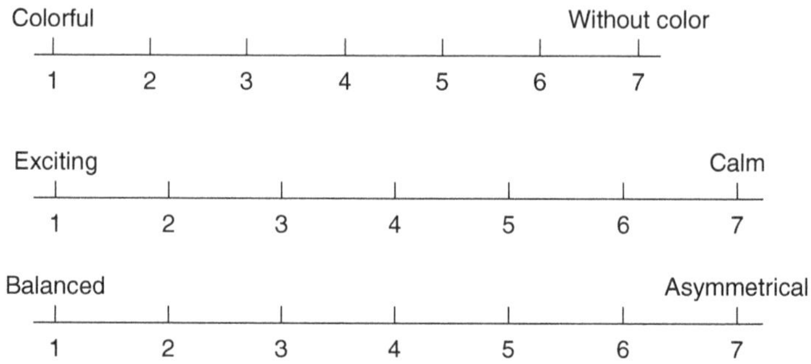

FIGURE 29.5 Rating scales for famous paintings.

b. Provide the remaining members of the class with the ratings (but not the paintings), and have them rate the paintings on the scale in Figure 29.6:

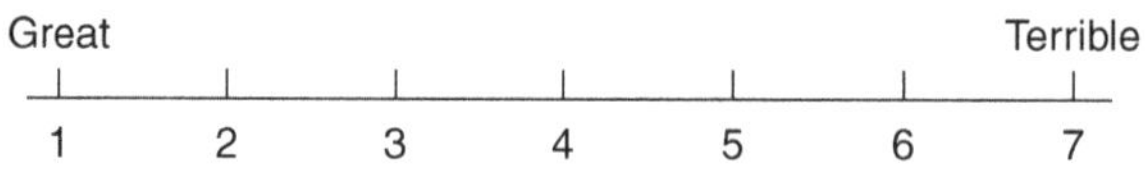

FIGURE 29.6 Overall rating scale system for famous paintings.

c. Develop a regression equation for each participant in part b that relates the final judgment to the three predictors of part a.

d. Can you develop a methodology for capturing the number of distinct judgmental policies in part b?

3. Consider how to evolve a statistical classification rule that would classify an individual into one of two populations on the basis of a set of measurements on the individual. A summary of the steps follows:

Step 1: The first requirement is to define the vector of observable variables that will be used in making the classification. Statistical criteria can be used to determine whether variables are useful for classification purposes. For example, the sample means can be obtained for the two populations, and the differences between the means relative to their standard deviations can be checked to select variables having large relative mean differences. In addition, one can also look at the correlation between the variables being considered and eliminate those that are too highly correlated with variables already selected. The cost of measuring the variable should also be compared with the additional classification power that the variable provides. Variable selection requires good judgment and should be tested by actually determining the operational effectiveness of the resulting classification rule.

Step 2: Once the set of variables is determined, their distribution should be checked to see how close the joint distribution of the variable is to being a multivariate normal distribution. The efficiency of our linear classification rule depends upon how good the normal assumption is. If a significant lack of normality is apparent, or if the covariance matrices are not close to being the same, we can use transformations on the variables to try to improve the situation.

Step 3: After the variables to be used have been determined, we need a classification statistic based on the variables to be used in the classification rule. We introduced the linear classification statistic

$$W = b_1 y_1 + b_2 y_2 + \ldots + b_p y_p$$

and justified its use from the fact that if the variables are normally distributed with equal covariance matrices and the coefficients are properly determined, the linear function given above minimizes the probability of making a misclassification. If the normality assumptions are not met, then some other function of the observable variables may be required.

Step 4: Given the linear classification function, one must determine the coefficients by solving the system of simultaneous equations as given in our illustration.

Step 5: The final step in developing the classification rule is to determine the classification constant. We introduced three criteria for determining this constant.

Step 6: Finally, one should make an evaluation of the operational effectiveness of the resulting classification rule. The statistician does this by considering the probabilities of making different types of errors and determining the expected loss if losses and prior probabilities are available. Although we restricted ourselves to the two-population case, the approach presented can be generalized by considering many populations. One approach to the multipopulation problem is to perform sequential classifications by first aggregating the populations into two groups. Each group is then considered as a super population, and the two-population technique can then be used to obtain a rule for classifying an individual into one of the super populations. Each super population is then broken down into two smaller aggregates, another classification rule evolved, and a classification made. This is repeated until the final classification rule places an individual into a single population. If one knows the distribution of the vector of variables for each population $f_k(y)$, the likelihood ratio classification rule can be used to classify an individual into one of the populations. This rule states that

If for all $f_k(y) > f_i(y)$ *for all j not* $= k$ classify the individual into population π_k.

To apply this rule, we need to assume an appropriate distribution model for each $f_k(y)$, estimate the parameters from the available samples, compute each $f_k(y)$ for the observed y, and classify the observation into the population associated with the maximum $f_k(y)$.

These multiple population approaches are, however, beyond the scope of this text, and thus we have presented only a two-population technique. Complete the classification methodology for a personnel manager who is considering whether to employ an applicant as a salesman given the data in Table 29.9.

TABLE 29.9 Applicant salesman data.

π_1 = Successful Salesman				π_2 = Unsuccessful Salesman			
I	Age Y_1	Expenses Y_2	Expected Income Y_3	I	Age Y_1	Expenses Y_2	Expected Income Y_3
1	42	600	800	1	41	300	350
2	48	300	500	2	40	500	300
3	42	350	500	3	50	500	700
4	42	500	600	4	41	100	500
5	39	250	450	5	37	300	600
6	42	500	600	6	33	500	500
7	56	1200	1150	7	42	650	650
8	49	500	600	8	52	750	700
9	43	600	600	9	45	500	500
10	41	700	700				
11	49	350	500				
12	51	300	550				

4. Develop a classification rule for the employment decision, using only the age variable Y, whose sampled values are given in Table 29.9. Find the classification constant *to make the two probabilities of misclassification* equal, and estimate the probabilities using the normality assumption. Check this probability by observing how many of the individuals in the two samples would have been incorrectly classified if the rule had been applied to them.

5. Work problem 4 using the two variables Y_1 and Y_2.

6. On the basis of the work done in problems 4 and 5, how much would you be willing to pay to obtain Y_3 for an individual being considered for employment?

7. Write a program to develop the three variable classification rules. Test the program using the data given in Table 29.9.

CHAPTER 30

MULTIPLE LINEAR REGRESSION: HANDLING VIOLATIONS OF RESTRICTIONS

OVERVIEW AND LEARNING OBJECTIVES

In This Chapter

30.1 Introduction

Sometimes, the restrictions (such as predictor variables being quantitative or the dependent variable not being a linear function of the predictors) for using multiple linear regression cannot be met. We will investigate how to handle categorical predictors, curvilinear regression, and transformations to apply when errors are not normally distributed or the variance is not homogeneous. The chapter concludes with a discussion of the problem of outliers and a list of common mistakes made in regression analyses.

30.2 Visual Tests for Verifying the Regression Assumptions

In deriving the expressions for regression parameters, we assumed:

- The predictor variables are linearly related to the response variable.
- Each predictor variable is not a random variable and they are measured without any error.
- The errors are statistically independent.
- The errors are normally distributed with 0 mean and a constant standard deviation.

If any of these assumptions are violated, the conclusions based on the regression are suspect. Often these assumptions are tested visually, as explained in the following sections.

30.2.1 Test for linearity

Preparing a scatter diagram of Y versus each predictor variable X can be helpful. Each graph should have a linear relationship. Any outlier points on a graph should have their value rechecked, as they might be due to some measurement error. If the graph is nonlinear, then a curvilinear regression should be used instead of a linear regression.

30.2.2 Test for independent errors

This test requires the construction of a scatter diagram of the errors, $\varepsilon_i = (Y_i - \hat{Y}_i)$, versus the predicted response $\hat{Y}_i$. Notice that the term *residual* is used instead of error. Any visible trends in this scatter diagram would indicate a dependence of errors on the predictor variable. Any visible trend is indicative of an inappropriate model. It would then be possible that a linear model is not appropriate for this case, and either a curvilinear model or a transformation be applied.

A plot of the residuals as a function of the observation number, where the numbers are given in the order in which the observations were made, can be implemented. Any trend up or down in such a graph indicates the presence of other factors, or side effects, that varied from one experiment to the next and affected the response. Causes for these trends should be identified.

Notice that there isn't a foolproof test for independence. We can only look at each potential type of dependence individually, as illustrated in the previous two cases. Dependence can be illustrated, but independence cannot be easily proven.

30.2.3 Test for normally distributed errors

For this test, a scatter diagram of the z scores for the response variable versus the z scores for the errors can be made. If the plot is approximately linear, then the assumption for normally distributed errors is satisfied.

30.2.4 Test for homoscedasticity

Homoscedasticity simply ensures that there is a constant standard deviation of errors. A scatter diagram of errors versus predicted response can be used for this test. If the spread in one part of the graph seems significantly different than that in other parts, then homoscedasticity is not present.

30.3 The Problem of Multicollinearity

Two lines are collinear if they have the same slope and same intercept. These two lines can be represented in just one dimension rather than the two dimensions required for lines that are not collinear. This means that two collinear lines are dependent, not independent. Similarly, when two predictor variables are linearly dependent, they are called collinear. The linear dependence between variables is measured by their correlation. Therefore, if the correlation between two predictor variables is nonzero, they are collinear. The problem of linear dependence among many predictor variables is called the problem of multicollinearity. When multicollinearity is present, the regression analysis may generate contradictory results.

Suppose that we are investigating the following model:

$$Y_i = \beta_0 + \beta_1 X_1 + \beta_2 X_2 + \varepsilon_i$$

where there is a perfect correlation between X_1 and X_2. If a change in X_1 occurs simultaneously with a change in X_2, then we cannot tell which of the independent variables actually is related to the change in Y. If we want to assess the separate effects of independent variables, it is essential that they not move exactly in unison through the experiment. The standard assumptions for multiple regression analysis exclude cases of perfect correlation between independent variables.

There are many indicators of multicollinearity. The correlation matrix for the regression analysis can be investigated. Also, the problem is present if a set of predictors as a group influence the response variable, but individually the same predictors demonstrate little influence on the response variable.

Another related problem occurs if redundant or irrelevant predictor variables are included in a model. If these unnecessary variables are correlated with the other predictor variables, then the variance of the coefficient estimates for the important variables will be increased. As a result, the overall efficiency of the coefficient estimates will be reduced.

The methodology for performing a multilinear regression analysis given in Chapter 29, "Multiple Linear Regression: Case Study," is not based upon investigation and interpretation of the coefficients; rather, the method simply looks at drops in the predictive efficiency of the reduced model from the full model. As a result, this methodology is not as concerned about multicollinearity.

30.4 Ridge Regression

When a researcher finds that the explanatory variables that he is working with are highly correlated with each other, he should use ridge regression if he insists on interpretation of the beta coefficients' values as indicators of relative predictive importance.

In theory, ridge regression enables the researcher to obtain such an estimate by minimizing mean square error risk. In practice, the large number of criteria that the researcher must consider makes it difficult to select a particular set of coefficients. Nonetheless, ridge regression provides a useful diagnostic to check for sign stability.

30.5 Categorical Predictors

Many times, when designing a regression model, the researcher must include variables which designate membership to a specific group. Consider a study where a categorical predictor defines two variates distinguishing programming language paradigm into functional and object oriented groups. In this instance we could define:

$$X_1 = \begin{cases} 1, & if\ a\ functional\ language\ paradigm \\ 0, & otherwise \end{cases}$$

$$X_2 = \begin{cases} 1, & if\ an\ object\ oriented\ language\ paradigm \\ 0, & otherwise \end{cases}$$

Then the model for predicting software development time, Y, would be $E(Y) = \beta_0 + \beta_1 X_1 + \beta_2 X_2$

$$= \beta_0 X_0 + \beta_1 X_1 + \beta_2 X_2, \quad \text{where} \quad X_0 = 1, \text{ for all cases}$$

Note that the normal equations encounter difficulty here since X^{Tr} X does not have an inverse, and the first column of X^{Tr} X equals the sum of the last two columns. Then no unique estimates off the regression coefficients can be found and the researcher would be restricted to executing a setwise regression analysis.

If the researcher needs to be able to interpret the regression coefficients the problem can be resolved by dropping one of the categorical predictor variables. But then the researcher might also desire to include the interaction. Then the function used would be:

$$E(Y) = \beta_0 + \beta_1 X_1 + \beta_2 X_2 + \beta_3 X_1 X_2.$$

This implies that when interaction is present, the effect of the categorical predictor variable should only be studied by comparing the regression functions for each class of the categorical predictor variable.

Consider the case where the predictor variable has more than two classes, then we require additional categorical predictor variables in the regression model. Again, dropping one categorical predictor variable leaves a set of linearly independent set of categorical predictor variables, thereby allowing for interpretation of the β_i's of the regression equation.

In Chapters 31 through 34, we will study analysis of variance models. Essentially these are models in which all independent variables are categorical predictor variables. Models in which there are some quantitative and some categorical predictor variables can represent covariance models. In these models the quantitative variables are included in the model to reduce the variance of the error terms.

Notice that the researcher could use one independent predictor variable with allocated cases:

Programming Language Class	X_1
functional	3
object oriented	2
procedural	1

In this design the model is $Y = \beta_0 + \beta_1 X_1 + \varepsilon$. Then we encounter the fact that the function allocated codes define a metric for the classes of the categorical predictor variable, which may or may not make sense.

Sometimes a researcher may wish to breakdown a quantitative variable with categorical predictor variables. Suppose we have a software development time as a quantitative independent variable and we express it in terms of the classes:

under 1 day
1 day to 1 week
a week to a month
etc.

At first sight this might seem to be a foolish approach because the researcher has actually thrown away information in the transformation. However, the researcher might be in doubt about the shape of the regression function, and hence may prefer the categorical predictor variable approach. This method allows the researcher to obtain information about the shape without making any assumptions about the functional form of the regression model.

Finally, the researcher might design a discriminate regression model. This is where

$$Y_i = \beta_0 + \beta_i X_i + \varepsilon_i,\ Y_i = 0,1.$$

This model is used to predict which class a case "belongs to" based upon the value of the quantitative variable X. This model cannot have normally distributed error terms, since Y can take on only two values, 0 or 1. Additionally, the error terms do not necessarily have equal variances. These are some of the difficulties encountered with the regression function when the dependent variable is a categorical predictor variate.

30.6 Curvilinear Regression

Both the simple linear and multiple linear regression models assume that a linear relationship exists. But many applications deal with nonlinear data, such as an exponential growth instance or a polynomial instance. Suppose that the researcher determines, by studying the scattergram between two variates that the exponential growth model should be applied, then

$$y = bx^a,$$

We can simply study the following relationship, which is linear:

$$(\ln y) = (\ln b) + a(\ln x),$$

and becomes

$$Y = c + aX,$$

where $Y = \ln y$, c = a constant = $\ln b$, and $X = \ln x$.

Other examples of curvilinear regression functions are listed with their conversions in Figure 30.1.

Nonlinear function	Linear conversion function
$Y = A + B/X$	$Y = A + B(1/X)$
$Y = 1/(A + B X)$	$(1/Y) = A + BX$
$Y = X/(A + BX)$	$(X/Y) = A + BX$
$Y = AB^X$	$(\ln Y) = (\ln A) + (\ln B) X$
$Y = A + BX^n$	$Y = A + B (X^n)$

FIGURE 30.1 Commonly used curvilinear function conversions.

ON DVD All figures and tables in this chapter appear on the companion DVD.

In curvilinear regressions, if a predictor variable appears in more than one transformed predictor variable, the transformed variables are likely to be correlated, causing the problem of multicollinearity.

30.7 Transformations

Often it is as easy or easier to transform the data as it is to modify the regression model as discussed in the section on curvilinear regression. Transformations can be used if the homoscedasticity assumption of the residuals is violated.

The transformations in Figure 30.2 have been found useful in practice.

Transform	Function	When Useful
Log	$h(y) = a \ln y$	if $s = a\bar{y}$; good if y_{max}/y_{min} is large
Square Root	$w = \sqrt{y}$	if $s = \sqrt{y}$; true for a Poisson process, helps stabilize the variance
Arc Sine	$\sin^{-1}\sqrt{y}$	y is a proportion or percentage
Omega	$w = 10\log_{10}\left(\frac{y}{1-y}\right)$	y is a proportion
Power	$w = y^a$	when testing homogeneity of variance: $s_e \propto \hat{y}^{1-a}$

FIGURE 30.2 Useful transformations.

30.8 Outliers

An *outlier* is any observation that is atypical of the remaining observations. Excluding an outlier for the analysis may lead to a misleading conclusion if the outlier in fact represents a correct observation of the system behavior. The easiest way to identify outliers is to look at the scatter plot of the data. Any value significantly away from the remaining observations should be investigated for possible experimental errors. Once the possibility of errors in the experiment has been eliminated, the researcher may decide to include or exclude the suspected outlier based on intuition. One alternative is to repeat the analysis with and without the outlier and state the results separately. Another alternative is to divide the operating region into two or more subregions and obtain a separate model for each subregion.

SIDE BAR

Commonly Made Mistakes in Regression

- Failing to verify that the relationship is linear
- Not paying attention to visual verification methods
- Not specifying confidence intervals for the regression parameters
- Failing to report the coefficient of determination

SIDE BAR

- Confusing the coefficient of determination with the coefficient of correlation
- Using highly correlated predictor variables, which leads to multicollinearity problems
- Application of a regression equation beyond the measured range
- A model having too many predictor variables
- Assuming a good predictor variable is a good control variable
- Measuring only a small subset of the complete range of operation

30.9 Exercises

Interpretation Exercises

1. Consider the data file in Table 30.1:

TABLE 30.1 Sample data file by case number.

Case	X_1	X_2	Y
1	2	6	23
2	8	9	83
3	6	8	63
4	10	10	103

a. Find a least squares equation to predict Y from X_1 and X_2, and its associated value of R^2.

b. Show that the equations $Y = -7 + 9X_1 + 2X_2$ and $Y = -87 + X_1 + 18X_2$ are just as good as the one found in part a by computing their coefficients of determination.

c. Why are there several equally good regression equations? (Hint: Consider $R_{X1,X2}$.)

Computational Exercises

2. Consider the following data file:

X	1	1	2	2	2	3	3	3	4	4
Y	1	1	3	4	4	8	9	10	15	16

a. Plot these data pairs on a scatter diagram with Y as the dependent variable.

b. What type of curve might fit these points?

c. Use simple linear regression to find the equation relating Y to X. (Hint: If $Y = AX^B$, then $\ln(Y) - \ln(A) + B\ln(x)$.)

Excel Exercises

3. A company needs to determine the relationship between cost per unit produced (Y) and the number of units produced (X). They collect the data shown in Table 30.2:

TABLE 30.2 Number of units by cost per unit.

Number of Units (X)	Cost per Unit (Y)
10	509.2
10	502.1
20	573.2
20	569.4
30	652.7
30	658.0
50	755.2
50	760.0
60	789.2
60	791.2
70	841.2
70	839.7
80	855.1
80	872.4

a. Plot the scatter diagram of this data with Y the dependent variable.

b. Use multiple regression to fit an equation of the form $Y = b_0 + b_1X + b_2X^2$ to the data.

c. What cost per unit would be expected from a production of 45 units?

4. For the data given in Table 30.3:

TABLE 30.3 Sample bivariate data file.

X	Y	X	Y	X	Y
-3.0	1.8	-0.6	0.0	2.0	5.0
-3.0	1.5	0.0	0.0	2.0	4.6
-2.6	1.2	0.4	0.2	2.0	5.4
-2.6	0.4	0.4	0.2	2.6	7.4
-1.8	0.4	0.4	1.0	2.6	7.0
-1.8	0.2	1.0	1.5	2.6	6.6
-1.0	0.0	1.0	0.8	3.0	8.0
-1.0	0.5	1.0	2.0	3.0	7.8
-1.0	0.6	1.6	4.6	3.0	7.5
-0.6	0.2	1.6	3.6	3.0	6.6

a. Plot a scatter diagram of this data with Y as the dependent variable.

b. State a type of curve that would fit the data.

c. Use multiple linear regression to identify the appropriate curve and state its equation.

d. What value of Y would be expected to correspond to $X = 2.3$?

SIDE BAR

Ridge Regression Project

1. Describe the symptoms that may indicate a collinearity problem in a regression study.
2. What is meant by the "variance inflation factor"? What does it measure?
3. What is the fundamental problem with collinearity?
4. Perform a literature review on ridge regression and fully discuss how it can resolve the collinearity problem.

Curvillinear Regression Project

1. For a random of students preparing for a given upcoming test, record the time spent in preparation for the test and the individual's score on the test.

   a. Fit a least-squares linear regression equation to these data points. Find R^2 to measure the predictive accuracy of this equation.
   b. Find the least squares quadratic regression equation for this data ($Y = A + BX + CX^2$). Find the value of R^2 for this regression equation.
   c. Determine which regression model best fits the data, a or b. Justify your answer.
   d. Consider the following data file:

X	1	1	2	2	2	3	3	3	4	4
Y	1	1	3	4	4	8	9	10	15	16

      I. Plot these data points on a scatter diagram with Y as the dependent variable.
      II. What type of curve might fit these points?
      III. Use simple linear regression to find the equation relating Y to X.
      IV. What value of Y would be expected to correspond to $X = 3.5$?

   e. Summarize the methodology for curvillinear regression illustrated by parts a–d in this problem.
   f. Perform a background review on Amdahl's law for I/O's in computer systems. Provide specific examples based upon regression models.
   g. Perform a background review on garbage collection times for a particular garbage collection algorithm resolved with regression models.

UNIT 7

EXPERIMENTAL DESIGNS

Let's study *Winning Database Configurations: An IBM Informix Database Survey*, by Marty Lurie, IBM Software Group at *www.ibm.com/developerworks/db2/zones/informix/library/techarticle/lurie/0201lurie.html* one more time. Lurie uses a regression model, at first, to look for any statistical relationship between 7.x and 9.x servers. Study the ANOVA analysis.

1. In this study how was ANOVA employed? What were the results?
2. Design a two-way ANOVA design and data table for this study. What would you choose for the dependent variable? Why? Describe your test for interaction.
3. Is there an appropriate ANOCOVA design to consider?
4. What type of sampling method was employed in this study? Does the answer to this question have an impact on what type of experimental design should be utilized?
5. For you designs in problems 2, 3, and 4: develop linear regression models which will generate equivalent analyses as the ANOVA designs.
6. Consider experimental designs which would control for indexing, partitioning, the NOAGE parameter, forced residency, and/or the fill factor parameter.

CHAPTER 31

ONE-WAY ANALYSIS OF VARIANCE

OVERVIEW AND LEARNING OBJECTIVES

In This Chapter

31.1 Introduction

Previously we developed the t-test to decide, using independent samples, whether the means of two populations are equal. To compare the means of three or more populations, we employ a technique that compares the variability of the values of the dependent variable in several samples.

In linear regression, the total variability of the values of the dependent variable is shown to be the sum of the variation associated with the independent variables and the unexplained variation not associated with the predictors. This idea, examining the sources of the variation of the dependent variable, is at the heart of a technique that can be thought of as comparing the means of several populations or as regression when the predictors are nominal rather than at least ratio. Variation in the values of the dependent variable is broken into components attributed to the independent variables and other, unexplained variation. Ratios of these variations, or sums of squared deviations, are used to make inferences about the population means. For this reason, the technique is called *analysis of variance*, or *ANOVA*.

31.2 Sample Study: (*Winning Database Configurations, Continued*)

Let's pick out the details of the ANOVA analysis in *Winning Database Configurations: An IBM Informix Database Survey* by Marty Lurie of the IBM Software Group, available at *http://www.ibm.com/developerworks/db2/zones/informix/library/techarticle/lurie/0201lurie.htm*, one more time. Lurie uses a regression model, at first, to look for any statistical relationship between 7.x and 9.x servers.

- Why is ANOVA a better statistical tool for completing the server comparisons?
- What were the results for a one-way ANOVA?
- Could a better experimental design be used for analyzing the server comparisons? Defend your answer.
- Which servers are driving more data? Why is this important?
- What are models and how should a DBA utilize them?
- What is the relationship between regression and the analysis of variance?
- How do variation and variance relate?

Reader's should be able to answer these types of questions after mastering the material discussed in the chapter.

31.3 Introduction to ANOVA

We begin with the case, corresponding to simple linear regression, in which the value of the dependent variable is influenced by a single *qualitative* predictor. This is called *one-way* analysis of variance. The predictor is called a *factor* and its values, called *levels* or *treatments*, divide the observed cases into several groups. The corresponding groups of values of the dependent variable are then samples from several populations. The levels of the factor are *fixed* if all possible levels are included in the collected data or if inferences are to be drawn about only the observed values. If the observed levels of the factor are a sample from the population of all possible levels and inferences are to be drawn over all levels, whether represented in the data or not, the levels are *random*. Consider, for example, drawing conclusions about all possible fertilizing patterns for wheat based on a study involving four such patterns.

In our first example, the levels of the factor are fixed, since conclusions are to be drawn only over the levels mentioned in the data.

EXAMPLE 31.1

The Acme Publishing Company has released a paperback book in three differently colored covers. Each of 20 bookstores in similar markets receives copies in only one color. The sales totals, as shown in Table 31.1, are for the first month after the book's publication.

TABLE 31.1 Sales totals for one month for a paperback book.

	Cover Color		
	Red	**Blue**	**Green**
Number of Copies Sold	145	255	226
	210	190	240
	158	276	185
	205	222	175
	182	248	206
		206	221
		269	231
		230	

All tables in this chapter appear on the companion DVD.

ON DVD

The publishers wish to know whether the mean sales for the book are the same for each cover. *Sales* is the dependent variable, and *cover color* the factor, with *treatments* (*levels*) being red, blue, and green. The three groups of sales values are independent samples from three populations, and we must decide if the means of these populations are not equal. These hypotheses are being tested:

$$H_0: \mu_{\text{red}} = \mu_{\text{blue}} = \mu_{\text{green}}$$

H_a: At least one mean differs from the other two

Solution:

The test statistic will be based on a comparison of sums of squared deviations and will have an F distribution when the null hypothesis is true. We now consider this analysis in general, before completing the example later in this chapter.

The one-way ANOVA problem is a test of these hypotheses:

$H_0: \mu_1 = \mu_2 = \ldots = \mu_k$

H_a: At least one population mean differs from the others

The data for the test is organized as shown in Table 31.2.

TABLE 31.2 Data for one-way ANOVA.

	Treatment 1	**Treatment 2**	**Treatment *k***
Values of the Dependent Variable	x_{11}	x_{12}	x_{1k}
	x_{21}	x_{22}	x_{2k}
	.	.	.
	.	.	.
	.	.	.
	x_{n1}	x_{n2}	x_{nk}

The observations of the values of the dependent variable X_{ij} form samples of sizes n_j from k populations, as determined by the treatments. The populations are assumed to have a common variance σ^2. Let $\bar{X}_{\cdot j}$ be the mean of the sample of treatment j, and let $\bar{X}_{\cdot\cdot}$ be the grand mean of all the values of X_{ij}. Then if $N = \sum_{j=1}^{k} n_j$ is the total number of observations,

$$\bar{X}_{\cdot j} = \frac{1}{n_j}\sum_{j=1}^{n_j} X_{ij} \quad \text{and} \quad \bar{X}_{\cdot\cdot} = \frac{1}{N}\sum_{j=1}^{k}\sum_{i=1}^{n_j} X_{ij}.$$

The total variation of the values of the dependent variable is measured, as in linear regression, by the *total sum of squares*, SS_{tot}:

$$SS_{tot} = \sum_{j=1}^{k}\sum_{i=1}^{n_j}\left(X_{ij} - \bar{X}_{\cdot\cdot}\right)^2$$

Each value of the dependent variable contributes one term to this sum.

It can be shown with some tedious algebra that

$$SS_{tot} = \sum_{j=1}^{k}\sum_{i=1}^{n_j}\left(X_{ij} - \bar{X}_{\cdot\cdot}\right)^2$$

$$= \sum_{j=1}^{k} \sum_{i=1}^{n_j} \left[\left(X_{ij} - \bar{X}_{\cdot j} \right) + \left(\bar{X}_{\cdot j} - \bar{X}_{\cdot\cdot} \right) \right]^2$$

$$= \sum_{j=1}^{k} \sum_{i=1}^{n_j} \left(X_{ij} - \bar{X}_{\cdot j} \right)^2 + \sum_{j=1}^{k} n_j \left(\bar{X}_{\cdot j} - \bar{X}_{\cdot\cdot} \right)^2 + 2 \sum_{j=1}^{k} \sum_{i=1}^{k_j} \left(X_{ij} - \bar{X}_{\cdot j} \right) \left(\bar{X}_{\cdot j} - \bar{X}_{\cdot\cdot} \right)$$

$$= \sum_{j=1}^{k} \sum_{i=1}^{n_j} \left(X_{ij} - \bar{X}_{\cdot j} \right)^2 + \sum_{j=1}^{k} n_j \left(\bar{X}_{\cdot j} - \bar{X}_{\cdot\cdot} \right)^2 + 2 \left[\sum_{j=1}^{k} \left(\bar{X}_{\cdot j} - \bar{X}_{\cdot\cdot} \right) \left[\sum_{i=1}^{nj} \left(X_{ij} - \bar{X}_{\cdot j} \right) \right] \right].$$

But $\bar{X}_{\cdot j}$ is the mean of the values of the *j*th treatment group, so

$$\sum_{i=1}^{n_j} \left(X_{ij} - \bar{X}_{\cdot j} \right) = 0$$

for all *j*. Therefore,

$$SS_{tot} = \sum_{j=1}^{k} \sum_{i=1}^{n_j} \left(X_{ij} - \bar{X}_{\cdot j} \right)^2 + \sum_{j=1}^{k} n_j \left(\bar{X}_{\cdot j} - \bar{X}_{\cdot\cdot} \right)^2.$$

The double summation is the sum of the squared deviations of each value from its treatment group mean, and is a measure of the variation observed *within* the groups. It is called the *sum of squares within*, SS_{within}. The second sum is the total of the squared deviations of the group means from the grand mean, a measure of the variation *between* groups. It is called the *sum of squares between*, $SS_{between}$, so that

$$SS_{tot} = SS_{within} + SS_{between}.$$

We have partitioned the total variability observed in all the values of the dependent variable into two quantities, a measure of the variability within each group and a measure of the variability between the groups. From these we derive a test statistic.

If the null hypothesis of the test is true and the population means are equal across the treatments, then $\bar{X}_{\cdot\cdot}$ and all the $\bar{X}_{\cdot j}$ are unbiased estimates of the common population mean μ. Further, estimators of the common population variance σ^2 can be developed from the sums of squared deviations from these sample means.

For example, let

$$s_j^2 = \frac{1}{n_j - 1} \sum_{i=1}^{n_j} \left(X_{ij} - \bar{X}_{\cdot j} \right)^2$$

be the variance of treatment group j, so that $E(S_j^2) = \sigma^2$. Then

$$\begin{aligned}
E(SS_{within}) &= E\left(\sum_{j=1}^{k}\sum_{i=1}^{n_j}\left(X_{ij} - \bar{X}_{\cdot j}\right)^2\right) \\
&= E\left(\sum_{j=1}^{k}\left(n_j - 1\right)s_j^2\right) \\
&= \sum_{j=1}^{k}\left(n_j - 1\right)E(s_j^2) \\
&= \sum_{j=1}^{k}\left(n_j - 1\right)\sigma^2 \\
&= \sigma^2\left[\sum_{j=1}^{k} n_j - \sum_{j=1}^{k} 1\right] \\
&= \sigma^2 (N - k).
\end{aligned}$$

Therefore, $E(SS_{within}/(N - k)) = (N - k)\,\sigma^2/(N - k) = \sigma^2$. The quantity $SS_{within}/(N - k)$ is called the *mean square error within groups*, MS_{within}, and is an unbiased estimator of the common population variance σ^2.

Similarly, it can be shown that

$$E(SS_{between}) = (k-1)\sigma^2 + \sum_{j=1}^{k} n_j\left(\mu_j - \mu\right)^2,$$

where μ is the mean of the population corresponding to treatment j and $\mu_j = (1/N)\sum_{j=1}^{k} n_j\mu_j$. We define the mean square error between groups to be

$$MS_{between} = SS_{between}/(k - 1),$$

so that, in general,

$$E(MS_{between}) = \sigma^2 + \frac{1}{k-1}\sum_{j=1}^{k} n_j(\mu_j - \mu)^2.$$

When the null hypothesis is true, $\mu_j = \mu$ for all j, and the second term of this sum is 0, so that $E(MS_{between}) = \sigma^2$. In this case, the observed values of MS_{within} and $MS_{between}$ should be approximately equal, and their *ratio* should be near 1. If H_0 is false, some of the population means are unequal and $\sum_{j=1}^{k} n_j\left(\mu_j - \mu\right)^2 > 0$, so that $E(SS_{between}) > \sigma^2$. We would expect $MS_{between}$ to be larger than MS_{within}.

The test statistic, then, is $F = MS_{between}/MS_{within}$, whose sampling distribution, if H_0 is true, is an F distribution with $(k - 1)$ degrees of freedom in the numerator and $(N - k)$ degrees of freedom in the denominator. The critical value for this upper-tail test can be found in Table A–4, Critical values of the F distribution, in Appendix A. If F exceeds the critical value, we reject the null hypothesis and accept the alternative, concluding that some of the population means differ and that the treatments do affect the dependent variable.

ON DVD

The information used in a one-way analysis of variance is often summarized in an ANOVA table, as shown in Table 31.3.

TABLE 31.3 One-way ANOVA table.

Source	Degrees of Freedom	Sum of Squares	Mean Sum of Squares	F
Between Groups	k - 1	$SS_{between}$	$MS_{between}$	$MS_{between}/MS_{within}$
Within Groups	N - k	SS_{within}	MS_{within}	
Total	N - 1	SS_{total}		

The following assumptions are necessary for a one-way analysis of variance:

- The variances of the populations from which the treatment groups are drawn must be equal.
- The effect of the factor on the dependent variable is additive:

$$X_{ij} = \mu + \alpha_j + \varepsilon_{ij},$$

where α_j is the effect of the jth treatment, and ε_{ij} is the random variation of observation X_{ij}.

It is also generally assumed that the ε_{ij} are $N(0, \sigma^2)$, and that the ε_i sum to 0, since this simplifies the model.

To complete Example 31.1, we perform the calculations shown in Table 31.4.

CONTINUATION OF EXAMPLE 31.1

Solution:

TABLE 31.4 Completion of calculations of Table 31.1.

	Cover Color		
	Red	**Blue**	**Green**
Number of Copies Sold	145	255	226
	210	190	240

(Continued)

CONTINUATION OF EXAMPLE 31.1

TABLE 31.4 Continued.

	Cover Color		
	Red	**Blue**	**Green**
Number of Copies Sold	158	276	185
	205	222	175
	182	248	206
		206	221
		269	231
		230	

Calculations:

$$\bar{X}_{.1} = 180 \quad \bar{X}_{.2} = 237 \quad \bar{X}_{.3} = 212$$

$$\bar{X}.. = \frac{900+1806+1484}{20} = 214$$

$$SS_{\text{between}} = \sum_{j=1}^{k} n_j \left(\bar{X}._j - \bar{X}..\right)^2 = 5(180-214)^2+8(237-214)^2+7(212-214)^2 = 10040$$

$$MS_{\text{between}} = 10040/(3-1) = 5020$$

$$SS_{\text{within}} = \sum_{j=1}^{k}\sum_{i=1}^{n_j}\left(X_{ij} - \bar{X}._j\right)^2 = (145-180)^2 + (210-186)^2 + (231-212)^2 = 13228$$

$$MS_{\text{within}} = 13228/(20-3) = 778.12$$

$$F = MS_{\text{between}}/MS_{\text{within}} = 5020/778.12 = 6.45.$$

The ANOVA table in Table 31.5 summarizes the calculations of Table 31.4.

TABLE 31.5 ANOVA summary for Example 31.1.

Source	**Degrees of Freedom**	**Sum of Squares**	**Mean Sum of Squares**	**F**
Between Groups	3 – 1 = 2	10040	5020	6.45
Within Groups	20 – 3 = 17	13228	778.12	
Total	20 – 1 = 19	23268		

At the 5% significance level, the critical value of the F statistic, corresponding to 2 and 17 degrees of freedom, is 3.59. The observed value of the statistic exceeds the critical value, so we reject H_0 and conclude that at least one of the population means differs from the others. The color of the book cover does affect sales.

EXCEL

EXCEL ONE-WAY ANOVA FUNCTION

The ANOVA: Single Factor function is a one-way ANOVA function in Excel. This function requires that only numerically coded data values are in the data file. Therefore, a simple application of the IF() function can resolve the recoding requirement.

ON DVD

In the American Cities database in Appendix B, the variable X1 (region) is nominal, and divides the cases of the file into three groups. Table 31.6 shows the file for the X1 and X2 variables after application of the IF() function.

TABLE 31.6 Data for X1 by X2.

X1	X2		X1	X2
E	10.700	X1	1	10.700
E	8.800		1	8.800
E	10.900		1	10.900
E	12.700		1	12.700
E	7.100		1	7.100
E	0.000		1	0.000
E	3.300		1	3.300
E	11.900		1	11.900
E	13.300		1	13.300
E	12.800		1	12.800

We utilized the functions shown in Table 31.7.

TABLE 31.7 IF(_,_,_) command for classifying.

X1	X2
=IF(J2=“E”,1,IF(J2=“C”,2,IF(J2=“W”,3)))	10.7
=IF(J3=“E”,1,IF(J3=“C”,2,IF(J3=“W”,3)))	8.8
=IF(J4=“E”,1,IF(J4=“C”,2,IF(J4=“W”,3)))	10.9
=IF(J5=“E”,1,IF(J5=“C”,2,IF(J5=“W”,3)))	12.7
=IF(J6=“E”,1,IF(J6=“C”,2,IF(J6=“W”,3)))	7.1
=IF(J7=“E”,1,IF(J7=“C”,2,IF(J7=“W”,3)))	0
=IF(J8=“E”,1,IF(J8=“C”,2,IF(J8=“W”,3)))	3.3
=IF(J9=“E”,1,IF(J9=“C”,2,IF(J9=“W”,3)))	11.9
=IF(J10=“E”,1,IF(J10=“C”,2,IF(J10=“W”,3)))	13.3
=IF(J11=“E”,1,IF(J11=“C”,2,IF(J11=“W”,3)))	12.8

We will test the null hypothesis that the mean changes in department store sales (X2) are equal across the three regions against the alternative hypothesis that at least one regional mean differs from the other two.

The output from the ANOVA: Single Factor function is shown in Table 31.8.

TABLE 31.8 Single factor ANOVA analysis.

ANOVA: Single Factor						
SUMMARY						
Groups	**Count**	**Sum**	**Average**	**Variance**		
X1	74	138	1.865	0.584		
X2	74	711.6	9.616	47.615		
ANOVA						
Source of Variation	**SS**	**Df**	**MS**	**F**	**P-value**	**F crit**
Between Groups	2223.088	1	2223.088	92.247	0.000	3.906
Within Groups	3518.509	146	24.099			
Total	5741.597	147				

If the null hypothesis is true, the probability that the F statistic will be at least 3.906 is 0.000. This is less than 5%, a reasonable significance level, so we reject the null hypothesis. We conclude that the mean change in department store sales varies across the three regions.

31.4 Tests of Homogeneity of Variance

To test the uniformity of several population means by an analysis of variance, it is necessary that the variances of the populations be the same. For samples from normally distributed populations, several tests of homogeneity of variance have been developed. For a study involving k samples, each generates a statistic to test these hypotheses:

$$H_0: \sigma_1^2 = \sigma_2^2 = \ldots = \sigma_k^2$$

H_a: At least one variance differs from the others

The simplest test of homogeneity of variance is the *Hartley test*, which may be used when all the samples are of the same size n. The test statistic F_{max} is the ratio of the largest sample variance to the smallest:

$$F_{max} = \frac{S_{max}^2}{S_{min}^2}$$

If the population variances are uniform, the statistic will tend to be near 1; if F_{max} is large enough, we have evidence that the variances of the several populations are not equal and we reject the null hypothesis of the test. Critical values of this test correspond to $(n - 1)$ degrees of freedom and to the number k of sample variances. For the 5% and 1% significance levels, critical values of the Hartley test are given in Table A-23, Critical values for the Hartley test of homogeneity of variance, in Appendix A.

ON DVD

ON DVD

The Hartley test loses power when the number of samples (treatment groups) grows beyond five. Another test of homogeneity of variance that requires uniform sample sizes and is nearly as simple as the Hartley test is the *Cochran test*. The variance for each sample is found, and the test statistic is

$$C = \frac{S^2_{max}}{\sum_{i=1}^{k} S_i^2}.$$

Critical values of C correspond to n and k, and are given for the 5% significance level found in Table A-24, Critical Values for the Cochran test of homogeneity of variance, in Appendix A.

Suppose that in a one-way analysis of variance involving five treatment groups, each of size 10 and selected from normally distributed populations, the following variances are computed:

$$S_1^2 = 16.5;\ S_2^2 = 37.8;\ S_3^2 = 14.1;\ S_4^2 = 22.6;\ S_5^2 = 42.8.$$

The value of Hartley's statistic for these values is:

$$F_{max} = \frac{S^2_{max}}{S^2_{min}} = \frac{42.8}{14.1} = 3.04.$$

Corresponding to $(n-1) = 9$ and $k = 5$, the critical value of F_{max} at the 5% significance level is 7.11. The value of the statistic is less than 7.11, so the null hypothesis of the test is not rejected; it cannot be concluded that the population variances are unequal.

For the same sample variances, the value of the Cochran statistic is:

$$C = \frac{S^2_{max}}{\sum_{i=1}^{5} S_i^2} = \frac{42.8}{16.5+37.8+14.1+22.6+42.8} = 0.320.$$

At the 5% significance level, the critical value of C corresponding to $n = 10$ and $k = 5$ is 0.424. The observed value of the statistic does not exceed the critical value, and again the null hypothesis of equal variances is not rejected.

A test of homogeneity of variance that does not require equal sample sizes, though its calculations are relatively involved, is the *Bartlett test*. For k samples of size n_j, $j = 1, 2, \ldots, k$, from normally distributed populations, the value of the Bartlett statistic is:

$$B = \frac{1}{C}\left[(N-k)\ln\left(\frac{SS_{within}}{N-k}\right) - \left[\sum_{j=1}^{k}\left[(n_j-1)\ln S_j^2\right]\right]\right],$$

$$\text{where } N = \sum_{j=1}^{k} n_j \quad \text{and} \quad C = 1 + \frac{1}{3(k-1)}\left[\sum_{j=1}^{k}\left(\frac{1}{n_j-1}\right) - \frac{1}{N-k}\right].$$

ON DVD

For sufficiently large sample sizes ($n_i \geq 5$), the sampling distribution of B is approximately a chi-square distribution with $(k-1)$ degrees of freedom. Large values of B tend to support the alternative hypothesis of unequal variances, and critical values of B may be found in Table A-4b, Critical values of the F distribution, in Appendix A.

Consider the data given in Table 31.1 at the beginning of this chapter. We perform the Bartlett test of homogeneity of variance for the populations from which the three treatment groups were selected with these calculations:

$$S_1^2 = 809.5;\ S_2^2 = 919.1;\ S_3^2 = 592.7$$

$$N = 20 \quad SS_{within} = 13228$$

$$C = 1 + \frac{1}{3(3-1)}\left[\left(\frac{1}{5-1} + \frac{1}{8-1} + \frac{1}{7-1} - \frac{1}{20-3}\right)\right] = 1.083$$

$$B = \frac{1}{1.083}\left[(20-3)\ln\left(\frac{13228}{17-3}\right) - \sum_{k-1}^{k}\left[\left(n_j - 1\right)\ln s_j^2\right]\right]$$

$$= \frac{1}{1.083}\left[17 * 6.657 - (5-1)6.96 - (8-1)6.823 - (7-1)6.358\right]$$

$$= 0.285$$

Note that the C used to find B is not the C of the Cochran test.

Corresponding to 3 – 1 = 2 degrees of freedom, the critical value of B at the 5% significance level is 5.99. The value of the statistic is less than this, so we do not reject the null hypothesis of the test. We do not conclude that the variances of the respective populations of book sales are unequal.

The tests of homogeneity of variance just described are all very sensitive to violations of the assumption of normally distributed populations. Box examined the effects of departures from normality on the Bartlett test and developed from it a more robust one. Analysis of variance is less sensitive to departures from normality than are these tests of homogeneity of variance, so when non-normality is suspected in the underlying populations, a small significance level may be used with the tests of homogeneity of variance.[1]

31.5 Multiple Comparisons

When the ANOVA F test results in the rejection of the null hypothesis that all treatment means are equal, this does not indicate which mean differs from another. Some pairs of means may exhibit significant differences, while others may not. One category of procedures for making *multiple comparisons* among pairs of treatment means is called *post hoc* or *a posteriori*. Such procedures make all possible comparisons between treatment means; if there are k treatments, there will be $k(k-1)/2$ comparisons. We briefly describe two such tests.

The Scheffé test calculates an F statistic for each comparison being performed, and compares each to a single critical value. If $\bar{X}_i$ and $\bar{X}_j$ are the means of treatment groups i and

j containing n_i and n_j values, respectively, and if there are k treatment groups in the experiment, the corresponding F statistic is:

$$F = \frac{(\bar{X}_i - \bar{X}_j)^2}{MS_{within}\left(\frac{1}{n_i} + \frac{1}{n_j}\right)(k-1)},$$

where MS_{within} is the mean sum of squares within groups from the ANOVA table. If there are N values of the dependent variable in the experiment, then the numbers of degrees of freedom associated with the F statistic are $k-1$ and $N-k$, as in the original analysis of variance.

CONTINUATION OF EXAMPLE 31.1

Solution:

The Scheffé F statistic for the first two treatments is:

$$F = \frac{(180-237)^2}{778.12\left(\frac{1}{5} + \frac{1}{8}\right)(3-1)} = \frac{3249}{505.778} = 6.42.$$

Similar computations complete this table of Scheffé F statistics for the three comparisons possible in this example:

Groups	F
1.2	6.42
2.3	1.50
1.3	1.92

Corresponding to $3-1=2$ and $20-3=17$ degrees of freedom, $F_{0.05}=3.59$. At the 5% significance level, only the difference between $\bar{X}_1$ and $\bar{X}_2$ is significant.

The Scheffé test is very rigorous with respect to Type I errors and will show fewer significant differences than other multiple comparison techniques. The use of a larger significance level, perhaps 10% instead of 5%, is often recommended.

Tukey's test may be applied when the treatment groups are of equal size. It is based on the statistic

$$q = \frac{\bar{X}_i - \bar{X}_j}{(MS_{within}/n)^{1/2}},$$

where n is the size of the treatment groups. Critical values for this test correspond to $N-k$ degrees of freedom, where k is the number of treatment groups and $N = nk$, and are found in Table A-12, Critical values of the correlation coefficient. A difference

ON DVD

between two means is significant if the value of q for that pair is more extreme than the table's critical value for the chosen significance level.

Note that comparisons made with the Scheffé technique are necessarily two-tailed, while the Tukey test may be one- or two-tailed.

31.6 Exercises

1. Five salespeople report these monthly sales amounts (given in thousands of dollars):

	Bob	Joe	Mary	Bill	Alice
Jan	4	5	15	35	17
Feb	7	6	18	27	26
Mar	9	12	21	29	17
Apr	9	7	26	30	20
May	14	12	20	35	12

a. Calculate the mean sales for each salesperson and the variances of each salesperson's sales amounts. At the 5% significance level, does either the Hartley test or the Cochran test of homogeneity of variance convince us that the variances are unequal?

b. Perform the Bartlett test of homogeneity of variance on the five sample sales amounts. Use the 5% significance level.

c. What is the relevance of tests of homogeneity of variance to analysis of variance?

d. Perform a one-way analysis of variance to determine if it can be concluded at the 5% significance level that mean sales amounts differ among the five salespeople.

e. Use the Scheffé test to determine which pairs of treatment group means are different at the 5% significance level.

f. Apply the Tukey test at the 5% significance level for the differences between the means of groups 3 and 5, and 4 and 5. What hypotheses are being tested, and what conclusions are reached?

2. Four technicians are each given six solutions in which they measure the concentration of an organic compound. All the solutions are in fact identical; we are concerned with possible differences in mean measured level of the organic compound among the technicians. The ANOVA table in Table 31.9 summarizes the analyses:

TABLE 31.9 A one-way ANOVA table.

Source	Degrees of Freedom	Sum of Squares	Mean Sum of Squares	F
Between groups	3	0.004235	0.001411	
Within groups	20	0.004625	0.000231	6.10
Total	23	0.008860		

a. What hypotheses are being tested here?

b. Interpret the results.

3. In general, how does the F statistic as an analysis of variance distinguish between the null and alternative hypotheses?

4. A study is completed on the software development time for implementing an algorithm programmed by 24 programmers using four different programming languages. The research question is "Does the choice of programming language affect software development time?" The 24 programmers were randomly assigned to program the algorithm in four different languages (A, B, C, and D). The software development times, in minutes, were taken and tested in random order. The analysis in Table 31.10 is exactly justified if the data are random samples from four normal populations. Note that test order is listed in parentheses.

TABLE 31.10 Programmer software development times by treatments.

Programming Language (Treatment)			
A	B	C	D
63 (12)	62 (20)	68 (16)	56 (23)
67 (9)	60 (2)	66 (7)	62 (3)
71 (15)	63 (11)	71 (1)	60 (6)
64 (14)	59 (10)	67 (17)	61 (18)
65 (4)		68 (13)	63 (22)
66 (8)		68 (21)	64 (19)
			63 (5)
			59 (24)

a. State the formal hypothesis being tested.

b. Are the differences between the treatment averages greater than expected given the variation within treatments? (Hint: Run a one-way ANOVA.)

c. Plot the t points for the four treatments on a common t_{20} distribution plot. Using this plot as a basis, could these four t values be drawn randomly from a t_{20} distribution plot?

5. Given the data in Table 31.11:

TABLE 31.11 Artificial data for one-factor model.

Levels of Factor A			
1	**2**	**3**	**4**
32.72	32.95	33.01	36.41
29.15	35.96	29.30	37.58
29.73	32.26	35.98	39.86
27.10	33.11	34.70	36.86
32.14	34.65	33.51	35.99
31.53	34.79	31.71	33.47
31.41	31.83	33.98	35.79
32.73	35.22	33.13	38.89
29.97	37.01	31.67	35.68
30.67	33.46	33.07	35.93

a. Build the ANOVA table for this model.

b. Build the table of means and main effects for this model.

c. State the hypotheses being tested.

d. If $\alpha = 0.05$, what are the results?

6. Given the data shown in Table 31.12 for a one-factor design model,

TABLE 31.12 One-factor design model data.

Levels of Factor A				
1	**2**	**3**	**4**	**5**
19.33	19.51	17.50	17.67	19.95
18.41	19.37	18.79	13.83	16.25
20.71	16.45	19.81	15.35	16.11
19.73	18.83	18.36	18.62	20.69
21.67	19.46		14.96	15.22
20.38	19.98		17.54	16.54
	16.16		14.37	17.49
	17.59		15.77	15.04
			16.49	18.28

(Continued)

TABLE 31.12 Continued.

Levels of Factor A				
1	2	3	4	5
			15.51	18.74
			18.18	
			19.34	

a. Find the ANOVA table.

b. For $\alpha = 0.05$, test H_0: $a_1 = a_2 = \ldots = a_s$ versus H_1: at least one equality is an inequality. Set a 95% confidence interval on $a_1 - a_2$ and on $\mu + a_1$.

7. Consider the problem of how to test which form of advertising is best to promote the sale of a product:

A_1 = newspaper
A_2 = radio
A_3 = TV

A 15-week test period is selected for the test, with each form of advertising being used for five different weeks and the increase in the total weekly sales compared. The particular weeks used for each type of advertising have been selected randomly from the 15 possible weeks. The gain of sales is determined by subtracting the sales from the expected sales for the week if there had been no advertising. This difference is the response variable Y of our experiment. We can tabulate the results of the experiment in Table 31.13, which is a one-way table:

TABLE 31.13 Total weekly sales by advertising treatments data.

A_1	A_2	A_3
790	640	721
853	832	834
940	734	737
820	969	931
830	742	923

Using this data, test the hypothesis that the average response is the same for each form of advertising. If you discard this null hypothesis, use the least significant difference technique to determine if one form of advertising is superior to either of the other two.

8. The personnel officer of a company is interested in what type of relationship exists between the educational background of 507 employees and their reading speed. In making a study of this problem, he uses the following educational classes: A_1: high school not completed; A_2: high school completed but no college; A_3: some college but no degree; A_4: a college degree. His experiment consists of selecting samples of 10 employees from each educational class and measuring their reading speed. The data obtained is given in Table 31.14. Each observation is represented by Y_{ij}, where *i* denotes the educational class and *j* denotes the *k*th observation in that class.

TABLE 31.14 Reading time speed by educational background data.

A_1 High School not Completed	A_2 Completed High School	A_3 Some College	A_4 College Degree
7.2	5.7	5.8	3.7
6.7	5.7	4.1	6.3
6.5	7.9	4.2	5.6
7.4	5.2	5.0	5.7
6.4	6.6	7.6	5.1
8.4	4.6	4.6	3.6
7.5	6.3	6.0	4.4
7.0	7.0	6.0	5.1
5.5	4.8	5.2	6.4
5.7	5.7	5.4	6.2

Test the hypothesis that the reading speeds are the same for all levels of education. If this hypothesis is discarded, divide the four levels into as many different groups as the least significant different approach might indicate.

Reference

1. Box, G.E.P., "Nonnormality and Tests of Variances," *Biometrika* 40 (1953), 318–335.

CHAPTER 32

TWO-WAY ANALYSIS OF VARIANCE

OVERVIEW AND LEARNING OBJECTIVES

In This Chapter

32.1 Introduction

In Chapter 31, we extended regression from the simple linear case, in which the dependent variable is related to a single predictor, to multiple linear regression, which involves two or more predictors. Similarly, we perform analysis of variance when a dependent variable is affected by two or more factors. In this chapter, we investigate two-way analysis of variance and discuss how factors can be employed as an experimental control.

The following example involves an analysis of variance in which the dependent variable is affected by two factors.

EXAMPLE 32.1

A manufacturer of soap products wants to find both the most attractive color for the container of a new shampoo and the most effective advertising campaign for its introduction. The package is produced in four colors, and three advertising campaigns are constructed; these are used in all possible combinations in 36 similar market areas. Table 32.1 lists the sales in those areas for the first month of the test period:

TABLE 32.1 Advertising campaign/package color data.

		Package Color (Sales in $1000)			
		Yellow	**Red**	**Blue**	**Green**
Advertising Campaign	A_1	2.6	1.9	3.9	3.7
		3.0	1.6	3.5	2.6
		2.0	1.3	3.0	3.3
	A_2	3.6	3.4	4.6	3.4
		3.0	2.6	4.1	3.7
		3.9	2.3	3.7	4.3
	A_3	4.9	3.8	4.9	4.7
		3.9	3.1	4.3	5.0
		4.2	2.8	5.4	4.1

All figures and tables in this chapter appear on the companion DVD. ON DVD

Solution:

From this data, the manufacturer can derive answers to three separate but related questions. Does the package color affect sales? Does the choice of advertising campaign affect sales? Do the two factors act together, that is, *interact*, to affect sales? (For example, a highly visual

EXAMPLE 32.1

television campaign may more successfully promote a brightly colored package.) In this case, these three null hypotheses are being tested simultaneously:

H_0: Mean sales are equal across package colors.

H_0: Mean sales are equal across advertising campaigns.

H_0: The factors do not interact to affect sales.

This example will be continued later in the chapter.

Interaction effects occur when the response of the dependent variable X to the levels of factor A changes with the levels of factor B. This is illustrated in Figure 32.1. In graph (a), the mean value of X changes over the levels of factor A in the same way for each level of factor B; that is, factors A and B do not interact. In graphs (b) and (c), the nature of the change of $\overline{X}$ with A depends on the level of factor B; A and B interact.

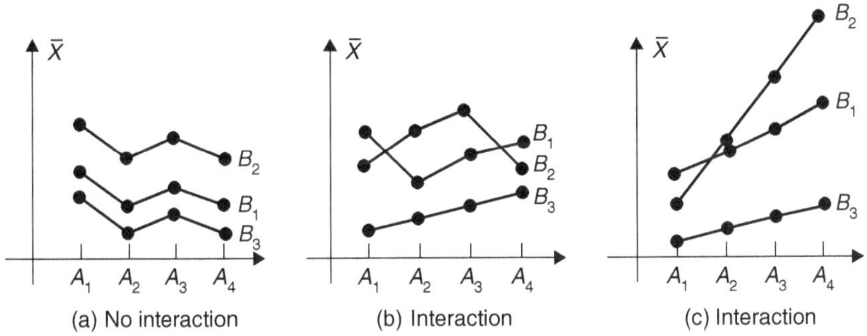

FIGURE 32.1 Illustrations of interaction effects in two-way analysis of variance.

In general, with n observations of the dependent variable for each pair of factor levels, the data may be organized as shown in Table 32.2:

TABLE 32.2 Factorial model data file.

Rows of Factor A		Columns of Factor B				
		B_1	B_2		B_c	Row Means
	A_1	X_{111}	X_{121}		X_{1c1}	
		X_{112}	X_{122}		X_{1c2}	
		..	..	......	..	$\overline{X_{1..}}$
		X_{11n}	X_{12n}		X_{1cn}	

(Continued)

TABLE 32.2 Continued.

		Columns of Factor B				
		B_1	B_2		B_c	**Row Means**
Rows of Factor A	A_2	X_{211}	X_{211}		X_{2c1}	
		X_{212}	X_{222}		X_{2c2}	
		..	..	......	..	$\overline{X_{2..}}$
		X_{21n}	X_{22n}		X_{2cn}	
		..	..		..	
		..	..		..	
	A_r	X_{r11}	X_{r21}		X_{rc1}	
		X_{r12}	X_{r22}		X_{rc2}	$\overline{x_{r..}}$
		..	..	......	..	
		X_{r1n}	X_{r2n}		X_{rcn}	
Col Means		$\overline{X_{.1.}}$	$\overline{X_{.2.}}$	......		$\overline{X_{.c.}}$

The mean of the values in cell *ij* is $\bar{X}_{ij}$, the grand mean of all the observations is $\bar{X}...$, and there are a total of $N = nrc$ observations. We assume again that the populations of values corresponding to the samples listed in the cells have common variance σ^2.

It can be shown that the total sum of squares for such a two-way analysis of variance can be partitioned in this way:

$$\begin{aligned} SS_{tot} &= \sum_{i=1}^{r}\sum_{j=1}^{c}\sum_{k=1}^{n}\left(X_{ijk} - \bar{X}...\right)^2 \\ &= cn\sum_{i=1}^{r}\left(\bar{X}_{i..} - \bar{X}...\right)^2 + rn\sum_{j=1}^{c}\left(\bar{X}_{.j.} - \bar{X}...\right)^2 \\ &+ n\sum_{i=1}^{r}\sum_{j=1}^{c}\left(\bar{X}_{ij.} - \bar{X}_{i..} - \bar{X}_{.j.} + \bar{X}...\right)^2 \\ &+ \sum_{i=1}^{r}\sum_{j=1}^{c}\sum_{k=1}^{n}\left(X_{ijk} - \bar{X}_{ij.}\right)^2 \end{aligned}$$

On the right-hand side of this equation, the first sum of squares, SS_{row}, measures the effect of the row factor (A) on the dependent variable. The second term, SS_{col}, measures the effect of the column factor (B). The third, SS_{int}, measures the effect of interaction of the two factors, and the last, SS_{within}, is due to variation within the cells of the table.

There are $(r-1)$ and $(c-1)$ degrees of freedom, respectively, associated with the row and column sums of squares. $(r-1)(c-1)$ degrees of freedom are associated with the interaction sum of squares, and $rc(n-1)$ degrees of freedom are associated with SS_{tot}. Note that $(r-1) + (c-1) + (r-1)(c-1) + rc(n-1) = nrc-1$.

If all the null hypotheses stated above are true, then each sum of squares divided by its number of degrees of freedom, or each mean sum of squares, is an unbiased estimate of the common variance σ^2 of the X_{ijk}. Ratios of mean sum of squares form F statistics with which to test the hypotheses, as shown in the two-way ANOVA table in Table 32.3. Computation of the F statistics for row and column effects depends on whether the row and column factor levels are random or fixed.

TABLE 32.3 Two-way ANOVA table.

Source of Variance	Sum of Squares	Degrees of Freedom	Mean Sum of Squares (MS)	F (fixed effects)	F (random effects)
Rows	$SS_{row} = cn\sum_{i=1}^{r}(\bar{X}_{i..}-\bar{X}_{...})^2$	r − 1	$\frac{SS_{row}}{(r-1)}$	$\frac{MS_{row}}{MS_{within}}$	$\frac{MS_{row}}{MS_{int}}$
Columns	$SS_{col} = rn\sum_{j=1}^{c}(\bar{X}_{.j.}-\bar{X}_{...})^2$	c − 1	$\frac{SS_{COL}}{(c-1)}$	$\frac{ms_{COL}}{MS_{within}}$	$\frac{MS_{COL}}{MS_{int}}$
Interaction	$SS_{int} = \sum_{i=1}^{r}\sum_{j=1}^{c}\overline{X_{ij.}}-\overline{X_{i..}}-\overline{X_{.j.}}+\overline{X_{...}})^2$	(c − 1)(r − 1)	$\frac{SS_{int}}{(c-1)(r-10)}$	$\frac{ms_{int}}{MS_{within}}$	$\frac{MS_{int}}{MS_{within}}$
Within Groups	$SS_{within} = \sum_{i=1}^{r}\sum_{j=1}^{c}\sum_{k=1}^{n}(X_{ijk}-\bar{X}_{ij.})^2$	rc(n − 1)			
Total	$SS_{total} = \sum_{i=1}^{r}\sum_{j=1}^{c}\sum_{k=1}^{n}(X_{ijk}-\bar{X}_{...})^2$	n − 1			

Note that the calculations to construct this ANOVA table require equal cell frequencies in the data. For unequal cell frequencies, the calculations become more complicated, but the approach and interpretation remain the same.[1]

CONTINUATION OF EXAMPLE 32.1

Solution: See Table 32.4.

TABLE 32.4 ANOVA table for Example 32.1.

Source of Variance	Sum of Squares	Degrees of Freedom	Mean Sum of Squares (MS)	F (Fixed effects)
Rows	14.64	3 − 1=2	$\frac{14.64}{2}=7.32$	$\frac{7.32}{0.24}=30.50$
Columns	28.64	4−1=3	$\frac{28.48}{3}=9.49$	$\frac{9.49}{0.24}=39.53$
Interaction	0.45	(3 −1)(4 −1)=6	$\frac{0.45}{6}=0.08$	$\frac{0.08}{0.24}=0.33$
Within Cells	5.76	$3\times4\times(3-1)=24$	$\frac{5.76}{24}=0.24$	
Total	49.31	36 − 1=35		

Both factors in this example are fixed since no other package colors or ad campaigns are being considered.

If there is significant interaction between the factors, we cannot draw reliable conclusions about their individual effects on the dependent variable. Therefore, we look first at the F statistic associated with the interaction sum of squares. At the 5% significance level, the critical F value corresponding to 6 degrees of freedom in the numerator and 24 in the denominator is 2.51. The F statistic for interaction is 0.33, less than the critical value, so at $\alpha = 5\%$ we cannot conclude that the two factors interact to affect sales. We examine the other F statistics to test the row and column effects.

Corresponding to 2 and 24 degrees of freedom, $F_{0.05} = 3.40$, $MS_{row}/MS_{within} = 30.50 > 3.40$, so we conclude that the choice of advertising campaign affects sales. Similarly, for 3 and 24 degrees of freedom $F_{0.05} = 3.01$, $MS_{col}/MS_{within} = 39.53 > 3.01$; package color also affects sales.

EXCEL

TWO-WAY ANOVA IN EXCEL OR A STATISTICAL PACKAGE?

The two-way ANOVA: Without Replication function is available under Data Analysis as a selection under the Tools menu. This function does not test for interaction and therefore, the author recommends executing two-way ANOVA problems utilizing a standard statistical package, such as SPSS.

Analyses of variance involving two or more factors can be performed in SPSS with the procedure ANOVA (which also performs one-way tests). This example program using the American Cities database in Appendix B groups the values of X7 in order to use it as a factor, then performs a two-way analysis of variance to determine if X3 (unemployment) is affected by X1 and X7.

ON DVD

TABLE 32.5 EXCEL two-way ANOVA printout.

```
10.     GET FILE
10.005  USNEWS
20.     RECODE
20.005  X7 (– 20 THRU 0 = 1)(1.01 THRU 20 = 2)
20.010  (20.01 THRU HI = 3)
30.     ANOVA
30.005  X3 BY X1(1,3),X7(1,3)
```

The output of the program is this:

```
ANOVA TABLE

     X3       UNEMPLOYMENT
  BY X1       CITY REGION
     X7       CHANGE IN CONSTRUCTION ACTIVITY
****************************************************************************
```

SOURCE OF VARIATION	SUM OF SQUARES	DF	MEAN SQUARE	F	SIGNIF OF F
MAIN EFFECTS	2.933	4	.733	.255	.905
X1	2.595	2	1.297	.452	.639
X7	.110	2	.055	.019	.981
2-WAY INTERACTIONS	9.006	4	2.252	.784	.540
X1 X7	9.006	4	2.252	.784	.540
EXPLAINED	11.939	8	1.492	.520	.837
RESIDUAL	175.195	61	2.872		
TOTAL	187.135	69	2.712		

74 CASES WERE PROCESSED.
4 CASES (5.4 PCT) WERE MISSING.

None of the computed F statistics is significant, so we do not reject any of the null hypotheses in the three tests that this analysis performs. We cannot conclude that the mean values of X3 change across X1 or X7, or that X3 is affected by an interaction of X1 and X7. Note that the main effects are the row and column effects.

Excel's two-way ANOVA function generates the output in Table 32.6 after recoding X7:

TABLE 32.6 Recoding of X7.

X1	X7	X3
=IF(F2=“E”,1,IF(F2=“C”,2,IF(F2=“W”,3)))	=IF(G2<=0,1,IF(G2<=20,2,3))	4.7
=IF(F3=“E”,1,IF(F3=“C”,2,IF(F3=“W”,3)))	=IF(G3<=0,1,IF(G3<=20,2,3))	0
=IF(F4=“E”,1,IF(F4=“C”,2,IF(F4=“W”,3)))	=IF(G4<=0,1,IF(G4<=20,2,3))	6.2
=IF(F5=“E”,1,IF(F5=“C”,2,IF(F5=“W”,3)))	=IF(G5<=0,1,IF(G5<=20,2,3))	6.5
=IF(F6=“E”,1,IF(F6=“C”,2,IF(F6=“W”,3)))	=IF(G6<=0,1,IF(G6<=20,2,3))	9.7
=IF(F7=“E”,1,IF(F7=“C”,2,IF(F7=“W”,3)))	=IF(G7<=0,1,IF(G7<=20,2,3))	6.4
=IF(F8=“E”,1,IF(F8=“C”,2,IF(F8=“W”,3)))	=IF(G8<=0,1,IF(G8<=20,2,3))	6.1
=IF(F9=“E”,1,IF(F9=“C”,2,IF(F9=“W”,3)))	=IF(G9<=0,1,IF(G9<=20,2,3))	6.3

The coded values are shown in Table 32.7:

TABLE 32.7 Recoded X7 data file.

X1	X7	X3
1	2.000	4.700
1	2.000	0.000
1	3.000	6.200
1	2.000	6.500
1	3.000	9.700
1	3.000	6.400
1	2.000	6.100
1	2.000	6.300
1	2.000	3.900

The resultant ANOVA table in Excel is shown in Table 32.8:

TABLE 32.8 Two-way ANOVA table for X3 by X1 and X7.

Anova						
	SS	**df**	**MS**	**F**	**P-value**	**F crit**
Rows	170.600	73	2.337	1.028	0.454	1.473
Columns	480.240	1	480.240	211.190	0.000	3.972
Error	166.000	73	2.274			
Total	816.840	147				

Since there is an absence of an F test for interaction, any attempt to analyze the problem further is not wise.

32.2 Two-Way ANOVA with One Entry Per Cell

In the two-way ANOVA of the previous section, in which each cell of the data table contained several entries, we tested for an interaction effect by computing $F = MS_{int}/MS_{within}$, where the denominator of the fraction was computed from the squared deviations of the observed values from their means *within* each cell. When there is only one entry per cell, however, this value is always 0; we cannot test for an interaction effect, nor can we use MS_{within} to form statistics to test for treatment effects. To test for treatment effects in this case, we assume that there is no treatment interaction and use MS_{tot} in the denominators of the row and column F statistics. Table 32.9 indicates the appropriate calculations for two-way ANOVA with one entry per cell.

TABLE 32.9 Calculations for two-way ANOVA with one entry per cell.

Source of Variance	Sum of Squares	Degrees of Freedom	Mean Sum of Squares	F
Rows	$c\Sigma_{i=1}^{r}(\overline{X}_{i.}-\overline{X}..)^2$	r − 1	$\frac{SS_{row}}{(r-1)}$	$\frac{MS_{row}}{MS_{INT}}$
Columns	$r\Sigma_{j=1}^{c}(\overline{X}_{.j}-\overline{X}..)^2$	c − 1	$\frac{SS_{COL}}{(c-1)}$	$\frac{ms_{COL}}{MS_{within}}$
Interaction	$\Sigma_{i=1}^{r}\Sigma_{j=1}^{c}(X_{ij}-\overline{X_{i.}}-\overline{X_{.j}}+\overline{X...})^2$	(c − 1)(r − 1)	$\frac{SS_{int}}{(c-1)(r-10)}$	
Total	$\Sigma_{j=1}^{c}\Sigma_{i=1}^{r}(X_{ij}-\overline{X}..)^2$	(rc−1)		

EXAMPLE 32.2

Consider the example in Table 32.10, which is simplified from Table 32.1 by using only the first entry in each cell.

TABLE 32.10 Data table with one entry per cel.

		PACKAGE COLOR				
		Yellow	Red	Yellow	Green	
ADV	A_1	2.6	1.9	3.9	3.7	3.03
	A_2	3.6	3.4	4.6	3.4	3.75
CAM	A_3	4.9	3.8	4.9	4.7	4.58
		3.70	3.03	4.47	3.93	$\overline{X}..= 3.78$

EXAMPLE 32.2

Solution:

Note the row, column, and grand means in the margins. The ANOVA table derived from this data is shown in Table 32.11.

TABLE 32.11 ANOVA derived from simplified table (Table 32.10).

Source of Variance	Sum of Squares	Degrees of Freedom	Mean Sum of Squares	F
Rows (AD CAMPIGN)	4.81	3−1=2	48.1/2=2.41	2.41/3022=10.95
Columns (COLOR)	3.29	4−1=3	3.29/3=1.10	1.10/0.22=5.00
Interaction	1.31	(3−1)(4−1)=6	1.31/6=0.22	
Total	9.30	12−1=11		

Corresponding to 2 and 6 degrees of freedom, $F_{0.05} = 5.14$; for 3 and 6 degrees of freedom, $F_{0.05} = 4.76$. In both tests, the F statistic is greater than the critical value, and we reject the null hypothesis. Both package color and advertising campaign affect sales of the new shampoo.

32.3 Randomized-Block Designs

In the ANOVA designs considered thus far, the experimental units have been randomly assigned to each factor level, and we have been interested in the effects on the dependent variable of all the factors in the experiment. Such experimental designs are called *completely randomized.* If an experiment includes two factors of which only one is to be investigated, the effect of the "uninteresting" factor can be controlled through the use of a *randomized-block design*.

In the example in Table 32.12, a researcher investigates the mean lifetimes of spark plugs of several brands (with the mean lifetimes the factor of interest), while controlling for the types of cars used in the tests.

TABLE 32.12 Randomized block design data.

		Spark Plug Type (Treatment)				
		A	B	C	D	
		14.1	15.7	13.7	19.4	
		17.5	12.5	12.0	19.3	
Auto Type (Control)	Y	18.0	17.8	12.8	13.1	Thousands of miles before failure
		18.2	11.2	18.4	21.8	

(Continued)

TABLE 32.12 Continued.

		Spark Plug Type (Treatment)			
		A	**B**	**C**	**D**
Auto Type (Control)	Z	15.2	15.2	12.4	21.5
		18.7	11.8	11.7	20.1
		18.5	18.2	11.8	15.0
		15.9	11.2	18.1	21.0

The model is the same as that for two-way ANOVA with multiple entries per cell (as shown in Table 32.1), except for the computation of certain F statistics, as shown in Table 32.13.

TABLE 32.13 ANOVA table for a randomized-block design.

Source of Variance	Sum of Squares	Degrees of Freedom	Mean Sum of Squares	F
Treatments	$SS_{Tr} = bn\sum_{j=1}^{B} (\overline{X_{.j.}} - \overline{X_{...}})^2$ = 162.5640	4 − 1 = 3	$\frac{SS_{Tr}}{(T-1)}$ =54.1880	$\frac{MS_{Tr}}{MS_{INT}}$ =627.900
Blocks	$SS_{Bl} = Tn\sum_{j=1}^{T} (\overline{X_{.j.}} - \overline{X_{...}})^2$ =0.2588	2 − 1 = 1	$\frac{SS_{Bl}}{(B-1)}$ =0.1360	$\frac{ms_{Bl}}{MS_{INT}}$ =15.76
Interaction	$SS_{int} = \sum_{i=1}^{T}\sum_{j=1}^{B} (\overline{X_{ij.}} - \overline{X_{i..}})^2$ =0.2588	(4 − 1) (2 − 1) = 3	$\frac{SS_{int}}{(T-1)(B-10)}$ =0.0863	$\frac{MS_{int}}{MS_{res}}$ =0.0012
Residual (res)	$SS_{res} = \sum_{i=1}^{T}\sum_{j-1}^{B}\sum_{k=1}^{n} (X_{ijk} - \overline{X_{ij.}})^2$ =178.2352	4 x 2 x (3 − 1) = 16	$\frac{SS_{res}}{(n-1)TB}$ =7.4265	
Total	$SS_{tot} = \sum_{i=1}^{T}\sum_{j=1}^{B}\sum_{k-1}^{n} (X_{ijk} - \overline{X_{ij.}})^2$ =341.1944	3 x 2 x 3 − 11 = 23	(nTB − 1)	

Note in the table that we refer to treatments and blocks rather than columns and rows, with T the number of treatments and B the number of blocks.

Comparing the computed F statistics to the appropriate critical values from Table A-4b, Critical values of the F statistic, in Appendix A, we find that we can reject the null hypothesis ON DVD that the mean spark plug lifetimes are equal across the treatments; we conclude that there is a difference in mean lifetimes among the four brands. We do not reject the null hypotheses of no interaction between treatments and blocks and of no effect by the blocking factor.

32.4 Latin Square Design

We have seen that a factor whose effect on the dependent variable we wish to control can be treated as a blocking factor. This technique can be effectively extended to two blocking factors by using an elegant design called a *Latin square.*

A Latin square is an array whose elements each appear exactly once in each row and column of the array. By creating a rectangular arrangement of the levels of the two blocking factors, we can assign treatment levels in order to form a Latin square and thereby control the effects of both blocking factors, as shown in Figure 32.2.

X	Y	Z
Z	X	Y
Y	Z	X

FIGURE 32.2 A 3 x 3 Latin square.

EXAMPLE 32.3

Suppose we wish to find if differences exist among the mean lifetimes of four brands of tires, T_1, T_2, T_3, and T_4. We test the tires in several climates (A) and on several road surfaces (B). We are not directly concerned with the effect of climate or road surface on tire life in this experiment, so A and B are blocking factors.

Solution:

If the blocking factors each have four levels, we could arrange them with the treatment levels in $4 \times 4 \times 4 = 64$ possible combinations, and measure tire lifetime in each case. It is possible, however, to pair each climate with each road condition just once, and assign the treatments in a Latin square to obtain the design shown in Table 32.14:

TABLE 32.14 Latin square design data file.

	Road Conditions				
Climate	**B_1**	**B_2**	**B_3**	**B_4**	
A_1	T_1	T_2	T_3	T_4	Treatment level (Brand of Tire)
A_2	T_2	T_3	T_4	T_1	
A_3	T_3	T_4	T_1	T_2	
A_4	T_4	T_1	T_2	T_3	

Each treatment is paired with each level of both blocking factors, and only 16 experimental measurements are required (though multiple entries per cell are possible). Note that to construct an experiment of this design, the number of levels of the blocking factors must equal the number of treatments.

Let the measurements of the dependent variable (corresponding to the kth treatment in the ith row and jth column) be indicated by X_{ijk}, the treatment means by $\bar{X}_{..k}$, the row means by $\bar{X}_{i..}$, the column means by $\bar{X}_{.j.}$, and the grand mean by $\bar{X}_{...}$. The ANOVA table for a Latin square design with n levels for all the factors is shown in Table 32.15.

TABLE 32.15 ANOVA table for a Latin square design.

Source of Variance	Sum of Squares	Degrees of Freedom	Mean Sum of Squares	F
Rows (A)	$n\sum_{i=1}^{n}(\overline{X_{i..}} - \overline{X_{...}})^2$	(n − 1)	$\frac{SS_{row}}{(n-1)}$	$\frac{MS_{row}}{MS_{INT}}$
Columns (B)	$n\sum_{i=1}^{n}(\overline{X}_{.i.} - \overline{X_{...}})^2$	(n − 1)	$\frac{SS_{col}}{(n-1)}$	$\frac{ms_{col}}{MS_{INT}}$
Interaction (T)	$n\sum_{k=1}^{n}(\overline{X}_{..k} - \overline{X}_{...})^2$	(n − 1)	$\frac{SS_{Tr}}{(n-1)}$	$\frac{MS_{Tr}}{MS_{int}}$
Residual	$\sum_{I=1}^{n}\sum_{J=1}^{n}\sum_{K=1}^{n}(X_{ijk} - \overline{X_{I..}} - \overline{X}_{.J.} - \overline{X}_{..K} + 2\overline{X}_{...})^2$	(n − 1)(n − 2)		
Total	$\sum_{i=1}^{n}\sum_{j=1}^{n}\sum_{k-1}^{n}(X_{ijk} - \overline{X_{...}})^2$	$n^2 - 1$		

In the case of multiple entries per cell, $\bar{X}_{ijk}$ is the mean of the values corresponding to treatment k in position ij.

The statistic $F = MS_{tr}/MS_{res}$ is used to perform a test of these hypotheses:

> H_0: The dependent variable means are equal across the treatments.
>
> H_a: The dependent variable means are unequal across the treatments.

32.5 Exercises

1. The curriculum committee is considering whether students in a college should be required to take a speed-reading course. It was felt that the benefit, as measured by Y = gain in reading speed six months after completing the course, would be affected by the student's year in school as well as by the type of course load he is carrying. An experiment is performed in which five students from each of the four undergraduate years and for each of three colleges are given the course. Six months later their reading speed is tested. The results are recorded in the two-way table shown in Table 32.16:

TABLE 32.16 Two-way data table.

College	Freshman	Sophomore	Junior	Senior
Arts & Sciences	110, 105, 101, 88, 101	115, 133, 117, 121, 113	112, 93 93, 108, 114	76, 81, 84, 90, 75
Engineering	142, 129, 135, 139, 137	121, 90, 120, 105, 103	117, 108, 86, 102, 106	103, 86, 80, 106, 94
Business	161, 148, 140, 143, 160	100, 97, 117, 101, 111	122, 125, 119, 137, 120	118, 108, 109, 123, 122

Test the null hypothesis relative to the differences in the two main effects and their interaction.

2. Given the data in Table 32.17:

TABLE 32.17 Artificial data from two-factor design with interaction.

Levels of Factor A	Levels of Factor T			
	1	**2**	**3**	**4**
1	6	2	5	3
2	2	0	1	1
3	5	2	4	2

Compute the ANOVA table for testing $H_0: \mu_1 = \mu_2 = \mu_3 = \mu_4$.

3. Redo problem 2 using Excel.

4. Given the data in Table 32.18:

TABLE 32.18 Artificial data.

Levels of Factor A	Levels of Factor T		
	1	**2**	**3**
1	6 4	3	3
2	3 5	5	5 7

a. Build a table for the number of cell values.

b. Build a table for the cell means.

c. Build the ANOVA table for testing $H_0: \mu_1 = \mu_2 = \mu_3$.

d. Do this problem computationally.

5. Redo problem 4 using Excel.

6. The data in Table 32.19 is assumed to be the result of an investigation of four drugs for controlling pain. Each drug is administered by three different methods. The model is assumed to be a two-factor design with no interaction and one observation per cell. The observation yy is the amount of a certain substance in the blood of a person two hours after receiving the drug.

TABLE 32.19 Contingency data table.

Method of Administering	Drug			
	A	B	C	D
1	12.6	14.1	10.4	11.4
2	11.3	12.2	21.0	11.8
3	15.1	16.0	21.1	14.1

Conduct a two-way ANOVA, complete with post hoc comparison of group means.

7. An investigation was made to determine how the speed of drill bits and the viscosities of a lubricant affect the speed with which a drill can penetrate granite. Records were examined for wells that had been drilled using four viscosities of a lubricant and three bit speeds. Information on three wells for each speed-viscosity combination was recorded and the data is given in Table 32.20.

TABLE 32.20 Speed of drill by viscosity of lubricant data file.

Speed of Drill (Factor A)	Viscosity of Lubricant (Factor T)			
	1	2	3	4
1	16.1 15.8 15.1	12.7 15.3 11.2	8.1 9.3 6.2	14.4 15.8 17.3
2	20.2 18.7 16.3	11.2 14.6 13.0	10.7 12.1 15.4	16.2 17.3 19.1
3	12.1 18.1 15.9	16.1 14.2 13.9	13.5 12.7 10.1	12.8 14.1 15.5

a. Find the best unbiased estimates of the main effects of the viscosities of the lubricant.

b. Find 95% one-at-a-time confidence intervals on the main effects of the speeds of the drill.

c. Find 95% simultaneous confidence intervals on the differences of all pairs of viscosities of the lubricant.

d. Exhibit an ANOVA table.

8. Exhibit all distinct 2 x 2 tables that can be derived from the two-way table of μ_{ij}, s shown in Table 32.21.

TABLE 32.21 Hypothetical two-factor data table.

Levels of Factor A	Levels of Factor T			
	1	2	3	4
1	3	1	4	9
2	–1	–3	0	4
3	5	3	6	11

a. Find the interaction of all 2×2 tables.

b. Do factors A and T interact?

9. A consumer preference study of three package designs (treatments) was organized as a randomized-block design among four supermarkets (blocks). The ANOVA table in Table 32.22 summarizes the results of that study, in which sales of each design in each store were recorded (thus, one case per cell).

TABLE 32.22 Randomized block ANOVA table.

Source	Degrees of Freedom	Sum of Squares	Mean Sum of Squares	F
Blocks	3	348.00	116.00	15.30
Treatments	2	547.17	273.58	36.09
Residual	6	45.50	7.58	
Total	11	940.67		

a. Why does the ANOVA table contain no entries for interaction effects?

b. What two pairs of hypotheses are being tested here? In which test are we really interested?

c. Using the 5% significance level for all tests, what conclusions are reached? What do they mean?

10. Factors A and B interact to affect the dependent variable X. What does this mean?

11. Which of the graphs in Figure 32.3 illustrate interaction of factors A and B in their effect on X?

12. True or False:

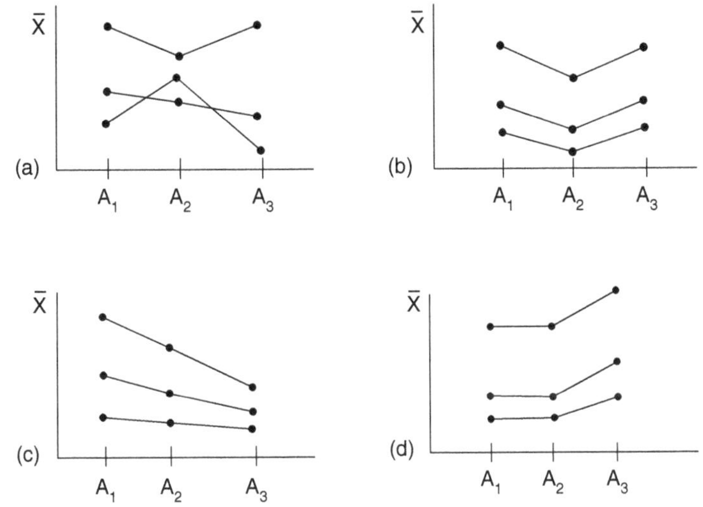

FIGURE 32.3 Interaction.

a. Analysis of variance analyzes the variation of the dependent variable in an attempt to apportion that variation among qualitative independent variables ________.

b. In one-way ANOVA, when the null hypothesis is false, $MS_{between}$ will probably be less than * MS_{within} ________.

c. To measure interaction, there must be more than one entry per cell ________.

d. The design of an experiment and the inferences to be drawn must be chosen in advance of the experiment ________.

e. Multiple comparisons are designed to compare all pairs of treatment group means ________.

f. The Scheffé test is less rigorous than other multiple comparison techniques with respect to Type I errors ________.

13. In testing end user satisfaction for a new release of a software system, three individuals were trained to evaluate software products consistently. Four components of the new product were presented to these evaluators to evaluate. The end user satisfaction experiment therefore considered the following variables (and is focused on answering the indicated questions and the measure of performance (response) is based upon the evaluators rating on a scale of 0 to 14):
Components Tested: 4 levels (Is there a difference in end user satisfaction between components?)

Evaluators: 3 levels (Are they consistent with one another?)

Several tests of each component were given to the evaluators to run in random order. The results are shown in Table 32.23.

TABLE 32.23 Components tested.

	A	B	C	D	
1	7.6 6.5 7.2	11.4 7.6 9.5	6.1 7.9 6.8	2.7 3.1 1.7	$\overline{Y_1} = 6.52$
2	7.2 13.6 10.7	12.9 12.4 10.7	11.1 6.9 9.0	3.3 1.9 2.3	$\overline{Y_2} = 8.50$
3	7.0 10.2 8.3	10.2 8.1 8.7	3.8 9.2 11.0	3.7 2.2 3.1	$\overline{Y_3} = 7.38$
	$\overline{Y}_A = 8.70$	$\overline{Y}_A = 0.16$	$\overline{Y}_A = 8.33$	$\overline{Y}_A = 2.67$	$\overline{\overline{Y}} = 7.47$

a. Run a two-way analysis and interpret the results.

b. Plot the model residuals as a function of time and interpret the results.

14. In a study of the effect of the experimenter on the responses obtained, 20 male and 20 female personnel officials were shown a neutral photograph of a human face and asked to rate it on a scale from –10 (extreme failure) to + 10 (extreme success). For half of the officials of each sex, selected at random, the experimenter looked at the official as often as possible while delivering his instructions. For the other half of the officials, the experimenter did not look at the official. The results are given in Table 32.24:

TABLE 32.24 Hypothetical data file of specific official to eye contact.

Official	**High Eye Contact Low Eye Contact**			
	Male	**Female**	**Male**	**Female**
1	+1	+6	–3	0
2	–3	+2	+2	+3
3	+2	+4	–4	–1
4	+4	+1	1	+6
5	+1	+5	0	+4

a. State the model you will use to analyze the effects of the experimenter's eye contact and the sex of the official on the response.

b. Test whether the factors have any main effects and whether they interact. Use a level of significance of 0.01 for each test. Summarize your results.

c. What should be the next step in the analysis of the study results?

15. Refer to problem 14.

   a. Set up the regression model counterpart to the ANOVA model.

   b. State the correspondences between the regression model parameters and the ANOVA model parameters.

   c. Obtain the statistics for testing for the presence of main effects and interactions by means of the regression approach.

   d. Conduct the tests, each with a level of significance of 0.01.

   e. Obtain an interval estimate by the regression approach for the contrast:

32.6 TWO-WAY ANOVA PROJECT I

Obtain and study the article: An Experimental Approach to Integrating ANOVA Concepts by LeRoy A. Franklin in the *Journal of Statistics Education*, Volume 10, Number 1 (2002) found at *http://www.amstat.org/publications/jse/v10n1/datasets.franklin.html*.

This article describes a large manufacturing company that built industrial fans. As often happens, a smaller competitor gave up and sold their business to the larger one. Even though the finished product appeared to be the same, each one claimed to have the best. It was a dilemma for the engineers to research and view as an opportunity to truly seek the most advantageous method for the newly merged corporation. Research students were asked to assist in the design and analysis of the experiments necessary for the research studies. The goal of these research experiments would result in standardizing processes requiring long-term manufacturing and financial ramifications.

The decision-making engineers expressed to the students that three factors would determine which method to choose: the type of hole in the spider (blades), the type of barrel onto which the spider was placed, and the type of assembly method for these components.

You [the student] has been contacted to assist in the design and analysis of the experiments.

1. Using SPSS, execute an ANOVA for a sample size of eight on the fadata.
   dat raw data

   a. Compute estimates of each of the model terms.

   b. Provide an estimate of the error term with 1 degree of freedom.

   c. Using the calculated values for F or the p-values, what conclusion do you reach?

   d. What hypothesis is being evaluated? How can you control for a Type I error? When would a Type II error be committed? What is the power for the hypothesis test?

2. Rerun the experiment with the sample size doubled to 16 observations, 2 at each of the factor combinations.

   a. How does the mean squared error term compare to the previous experiment?

   b. What are the results for this experiment?

   c. How do the estimates for observed power compare for the two experiments?

3. Rerun the experiment with a sample size of 32, 4 observations at each treatment combination.

   a. What impact does the larger sample size have on the degrees of freedom for the error sum of squares?

   b. What effect does overall Type I probability have on making several hypothesis tests?

   c. Review the results of the four experiments to determine what your conclusions would have been using $\alpha = 0.01$.

Interpret the results of the experiments when looking at the estimated means for the various treatment combinations. What ultimately helps you identify a combination of factor settings that substantially improve the strength of the industrial fans?

Reference

1. Winer, B.J. (1971). *Statistical Principles in Experimental Design*, (2nd Ed.) McGraw-Hill, Inc, New York.

CHAPTER 33

ANALYSIS OF COVARIANCE

OVERVIEW AND LEARNING OBJECTIVES

In This Chapter

33.1 Introduction

In many situations, the effect of an "uninteresting" factor cannot be controlled in the design of the experiment. In such cases, linear regression can be used to *adjust* the values of the dependent variable to remove the unwanted effect. Analysis of variance is then applied to the adjusted values to determine the effects on the dependent variable of the treatments of interest. This procedure, which we will study in this chapter, is called *analysis of covariance*, or ANCOVA.

Just like blocks can be used for experimental control in a randomized block design, (a) quantitative variable(s) can be used in a covariance study for statistical control. Consider the following example.

EXAMPLE 33.1

Annoying Novelties, Inc. investigates the relationship between sales of its products and three distinct package designs. Knowing that advertising also affects sales, they have recorded advertising expenditures as well as sales totals for several months for each package design. The data is listed in Table 33.1.

TABLE 33.1 Advertising expenditures and sales of three package designs of Annoying Novelties, Inc.

Package Design					
P_1		P_2		P_3	
Advertising (in \$1000)	Sales (in \$1000)	Advertising (in \$1000)	Sales (in \$1000)	Advertising (in \$1000)	Sales (in \$1000)
4.9	42.1	2.9	44.8	8.6	91.2
5.7	69.8	2.9	44.7	8.5	88.8
4.7	44.5	2.7	38.5	8.8	95.1
5.4	61.6	2.8	38.4	7.4	80.9
4.2	51.7	3.8	50.0	9.5	102.4
6.0	69.0	1.8	25.6	7.4	78.4
5.3	66.9	1.7	26.5	8.6	88.9
4.0	42.9	1.0	26.5	7.0	74.7
4.0	43.1	1.5	29.3	8.8	83.4
6.2	71.6	3.9	47.0		
6.5	65.4	1.9	29.5		
		1.0	18.2		

All figures and tables in this chapter appear on the companion DVD.

ON DVD

EXAMPLE 33.1

A one-way analysis of variance of sales by package design to test these hypotheses:

H_0: Mean sales are equal across package designs.
H_a: Package design affects mean sales.

produces, using standard statistical software system, the output in Table 33.2:

TABLE 33.2 Sales and package design ANOVA table.

Sales by Package Design					
Analysis of Variance					
Source	**Df**	**Sum of Sq**	**Mean Sq**	**F Ratio**	**F Prob**
Between Groups	2	13820.249	6910.124	61.284	.000
Within Groups	29	3269.906	112.775		
Total	31	17090.155			

The negligible probability associated with the F statistic seems to indicate that mean sales do vary with package design. However, this contention ignores the possible influence of advertising on sales, which might be causing the observed variation.

Let advertising expenditures be X and sales Y. Figure 33.1 is a scatter diagram of the data, with packaging technique also indicated.

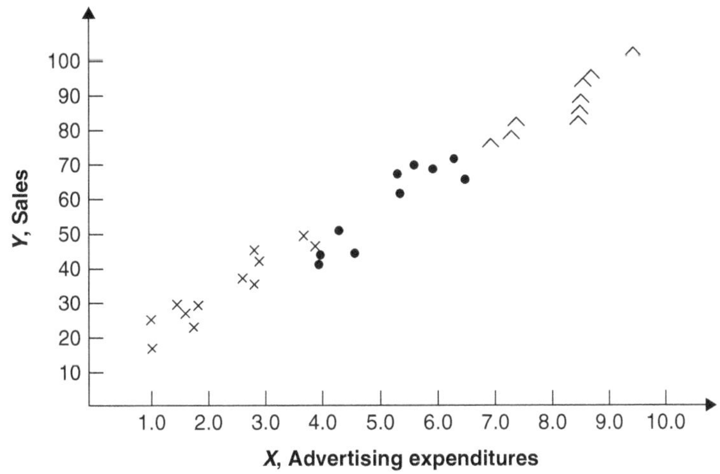

FIGURE 33.1 • = Package design P1, x = Package design P2, and ^ = Package design P3.

There is a linear relationship between X and Y both within each of the treatment groups and across the groups. We must adjust the values of Y (sales) to remove the effects of these relationships. We begin by considering each treatment group separately.

EXAMPLE 33.1

For the cases in treatment group P_1, the least squares regression equation is:

$$\overline{Y}_x = A_1 + B_1 X = -4.26224 + 11.87144X,$$

and the coefficient of determination $R^2 = 0.73244$ is significant at $\alpha = 1\%$. The mean value of X for the first treatment group is $\overline{X}_1 = 5.173$, and we adjust the values of Y in this group by computing

$$Y_{1j}^* = Y_{1j} - B_1\left(X_{1j} - \overline{X}_1\right) = Y_{1j} = 11.871(X_{1j} - 5.173).$$

For the first case in treatment group P_1,

$$Y_{1j}^* = 42.1 - B11.871(4.9 - 5.1723) = 45.3.$$

For group P_2, $B_2 = 9.88342$ and $\overline{X}_2 = 2.235$, and for group P_3, $B_3 = 9.84645$ and $\overline{X}_3 = 8.267$. Calculations like those just given produce the table of adjusted values shown in Table 33.3.

TABLE 33.3 ANOCOVA data file for package design controlled by advertising expenditures to predict mean sales.

Package Design					
P_1		P_2		P_3	
X	Y	X	Y	X	Y
4.9	45.3	2.9	38.2	8.6	87.9
5.7	63.5	2.9	38.1	8.5	83.5
4.7	50.1	2.7	33.9	8.8	89.9
5.4	58.9	2.8	32.8	7.4	89.4
4.2	63.3	3.8	34.5	9.5	90.3
6.0	59.2	1.8	29.9	7.4	86.9
5.3	65.4	1.7	31.8	8.6	85.6
4.0	56.8	1.0	38.7	7.0	87.2
4.0	57.0	1.5	36.6	8.8	80.1
6.2	59.4	3.9	30.5		
6.5	49.6	1.9	32.8		
		1.0	30.4		

As you can see from the scatter diagram in Figure 33.2, we have adjusted for the linear relationship between sales and advertising *within* each of the three treatment groups, but a linear relationship between these quantities remains *across* the groups.

EXAMPLE 33.1

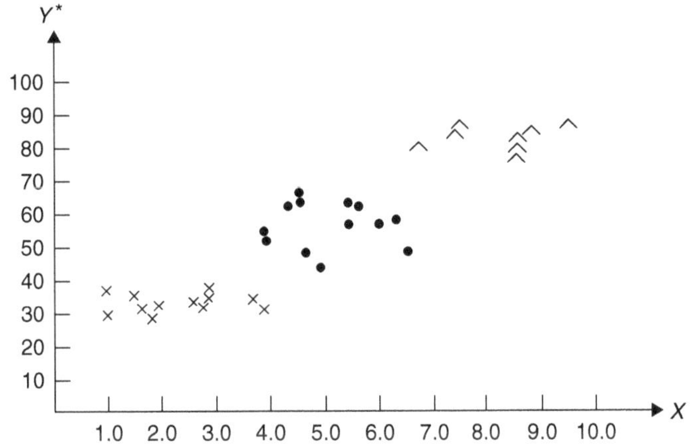

FIGURE 33.2 • = Package design P1, x = Package design P2, and ^ = Package design P3.

Assuming that the three regression slopes B_1, B_2, and B_3 are equal, we can find the least squares regression line for X and the adjusted values:

$$Y_X^* = A + BX = 18.896 + 7.687X.$$

The mean of all the values of X is $\overline{X} = 4.975$, and the final adjustment of the Y values takes this form:

$$\hat{Y}_{ij} = Y_{ij}^* - B(X_{ij} - \overline{X}) = Y_{ij}^* - 7.687(X_{ij} - 4.975).$$

For the first case in treatment group P_1, for example,

$$\hat{Y}_{11} - 45.3 - 7.687(4.9 - 4.975) - 45.9.$$

Similar calculations complete the table of fully adjusted values shown in Table 33.4.

TABLE 33.4 ANOCOVA data table with fully adjusted mean sales values.

Package Design					
P_1		P_2		P_3	
X	Y	X	Y	X	Y
4.9	45.9	2.9	54.2	8.6	60.0
5.7	57.9	2.9	54.1	8.5	56.4
4.7	52.2	2.7	51.4	8.8	60.5
5.4	55.6	2.8	49.5	7.4	70.8
4.2	69.2	3.8	43.5	9.5	55.5
6.0	51.3	1.8	54.3	7.4	68.3

(Continued)

EXAMPLE 33.1

TABLE 33.4 Continued.

Package Design					
P_1		P_2		P_3	
X	Y	X	Y	X	Y
5.3	62.9	1.7	57.0	8.6	57.7
4.0	64.3	1.0	69.2	7.0	71.6
4.0	64.5	1.5	63.3	8.8	52.2
6.2	50.0	3.9	38.8		
6.5	37.9	1.9	56.4		
		1.0	61.0		

The coefficient of determination of X and the adjusted values $\hat{Y}$ is essentially 0, as illustrated in the scatter diagram in Figure 33.3.

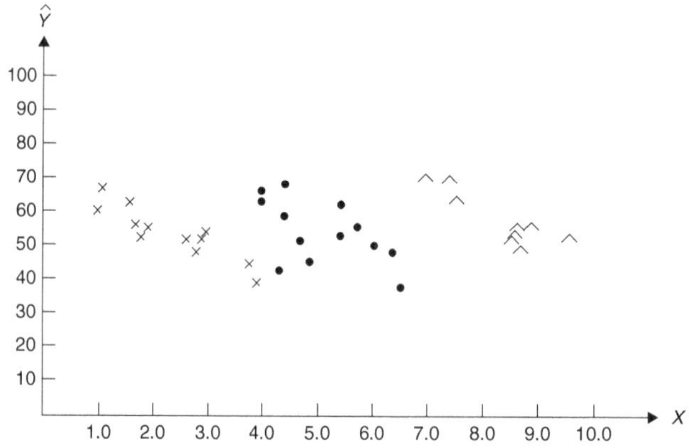

FIGURE 33.3 • = Package design P_1, x = Package design P_2, and ^ = Package design P_3.

It also appears that there is no difference in sales, when adjusted for the effect of advertising, across the three treatment groups.

We test this last contention by performing the following analysis of variance, which uses the fact that variation in the values of the dependent variable, Y, is due to:

- Overall linear dependence between Y and X, the *covariate*
- The treatments, P
- Differences in the linear relationships between X and Y across the treatment groups
- Random variation within the treatment groups

EXAMPLE 33.1

This analysis is performed on the original data values, organized as shown in Table 33.5.

TABLE 33.5 One-way ANOCOVA design table.

Treatment Groups						
1		**2**		**...**	**K**	
X	**Y**	**X**	**Y**	**...**	**X**	**Y**
X_{11}	Y_{11}	X_{12}	Y_{12}		X_{1k}	Y_{1K}
X_{21}	Y_{21}	X_{22}	Y_{22}	...	X_{2k}	Y_{2k}
...	...	...	...	...	...	...
$X_{n_1 1}$	$Y_{n_1 1}$	$X_{n_2 2}$	$Y_{n_2 2}$	...	$X_{n_k k}$	$Y_{n_k k}$

Various sums of products and squares are computed in these ways:

$$SP_{(j)xy} = \sum_{i=1}^{n_i} \left(X_{ij} - \overline{X}_{.j}\right)(Y_{ij} - \overline{Y}_{.j})$$

$$SP_{(j)xx} = \sum_{i=1}^{n_i} (X_{ij} - \overline{X}_{.j})^2$$

$$SP_{yy} = \sum_{j=1}^{k} \sum_{i=1}^{n_i} (Y_{ij} - \overline{Y}_{-})^2$$

$$SP_{xy} = \sum_{j=1}^{k} SP_{(j)xy}$$

$$SP_{xx} = \sum_{j=1}^{k} SP_{(j)xx}$$

$$SS_{xy} = \sum_{j=1}^{k} \sum_{i=1}^{n_i} (X_{ij} - \overline{X})(Y_{ij} - \overline{Y})$$

$$SS_{xx} = \sum_{j=1}^{k} \sum_{i=1}^{n_i} (X_{ij} - \overline{X})^2$$

$$SS_{yy} = \sum_{j=1}^{k} \sum_{i=1}^{n_i} (Y_{ij} - \overline{Y})^2$$

Note that these values are related to the variances and covariances of the variables within and across the treatment groups. The ANOVA table for this analysis is shown in Table 33.6.

EXAMPLE 33.1

TABLE 33.6 ANOVA table for analysis of covariants.

Source of Variance	Degrees of Freedom	Sum of Squares	Mean Sum of Squares	F
Overall linear dependence	1	$SS_{lin\ dep} = \frac{S_{xy}^2}{S_{xx}}$	$MS_{lin\ dep} = \frac{SS_{lin\ dep}}{1}$	
Treatments	$k-1$	$SS_{tr} =$	$MS_{tr} = \frac{SS_{tr}}{(k-1)} \frac{MS_{tr}}{MS_{within}}$	
			$(S_{yy} - \frac{S_{xy}^2}{S_{xx}})(SP_{yy} - \frac{SP_{xy}^2}{SP_{xx}})$	
Difference in slopes	$k-1$	$SS_{diff} =$	$MS_{diff} = \frac{SS_{diff}}{(k-1)} \frac{MS_{diff}}{MS_{within}}$	
			$\sum_{j=1}^{k} \frac{SP_{(j)xy}^2}{SP_{(j)xx}^2} - \frac{SP_{xy}^2}{SP_{xx}}$	
Within groups	$N-(k-1)$	$SS_{within} =$	$MS_{within} = \frac{SS_{within}}{N-(k-1)}$	
			$SP_{yy} - \sum_{j=1}^{k} \frac{SP_{(j)xy}^2}{SP_{(j)xx}}$	
Total	$N-1$	$SS_{total} = S_{yy}$		

The F statistic MS_{diff}/MS_{within}, with $k - 1$ and $N-(k - 1)$ degrees of freedom, is used to test the null hypothesis that the regression slopes are equal, while MS_{tr}/MS_{within}, also with $k - 1$ and $N-(k - 1)$ degrees of freedom, tests the hypothesis that the mean value of the dependent variable does not vary across the treatment groups.

In our example, we compute the following sums:

$SP_{(1)xy} = 93.14$	$SP_{(20)xy} = 105.94$	$SP_{(3)xy} = 52.19$
$SP_{(1)xx} = 7.84$	$SP_{(2)xx} = 10.72$	$SP_{(3)xx} = 5.30$
$SP_{yy} = 3269.84$	$SP_{xy} = 251.27$	$SP_{xx} = 23.86$
$SP_{yy} = 17090.15$	$SS_{xy} = 1835.43$	$SS_{xx} = 206.09$

EXAMPLE 33.1

From these, we construct the ANOVA table shown in Table 33.7.

TABLE 33.7 ANOVA table.

Source of Variance	Degrees of Freedom	Sum of Squares	Mean Sum of Squares	F
Overall Linear	1	16346.3	16346.3	
Dependence				
Treatments	3−1=2	120.2	60.1	2.99
Difference in Slopes	3−1=2	21.3	10.6	0.53
Within Groups	32−(3−1)=30	602.5	20.1	
Total	32−1=31	17090.2		

Corresponding to 2 and 30 degrees of freedom, $F_{0.05} = 3.32$. The treatment's F statistic is not greater than this value, so we cannot conclude that sales are influenced by the choice of package design. The observed variation in the values of Y (sales) was due to the influence of the covariate X (advertising). Note the magnitude of $MS_{lin\ dep}$. Finally, the F statistic associated with differences in regression slopes is not greater than the critical value, so we do not conclude that the three regression slopes were different.

33.2 Exercises

1. To predict future equipment needs, a facilities manager evaluates three methods of data entry by means of this experiment. Eighteen newly hired trainees were given an aptitude test; scores on that test are the values of the variable X below. The trainees were divided into three groups, and for five weeks, each group entered data by one entry method. At the end of that time, their efficiency was evaluated. Their ratings are the values of Y.

TABLE 33.8 Data entry method data set.

Data Entry Method								
E-mail			FTP			Internal E-mail		
Trainee	X	Y	Trainee	X	Y	Trainee	X	Y
A	32	62	G	21	72	M	38	65
B	46	69	H	24	85	N	45	88
C	27	64	I	18	61	O	52	104
D	35	48	J	32	87	P	48	86
E	31	57	K	35	69	Q	41	72
F	40	74	L	26	74	R	37	63

a. Plot Y_{ij} against X_{ij} for all the values.

b. For each group of values, find $\overline{X}_{\cdot j}$, and calculate the values $Y_{ij}^* = Y_{ij} - B_j(X_{ij} - \overline{X}_j)$, where B_j is the slope of the regression line of The values X_{ij} and Y_{ij}, *for the jth group*. Plot the adjusted values Y_{ij}^* against X_{ij}.

c. Find the slope of B of the regression line of Y *with* X_r^* X and $\overline{X}$, the mean of the X values. For each Y value, find $\widehat{Y}_{ij} = Y_{ij}^* - Bj\,(Xij - \overline{X}j)$, and plot a scatter diagram of $\widehat{Y}$ with X.

d. What conclusions can you reach from parts (a), (b), and (c)?

e. Perform an analysis of variance of $\widehat{Y}$ by data entry method, and compare the results with those of an analysis of covariance of Y by data entry method with aptitude.

2. A one-way classification with three treatments produced responses Y and associated observations X as shown in this table:

TABLE 33.9 One-way ANCOVA data set.

Treatments					
1		**2**		**3**	
X	Y	X	Y	X	Y
6	41	3	42	8	39
8	41	9	65	9	38
2	23	10	70	12	61
7	42	6	44	4	22
9	56	4	29	7	28
4	47	5	36	1	3
8	65	8	42	8	33
7	51	4	68	6	17
6	41			9	14
5	39			4	22
4	57				

a. At the 5% significance level, must we conclude that the slope of the regression line of Y with X varies across the treatment groups? Use $a = 5\%$.

b. Complete the analysis of covariance required to test for treatment effects on Y at the 5% significance level, while controlling for X.

3. Using the data given in exercise 1, create a raw data file and use a statistical software package to perform an analysis of covariance of efficiency score by data entry method while controlling for aptitude test score.

4. Create a raw file of the data given for problem 2 and using a statistical software package to perform an analysis of covariance of Y by the treatment levels while controlling for X. Use the 5% significance level.

5. Using the American Cities file and a statistical software package, perform an analysis of covariance of X3 by X1 while controlling for X7. What conclusions, if any, do you reach?

CHAPTER 34

EXPERIMENTAL DESIGNS

OVERVIEW AND LEARNING OBJECTIVES

In This Chapter

34.1 Introduction

We have examined a number of ways of analyzing the data generated by an experiment, but have not dealt with creating the experiment. How does a researcher construct an experiment? What factors are important in its design? These are the fundamental questions in the area of experimental design.

> *Experimental design* is the area of investigation dealing with the structure and planning of experiments.

34.2 Classification of Designs

The purpose of an experiment is to discover the effect of certain independent variables on a dependent variable. The outcome of the experiment will be measured in terms of variation in the value of the dependent variable.

Often, an effort will be made to control the possible effects of (independent) variables not intended to be part of the experiment whose presence might influence the value of the dependent variable and thereby muddy the results. A *completely randomized block design* maintains such control by appropriate construction of the experiment, while the *analysis of covariance* model is a statistical technique to control the effects of extraneous variables or concomitant examples.

Another goal in experimental design is the maintenance of a balance between replicability of the research results and the ability to generalize them. Certainly it is desirable that the researcher be able to extend the results of his study to individuals, cases, or situations that were not part of the experiment, but it should also be possible for others to replicate his results.

Before we define types of experimental designs that might meet these goals, we first distinguish between two approaches to applied research:

> *Experimental research* occurs when the experimenter manipulates certain independent variables, while controlling the effects of others.
>
> *Ex post facto research* occurs when the experimenter *observes* values of the independent variables whose effects on the dependent variable are being studied, while controlling the effects of others.

Note that in *experimental research*, the experimenter *determines* values of the independent variables being examined, while in *ex post facto research*, he *observes* values of the independent variables but does not influence them.

The statistical analyses applied to the data are not determined by the type of research employed. They will be the same in either case, and the following algorithm describes the analytical process:

Step 1: Observe the system to be studied.

Step 2: If the problem is defined, continue; ELSE go to Step 1.

Step 3: If variables can be identified, continue; ELSE go to Step 1.

Step 4: If hypotheses can be formulated, continue; ELSE go to Step 2.

Step 5: Establish the research design.

Step 6: Determine the sampling procedures.

Step 7: Collect the data.

Step 8: Perform the statistical analysis.

Step 9: Interpret the results.

Step 10: If results correspond to the problem, continue; ELSE go to Step 5.

Step 11: Report the results.

When we identify ways to control the variation of extraneous independent variables, we have completely described the components of the system under study: the dependent variable, the independent variables whose effects on the dependent variable concern us, and other independent variables whose influence on the dependent variable we wish to control. There are four major methods for controlling the impact of these *control variables*:

- Holding such variables constant
- Randomizing their effects
- Matching experimental units with respect to values of the control variables
- Controlling their impact statistically

We have used the technique of holding all the independent variables constant except the one being studied when we observed the drop in R^2 from the full to a reduced model in multiple linear regression. Randomization is possibly the "best" method of control; experimental units are assigned randomly to the various experimental treatments (values of the independent variable being studied) with respect to values of the control variables. There are several ways to match experimental units to control the impact of an independent variable. In one, the elements of the experimental group are stratified by values of the control variable, and these strata are then distributed among the treatments. In another, the sample is divided into equal parts corresponding to the number of values of the control variable, and the experiment is repeated several times. Finally, control can be achieved statistically whenever the control variable is correlated with the dependent variable, since the control variable then accounts for part of the variation of the dependent variable. The researcher need only eliminate that portion of the variation from the analysis.

34.3 Experimental Design Definitions

Classical experimental design:

- Two groups are formed either by random assignment or by matching.
- Two groups are identical before the experiment begins.
- During the experiment, one group (the *treatment group*) receives the experimental treatment and the other (the *control group*) does not.
- After the experiment, differences in the values of the dependent variable from the control to the treatment group can thus be attributed to the experimental treatment.

Often, practical limitations—cost, time, etc.—prevent this design from being fully realized. In some cases, for example, a similar but nonequivalent control group is used. Or the researcher will construct a comparison group by dividing an intact group into treatment and comparison components, according to the value of some measure taken before administration of the experimental treatment. Designs with nonequivalent control groups are called *quasi-experimental designs.*

The simplest experimental design is called the *simple randomized design*:

Simple randomized design:

- Only one independent variable and the dependent variable are included in the analysis.
- Control is achieved by randomization, and possibly by holding other variables constant.

One-way analysis of variance is an example of such a design when the dependent variable is interval level. When the dependent variable is nominal, we have used the chi-square test.

The inclusion of more independent variables in the analysis leads to *factorial designs* in which the researcher conducts several experiments simultaneously with the same experimental units, and the possible effects of interactions between independent variables can be studied.

We make the following definition.

Factorial designs:

- Two or more independent variables, called *factors*, are studied simultaneously.
- No provisions are made for controlling the effects of other independent variables on the dependent variable other than randomization and possibly holding control variables constant.

- Given the factors A and B (refer to Figure 34.1), the following tests are performed:
  - H_0: Row means are equal $(\mu_{A_1} = \mu_{A_2} = \ldots = \mu_{A_l})$
    H_1: Row means are unequal;
  - H_0: Column means are equal $(\mu_{B_1} = \mu_{B_2} = \ldots = \mu_{B_k})$
    H_1: Column means are unequal;
  - H_0: Cell means have equal trends across treatments (values of B)
    H_1: Cell means have unequal trends across treatments.

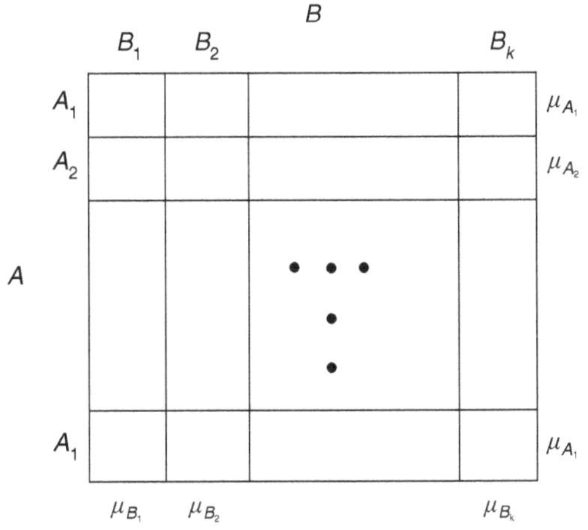

FIGURE 34.1 Factorial design (A and B are factors).

All figures and tables in this chapter appear on the companion DVD.

Factorial designs have been illustrated by multiple analysis of variance, but they may also be analyzed with multiple regression.

Simple randomized designs can be improved by using samples matched with respect to the *blocking variable.* Such designs are called *randomized block designs*.

Randomized block designs:

- The population is stratified by values of the blocking variable.
- Subsamples are drawn from each stratum in such a way that the distribution of values of the blocking variable in the entire sample grossly matches that of the population.
- Given the factor A and blocking variable B, this hypothesis is tested (see Figure 34.2):

  H_0: Column means are equal $(\mu_{A_1} = \mu_{A_2} = \ldots = \mu_{A_{kl}})$
  H_a: Column means are unequal

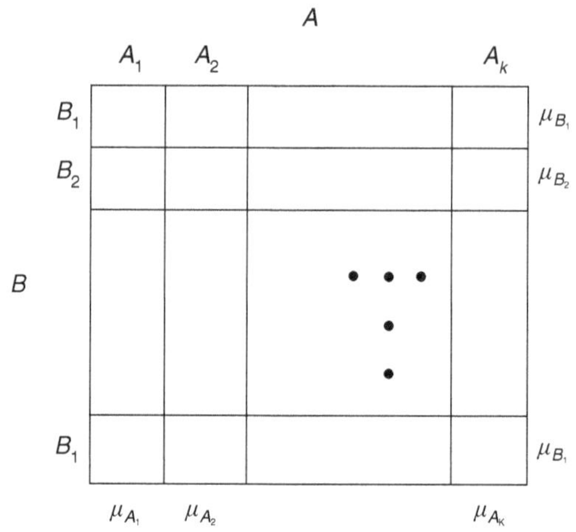

FIGURE 34.2 Randomized block design (A is a factor and B is a blocked factor).

Whether the hypotheses of equal row means or interaction are to be tested will be a function of the particular experiment.

Such a design has three distinct advantages over a simple randomized design:

- The stratification gives some control over the blocking variable.
- Because of the stratification, the analysis is more sensitive to possible relations between the dependent variable and the factor.
- Data for the independent variable can be studied separately.

Both analysis of variance and multiple regression can be used with the randomized block design.

If the effects of all treatments are temporary, the researcher may use a *repeated measure design*.

Repeated measure design:

The experimental treatments are administered in order to each experimental unit.

The statistical techniques applied to data generated by such a design are the same as those used with a randomized block design.

Sometimes it is impossible to obtain experimental units for every cell of a factorial design, as shown in Figure 34.3.

	A_1	A_2	A_3
B_1	Not available	Not available	
B_2	Not available	Not available	
B_3		Not available	Not available

FIGURE 34.3 Experimental design with nested variable B in A.

Note that only one cell in each row contains data. Such experiments are generally analyzed as simple randomized designs, and are called *experimental designs with nested variables*. The factor B is nested in A in Figure 34.3.

Analysis of covariance extends the classical experimental designs described above by combining techniques of prediction with the analysis of variance. Prediction of the dependent variable from the control variables allows the adjustment of the values of the dependent variable. These adjusted values are then analyzed using the usual experimental designs.

34.4 Avoiding Pitfalls

In any research endeavor, it is wise to follow these guidelines:

- Select a meaningful dependent variable.
- When using a pre-test, do not use both selection and control experimental units with the same measurements for the dependent variable.
- If only one set of measurements of the dependent variable is used, when the selection is made it will alter the results of the control mechanism by what is called the *regression effect*.
- Use the same instrument for pre- and post-testing.
- Make sure the treatment and control groups are comparable.
- When the treatment and control groups are nonequivalent, use the appropriate statistical adjustments. In particular, avoid the use of *raw gain scores*, which are the differences between pre- and post-test scores, unless the standard deviations of the pre- and post-test scores are equal.
- Do not construct a matched control group after the treatment group has been selected. If this is done, regression will act differently on the two groups and will inflate apparent differences. Be careful in the collection of data.
- Avoid missing data. The analysis should be based on experimental units present in both control and experimental groups.
- Do not generalize beyond the information extracted from the experiment and the results of the statistical analysis.

34.5 Experimental Goals

Upon completion of his research, the researcher should be able to answer the following questions:

- Is there a significant difference in the value of the dependent variable that can be attributed to one or more of the factors?
- Is this difference statistically significant?

- Is the effect of factors on the dependent variable significant in terms of the problem?
- Can the treatments of the experiment be applied in another setting with a reasonable expectation of similar results?
- How likely is it that the observed effects are the results of the treatments?
- Are the results believable and interpretable?

34.5.1 Experimental design in practice

What does validation consist of? Providing dramatic examples of the advantages of relying on something more than plausibility. As new researchers, we often set out to use our background in experimental design to solve the problems of various organizations. Always mindful of the model of a controlled experiment, we then find ourselves constantly assigned research projects where we have to deviate from the ideal model. Consider the situation described in Table 34.1:

TABLE 34.1 Traditional experiment.

Groups	Time Period	
	Before	**After**
Experimental Group	x_1	x_2
Control Group	x'_1	x'_2

In this type of study, the objective is to test whether a difference is attributable to what we think it was attributable to, or whether $d = x_1 - x_2$ is significantly larger than $d' = x'_1 - x'_2$.

In practice, we often do not have the patience to conduct a full-scale investigation or have pragmatic constraints placed upon the study that lead to a design not containing all four cells. Sometimes we have only two cells, especially in post hoc studies, as shown in Table 34.2:

TABLE 34.2 Before/after comparative design.

Before	After
X_1	X_2

In this situation, we have two observations of the same individuals or groups taken at different times. Note that what factors ultimately were most responsible for the results could only be inferred indirectly with this design.

Next we encounter the design shown in Table 34.3:

TABLE 34.3 Virtual before/after comparative design.

Before	After
X_1	
	x'_2

Here, at one point in time we have one sample, and at a latter point in time we have another sample. A major problem with this design occurs if the populations in the two samples are initially different. In this case, the differences might not be attributable to conditions taking place in the time interval between the two observations.

To strive for a controlled experiment without paying the price, the design in Table 34.4 might be utilized (as in correlation analysis):

TABLE 34.4 Weak before/after design.

X_1
X_2

Here, we have two or more groups that we study at the same point in time. In this design, there is the possibility that other uncontrolled variables can contribute.

In the worst-case scenario, we use one cell, as shown in Table 34.5:

TABLE 34.5 Worst case design.

X_2

Here, we simply filled in a previous time cell with hypothetical data, as shown in Table 34.6:

TABLE 34.6 Hypothetically modified worst case design.

X_1	X_2

This design occurs in practice more often than we probably would be willing to admit.

Even worse would be the design shown in Table 34.7:

TABLE 34.7 Ultimate worst case design.

X_1	X_2
x'_1	x'_2

In conclusion, we should strive to set forth the model of the controlled experiment. But we should realize that in practice, often some cells may be missing from the ideal model, yet more economy and orderliness are still possible. We accomplish this by using theory in our planning to help us decide whether a particular inquiry would be important if we made it by narrowing down the number of variables and by making sure we can classify our data through a particular continuum, even if only provisionally.

34.6 Exercises

1. In an analysis of variance design, the number of observations of the dependent variable may not be the same for each cell of the design. State two possible explanations for this phenomenon and how to deal with each.

2. In a design that compares three treatment groups, what can be done to increase the precision and reliability of the test?

3. In what way are the randomized block design and the covariance design alike? How are they different?

4. In an analysis of covariance, what should we do if equality of regression slopes is rejected?

5. Describe how control is achieved in a one-factor experiment with repeated measurements.

6. In a three-way analysis of variance with cell frequencies greater than 1, into how many parts is the total sum of squares partitioned? Explain.

7. Why is a randomized block design more precise in general than a simple randomized design?

8. When interaction is found to be statistically significant in a randomized block two-factor design, what should be done?

9. Why are equal or proportional groups of measurements per treatment used in factorial designs?

10. In certain data processing tasks performed by a computer center, data is run through several programs, and each step of the process is documented. The manager of the center wants to assess the effects on the information ultimately produced of five different organizations of the documentation steps. He also wants to control the effects of two blocking factors—the particular data file being processed and the programmer writing the task.

Therefore he constructs a Latin square design, using five levels for each factor. The five organizations of documentation steps are indicated by the letters A, B, C, D, and E, and the results, which are numerical evaluations of the final product, are shown in this table.

		Data File				
		1	2	3	4	5
Programmer	1	A 24	B 16	C 22	D 16	E 13
	2	B 20	C 20	D 15	E 15	A 14
	3	C 29	E 16	B 10	A 17	D 10
	4	D 23	A 10	E 10	C 23	B 16
	5	E 20	D 23	A 20	B 19	C 26

a. In what hypothesis test is the manager interested?

b. Calculate the row, column, treatment, and grand means.

c. Calculate the four sums of squares and construct the ANOVA table.

d. What conclusions can be reached from these data?

11. Why are equal or proportional groups of measurements per treatment used in factorial designs?

12. Why are matched samples often used in classical experimental design?

13. State the linear model for the two-factor design. What assumptions are made when we employ this model, and how critical are they?

14. Consider the Latin square design on a study of MPG given in Table 34.8:

TABLE 34.8 Latin square design table.

Variable	1st level	2nd	3rd	4th
Car	1	2	3	4
Driver	I	II	III	IV
Fuel Additive	A	B	C	D

Driver(i)	1	2	3	4
I	A 21	B 26	D 26	C 25
II	D 23	C 26	A 26	B 27
III	B 15	D 14	C 16	A 16
IV	C 17	A 15	B 20	D 20

a. State the null hypothesis.

b. State the model form.

c. Using Excel, generate the two-way ANOVA table and interpret the results.

d. Generate and interpret the residual plots.

EXPERIMENTAL DESIGN PROJECT

In an excellent article by Ho Siong Lin of Ngee Ann Polytechnic in the Republic of Singapore, found at htpp://stat.auckland.ac.nz/~lase/publications/2/topic2o.pdf, the author explores the effectivenss of using a catapult for enhancing the understanding of design of experiments in the classroom. Mainly, the author recommends using factorial design to analyze the experimental data. An effort is made to determine the key input factors that significantly affect the output responses. After careful study of this article, either obtain a suitable catapult or build your own catapult. Finish this project by completing the activities found in: Teaching Basic Experimental Design with Catapults and other Hands-On Activities by Mark G. Haug, PhD, JD, University of Kansas School of Business given at USCOTS 2007 Breakout Session found at *http://www.causeweb.org/uscots/uscots07/program/files/breakout3_4.pdf* and the Satapult - Study discussed at *www.launsby.com/Documents/Statapult4download.doc*.

UNIT 8

NONPARAMETRIC TESTS AND COMMONLY USED DISTRIBUTIONS

Study *Detection of intrusions in information systems by sequential change point methods* by Tartaakovsky and etc. at *academic.research.microsoft.com/Paper/4671956.aspx* and answer the following questions. Basically, this article describes and test abstract sequential multi-chart detection procedures for detecting changes in multichannel sensor systems. In the case of complete information on pre-change and post-change distributions, the detection algorithm, represents a likelihood ratio-based multichannel, generalization of Page's cumulative, sum (CUSUM)test. The CUSUM test is applied to general stochastic models, that may, include correlated and nonstationary observations. The authors main motivation for doing the study was for network security: rapid anomaly, detection for an early detection of attacks in computer networks that lead to changes in network traffic. This kind of application encourages the development of a nonparametric multichannel detection test that does not use exact pre-change (legitimate) and post-change (attack) traffic models. The proposed nonparametric method can be effectively applied to detect a wide variety of attacks such as denial-of-service attacks, worm-based attacks, port-scanning, and man-in-the-middle attacks. In addition, a multichannel CUSUM procedure is proposed that is based on binary quantized data.

1. What type of detection algorithms were proposed? Why?
2. Why was a Monte Carlo simulation run?
3. Which method evaluated might perform best in practice? Why is this a surprising result?
4. Explain what is meant by an asymptotic formula.
5. What type of practical applications in an identically independently distributed environment could utilize the CSUM stopping time under the 'no change' hypothesis? Why?

One way of dealing with the complexities of the world is to build a model to simulate a situation or environment. The model then can be tested and manipulated. These results can be used to predict what will happen in the world. An airplane company wants to know how a proposed new plane will fly, so it builds a model and places the model in a wind tunnel and runs tests. Or, it creates a computer program that allows engineers to specify the plane on the screen and then mathematically simulate the plane flying under different conditions.

Of course, the effectiveness of using a model to predict what really will happen depends on how well the relevant aspects of the situation are captured in the model. One certainly can imagine an airplane whose design worked great on the computer but that crashes when it is first flown. There is a saying in the computer field, "Garbage In, Garbage Out," or "GIGO". If you put nonsense into the computer, you will get nonsense out. Don't believe everything that comes out of the computer. Just because "The computer says" something doesn't mean it is right.

People have attempted to create computer models for all kinds of situations. There are computer models of the weather. Enter in the temperature, wind velocity, and so on at many different locations on the earth. The computer predicts what the weather will be like tomorrow or next week. There are computer models of the economy. Suppose we raise the income tax rate. What will be the effect on the economy a year from now or five years from now?

We have seen how Excel can be used as a simulation tool. "What if analysis is computer simulation. To predict the profitability of a new enterprise, create an Excel worksheet to model the projected finances and try it out under different scenarios.

Some situations are best modeled probabilistically. One third of the time this will happen; two thirds of the time something else will happen. These situations can be modeled on the computer with the use of **random numbers**. Simulations that make use of random numbers often are called **Monte Carlo simulations**. Monte Carlo is a city on the Mediterranean Sea that is the site of a well-known gambling casino.

In this unit we will discuss random numbers in Excel and ten their use in simulations will be presented in unit ten.

CHAPTER 35

RANDOM NUMBER GENERATION

OVERVIEW AND LEARNING OBJECTIVES

In This Chapter

35.1 Introduction

35.2 Random Number Generation

35.3 Desired Properties of a Good Generator

35.4 Linear Congruential Generators

35.5 Multiplicative Linear Congruential Generators

35.6 Extended Fibonacci Generators

35.7 Combined Generators

35.8 Seed Selection

35.9 Myths about Random Number Generation

35.10 Exercises

Random Number Project I

Random Number Project II

35.1 Introduction

In almost all our work so far, we have paid careful attention to the shapes of population distributions in assessing the applicability of our statistical analyses and tests. In this chapter, we examine *random number generation*.

35.2 Random Number Generation

One of the key steps in performing a statistical analysis for a simulated study is often to have a routine to generate random values for variables with a specified random distribution; for example, exponential and normal. This is done in two steps. First, a sequence of random numbers distributed uniformly between 0 and 1 is obtained. Then the sequence is transformed to produce random values satisfying the desired distribution. The first step is called *random number generation* and the second *random variate generation*.

35.3 Desired Properties of a Good Generator

Most random number generators are defined as a recursive relation in which the next number in the sequence is a function of the last one or two numbers. For example,

$$X_n = (5X_{n-1} + 1)\mathrm{mod}16$$

defines, starting with $X_0 = 5$, the random sequence:

5, 10, 3, 0, 1, 6, 15, 12, 13, 22, 11, 8, 9, 14, 7, 4, 5.

Note that $X_1 = (5(5) + 1)\mathrm{mod}\ 16 = 26 \mathrm{\ mod\ } 16 = 10$. For this random number generator, all the X_i values are integers between 0 and 15. Also take note that the sequence regenerates 5. The fact that the number 5 is regenerated means the sequence can be regenerated any time provided the starting value X_0 is given. The value that is used to begin the sequence is called the *seed*.

Given the seed, we can tell with a probability of 1 the numbers in the sequence that will be generated, or we say the function representing the random number generator is *deterministic*. Yet the numbers are random in the sense that they would pass statistical tests for randomness, to be discussed in Chapter 37, "Testing for Randomness." These numbers are therefore only partially random and are called *pseudorandom numbers*.

In our example, only 16 numbers are unique within the sequence, and then these 16 numbers are repeated. In other words, this random number generator has a *cycle length* of 16. If a generator does not repeat an initial part of the sequence, then that part is referred to as the *tail*. The *period* of the generator is the sum of the tail length and the cycle length.

The desirable properties for a random number generator are:

- The related software should execute efficiently.
- The sequence should possess a large period.
- The numerical values in the sequence should be independent and uniformly distributed.

The last property requires a battery of statistical tests to validate randomness.

Some of the random number generators commonly used include:

- Linear congruential generators
- Extended Fibonacci generators
- Combined generators

A description of each of these approaches follows.

35.4 Linear Congruential Generators

The residues of successive powers of a number possess good randomness properties, where the nth number in the sequence is found by dividing the nth power of an integer a by another integer m and keeping the remainder. That is,

$$X_n = (aX_{n-1} + b)\,\text{mod}(m).$$

The parameter a is called the *multiplier* and m is called the *modulus*. Such generators are referred to as *Linear Congruential Generators* (*LCGs*).

In general, the choice of a, b, and m affects the period and autocorrelation in the resultant sequence. Generally,

- The modulus m should be large.
- For mod(m) computations to be efficient, m should be a power of 2.
- If b is nonzero, the maximum possible period m is obtained when:
  - m and b are relatively prime (they have no common factors other than 1).
  - Every prime number that is a factor of m is also a factor of $a - 1$.
  - $a - 1$ is a multiple of 4.

Notice that all of these conditions are met if $m = 2^k$, $a = 4c + 1$, and b is odd. It is assumed that c, b, and k are positive integers.

35.5 Multiplicative Linear Congruential Generators

A *multiplicative LCG* is of the form:

$$X_n = aX_{n-1} \bmod(m).$$

Multiplicative LCGs are more efficient than mixed LCGs. Additional efficiency can be obtained by choosing m to be a power of 2. However, the maximum possible period for a multiplicative LCG with modulus $m = 2^k$ is only one-fourth the full period possible, that is, 2^{k-2}. The latter period is achieved if the multiplier a is of the form $8i \pm 3$ and the initial seed is an odd integer.

A solution to the small period problem is to use a modulus m that is a prime number. In this case, with a proper choice of the multiplier a, it is possible to get a period of $m-1$, which is almost equal to the maximum possible length m. Notice that unlike a mixed LCG, X_n obtained from a multiplicative LCG can never be 0 if m is prime. The values of X_n lie between 1 and $m-1$, and any multiplicative LCG with a period of $m-1$ is called a *full period generator*.

Not all values of the multiplier are equally good. It can be shown that a multiplicative LCG will be a full period generator if and only if the multiplier a is a *primitive root* of the modulus m. By definition, a is a primitive root of m if and only if $a^n \bmod(m) \neq 1$ for $n = 1, 2, \ldots, m-2$.

One of the important cautions in implementing LCGs is that the properties are guaranteed only if the computations are done exactly and without any rounding errors. Another problem in implementing LCGs is that the product aX_{n-1} can exceed the largest integer allowed on the computer system, causing integer overflow.

35.6 Extended Fibonacci Generators

A Fibonacci sequence $\{X_n\}$ is generated by the following relationship:

$$X_n = X_{n-1} + X_{n-2}$$

Random numbers can be generated using the Fibonacci sequence in this way:

$$X_n = \left(X_{n-1} + X_{n-2}\right) \bmod(m)$$

However, this sequence does not possess good randomness properties as it has a high serial correlation. But the following relationship passes most statistical tests for randomness:

$$X_n = \left(X_{n-5} + X_{n-17}\right) \bmod(2^k)$$

35.7 Combined Generators

It is possible to combine two or more random generators to produce a "better" generator. The following are three such techniques:

- Adding random numbers obtained by two or more generators.
  Note that this modification will sometimes increase the period and randomness.[1]

- Exclusive-or random numbers obtained by two or more generators.[2]
- Shuffle. Shuffling is when one sequence is used as an index to decide which of several numbers generated by a second sequence should be returned.[3] Algorithm M by Marsaglia and Bray uses an array of size 100 that is filled with random numbers from a random sequence X_n. To generate a random number, generate a new Y_n (between 0 and $m-1$) and scale it to obtain an index $I = 1 + 100\, Y_n/m$. The value in the ith element of the array is returned as the next random number. A new value of X_n is generated and stored in the ith location. One problem with shuffling is that it is not easy to skip a long subsequence as is often required in applications.

35.8 Seed Selection

The following guidelines should be taken into consideration when selecting a seed for a random number generator:

- Do not use 0. This choice of value would make a multiplicative LCG stick at 0.
- Avoid even values. If you are using a multiplicative LCG, the seed should be odd.
- Do not subdivide one stream. Using a single stream for all variables is a common mistake that can result in a strong correlation between the variables.
- Use nonoverlapping streams. If the streams overlap, there will be a correlation between the streams, and the resulting sequences will not be independent.
- Reuse seeds in successive replications.
- Do not use random seeds. This causes two problems: First, the results are difficult to reproduce; and, second, it is not possible to guarantee that the multiple streams will not overlap.

35.9 Myths about Random Number Generation

The following are some common misconceptions about random number generation:

- A complex set of operations leads to a random number.
- A single test such as the chi-square test is sufficient to test the goodness of a random number generator.

Note that the sequence 0, 1, 0, 1, 0, 1, …, 0, 1 will pass the chi-square test, but clearly the sequence is not random (a pattern is present).

- Random numbers are unpredictable.

LCGs are deterministic.

- Some seeds are better than others.

In general, generators whose period or randomness depends upon the seed should not be used, since an unsuspecting user may not remember to follow the guidelines.

- Accurate implementation is not important.

The period and randomness properties of generators are guaranteed only if the generation formula is accurately implemented without any overflow or truncation.

- Bits of successive words generated by a random number generator are equally randomly distributed.

35.10 Exercise

1. A table of random numbers contains 15,050 digits, which should have been chosen at random from the digits 0 through 9. The following table gives the frequencies of each digit in the table. At the 5% significance level, should it be concluded from this data that the table is not random?

Digit	0	1	2	3	4	5	6	7	8	9
Frequency	1493	1491	1461	1552	1494	1454	1613	1491	1482	1519

RANDOM NUMBER PROJECT I

In a programming language of your choice evaluate the effectives of the language supplied uniform number generation. For the evaluation write a program, in a language of your choice, that conducts at least six or more of the test of randomness. Perform each test on randomly generated sets of 1000 numbers. Which test for randomness does the function pass? Which does it fail? Determine how the random numbers are generated. Check to see if the generator is performing in accordance with the procedure.

RANDOM NUMBER PROJECT II

Redo Project I but implement your own generators using the mixed congruential method, multiplicative congruential method, and additive congruential method. Additionally perform a comparative analysis of these congruential methods.

References

1. L'Ecuyer, P., "Efficient and Portable Combined Random Number Generators," *Communications of the ACM*, 31 (1988): 742–749 and 774.
2. Santha, M. and Vazirani, U.V., "Generating Quasi-Random Sequences from Slightly Random Sources," *Proc. 25th Annual Symposium on Foundations of Computer Science* (1984): 434–440.
3. Marsaglia, G. and Bray, T.A., "A Convenient Method for Generating Normal Variables," *SIAM Review*, 6 (1964): 260–264.

CHAPTER 36

RANDOM VARIATE GENERATION

OVERVIEW AND LEARNING OBJECTIVES

In This Chapter

36.1 Introduction

There are a number of methods to generate nonuniform variables. For a particular distribution, one method may be more efficient than the others. In this chapter, some of the commonly used methods are described. In Chapter 39, "Commonly Used Distributions," a list of common distributions and their generation algorithm is presented.

We will look at four methods for generating nonuniform variables: inverse transformation, rejection, composition, and convolution. These methods assume that we have already generated a sequence of random numbers distributed uniformly between 0 and 1, as discussed in Chapter 35, "Random Number Generation."

36.2 The Inverse Transformation

Suppose we want to generate random values from A to B from a uniform distribution defined over [A, B]. We recall that the cumulative distribution function (CDF) in this case is monotonically increasing and continuous over the interval and is defined as:

$$P(x) = \frac{x - A}{B - A}, \quad \text{for} \quad A \le x \le B.$$

Then in the *inverse method* we set r, a uniform value defined on [0, 1], equal to P(x), giving:

$$r = P(x) = \frac{x - A}{B - A}.$$

Since the cumulative distribution function is one to one, onto and continuous on [A, B], then the inverse function for P(x) exists. This "inversing" of variables gives the inverse method its name.

$$r = \frac{x - A}{B - A},$$

then

$$x = A + r(B - A).$$

The final equation is easy to use as a process generator for the uniform distribution. Given values for the parameters (A and B) and a random number on [0, 1], a sampled value (a variate) is obtained by substituting the appropriate values into the last equation. For example, if the distribution of interest is uniformly distributed with a lower limit of 5 and an upper limit of 10 and the random number 0.75 is generated, the sampled value would be

$$X = 5 + 0.75(10 - 5)$$

$$= 8.75.$$

Consider the need to sample a value from the negative exponential distribution. The first step is using the inverse transformation method to obtain the cumulative distribution function, which we know is:

$$P(x) = 1 - e^{-\lambda x}$$

Next, the uniform random number r is set equal to $P(x)$, and the expression is manipulated to express x as a function of r:

$$r = P(x) = 1 - e^{-\lambda x}$$

$$e^{-\lambda x} = 1 - r$$

Taking natural logarithms of each side, we get

$$-\lambda x = ln(1 - r)$$

$$x = \frac{-\ln(1 - r)}{\lambda}.$$

The latter equation is our random variate generator for the negative exponential distribution. Given a mean number of occurrences and a uniform random number, the number of intervals before the next occurrence is easily simulated. Keep in mind, however, that the sampled value is measured in intervals and must usually be transformed into minutes, hours, or whatever the appropriate unit of measurement may be before it becomes useful in an application. For example, if $\lambda = 1$ occurrence 5 minutes and $r = 0.40$, the sampled value is:

$$X = \frac{-\ln(0.6)}{1} = 0.52.$$

Measured in minutes, this is

$$(0.51 \text{ intervals}) * (5 \text{ minutes/interval}) = 2.55 \text{ minutes}.$$

36.3 The Rejection Method

A rejection technique can be used if another density function, the Probability Density Function (pdf), $g(x)$ exists so that $cg(x)$ *majorizes* the density function $f(x)$; that is, $cg(x) \geq f(x)$ for all values of x. If such a function can be found, then the following steps can be used to generate the random variate x with density $f(x)$:

1. Generate x with pdf $g(x)$.

2. Generate y uniform on $[0, cg(x)]$.

3. If $y \leq f(x)$, then output x and return. Otherwise, repeat from 1.

The algorithm continues *rejecting* the random variates x and y until the condition $y \leq f(x)$ is satisfied; hence, the name for the method.

Consider the problem of sampling values from a normally distributed random distribution. The inverse function does not exist; therefore the inverse transformation method does not meet our needs. However, the rejection method does solve the problem. In fact, this method works for any continuous distribution.

First enclose the continuous distribution in a rectangle, letting a rectangular distribution majorize the continuous distribution. Since the standard normal distribution is asymptotic to the z-axis, it is necessary to truncate the distribution at some reasonable point, as shown in Figure 36.1.

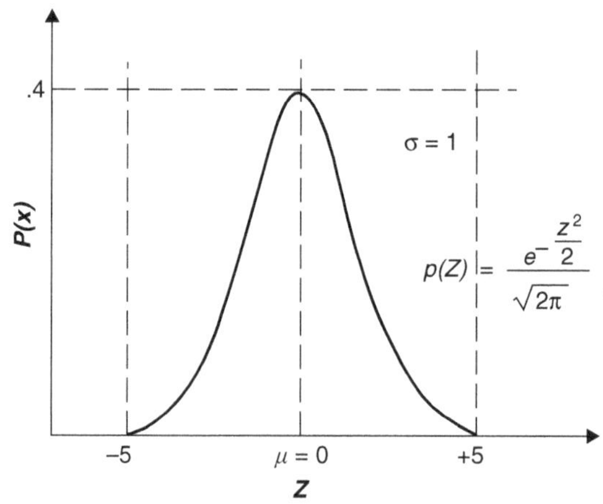

FIGURE 36.1 The majorizing of the standard normal distribution.

ON DVD All figures in this chapter appear on the companion DVD.

Points that are five standard deviations from the mean are chosen as the truncation points. The maximum height of the curve is 0.3989. Consequently, a rectangle with a height of 0.40 includes the entire distribution.

The next step is to randomly select a Z value between +5 and –5.

$$Z = -5 + 10r_1$$

After the Z value has been randomly selected, it is inserted in the equation for the standard normal distribution to determine if the height of the curve is at that point. Then a second random number is generated. It is transformed, using the uniform process generator, so that it randomly selects a point along the height of the rectangle.

$$H = 0.4r_2$$

Now comes the rejection part of the algorithm. If the value of H falls below the computed height of the curve at the randomly selected Z value, the Z value is treated as if it comes from the standard normal distribution. If, however, the value of H falls above the curve, the Z value under consideration is rejected and the process is repeated until an appropriate value is found.

36.4 The Composition Method

The composition method can be used if the desired Cumulative Distribution Function (CDF) can be expressed as a weighted sum of *n* other CDFs. That is,

$$F(x) = \sum_{i=1}^{n} p_i F_i(x)$$

where, $p_i \geq 0, \sum_{i=1}^{n} p_i = 1$, and the $F_i(x)$'s are distribution functions. The technique can also be used if the density function f(x) can be expressed as a similar weighted sum of *n* other density functions.

In either case, the steps to generate *x* are as follows:

1. Generate a random integer where

$$P(I = i) - p_i$$

   This can easily be done using the inverse transformation method.

2. Generate *x* with the *i*th $f_i(x)$ and return.

The pdf for a Laplace distribution is a composition of two exponential pdfs. Figure 36.2 illustrates the Laplace distribution given by the pdf:

$$f(x) = \frac{1}{2a} e^{\frac{-|x|}{a}}$$

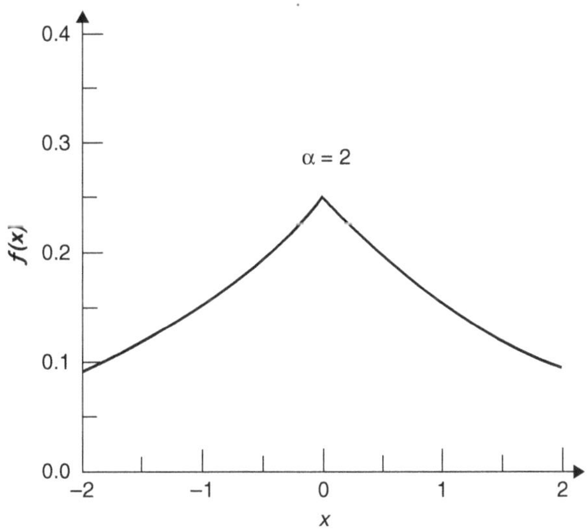

FIGURE 36.2 Laplace density function.

The probability of *x* being positive is half, and the probability of *x* being negative is half.

By the composition technique, Laplace variates can be generated in the following way:

1. Generate two uniform (0, 1) values r_1 and r_2.
2. If $r_1 < 0.5$, return $x = -a\ln(r_2)$; otherwise, return $x = a\ln(r_2)$.

36.5 The Convolution Method

The convolution method can be used if the random variable x can be expressed as a sum of n random variables $y_1, y_2, \dots, y_n$ that can be easily generated; that is,

$$x = y_1 + y_2 + \dots + y_n.$$

Notice the difference between composition and convolution. The former method is used when the pdf or CDF can be expressed as a sum of other pdfs or CDFs. The latter technique is used when the random variable itself can be expressed as a sum of other random variables. The following list of examples illustrate this summation process:

- A binomial variate with parameters n and p is a sum of n Bernoulli variates with success probability p.
- The chi-square distribution with υ degrees of freedom is a sum of squares of υ unit normal N(0, 1) variates.
- The sum of a "large" number of variates from any distribution has a normal distribution.
- The sum of two uniform variates has a triangular density.

Reconsider the normal distribution. Begin by obtaining 12 uniformly distributed variates, computing their sum, and then subtracting 6:

$$\sum_{i=1}^{12} (r_i) - 6.$$

Mathematically these sample sums are normally distributed with a mean of 0 and a standard deviation of 1:

$$\mu_{\Sigma} = 0.50(12) - 6 = 0$$
$$\sigma_{\Sigma} = 1/\sqrt{12S}\left(\sqrt{12}\right) = 1.$$

Depending on your mathematics skills, the derivation processes discussed in this chapter may or may not have been easy to follow. In most instances, however, knowledge of the final result of the derivation is all that is needed.

36.6 Exercises

1. Let a sequence of random numbers R_1, R_2, R_3, R_4, R_5, be 0.45, 0.37, 0.89, 0.11, and 0.66. Extend the sequence through R_{10} using the additive congruential method, where $m = 100$.
2. Determine whether the historical linear congruential generators shown below can achieve a maximum period. Conduct an Internet search. Also, state restrictions on X_0 to obtain this period.

a. In SIMSCRIPT for CDC; $a = 2{,}814{,}749{,}767{,}109$; $c = 59{,}482{,}661{,}568{,}369$; $m = 2^{48}$

b. IBM 360; $a = 69{,}019$; $c = 0$; $m = 2^{33}$

c. $a = 6507$; $c = 0$; $m = 1024$

3. Use the mixed congruential method to generate a sequence of three two-digit random numbers with $X_0 = 37$, $a = 7$, $c = 29$, and $m = 100$.

4. Use the mixed congruential method to generate a sequence of three two-digit random integers between 0 and 24 with $X_0 = 13$, $a = 9$, and $c = 35$.

5. Consider the multiplicative congruential generator under the following circumstances:

a. $a = 11$, $m = 16$, $X_0 = 7$

b. $a = 11$, $m = 16$, $X_0 = 8$

c. $a = 7$, $m = 16$, $X_0 = 7$

d. $a = 7$, $m = 16$, $X_0 = 8$

Generate enough values in each case to complete a cycle. What inferences can be drawn? Is maximum period achieved?

6. Generate five random observations from a *uniform distribution* between –10 and +40.

7. Suppose that random observations are needed from the *triangular distribution* whose probability density function is

$$f(x) = \begin{cases} 2x, \ if \ 0 < x < 1 \\ 0, \ otherwise \end{cases}$$

a. Derive an expression for each random observation as a function of the random decimal number *r*.

b. Generate five random observations.

CHAPTER 37

TESTING FOR RANDOMNESS

OVERVIEW AND LEARNING OBJECTIVES

In This Chapter

37.1 Introduction

In almost all our work so far, we have paid careful attention to the shapes of population distributions in assessing the applicability of our statistical analyses and tests. For example, to apply a t-test when using small samples, we must establish the normality of the underlying population(s); in linear regression, we assume normality and homoscedasticity. That is, our tests thus far have been parametric, depending on the characteristics of the underlying distributions. In this chapter, we examine tests for randomness and goodness-of-fit.

We sometimes wish to determine if a sequence of values is random, with no pattern or relationship among the values, as if each had been chosen at random from some range of values, or is not random, showing some pattern. Such investigations take the form of hypothesis tests with these hypotheses:

H_0: The values are random.

H_a: The values are not random.

The decision rule may be formulated in a variety of ways, of which we will examine four.

37.2 The Frequency Test

Suppose that a sequence of n values is to be chosen, at random and one at a time, from a list of k candidate values. At each selection of a random value to be included in the sequence, each candidate value has probability k^{-1} of being selected. The number of times a given value appears has the binomial distribution $b(n, k^{-1})$, with mean $nk^{-1} = n/k$; we would expect each of the candidate values to appear in the sequence of n selected values approximately n/k times, *if the selection at each step is in fact random.* Significant deviations of observed frequencies of the values from the predicted frequencies would suggest that the values are not random, but that some values are more likely than others.

We have used, in Chapter 23, "Chi-Square Tests of Goodness-of-Fit and Missing Data," tests based on a statistic with a chi-square distribution to compare predicted and observed frequencies, and do so again here.

If the null hypothesis is true and the values are truly random, then the statistic

$$\chi^2 = \sum_{i=1}^{k} \frac{\left(O_i - n/k\right)^2}{n/k} = \frac{k}{n} \sum_{i=1}^{k} \left(O_i - \frac{n}{k}\right)^2,$$

where O_i is the observed frequency of each value, has a chi-square distribution with $k-1$ degrees of freedom.

If the value of the chi-square statistic is near 0, the observed and expected frequencies must be similar, and we cannot conclude that the values are not random. If χ^2 is large, the observed and expected frequencies must differ, and the values are not random. We may use the chi-square table in the usual way to find the critical value of the chi-square statistic above which we may reject H_0 and accept H_a, concluding that the values are not random.

For example, consider this list of 50 digits from 0 to 4:

4 3 2 3 3 2 2 1 3 1 1 0 2 0 2 1 3 0
1 4 0 0 1 4 0 2 3 2 1 1 0 2 3 4 0 0
2 4 4 2 2 1 4 4 0 3 1 1 2 1

At the 5% significance level, we test the null hypothesis that these values are random against the alternative that they are not.

The digits are selected from a list of five possible values—0, 1, 2, 3, and 4—so $k=5$, while $n=50$. Corresponding to $k-1=4$ degrees of freedom, the critical value is $\chi^2 = 9.488$.

Each expected frequency is 50 * ½ = 10, while the five observed frequencies are

$$O_0=10;\ O_1=12;\ O_2=12;\ O_3=8;\ O_4=8.$$

The value of the chi-square statistic is

$$\begin{aligned}\chi^2 &= \frac{5}{50}\sum_{i=0}^{4}(O_i-10)^2\\ &= \frac{1}{10}\left[(10-10)^2+(12-10)^2+(12-10)^2+(8-10)^2+(8-10)^2\right]\\ &= \frac{1}{10}[0+4+4+4+4]\\ &= \frac{16}{10}=1.6\end{aligned}$$

This value is less than the critical value, so we do not reject the null hypothesis; we cannot conclude that these values were randomly selected.

To run this problem in Excel, simply enter the data in cells A1 through A50 and then enter the formulas shown in Table 37.1:

TABLE 37.1 EXCEL functions to construct a chi-square test.

Values	Ob Freq	Ex Freq	Diff	Diff ^ 2
0	=COUNTIF($A$1:$A$50,E7)	10	=F7-G7	=H7*H7
1	=COUNTIF($A$1:$A$50,E8)	10	=F8-G8	=H8*H8
2	=COUNTIF($A$1:$A$50,E9)	10	=F9 G9	=H9*H9
3	=COUNTIF($A$1:$A$50,E10)	10	=F10-G10	=H10*H10
4	=COUNTIF($A$1:$A$50,E11)	10	=F11-G11	=H11*H11
	chi-sq =	=(1/10)*SUM(I7:I11)		

All figures and tables in this chapter appear on the companion DVD.

The following function in Excel provides the calculation of the probability of a value occurring at the 5% significance level:

prob =	= CHIDIST(1.6,4)

This generates the data found in Table 37.2:

TABLE 37.2 Data file for a goodness of fit test.

Values	Ob Freq	Ex Freq	Diff	Diff ^ 2
0	10	10	0	0
1	12	10	2	4
2	12	10	2	4
3	8	10	-2	4
4	8	10	-2	4
	chi-sq =	1.6		

Then complete the analysis as done above.

Also note that this is confirmed with the following probability, which is greater than $a = 0.05$:

prob =	0.80879

37.3 The Gap Test

A *gap* in a sequence of random numbers is the number of values *between* two identical values. For example, in the sequence 6 5 3 2 3 6 1 … , there is a gap of length 4 between the first and second occurrence of the digit 6, and a gap of length 1 between the first and second 3. In a sequence of digits chosen randomly from a set of k candidate values, the length of a gap after a given value is a random variable X with probability function

$$f(x) = P(X = x) = \left(1 - \frac{1}{k}\right)^x \frac{1}{k}, \; x = 0, 1, 2, \ldots$$

That is, a gap of length x occurs after a particular value γ when the next x values are not γ, and the $(x + 1)$st value is γ. This occurs with probability

$$\left(1 - \frac{1}{k}\right)^x \frac{1}{k}.$$

If a sequence of random values contains N gaps, the expected number of gaps of length x is:

$$N^* f(x) = N\left(1 - \frac{1}{k}\right)^x \frac{1}{k}.$$

To investigate the randomness of the sequence, we can compare the expected and observed numbers of gaps of each length using the chi-square statistic as in the frequency test. The associated number of degrees of freedom will be one less than the number of gaps considered.

Consider again the sequence of 50 digits from 0 to 4 tested in the previous section. There are N = 45 gaps in the sequence, where values were chosen from the digits 0, 1, 2, 3, and 4. Since the probability of a gap of, say, length 2 is

$$\left(1-\frac{1}{5}\right)^2 \frac{1}{5} = \frac{16}{125} = 0.28,$$

the expected number of gaps of length 2 is

$$\frac{16}{125} * 45 = 5.76.$$

Finding the other expected gap frequencies in the same way, and counting the observed numbers of gaps, we obtain Table 37.3.

TABLE 37.3 The gap test of randomness.

	Digits					Gap	
Length	**0**	**1**	**2**	**3**	**4**	**Expected**	**Observed**
0	2	3	2	1	2	10	9.0
1	1	2	1	1		5	7.2
2	1	1	2			4	5.76
3	3	1	1	1	3	9	4.61
4		2	1			3	3.69
5	1	1	2	1		5	2.94
6						0	2.36
7			1	1		2	1.89
8	1					1	1.51
9				1		1	1.21
10			1		1	2	0.97
11		1				1	0.77
12			1			1	0.62
13						0	0.49
14						0	0.40
15						0	0.32
16						0	0.25
17						0	0.20
18					1	1	0.10
						45	

We now compute the chi-square statistic to test the null hypothesis that the sequence of values is random against the alternative that it is not:

$$\chi^2 = \sum_{i=0}^{18} \frac{(O_i - E_i)^2}{E_i} = \frac{(10-9.0)^2}{9.0} + \frac{(5-7.2)^2}{7.2} + \ldots$$
$$+ \frac{(1-0.16)^2}{0.16}$$
$$= 17.114.$$

Corresponding to 19 – 1 = 18 degrees of freedom, the value of $\chi^2_{0.05}$ is 28.9. The value of the chi-square is less than this critical value, so again we cannot reject the null hypothesis that a given sequence of values is random at the 5% significance level.

Excel contains no procedure that performs the gap test.

37.4 The Poker Test

In Chapter 7, "Random Experiments, Counting Techniques, and Probability," we performed several examples and exercises of this general form: "What is the probability that a five-card poker hand will contain two of a kind?" We can think of generating a random sequence of numbers as dealing from an infinite, well-shuffled deck, and apply reasoning similar to that of the card problems to develop another test of randomness.

Given a sequence that we wish to test for randomness, we break up the sequence into 5-tuples, and use a chi-square statistic to compare the expected and observed frequencies of particular types of "hands": two of a kind, three of a kind, two pair, etc. Because our "deck" is essentially infinite, finding the probabilities of the types of hands will differ from the card examples, as shown in this redo of the first example in this chapter.

Broken into 5-tuples, the sequence of numbers is this:

(4 3 2 3 3)(2 2 1 3 1)(1 0 2 0 2)(1 3 0 1 4)
(0 0 1 4 0)(2 3 2 1 1)(0 2 3 4 0)(0 2 4 4 2)
(2 1 4 4 0)(3 1 1 2 1)

When selecting digits randomly from the five alternatives (0, 1, 2, 3, and 4), the probability that an arbitrary 5-tuple will contain exactly one pair of identical values is:

$$1 * \frac{1}{5} * \frac{4}{5} * \frac{3}{5} * \frac{2}{5} * \binom{5}{2} = \frac{4!5!}{5^4 2!3!} = \frac{48}{125} = 0.384.$$

There are 10 5-tuples in the sequence, so the expected number of hands containing exactly one pair of values is $10 * 0.384 = 3.84$. Similar calculations, and counting the occurrences of the various arrangements, yield Table 37.4.

TABLE 37.4 Poker test data file.

"Hand"	Frequencies		
	Probability	Expected	Observed
One Pair	0.384	3.84	3
Two Pair	0.288	2.88	4
Three of a kind	0.192	1.92	3
Full House	0.064	0.64	0
Four of a kind	0.032	3.32	0

We find the value of the chi-square statistic in the usual way:

$$\chi^2 = \sum_{i=1}^{5} \frac{(O_i - E_i)^2}{E_i} = \frac{(3-3.84)^2}{3.84} + \frac{(4-2.88)^2}{2.88} + \frac{(3-1.92)^2}{1.92} + \frac{(0-0.64)^2}{0.64} + \frac{(0-0.32)^2}{0.32} = 2.036$$

The associated number of degrees of freedom is one less than the number of "hands" considered. In this case, that value is 5 – 1 = 4, and the value of $\chi^2_{0.05}$ is then 9.488. The observed value of the chi-square statistic is less than this critical value, so we cannot conclude at the 5% significance level that the sequence is not random.

Again, this is not a test that Excel supports.

37.5 The Runs Test

If a group of 12 men and women form a line in this order:

F F F F F F M M M M M M

we would conclude immediately that their arrangement is not random, as we would if this arrangement were observed:

M F M F M F M F M F M F

Sequences composed of only two types of symbols can be tested for randomness by counting the number of *runs*, or sequences of the same symbol, in the entire sequence. For example, the sequence

0 0 1 1 1 0 1

contains a run of two, a run of three, and two runs of one, a total of four runs.

If the number of runs in a sequence is small, lack of randomness through *clustering* is indicated, as in the first example above, while *mixing*—the second example—is reflected in too many runs. Since randomness may fail for either of these two mutually exclusive reasons, and the number of runs distinguishes them, we can perform a two-sided test for randomness or either of two one-sided tests.

Two-sided test:

H_0: The sequence is random.

H_a: The sequence is not random.

One-sided tests:

lower-tail

H_0: The sequence is random.

H_a: The sequence tends to cluster.

upper-tail

H_0: The sequence is random.

H_a: The sequence tends to mix.

A sequence to be tested consists of n_1 symbols of one type and n_2 symbols of a second; the number of possible runs has a minimum of 2 and a maximum of $2n_1$ if $n_1 = n_2$ or $2n_1+1$ if $n_1 < n_2$. Consult Table A-22, Critical values for the number of runs, in Appendix A, which provides critical values of the two-sided test at the 5% significance level (or the one-sided tests at $a = 2.5\%$) for n_1 and $n_2 \leq 20$.

For example, this sequence of binary digits contains six runs and is composed of $n_1 = 10$ 0s and $n_2 = 12$ 1s.

0 0 0 1 1 1 1 0 0 0 0 1 1 0 0 0 1 1 1 1 1 1

From Table A-22, the critical values for the two-sided test at the 5% significance level are 7 and 17; the observed number of runs is less than 7, so we conclude that the sequence is not random.

More strongly, since the number of runs is less than the critical value for the lower-tail test at the 2.5% significance level, we conclude that the sequence tends to cluster.

As N increases, the sampling distribution of U, the number of runs, approaches the normal distribution with mean $\mu_U = 1 + (2n_1 n_2)/N$ and standard deviation

$$S_U = \left[\frac{2n_1 n_2 (2n_1 n_2 - N)}{N^2(N-1)}\right]^{1/2}.$$

The critical values of U for the two-sided test at the a significance level are:

$$U^* = \mu_U \pm Z_{a/2} S_U.$$

37.6 Runs Above and Below a Central Value

A sequence whose entries take on more than two values may be tested for randomness with a runs test in two different ways. In the first, a dichotomy is imposed on the values in the sequence by creating a new sequence whose entries reflect whether each entry in the original sequence is above (A) or below (B) some central value in the population from which the original entries were selected, typically the mean or median. The sequence of As and Bs can be tested with the runs already given. Such a test is said to be based on *runs above and below the mean or median*.

For example, to test this sequence of 20 digits from 0 to 9, we create the corresponding sequence of symbols A and B depending on whether each entry in the sequence is above or below the population median of 4.5:

3	4	4	1	8	0	1	0	9	5	7	6	8	2	5	9	3	0	8	7
B	B	B	B	A	B	B	B	A	A	A	A	A	B	A	A	B	B	A	A

There are eight runs in the sequence of 10 As and 10 Bs; from Table A-22, the critical values for the two-sided test at the 5% significance level are 6 and 16. The observed number of runs lies between the critical values, so we cannot conclude that the original sequence is not random.

Again, this is not a test that Excel supports.

37.7 Runs Up and Down

In dichotomizing a sequence, as was done in the previous section, information about the sequence is inevitably lost. Some of this can be preserved by considering not runs above and below a value, but directional runs, or runs up and down. Again, we create a new sequence of symbols from the original sequence, but here, every entry except the first receives a+ or a– depending on whether the entry is greater or less than its predecessor. The runs test is then applied to the sequence of +s and –s. (If any entry replicates its predecessor, a 0 is entered. Zeros are ignored in creating and counting runs.)

Applying this technique to the previous example, we generate this sequence of +s and –s:

3	4	4	1	8	0	1	0	9	5	7	6	8	2	5	9	3	0	8	7	
	+	0	–	+	–	+	–	+	–	+	–	+	–	+	+	–	–	+	–	–

In the sequence of nine +s and nine –s, we find U = 16 runs. From Table A-22 in Appendix A, the critical values for the two-sided test at a = 5% are 5 and 15. U is not between these critical values, so we can conclude from this test that the original sequence of values is not random. In particular, since U is greater than the upper critical value, we can conclude that the sequence is not random because the direction changes too often.

This procedure is not performed by Excel.

37.8 The Kolmogorov Goodness-of-Fit Test

One of the many applications of the chi-square distribution is the goodness-of-fit test, which is used to investigate whether a set of values might have come from a specified distribution. Another technique, developed by the Russian mathematician A.N. Kolmogorov, tests for goodness-of-fit by comparing the empirical *cumulative distribution function (CDF)* with the hypothesized CDF, using the largest absolute difference of the two functions as the test statistic, usually called D. As with the chi-square test, the null hypothesis is that the values come from the proposed distribution, while the alternative is that they come from some other distribution.

ON DVD

For the western cities in the American Cities database, consider the 16 nonmissing values of X5, income of the average factory worker. We will use the Kolmogorov test to compare the null hypothesis that these values come from a normal distribution with mean 45,000 and standard deviation 2,300 with the alternative hypothesis that they come from some other distribution at the 5% significance level.

For each of the 16 data values, the cumulative distribution function $F(x)$ of the hypothesized normal distribution has the value

$$F(x) = P(X \le x), \quad \text{where} \quad X \text{ is } N(45000, 2300).$$

For example, at 44,410, one of the data values, the value of $F(x)$ is

$$F(44410) = P(X \le 44410) = P\left(\frac{(X-45000)}{2300} \le \frac{44410-45000}{2300}\right)$$
$$= P(Z \le -0.26) = 0.5000 - 0.1026 = 0.3974$$

The value of the cumulative *empirical* distribution at any data point x is

$$S(x) = \frac{N(x)}{N},$$

where $N(x)$ is the number of values less than or equal to x, and N is the total number of data values (in our example, 16). At 44,410, the value of $S(x)$ is

$$S(44410) = \frac{9}{16} = 0.5625.$$

Similar calculations provide the remainder of the entries in Table 37.5, in which the values of X5 have been placed in ascending order.

TABLE 37.5 The Kolmogorov test of goodness-of-fit.

X5	S(x)	F(x)
41125	0.0625	0.0529
42708	0.1250	0.1660
42871	0.1875	0.1762

(Continued)

TABLE 37.5 Continued.

X5	S(*x*)	F(*x*)
43459	0.2500	0.2514
44083	0.3125	0.3446
44142	0.3750	0.3557
44228	0.4375	0.3669
44291	0.5000	0.3783
44410	0.5625	0.3974
44524	0.6250	0.4168
44768	0.6875	0.4602
44818	0.7500	0.4681
45953	0.8125	0.6591
47639	0.8750	0.8749
47985	0.9375	0.9032
49307	1.0000	0.9693

The relationship between S and F can be seen by graphing the functions on common axes, as shown in Figure 37.1. Note that F is continuous and increasing, while S is an increasing step function.

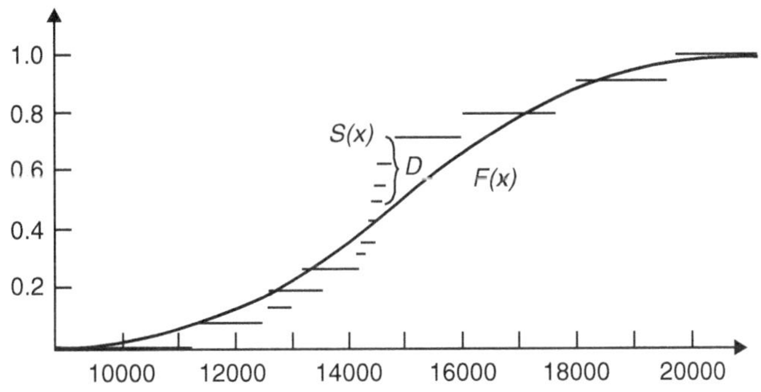

FIGURE 37.1 The empirical and theoretical distributions in the Kolmogorov test.

The Kolmogorov test statistic D is the smallest number greater than or equal to the absolute difference $|S(x)-F(x)|$ for all values of x. It can be seen from Figure 37.1 that F and S differ most in the vicinity of x = 44,818 (note that the values of the horizontal axis should all be increased by 30,000). In particular, at x = 44,818, this difference is $|0.7500-0.4681|=0.2819$.

In some situations, the statistic D is not so easily found. Consider a portion of this graph in a Kolmogorov test as shown in Figure 37.2.

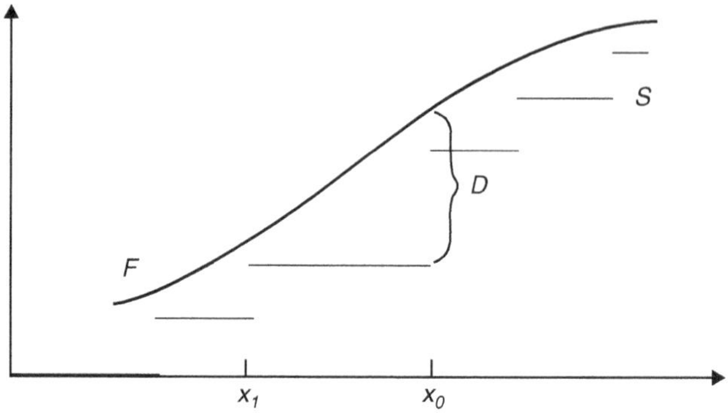

FIGURE 37.2 Finding D in the Kolmogorov test.

The largest difference between F and S is indicated, but it is not simply $|S(x_0) - F(x_0)|$; rather, it is the smallest number larger than $|S(x) - F(x)|$ as x approaches x_0. Since F is continuous, D is $|S(x_1) - F(x_0)|$.

ON DVD

Critical values for the two-sided Kolmogorov test, corresponding to the sample size, are given in Table A-10; for our sample of 16 values and a significance level of 10%, the critical value is 0.327. The test statistic, 0.2819, is less than the critical value, so we cannot conclude that the sample values did not come from the hypothesized normal distribution.

Note that Excel does not possess a Kolmogorov test statistic.

Two important observations should be made about the Kolmogorov test. First, the distribution of the test statistic D is not dependent on the nature of the underlying population distribution. Second, while the chi-square test of goodness-of-fit requires grouping possible values of a continuous distribution into classes, no such grouping is required here. Thus, it can be argued that the Kolmogorov test is better suited to situations involving continuous distributions.

37.9 The Kolmogorov-Smirnov Two-Sample Test

In a procedure very much like the Kolmogorov test, we can examine whether two independent samples might have come from identical distributions. This test was developed by another Russian, N.V. Smirnov, and generally carries the names of both men.

Given two independent samples, we construct the two empirical cumulative distribution functions $S_1(x)$ and $S_2(x)$, and use the statistic D=maximum $|S_1(x) - S_2(x)|$ to test the hypotheses:

H_0: The samples come from identical populations.
H_a: The populations are not identical.

ON DVD

The calculation of D is more straightforward here than in the Kolmogorov one-sample test, as shown by this example in which we compare the values of X5 from the American Cities database for 10 eastern cities and the 16 western cities used in the previous section.

Table 37.6 shows the empirical cumulative distribution functions $S_1(x)$ for the first 10 eastern cities and $S_2(x)$ for the western cities, and all the relevant differences between them.

TABLE 37.6 Empirical cumulative distribution functions for eastern cities, $S_1(x)$, and for western cities, $S_2(x)$.

X3				
Eastern	**Western**	$S_1(x)$	$S_2(x)$	$\lvert S_1(x)-S_2(x)\rvert$
40905		0.10	0	0.1000
	41275	0.10	0.0625	0.0375
41299		0.20	0.0625	0.1375
	42708	0.20	0.1250	0.0750
	42871	0.20	0.1875	0.0125
43002		0.30	0.1875	0.1125
43432		0.40	0.1875	0.2125
	43459	0.40	0.2500	0.1500
43979		0.50	0.2500	0.2500
	44083	0.50	0.3125	0.1875
44132		0.60	0.3125	0.2875
	44142	0.60	0.3750	0.2250
	44228	0.60	0.4375	0.1625
	44291	0.60	0.5000	0.1000
	44410	0.60	0.5625	0.0375
	44524	0.60	0.6250	0.0250
	44768	0.60	0.6875	0.0875
	44818	0.60	0.7500	0.1500
45784		0.70	0.7500	0.0500
	45953	0.70	0.8125	0.1125
46478		0.80	0.8125	0.0125
	47639	0.80	0.8750	0.0750
	47985	0.80	0.9375	0.1375
48362		0.90	0.9375	0.0375
48740		1.00	0.9375	0.0625
	49307	1.00	1.0000	0.0000

The value of D is simply the largest of the absolute differences; in this case, 0.2875. From Table A-22 in Appendix A, the critical value when the sizes of the samples are 10 and 16 is ½ for the 5% significance level. We cannot conclude that the distributions are different.

For sample sizes n_1 and n_2 larger than those given in the table, this formula approximates the critical value of the statistic D at the 5% significance level:

$$(1.36)\sqrt{\frac{n_1+n_2}{n_1 n_2}}.$$

37.10 Exercises

1. Apply the randomness test of runs above and below the median to this sequence of 15 values, using the 5% significance level:

   22, 2, 4, 12, 11, 15, 28, –5, 8, 4, –1, –10, –2, 25, 7

2. A programmer has developed a batch of 14 programs, some of which contain bugs. This sequence shows which contain bugs and which do not:

   B B NB NB NB B NB NB NB NB B NB B B

   At the 5% significance level, are the programs randomly arranged with respect to containing bugs? (Use the runs test.)

3. Apply the test of runs up and down to this sequence, at the 5% significance level:

   2 4 8 2 9 4 0 7 2 5

   7 5 9 0 7 7 5 9 6 0

4. Apply the gap test of randomness to this sequence, chosen from the digits, 0, 1, 2, and 3. Use the 10% significance level.

   1 0 1 1 1 1 2 0 0 3

   3 1 3 1 1 0 0 0 0 2

5. This sequence was selected from the digits 0 though 9. Apply the poker test of randomness to it at the 5% significance level.

   1 0 0 9 7

   7 6 5 2 0

   3 4 6 7 3

   8 0 9 5 9

   3 9 2 9 2

   3 2 5 3 3

   1 3 5 8 6

   5 4 8 7 6

   0 9 1 1 7

   7 4 9 4 5

CHAPTER 38

NONPARAMETRIC SUBSTITUTES FOR SOME FAMILIAR PARAMETRIC TESTS

OVERVIEW AND LEARNING OBJECTIVES

In This Chapter

38.1 Introduction

Our tests thus far have been parametric, depending on the characteristics of the underlying distributions. We now consider the class of nonparametric tests, in which we need make few if any assumptions about the underlying distributions. In general, such tests are more widely applicable than parametric tests but less powerful. We will examine nonparametric substitutes for several common parametric tests.

The conditions necessary for procedures such as the t-test or analysis of variance do not always hold in practice. For such situations we have a selection of nonparametric tests that do not require conditions on the underlying distributions.

38.2 The Mann-Whitney Test

The Mann-Whitney test considers the null hypothesis that two independent samples come from identical population distributions against the alternative that the underlying populations are different. The test is most sensitive to a difference in the *locations* of the two populations, so it is often thought of as a nonparametric substitute for the independent sample t-test (its effectiveness is not affected by the nature of the underlying distributions), and it is assumed that the underlying populations differ only in location.

To perform this test, we take two independent samples, then order *all* the sampled values together as one large group, creating order statistics and assigning the rank 1 to the smallest, 2 to the next smallest, and so on. Identical values are given the mean rank of the positions they would occupy were they slightly different. The test statistic is based on these ranks.

Let n_1 and n_2 be the sizes of the two samples, and let R_1 and R_2 be the sums of their ranks, respectively. Then for sample sizes of at least 8, if the null hypothesis is true and the two population means are equal, both R_1 and R_2, as random variables, have approximately normal distributions with means

$$\mu_{R_1} = \frac{n_1(n_1 + n_2 + 1)}{2} \quad \text{and} \quad \mu_{R_2} = \frac{n_2(n_1 + n_2 + 1)}{2}$$

and standard deviation

$$\sigma_{R_1} = \sigma_{R_2} = \left[\frac{n_1 n_2(n_1 + n_2 + 1)}{12}\right]^{1/2}.$$

Either R_1 or R_2 may be used as the test statistic, and the critical values for a two-sided test of the form

$$H_0: \mu_1 = \mu_2$$
$$H_a: \mu_1 \neq \mu_2$$

with significance level a will be $\mu_R \pm Z_{a/2}\sigma_R$. If μ_1 is not equal to μ_2, the values of sample 1 will tend to cluster at either the beginning or the end of the ranking, thus making the corresponding rank sum R_1 either small or large. If R_1 is not between the critical values, we may reject the null hypothesis and conclude that the population means are different.

When ties occur, the distributions of R_1 and R_2 are not exactly as described above, and correction factors are sometimes used. The difference is very small, however, unless the number of ties is large. (This is true of all tests based on sample ranks.)

For example, consider these two sets of test scores from students instructed by two distinct methods but given the same exam:

Group 1:	65	73	98	67	59	86	83	73	79	90
Group 2:	78	64	67	82	94	55	74	77	81	71

We order all the values and assign ranks, as shown in Table 38.1. Note how ties are resolved.

TABLE 38.1 Group rankings.

Group			Group		
1	**2**	**Rank**	**1**	**2**	**Rank**
	55	1		78	12
59		2	79		13
	64	3		81	14
65		4		82	15
67	67	5.5	83		16
	71	7	86		17
73		8.5	90		18
73		8.5	90		18
	74	10	98		20
	77	11			

All tables in this chapter appear on the companion DVD. ON DVD

The sum of ranks for group 1 is $R_1 = 112.5$ ($R_2 = 97.5$). The expected value of R_1 if H_0 is true is

$$\frac{10(10+10+1)}{2} = 105,$$

and σ_R is:

$$\left[\frac{n_1 n_2 (n_1 + n_2 + 1)}{12}\right]^{1/2} = \left[\frac{10 * 10(10+10+1)}{12}\right]^{1/2} = \left(\frac{2100}{12}\right)^{1/2} = 13.23.$$

At the 10% significance level, the critical values of R_1 are:

$$105 \pm 1.645 * 13.23 = 105 \pm 21.76 = (83.24, 126.76)$$

Since the test statistic R_1 does not lie outside the critical values, we cannot reject the null hypothesis; we cannot conclude that the test scores for the two groups differ significantly.

One-sided tests for the differences between two means can also be performed using the Mann-Whitney technique. Suppose, for example, that the values of sample 1 are in general less than those of sample 2. Then the rank sum R_1 will be reduced. If it is small enough, we may conclude that $\mu_1 < \mu_2$, and the critical value for this lower-tail test is $\mu_R - Z_a\sigma_R$. Similarly, the critical value for the upper-tail test with significance level a is $\mu_R + Z_a\sigma_R$.

38.3 The Wilcoxon Matched-Pairs Signed-Rank Test

When considering samples of matched-pair data (X_i, Y_i), we have used the t-test to examine the corresponding sample of *matched-pair differences* $D_i = X_i - Y_i$. In the Wilcoxon test for the difference of the central locations of two populations based on a sample of matched pairs, we construct the test statistic from the ranks and directions of the matched pair differences.

Consider, in an experiment of this kind, the population for all possible or hypothetical matched-pair differences, and let the median of this population be M_D. We assume that the differences are symmetrically distributed around M_D and state our hypotheses in terms of this parameter. If the two populations have the same central tendencies, then M_D will be near 0; if the populations differ in location, we will have $M_D \neq 0$. These are the hypotheses of the two-sided Wilcoxon test:

$$H_0: M_D = 0$$

$$H_a: M_D \neq 0.$$

In Unit 5, this example was solved with a matched-pairs t-test. We perform it here with the Wilcoxon test. Two tire-tread materials are being compared. Pairs of tires of the two materials are mounted on test cars and their lifetimes measured. The results are shown in Table 38.2.

TABLE 38.2 Comparison of two tire-tread materials.

Pair	1	2	3	4	5	6	7	8	9	10	11	12
Life of Material A (1000s of miles)	42.6	43.1	41.9	45.2	43.6	41.0	42.6	43.7	42.2	46.3	42.4	41.8
Life of Material B (1000s of miles)	40.3	42.7	42.0	42.1	43.4	39.5	41.5	43.4	41.4	43.1	41.6	40.1

From the values in Table 38.2, we assemble Table 38.3, which includes the matched-pair differences, their ranks, and their signs.

TABLE 38.3 Tire-tread differences, ranks, and signs.

Pair	A	B	D = A – B	D	Rank of D	Sign of D
1	42.6	40.3	2.3	2.3	10	+
2	43.1	42.7	0.4	0.4	4	+
3	41.9	42.0	-0.1	0.1	1	–
4	45.2	42.1	3.1	3.1	11	+
5	43.6	43.4	0.2	0.2	2	+
6	41.0	39.5	1.5	1.5	8	+
7	42.6	41.5	1.1	1.1	7	+
8	43.7	43.4	0.3	0.3	3	+
9	42.2	41.4	0.8	0.8	5.5	+
10	46.3	43.1	3.2	3.2	12	+
11	42.4	41.6	0.8	0.8	5.5	+
12	41.8	40.1	1.7	1.7	9	+

If a difference is 0, it is disregarded, and the sample size is reduced by 1; note again how ties are resolved in ranking the absolute values of differences. From Table 38.3, we compute two statistics, the sum of the ranks of the positive difference T^+ and the sum of the ranks of the negative difference T^-. Here, $T^+ = 77$ and $T^- = 1$.

Either can be used as the test statistic, and from Table A-11, Critical values of the Wilcoxon signed-rank test, in Appendix A we see that values this extreme would occur under the null hypothesis with almost negligible probability. Thus, we can reject the null hypothesis and conclude that there is a significant difference in the durability of the two materials. ON DVD

For tests in which the alternative hypothesis is of the form H_a: $M_D > 0$, we reject the null hypothesis and accept this alternative if T^+ is large enough (and T^- small enough); we accept the alternative H_a: $M_D < 0$ if T^- is sufficiently large.

For sample sizes larger than those treated in Table A-11, we can use the fact that as the sample size increases, the sampling distributions of both T^+ and T^- under H_0 approach the normal distribution with mean $n(n + 1)/4$ and standard deviation $[n(n + 1)(2n + 1)/24]^{1/2}$. ON DVD

38.4 The Kruskal-Wallis Test

In the parametric case, we used one-way analysis of variance to test the equality of several population means using independent samples. In this way, ANOVA was an extension of the independent samples t-test for the equality of two population means. The Mann-Whitney test is a nonparametric substitute for the t-test that compared the central tendencies of two populations; we now extend that idea to three or more populations with the Kruskal-Wallis test, which uses the ranks of the elements of independent samples from $k \geq 3$ populations to test these hypotheses:

H_0: The population distributions are identical.

H_a: The population distributions are not identical.

The test is most sensitive to differences in the locations, in particular the medians, of the populations, so it is generally thought of as a test of the equality of the population medians. Here, we are testing these hypotheses:

H_0: The population medians are identical.

H_a: At least one median differs from the others.

We demonstrate the Kruskal-Wallis test with the following exercise: A computer center has, for seven months, recorded the numbers of executions in four popular languages. The numbers of executions are shown in Table 38.4.

TABLE 38.4 Demonstration of the Kruskal-Wallis test.

Month	Basic	Pascal	Cobol	Fortran	
1	1	5	14	2	
2	5	9	18	7	
3	2	5	9	3	
4	6	11	21	10	**Hundreds of Jobs**
5	7	12	20	7	
6	3	5	12	5	
7	3	8	16	6	

We wish to test the null hypothesis that the median numbers of jobs in each language are the same against the alternative that at least two of the medians differ.

The Kruskal-Wallis test is based on the ranks of all the observations, so we begin by constructing a table of ranks, as shown in Table 38.5. Ties again receive an average rank value.

TABLE 38.5 Ranks for the Kruskal-Wallis test.

Month	Basic	Pascal	Cobol	Fortran
1	1	9	24	2.5
2	9	18.5	28	15
3	2.5	9	18.5	5
4	12.5	21	28	20
5	15	22.5	27	15
6	5	9	22.5	9
7	5	17	25	12.5

If for each sample n_i is the sample size and R_i is the sum of the ranks of the sample, the Kruskal-Wallis test statistic H is given by the formula:

$$H = \frac{12}{n(n+1)}\left[\sum_{i=1}^{k}\frac{R_i^2}{n_i}\right] - 3(n+1),$$

where k is the number of samples, and $n = \sum_{i=1}^{k} n_i$ is the total number of observations.

In our example, $R_1 = 50$, $R_2 = 106$, $R_3 = 171$, and $R_4 = 79$, so the value of H is:

$$H = \frac{12}{28(28+1)}\left[\sum_{i=1}^{4}\frac{R_i^2}{n_i}\right] - 3(28+1)$$

$$= \frac{12}{28*29}\left[\frac{50^2}{7}+\frac{106^2}{7}+\frac{171^2}{7}+\frac{79^2}{7}\right] - 3*29$$

$$= \frac{12}{812}*\frac{49218}{7} - 87 = 16.91.$$

If the null hypothesis is true and the population medians are equal, the sampling distribution of H is a chi-square distribution with $k - 1$ degrees of freedom. In our example, corresponding to $4 - 1 = 3$ degrees of freedom and a significance level of 5%, using a standard table on the critical values for the Kruskal-Wallis test in Appendix A is $\chi^2_{0.05} = 7.815$. The observed value of H is greater than the critical value, so we may reject H_0 and conclude that the population medians are unequal.

Note that the ability to reject H_0, as in analysis of variance, does not allow us to say with the same confidence that one particular median is greater than the others. Further tests must be performed to establish such a result.

38.5 The Spearman Rank Correlation Coefficient

The ranks of elements of samples can also be used to construct a measure of correlation.

Consider a sample of n pairs of the values (X_i, Y_i). We order the X values and the Y values separately, and rank the values in the usual way. Corresponding to the original pair of values, we have n pairs of ranks, and the Spearman rank correlation coefficient, named for the English psychologist C. Spearman, is computed from the sum of the squares of the differences between these ranks.

If there is a perfect correlation between the X and Y values, then the ranks in each pair will be identical, as shown in Table 38.6.

TABLE 38.6 Duplicating ranks.

Ranks	
X	Y
1	1
2	2
.	.
.	.
.	.
$n-1$	$n-1$
n	n

Each of the rank differences D_i will be 0, as well the sum of their squares.

If there is a perfect inverse correlation between the X and Y values, then the ranks of one will ascend, while those of the other will descend, as shown in Table 38.7.

TABLE 38.7 Inverse ranks.

Ranks	
X	Y
1	n
2	$n-1$
.	.
.	.
.	.
$n-1$	2
n	1

This situation produces the largest sum of the squares of the rank differences:

$$\sum_{i=1}^{n} D_i^2 = (n-1)^2 + \left[(n-1)-2\right]^2 + \ldots + \left[2-(n-1)\right]^2 + (1-n)^2$$

$$= 2\left[(n-1)^2 + (n-3)^2 + \ldots (1-n)\right]^2$$

$$= \frac{n\left(n^2-1\right)}{3}$$

The value of $\sum_{i=1}^{n} D_i^2$ must fall between 0, indicating perfect direct correlation, and $n(n^2 - 1)/3$, indicating perfect inverse correlation. We prefer that measures of correlation fall between –1 and +1, and a function of $\sum_{i=1}^{n} D_i^2$ that does this in a familiar way is defined to be the Spearman rank correlation coefficient R_S:

$$R_s = 1 - \frac{6\sum_{i=1}^{n} D_i^2}{n(n^2 - 1)}.$$

A formula more commonly used to calculate R_S is this:

$$R_s = \frac{12\sum_{i=1}^{n} R_{X_i} R_{Y_i}}{n(n^2 - 1)} - \frac{3(n+1)}{n-1},$$

where R_{X_i} and R_{Y_i} are the ranks of the values of X_i and Y_i. The following formula is equivalent to those above:

$$R_s = \frac{12\sum_{i=1}^{n} (R_{X_i} - \bar{R}_X)(R_{Y_i} - \bar{R}_Y)}{n(n^2 - 1)}.$$

That is, if the original data values are themselves ranks, R_S is equivalent to the Pearson correlation coefficient.

TABLE 38.8 Rating by natural gas usage.

Values		Ranks		
Area	**Gas 1000s of CU FT**	**Area**	**Gas**	**D**
1.4	31.6	3.5	3	0.5
1.7	40.3	6.5	6	0.5
2.1	52.0	13	13	0
1.9	41.5	10	7	3
2.4	50.7	15	12	3
1.8	43.8	8	8	0
1.7	44.0	6.5	9	-2.5
1.9	48.2	10	11	-1
1.3	30.5	2	1	1
1.2	30.7	1	2	-1
2.0	55.2	12	14	-2
1.9	47.6	10	10	0
2.2	57.1	14	15	-1
1.4	36.2	3.5	5	-1.5
1.6	34.5	5	4	1

Consider the values shown in Table 38.8. This example, from Chapter 24, lists floor area and natural gas use in January for 15 Midwestern houses.

The sum of the squares of the rank differences is $\sum_{i=1}^{15} D_i^2 = 36.0$, and the value of the Spearman coefficient is

$$R_s = 1 - \frac{6\sum_{i=1}^{15} D_i^2}{15(15^2 - 1)} = 1 - \frac{6 * 36}{15 * 224} = 1 - 0.064 = 0.936.$$

The value of R_S is near 1, indicating as before a strong correlation between gas use and house floor area.

38.6 Exercises

1. A sequence of 60 values is chosen from a distribution purported to be the discrete uniform on {0, 1, ..., 9}. These frequencies are observed:

   Digit 0 1 2 3 4 5 6 7
   Frequency 9 10 4 2 3 5 1 7 11

   a. Test the assertion of randomness from {0, 1, ..., 9} with the chi-square goodness-of-fit test at the 5% significance level.

   b. Perform the test in part a using the Kolmogorov test at the 5% significance level. Is this test appropriate here?

2. Employees at two computer centers are categorized in five job classifications as shown in Table 38.9.

   TABLE 38.9 Computer store center personnel data.

	Systems Engineers	Systems Analysts	Systems Programmers	Applications Programmers	Operators
Center A	2	0	12	14	2
Center B	4	9	18	22	7

   Using the Kolmogorov-Smirnov two-sample test, can it be concluded at the 5% significance level that jobs are distributed differently in the two centers?

3. Table 38.10 lists the thousands of kilometers of railroad track available in North and Central America and in South America. Use the Mann-Whitney test at the 5% significance level to test whether mean length of track is the same in the two regions.

TABLE 38.10 North versus south railway.

Group 1: North and Central America	1,000 km	Group 2: South America	11,000 km
Canada	70.1	Argentina	40.2
Costa Rica	0.6	Bolivia	3.4
Cuba	14.5	Brazil	23.9
El Salvador	0.6	Chile	9.0
Guatemala	0.8	Colombia	3.4
Honduras	1.1	Ecuador	1.1
Mexico	19.2	Paraguay	0.5
Nicaragua	0.3	Peru	2.1
Panama	0.7	Uruguay	3.0
United States	332.8	Venezuela	2.0

4. Perform the test of problem 3 using a t-test for independent samples. Which test is more appropriate here? Why?

5. A computer center director and a senior programmer each rated a group of 16 junior programmers on their software documentation skills. The ratings are provided in Table 38.11.

TABLE 38.11 Junior programmer skills ratings.

	Junior Programmers															
	1	2	3	4	5	6	7	8	9	10	11	12	13	14	15	16
Director	4	4	5	5	3	2	5	3	1	5	5	5	4	5	5	5
Senior Programmer	2	3	3	3	3	3	3	3	2	3	2	2	5	2	5	3

Use the Wilcoxon matched-pairs signed-rank test at the 5% significance level to test for a difference in the average rating by these two individuals.

6 a. Perform the test in exercise 5 using the t-test for matched-pairs.

b. What assumption does the Wilcoxon test make about the underlying distributions?

c. In the Wilcoxon test, pairs that tie are discarded. What is the problem with this procedure? Suggest an alternate procedure.

d. Is the Wilcoxon test or the t-test more appropriate to the data in problem 6?

7. Twenty-three COBOL programmers are randomly assigned to three study groups. The groups are given a programming assignment. Group 1 was told to use a nonstructured, bottom-up approach, Group 2 was to use a structured, top-down approach, and Group 3 was given no specific instructions. The times in hours required by each individual to complete the project are shown in Table 38.12.

TABLE 38.12 Group comparative study of problem solving methods.

	Times in Hours							
Group 1	9	8	8	6	5	4	4	4
Group 2	6	6	5	4	4	3	2	2
Group 3	8	8	7	7	7	6	5	

Use the Kruskal-Wallis test at the 5% significance level to decide if the average time varied over the three groups.

8 a. Perform the test in problem 7 using one-way analysis of variance.

b. What assumptions required by analysis of variance need not be satisfied when using the Kruskal-Wallis test?

c. How is the Kruskal-Wallis test affected by a large number of ties?

9. Sixteen software packages were ranked by two computer centers as shown in Table 38.13:

TABLE 38.13 Computer store rankings of software packages.

	Package															
	A	B	C	D	E	F	G	H	I	J	K	L	M	N	O	P
Center 1	1	2	3	4	5	6	7	8	9	10	11	12	13	14	15	16
Center 2	2	1	8	6	4	3	11	14	9	5	12	7	15	16	10	13

a. Calculate the Spearman rank correlation coefficient for this data. At $a = 5\%$, is it significant?

b. Calculate the Pearson product-moment correlation coefficient for the given values.

CHAPTER 39

COMMONLY USED DISTRIBUTIONS

OVERVIEW AND LEARNING OBJECTIVES

In This Chapter

39.1 Introduction

This chapter describes the random number generation algorithms for distributions commonly used in business applications. Distributions are listed in alphabetic order for easy reference. For each distribution, the key characteristics are summarized, algorithms for random number generation are presented, and examples of their application are provided.

39.2 The Bernoulli Distribution

The Bernoulli distribution is the simplest discrete distribution. A Bernoulli variate can take only two values, which are usually denoted as failure and success, or $x = 0$ and $x = 1$. If p denotes the probability of success, then $1 - p$ is the probability of failure. The experiments to generate a Bernoulli variate are called *Bernoulli trials*.

The key characteristics of the Bernoulli distribution are summarized in Table 39.1.

TABLE 39.1 Bernoulli (p) distribution.

Parameters:	p = probability of success $x = 1$; $0 \le p \le 1$
Range:	$x = 0, 1$ $1-p$, if $x = 0$
pmf:	$f(x) = p$, if $x = 1$ $f(x) = 0$, otherwise
Mean:	p
Variance:	$p(1-p)$

ON DVD All figures and tables in this chapter appear on the companion DVD.

The Bernoulli distribution is used to model the probability of an outcome having a desired class or characteristic, such as:

- A stock is up or down
- A sale does or does not reach completion
- A bit in a computer message is affected by noise and the message arrives in error

The inverse transformation method can be used to generate Bernoulli variates. Generate a uniform (0,1) value u. If $u \le p$, return 0; otherwise, return 1.

39.3 The Binomial Distribution

The number of successes x in a sequence of n Bernoulli trials has a binomial distribution. The key characteristics of the binomial distribution are summarized in Table 39.2.

TABLE 39.2 Binomial distribution: b(n, p).

Parameters:	p = probability of success in a trial, $0 < p < 1$ n = number of trials
Range:	$x = 0, 1, \ldots, n$
pmf:	$f(x) = \binom{n}{x} p^x (1-p)^{n-x}$
Mean:	np
Variance:	npq

The binomial distribution is used to model the number of successes in a sequence of n independent and identical Bernoulli trials, such as:

- The number of items in a batch that have a certain characteristic
- The number of sales that reach completion
- The number of bits in a computer message that are not affected by noise

The variance of the binomial distribution is always less than the mean. The Poisson distribution can be used in place of a binomial distribution if the variance is greater than the mean or equal to the mean.

Binomial variates can be generated in the following ways:

Generate n uniform (0,1) numbers. The number of random numbers that are less than p is b(n,p). This is the composition method.

Compute the cdf F(x) for $x = 0, 1, 2, \ldots, n$ and store in an array. For each binomial variate, generate uniform (0,1) value u and search the array to find x so that $F(x) \le u \le F(x+1)$; return x. This is the inverse transformation method.

39.4 The Chi-Square Distribution

The sum of squares of several unit normal variates has a distribution known as chi-square. The distribution was first derived by Karl Pearson in 1900. The key characteristics are summarized in Table 39.3.

TABLE 39.3 Chi-square distribution: $\chi^2(\nu)$

Parameters:	ν = degrees of freedom
Range:	$0 \le x \le \infty$
pmf:	$f(x) = \dfrac{x^{(\upsilon-2)} e^{-x/2}}{2^{\upsilon/2}\,\Gamma(\upsilon)}$

(Continued)

TABLE 39.3 Continued.

	Here, $\Gamma(v)$ is the gamma function as follows: $$\Gamma(v) = \int_0^{\infty} e^{-x} x^{v-1} dx$$
Mean:	v
Variance:	2^v

The chi-square distribution is used whenever a sum of squares of normal variables is involved, such as modeling sample variances or performing a goodness-of-fit test.

Chi-square variates can be generated in the following ways:

Generate v N(0,1) variates and return the sum of their squares.

Use the observation that the chi-square distribution $\chi^2(v)$ is the same as the gamma distribution $\gamma(2, v/2)$.

For v even,

$$\chi^2(v) = \frac{1}{2} \ln\left(\prod_{i=1}^{v/2} \mu_i\right).$$

For v odd,

$$\chi^2(v) = \chi^2(v-1) + [N(01)]^2.$$

39.5 The Exponential Distribution

The exponential distribution is used extensively in waiting line problems. The key characteristics of the distribution are summarized in Table 39.4.

TABLE 39.4 Exponential distribution.

Parameters:	a = mean, $a > 0$
Range:	$0 < x < \infty$
pmf:	$$f(x) = \frac{1}{a} e^{-x/a}$$
CDF:	$$F(x) = 1 - e^{-x/a}$$
Mean:	a
Variance:	a^2

The exponential distribution is the only continuous distribution with the memory-less property such that remembering the time since the last event does not help in predicting the time until the next event. If the interarrival times of customers at a shop are exponentially distributed with mean $\frac{1}{\lambda}$, the probability distribution function is:

$$F(\tau)=P(\tau<t)=1-e^{-\lambda t},\ t\geq 0$$

If we see an arrival and start our clock at $t=0$, the mean time to the next arrival is $\frac{1}{\lambda}$. Now suppose we do not see an arrival until $t=x$. The distribution of the time remaining until the next arrival is:

$$\begin{aligned}P(\tau-x<t|\tau>x)&=\frac{P(x<\tau<x+t)}{P(\tau>x)}\\&=\frac{P(\tau<x+t)--P(\tau<x)}{P(\tau>x)}\\&=\frac{\left(1-e^{-\lambda(x+t)}\right)-\left(1-e^{-\lambda x}\right)}{e^{-\lambda t}}\\&=1-e^{-\lambda x}\end{aligned}$$

This is identical to the situation at $t=0$. In particular, the mean time to the next arrival is still $\frac{1}{\lambda}$. Thus, the expected time to the next arrival is always $\frac{1}{\lambda}$ regardless of the time since the last arrival. Remembering the past history does not help.

The exponential distribution is used to model the time between successive events, particularly if the events are caused by a large number of independent factors, such as:

- The time between successive customers arriving at a shop
- The time between failures of a device

Exponential variates can be generated using the inverse transformation method. Generate a uniform (0,1) random number u and return $-a\ln(u)$ as the exponential variate value.

39.6 The F Distribution

The ratio of two chi-square variates has an F distribution. The key characteristics of the F distribution are summarized in Table 39.5.

TABLE 39.5 F distribution: F(n, m).

Parameters:	n = numerator degrees of freedom m = denominator degrees of freedom
Range:	$0\leq x\leq\infty$

(Continued)

TABLE 39.5 Continued.

pmf:	$f(x)=\frac{(n/m)^{n/2}}{\beta(n/2,m/2)}x^{(n-2)/2}\left(1+\frac{n}{m}x\right)^{-(n+m)/2}$ where $\beta(.)$ is the beta function $\beta(a,b)=\int_0^1 x^{a-1}(1-x)^{b-1}\,dx=\frac{\Gamma(a)\Gamma(b)}{\Gamma(a+b)}$
Mean:	$\frac{m}{m-2}$, provided $m>2$
Variance:	$\frac{2m^2(n+m-2)}{n(m-2)^2(m-4)}$, provided $m>4$

Notice that the two parameters n and m of the F distribution are called numerator degrees of freedom and denominator degrees of freedom. The names of the parameters are derived from the fact that F is equal to:

$$\frac{\left[\chi^2(n)/n\right]}{\left[\chi^2(m)/m\right]}.$$

In general:

$$F_{[1-a;n,m]}=\frac{1}{F_{[a;n,m]}}.$$

For example, this relationship can be used to compute $F_{0.05}$ from $F_{0.95}$.

The F distribution is used to model the ratio of sample variances, such as in the F-test for regression and analysis of variance.

We generate F variates by using the definition of an F variate; this method is called the *characterization method*. Generate two chi-square variates $\chi^2(n)$ and $\chi^2(m)$ and then compute

$$F(n,m)=\frac{\left[\chi^2(n)/n\right]}{\left[\chi^2(m)/m\right]}.$$

39.7 The Gamma Distribution

The gamma distribution, like the exponential distribution, is commonly used in waiting line problems. The key characteristics of the gamma distribution are summarized in Table 39.6.

TABLE 39.6 Gamma distribution: $\gamma(a, b)$.

Parameters:	a = scale parameter, $a > 0$ b = shape parameter, $b > 0$
Range:	$0 \leq x \leq \infty$
pmf:	$f(x) = \dfrac{(x/a)^{b-1} e^{-x/a}}{a\Gamma(b)}$
	where $\Gamma(.)$ is the gamma function
	$\beta(a,b) = \int_0^1 x^{a-1}(1-x)^{b-1}\,dx = \dfrac{\Gamma(a)\Gamma(b)}{\Gamma(a+b)}$
Mean:	ab
Variance:	a^2b

The gamma distribution is used to model service times of devices in waiting line problems and for repair times.

Gamma variates can be generated by finding the sum of b exponential variates when b is an integer. This gives the following expression for generating gamma variates from uniform (0,1) variates:

$$\gamma(a,b) \cong -a \ln\left[\prod_{i=1}^{b} u_i\right].$$

39.8 The Geometric Distribution

The distribution of number of trials up to and including the first success in a sequence of Bernoulli trials is called a *geometric distribution*. The key characteristics of the geometric distribution are summarized in Table 39.7.

TABLE 39.7 Geometric distribution: $G(p)$.

Parameters:	p = probability of success, $0<p<1$
Range:	$x = 1, 2, \ldots, \infty$
pmf:	$f(x) = (1-p)^{1-x}p$
CDF:	$F(x) = 1 - (1-p)^x$
Mean:	$1/p$
Variance:	$\dfrac{1-p}{p^2}$

The geometric distribution is a discrete equivalent of the exponential distribution. It is a memory-less distribution in the sense that remembering the results of past attempts does not help in predicting the future.

The geometric distribution is used to model the number of attempts between successive failures (or successes), such as:

- The number of local queries to a database between successive accesses to the remote database
- The number of successive error-free bits between in-error bits in a message received on a noisy transmission line
- Geometric variates can be easily generated using the inverse transformation method. Generate a uniform (0,1) random number u and compute

$$G(p)=\left\lceil \frac{\ln(u)}{\ln(1-p)} \right\rceil,$$

where $\lceil . \rceil$ represents rounding up to the next larger integer.

39.9 The Normal Distribution

The normal distribution was actually discovered by Abraham De Moivre in 1733. Gauss and Laplace rediscovered it in 1809 and 1812. The normal distribution N(0,1) with $\mu = 0$ and $\sigma =$ 1 is called the *unit normal distribution* or *standard normal distribution*. The key characteristics of the normal distribution are summarized in Table 39.8.

TABLE 39.8 Normal distribution: N(μ, σ).

Parameters:	μ = mean
	σ = standard deviation, $\sigma>0$
Range:	$-\infty \leq x \leq \infty$
pmf:	$f(x)=\frac{1}{\sigma\sqrt{2\pi}}e^{-(x-\mu)^2/2\sigma^2}$
Mean:	μ
Variance:	σ

The normal distribution is used whenever the randomness is caused by several independent sources acting additively, such as:

Errors in measurement:

- Errors in modeling to account for a number of factors that are not included in the model
- Sample means of a large number of independent observations from a given distribution

Normal variates can be generated through four methods:

Convolution method:

By generating the sum of a large number of uniform (0,1) random values, u_i, then:

$$N(\mu,\sigma) \cong \mu + \sigma \frac{\left(\sum_{i=1}^{n} u_i\right) - n/2}{(n/12)^{1/2}}.$$

Polar method:

Generate two uniform (0,1) variates u_1 and u_2.

Let $v_1 = 2u_1 - 1$, $v_2 = 2u_2 - 1$ and $r = v_1^2 + v_2^2$.

If $r \geq 1$, to back to step 2; otherwise let $s = \left[(-2\ln r)/r\right]^{1/2}$ and return:

$$x_1 = \mu + \sigma v_1 s$$
$$x_2 = \mu + \sigma v_2 s$$

x_1 and x_2 are two independent N(μ, σ) variates.

Rejection method:

Generate two uniform (0,1) variates u_1 and u_2.

Let $x = -ln\ u_1$.

If $u_2 > e^{-(x-1)^2/2}$, go back to step 1.

Generate u_3.

If $u_3 > 0.5$, return $u + \sigma x$; otherwise, return $\mu - \sigma x$.

Box-Muller method:[1]

Generate two uniform (0,1) random variates u_1 and u_2.

Let

$$x_1 = \mu + \sigma \cos(2\pi u_1)\sqrt{-2\ln(u_2)}$$
$$x_2 = \mu + \sigma \sin(2\pi u_1)\sqrt{-2\ln(u_2)}$$

There is some concern that if this method is used with us from an LCG, the resulting xs may be correlated.

39.10 The Poisson Distribution

The Poisson distribution is a limiting form of the binomial distribution and is used extensively in waiting line problems. The key characteristics of the Poisson distribution are summarized in Table 39.9.

TABLE 39.9 Poisson distribution.

Parameters:	λ = mean, $\lambda > 0$
Range:	$x = 0, 1, 2, \ldots, \infty$
pmf:	$f(x) = P(X = x) = \lambda^x \frac{e^{-\lambda}}{x!}$
Mean:	λ
Variance:	λ

The Poisson distribution is used to model the number of arrivals over a given interval, such as:

- The number of requests to a server in a given time interval t
- The number of failures per unit of time
- The number of queries to a database system over t seconds
- The number of typing errors per form

The Poisson distribution is particularly appropriate if the arrivals are from a large number of independent sources. Such arrival processes are also called *Poisson processes*.

Poisson variates can be generated by the inverse transformation method. Compute the cdf $F(x)$ for $x = 0, 1, 2, \ldots$ up to a suitable cutoff and store in an array. For each Poisson random variate, generate a uniform (0,1) value u, and search the array to find x such that $F(x) \leq u < F(x+1)$; return x.

39.11 The Student *t* Distribution

This distribution was derived by William S. Gosset (1867–1937), a chemist for a brewery whose owner did not appreciate his publishing. Gosset, therefore, published his paper under the pseudonym "Student." The symbol t was used to denote the variable and hence the name "Student's t distribution." The key characteristics of the t distribution are summarized in Table 39.10.

TABLE 39.10 Student's *t* distribution.

Parameters:	v = degrees of freedom
Range:	$-\infty \le x \le \infty$
pmf:	$f(x)=\frac{\{[\Gamma(v+1)/2]\}\left[1+(x^2/v)\right]^{-(v+1)/2}}{(\pi v)^{1/2}\,\Gamma(v/2)}$
Variance:	$v/(v-2)$, for $v > 2$,

If x is approximately a unit normal variate and y is approximately a chi-square variate with v degrees of freedom, then:

$$\frac{N(0,1)}{\sqrt{\chi^2(v)/v}} \cong t(v).$$

The pdf of a t variate is very similar to that of a unit normal. The distribution is bell shaped and is symmetrical about 0. For large degrees of freedom ($v>30$), a t distribution can be approximated by a unit normal.

The t distribution is used whenever the ratio of a normal variate and the square of a chi-square variable are involved and is commonly used in setting confidence intervals and in t-tests.

The t variates can be generated using characterization by generating x as N(0,1) and y as $\chi(v)$ and returning $\frac{x}{\sqrt{y/v}}$ as t(v).

39.12 The Continuous Uniform Distribution

The key characteristics of the continuous uniform distribution are summarized in Table 39.11.

TABLE 39.11 Continuous uniform distribution: U(a, b).

Parameters:	a = lower limit
	b = upper limit, $b>a$
Range:	$-a \le x \le b$
pmf:	$f(x)=\frac{1}{b-a}$
CDF:	$F(x)=\begin{cases}0, \text{ if } x<a\\ \frac{x-a}{b-a}, a\le x<b\\ 1, \text{ if } b\le x\end{cases}$
Mean:	$\frac{a+b}{2}$
Variance:	$(b-a)^2/12$

A uniform distribution is commonly used if a random variable is bounded and no further information is available, such as:

- Total service time in a simulation study for a server
- The amount of time for a sale completion
- To generate U(a,b), simply generate u as U(0,1) and return $a+(b-a)u$

39.13 The Discrete Uniform Distribution

This distribution is a discrete version of the uniform distribution. It takes a finite number of values, each with the same probability. The key characteristics of this distribution are summarized in Table 39.12.

TABLE 39.12 Discrete uniform distribution: $U_d(m, n)$.

Parameters:	m = lower limit n = upper limit, $n>m$; n and m integers
Range:	$x = m, m+1, m+2, \ldots, n$
pmf:	$f(x) = \frac{1}{n-m+1}$
CDF:	$F(x) = \begin{cases} 0, \text{ if } x < m \\ \frac{x-m+1}{n-m+1}, m \le x < n \\ 1, \text{ if } n \le x \end{cases}$
Mean:	$\frac{n+m}{2}$
Variance:	$[(n-m+1)^2-1]/12$

The discrete uniform distribution is used when it is believed that the value is equally likely over a bounded interval, such as:

- A clerk's service time in a shop
- The number of haircuts in a day per barber in a barbershop
- To generate $U_d(m,n)$, generate u as U(0,1) and return $[m+(n-m+1)u]$

Figure 39.1 presents a summary of the various distributions discussed in this chapter.

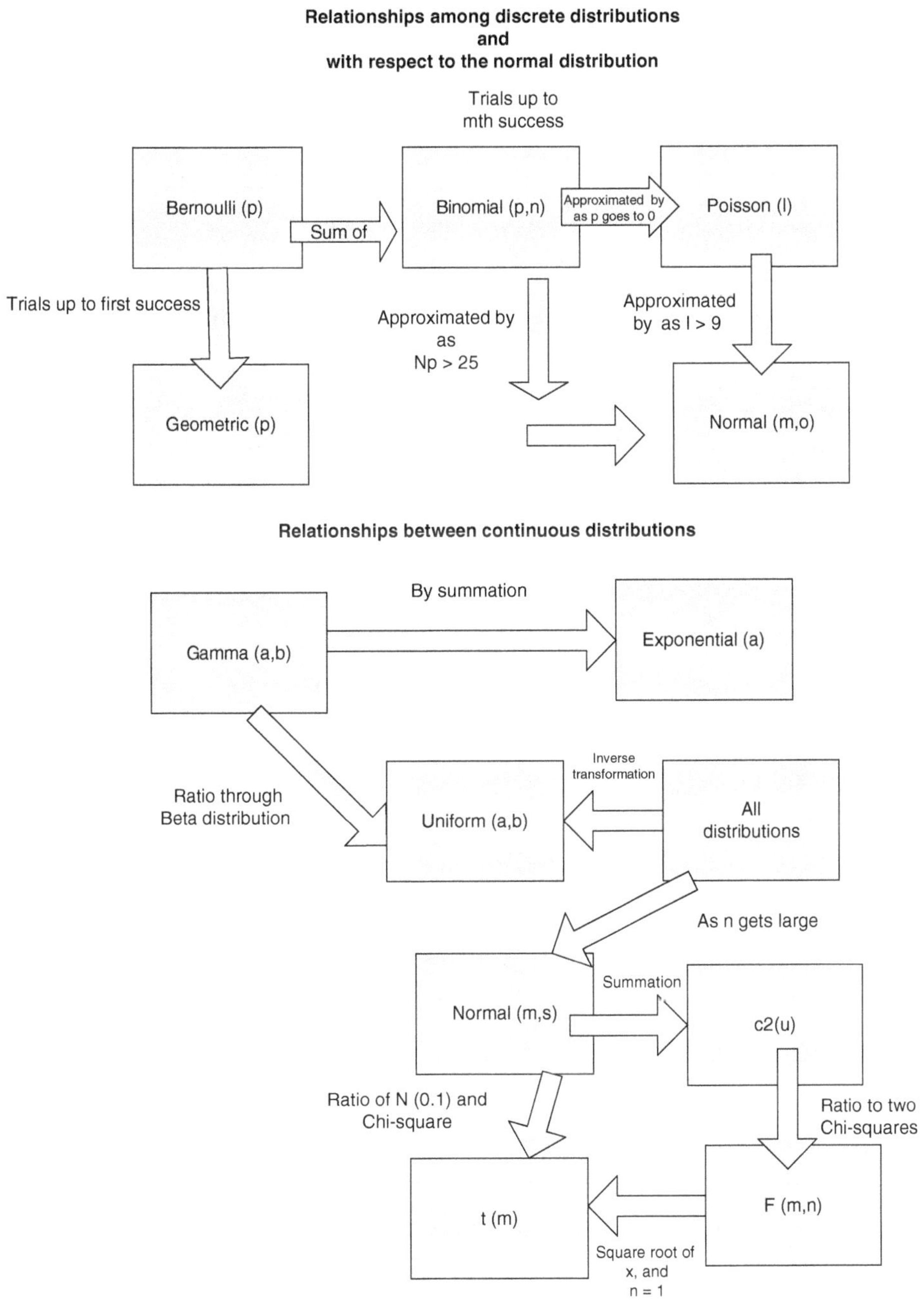

FIGURE 39.1 General summary for commonly used distributions.

39.14 Exercises

1. What distribution would you use to model the following?

   a. Number of requests between typing errors, given that each request has a certain probability of being in error

b. The mean of a large set of observations from a uniform distribution

c. Sample variances from a normal population

d. Ratio of two sample variances from a normal population

e. Time between successive arrivals, given that the arrivals are memory-less

f. Number of systems that are idle in a distributed system, given each system has a fixed probability of being idle

2. Let x, y, z, w be four unit normal variates. Find the distribution and 90th percentiles for the following quantities:

a. $(x+y+z+w)/4$

b. $x^2 + y^2 + z^2 + w^2$

c. $(x^2 + y^2)/(z^2 + w^2)$

d. $w/\sqrt{(x^2 + y^2 + z^2 + w^2)/4}$

Reference

1. Box, G.E.P. and Muller, M.E., "A Note on the Generation of Random Normal Deviates," *Annals of Mathematical Statistics*, 29 (1958), 610–611.

UNIT 9

RESEARCH METHODS

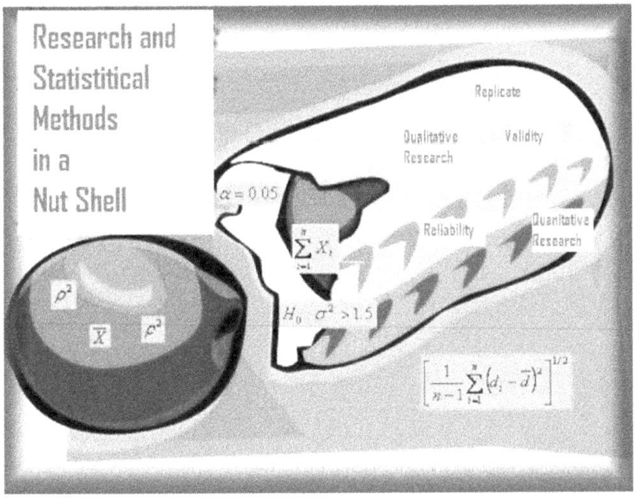

Study *Research Methods in Computing: What are they, and how should we teach them?* by Hilary J. Holz, Anne Applin, Bruria Haberman, Donald Joyce, Helen Purchase, and Catherine Reed found at *http://www.dcs.gla.ac.uk/publications/PAPERS/8445/sigsce-final.pdf*, an article printed in ITCSE '06 June 26-28, 2006. Bologna, Italy.

- What does SIGCSE-CSRM represent? Why is this important?
- What common characteristics of research methods in various domains of computing fields were identified?
- What standards did SIGCSE-CSRM decide to set? On what basis and process?
- Explain with illustrations the purpose of a general framework of learning activities suitable for teaching CSRM literacy.

Computing researchers and educators interested in teaching Computing Research Methods need a structure for collaborative exploration of this important topic. Our textbook, especially units nine and ten, is the author's attempt to present material on the content, pedagogy and curricular issues related to Computing Research Methods.

CHAPTER 40

A GUIDE TO RESEARCH

OVERVIEW AND LEARNING OBJECTIVES

In This Chapter

40.1 Introduction

The purpose of this chapter is to provide the reader with an overview of the research process which includes its logical and philosophical aspects, relevant ethics, and use of existing background. Hopefully, after studying this chapter, the reader will be able to answer the question: "Is my research project noteworthy?"

But what is research? How is research accomplished? How can a research project be evaluated? How can the research conducted be communicated? How and where can the results of the research be implemented and applied?

A good starting point to resolve these questions is to look at the role the background review, which contains and extends the literature review, plays in the entire research process.

Too often graduate students, many who are attempting to complete a research project for the first time, mechanically follow the traditional template that resembles:

1. Formulate a problem statement.
2. Perform a background review.
3. Develop a conceptual framework.
4. State the problem statement as a formal hypothesis.
5. Choose and/or implement a design and the procedures.
6. Obtain the results (analyze the data).
7. Interpret the results and make conclusions.

Sadly, students have a tendency to think and implement this template in a strictly sequential manner. As a result of their action, the student believes that the background review's primary purpose is to:

1. Begin by preparing a detailed annotated bibliography for the proposed problem.
2. Validate the originality of the research.

Unfortunately, the student perceives that every preceding step must be completed before working on the current step.

Formulating and finding answers to questions is the primary purpose of research. In this quest, the researcher must determine which questions are pertinent and which answers will extend knowledge in the research domain. These research efforts can be accomplished using exploration, description, predicting, and being explanatory. Finally, applications can then naturally follow.

40.2 Conceptual Framework

A researcher must prepare for research through a continuous background review to enable their ideas to discover noteworthy concepts in the research domain. Reading, searching, intercommunity communication, and filing activities enable identification of specific concepts pertinent to the proposed area of concern. These activities enable the formulation of key concepts for the purpose of communicating with key communities in the research domain of the topic. Eventually, the researcher compiles sufficient background to try their hand at creating a problem statement. The researcher needs to formulate the research problem at a level formally complete so that the findings from the proposed study have validity, and are related to other significant studies on the associated concepts.

This is the actual beginning point for the real background review. A continuous background review must be undertaken to find further studies on key concepts and their interrelationships. Additionally, organizational and sensitizing activity begins. Included in this continuous process is an attempt to find theoretical writings which may help in organizing.

The steps just described are the process of building a conceptual or theoretical framework. Researchers maintain a perpetual review program, as a result they are constantly prepared to identify problems for future research. As an idea occurs that is fruitful for future research they file it away for reference purposes. Throughout the research they must reveal what sets of concepts they are utilizing, concepts that assist them in defining problems, in collecting data, and in analyzing the results to problems. Pertinent questions to ask during the development and maintenance of a personal conceptual framework include:

1. Is the framework truly supportive in my research?
2. Does the framework enable a higher degree of efficiency in identifying meaningful research problems?

An excellent article on constructing and utilizing a conceptual framework is *Exploring the usefulness of a conceptual framework as a research tool: A researcher's self reflections* by Robyn Smyth in Issues In Educational Research, Vol 14, 2004 found at http//ier.org.au/iier14/smyth.html.

Basically the suggested approach by Smyth follows the flows found in the following diagram:

Notice that the background review and the conceptual framework are the bus structure for this research circuit. Many two way interactions exist for the researcher to find assistance in the fundamental research. Communication via the bus, the background review and conceptual framework, enable researchers to determine the relevancy of the particular problem under study. This quest for relevancy leads to dealing with the issues of reliability and validity.

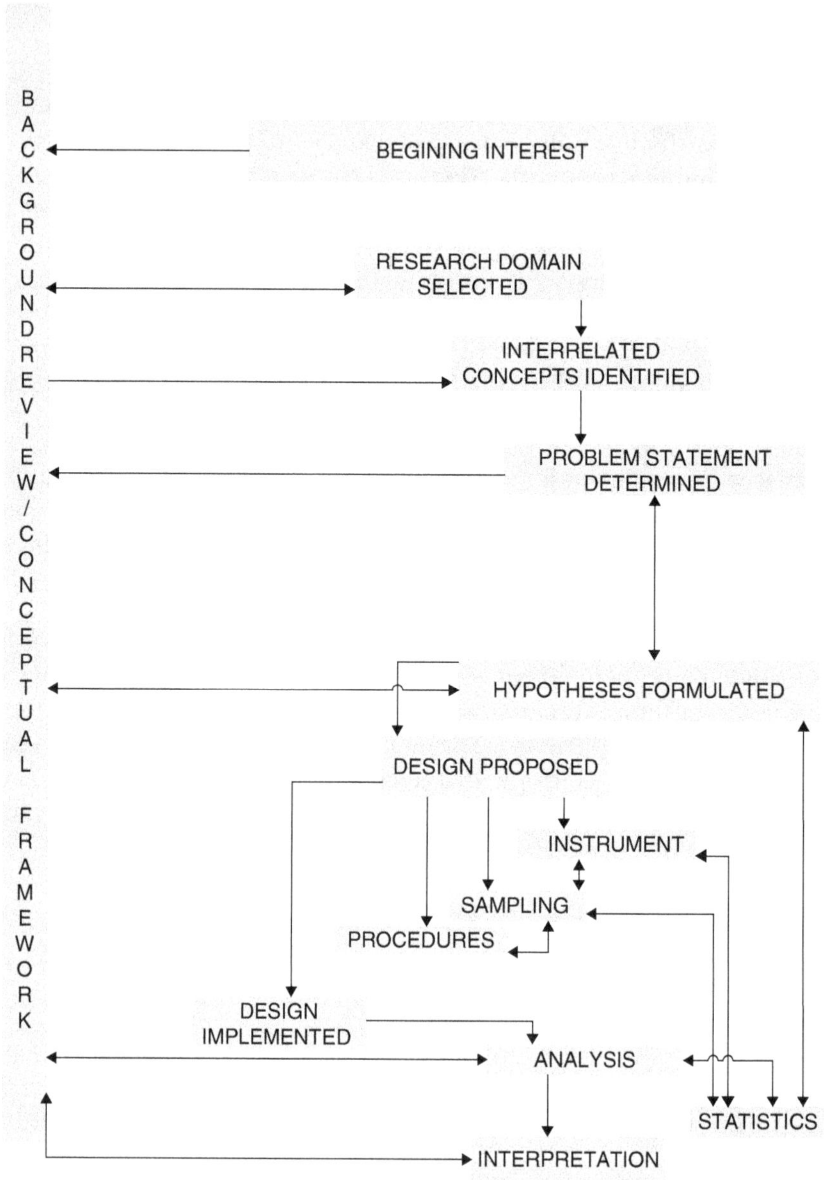

FIGURE 40.1 A Conceptual Framework Bus Architecture.

40.3 Reliability, Validity, Utility, and Usage

The reliability of a concept should be clearly distinguished from its validity. Reliability deals with how well a proposed evaluation of the concept measures the concept. For instance, will the evaluation of the concept consistently measure the concept. On the other-hand, validity measures how well the concept measures reflect what the concept actually is about.

The validity of a research study reflects the degree of correlation with some currently and widely accepted indicators in the domain of research under investigation. The reliability of a research study may be briefly defined as an index measuring the degree of agreement among re-observations of the same phenomena. Unreliability therefore implies variation among re-observations.

Two additional distinct concepts that should not be confused with reliability and validity are utility and usage. The utility of a concept may be defined as the correlation coefficient between the use of the concept and the efficiency of controlling its relevant phenomena. The definition of usage is to measure the proportion of a specified group which uses the concept or uses it in a specified way.

Note that it is possible to take usage as a criterion for validating a concept. Also utility of a concept actually means how well it contributes to our ability to predict and control phenomena. Many times in science, a concept is useful or useless in proportion to its ability to enable prediction and control of relevant phenomena better than with alternative concepts or absence of them.

In the physical science component of the field of computing, utility would seem to be the most important of the four properties. Reliability clearly would be next, while validity and usage are desirable but of less importance in the long run for scientific progress.

In summary the four properties imply that researchers should ask:

1. Are the results in the current research study useful for prediction?
2. Are the procedures employed reliable?
3. How valid are the results by specified criteria?
4. How widely used is the concept?

40.4 The Scientific Method

Any one who has taken a physical science class in their formal education has been exposed to the scientific method. Basically, it is a sequence of steps undertaken to find the answers to proposed questions. Physical scientist formulate questions and seek to find answers to these questions employing the following regimental set of steps:

1. Conduct research and observation of the phenomena.
2. Identify the problem.
3. State the formal hypothesis.
4. Conduct the experiment.
5. Analyze the results and draw conclusions.

But how does the following of these steps differ from solving the problem using common sense?

First: The scientific method systematically builds their theoretical structures, tests them for internal consistency, and subjects them to empirical testing. The average person may use conceptual schemes and theories, but usually only in a loose fashion.

Second: Scientists systematically and empirically test their theories and hypotheses. The average person also tests hypotheses, but does so in a selective fashion.

Third: There is a profound difference in the application of control. Scientists take deliberate steps, the average person would simply overlook, to control their explanations of observed phenomena in a systematic manner.

Fourth: Laymen simply invoke common sense for their explanations among phenomena. Scientists cultivate relations, they consciously and systematically pursue relations.

Fifth: When attempting to explain the observed phenomena the scientist rules out metaphysical explanations, which tend to be the layman's mainstay.

In summary, science is concerned with things that can be observed and tested.

40.4.1 Research

Often the first step in research in most graduate programs is to request that the student perform a background search to identify at least 3 to 6 potential questions. In other words, have the student begin to develop a research area. Clearly, the student as any professional researcher might be influenced by past experiences which could enhance their research studies. In particular, their past interests and experiences help them to identify particular concepts to consider in the actual pursuit of a new problem.

40.4.2 Problem

Usually the first draft of the problem is stated in ordinary prose and their rationale presented in the form of an argument. First, the researcher may experience doubt when observing the phenomena. Next, the researcher formulates the problem. They study the literature and internet accessible resources, seek the experiences of others, and draw upon their own experiences. Once the basic question is formulated, then the hypothesis is constructed and, ultimately, implications are deduced empirically. The hypothesis is tested by observation and experimentation. On the basis of the research evidence, the hypothesis is either rejected or it will fail to be rejected.

40.4.3 Project experimentation

In this phase of the research process the hypothesis is tested usually under the setting of a well controlled experiment. A design is either developed or chosen compatible with the research hypothesis and the question being tested, that controls or equalizes the effect of as many important extraneous variables as feasible. The design should also randomize the effects of all uncontrolled factors. Then the changes taking place in the dependent variable are observed by making changes in the independent variables.

The essential characteristics of a good project experimentation are:

1. An experimental design is used that insures that the observed dependent variable effects are unbiased estimates of the true effects.

2. A quantitative description of the precision of the dependent variable effects can be regarded as estimates of the true effects.
3. According to the broader purposes of the experiment the observed dependent variable effects will have whatever degree of precision that is required.
4. The hypothesis is a reliable and valid test.
5. The design can satisfy the requirements at minimum "cost."

During this phase researchers should ask:

What are the available resources?

What are the constraints which "time" may have for the completion of the study?

What is the actual "goodness" of a design, or are the internal and external validities of the experiment met?

40.4.4 Project conclusion

In the last part of the experiment the researcher is to interpret, conclude and make recommendations for further research. Results achieved should be compared with:

1. studies previously reviewed,
2. the conceptual framework within which the research was conducted, and
3. studies that may not have been previously considered but are suggested by the outcomes.

For more information on the scientific method the interested reader should consult *The Scientific Method* found at *http://oiscl.net/scientific_Method.pdf*.

40.5 Topic Research

How do researcher come up with an idea for a research project? The anthropologist often comes up with ideas by being actively engaged in field work. These ideas are a result of seeing what has been happening in the field.

Background reviews, as discussed, allow one to extend and refine previous research. Additionally, many organizations publish Requests for Proposals (RFPs). RFPs usually describe the problem, the context in which it operates, the desired approach for investigating the problem, and the amount the organization would be willing to pay for the research. Based upon the researcher's educational background, culture, and experiences, the researcher needs to check the proposed problem for whether or not the problem is feasible.

A good reference on topic research is found at *http://www.socialresearchmethods.net/kb/probfrm.php*.

40.6 Project Research

What is the purpose of a research report? Clearly it is used to tell the readers the problem investigated, the methods used to solve the problem, the results of the investigation, and the conclusions inferred from the results. As expeditiously and clearly as possible, the report should inform the reader what was done, why it was done, the outcome of doing it, and the investigator's conclusions. The reader should be able to reach his own conclusions regarding the adequacy of the research and the validity of the reported results and conclusions.

The general structure of the research report is the same as the structure of the research itself:

I. Problem

- **a.** Theory, hypotheses, definitions
- **b.** Previous research, the background review

II. Methodology [Data Collection]

- **a.** Sample and sampling method
- **b.** How hypotheses were tested [methodology], experimental procedures, instrumentation
- **c.** Measurement of variables
- **d.** Methods of analysis, statistics
- **e.** Pretesting and pilot studies

III. Results, Interpretations and Conclusions

In project research the problem section differs from that usually found in a thesis. In a thesis it is usually long and detailed, while in a research report it is kept to a minimum. The researcher needs to:

1. tell the reader what the research problem is about in a question form.
2. state the general problem not in a precise and operational manner, but simply set the general stage for the reader. Sub-problems should be precise and contain implications for testing. Some researchers state the general and specific hypotheses in an effort to inform the reader of the main area of investigation and the specific propositions that were tested.

All variables should be operationally defined. Every effort should be made to inform the reader not only of the variables but also what they mean. These operational terms must be given justification.

40.7 Scientific Writing

A research report must be written with clarity and accuracy. After all, the report must be informative. Efforts at generating emotion, to entertain, or to persuade are to be avoided. As a result the report should be written in a fashion similar to an military officer giving orders to the troops, a modified version for research report writing includes:

1. Keep it SIMPLE.
2. Operationally define all terms.
3. Start with the goal, the problem statement.
4. Use simple words which avoid usage of jargon.
5. Write in the third person with a passive voice.
6. Normally speak to the reader in past tense, except for usage of the present tense for specific directions to the reader.
7. Avoid exclusionary and stereotyping in writing.
8. Eliminate all unnecessary words, sentences, and paragraphs.
9. Find a peer reviewer for your report.
10. Write the abstract last. It includes primarily the main points but factually and briefly stated. This is not a summary.
11. Maintain the practice of critical examination of others writing. Practice, thought, and study of the 'good, bad, and ugly' of others writing is mandatory.
12. An exception to (2) is used infrequently to free up the dryness of the report by using the active voice in the first person and using the appropriate tense.

For additional information, for researchers in the computing field, about better writing consult the Chicago Style American Psychological Association (APA) publication manual.

An excellent list of questions for evaluating a research report is Guidelines for Critiquing Research Articles found at *http://www.sonoma.edu/users/n/nolan/n400/critique.htm.*

40.8 Matters of Ethical Concern in Research

Computer professionals are keenly aware of professional codes of ethics, behavioral guidelines, and professional responsibility since their professional organizations have codes of professional conduct as well as self-assessments on ethical issues in computing. The main ones are the ACM and IEEE Computer Society. Jointly they developed the Software Engineering Code of Ethics and Professional Practice. A Code of Ethics for Research emphasizes some of the same issues addressed by the latter codes of conduct: ethical issues of honesty and fairness, responsibility to respect confidentiality, maintenance of professional competence, being aware of relevant laws, and honoring contracts and agreements. Additional stress is placed upon: responsibility to respect property and protect privacy, avoid harming others, and respecting property rights.

The University of Pittsburgh offers the following guidelines, *GUIDELINES FOR RESPONSIBLE CONDUCT OF RESEARCH by the Office of Research Integrity, Office of Research Integrity (2007) found at http://www.pitt.edu/~provost/ethresearch.html.*

40.9 Exercises

1. Explain the relationship that exists between being able to generalize and replicate in a research project.
2. Complete a background search to operationally define face validity, content validity, criterion-related validity, translation validity, predictive validity, concurrent validity, convergent validity, and discriminant validity.
3. Complete a background search on how to operationally define ex-post facto experimentation. How does this approach differ from a true experiment?
4. Explain how reliability and validity are related.
5. Establish your own conceptual framework for research adapted to your field of interest in which you will carry out investigations. Try to demonstrate that your conceptual framework has potential as a meta-analysis tool by fulfilling the conditions set for it. By the criteria outlined in this chapter, determine if it can be judged to be a useful research tool in the context for which it was developed. These are the criteria that any researcher should satisfy regardless of the simplicity or complexity of the framework being evaluated.
6. Suppose you plan to test the hypothesis that a push information method to the sales force of your company enhances the overall company sales production by increasing the sales force's understanding of the client base. Outline an ideal research design to test this hypothesis, assuming you have complete command of the situation and plenty of money and help. Set up an alternate design as follows:
    - A design with complete randomization
    - A matching design with complete randomization

Now compromise the ideal. Set up alternative designs as follows:

- A matching design without randomization
- A compromise design

How do these four designs compare in their control power? Which design do you think should best test the hypothesis? In which design can you put the greatest faith? Why?

7. A college faculty group is to study a core curriculum program instituted during the previous year. All incoming freshmen are required to take this program. Discuss the problems and difficulties the researchers face. How much faith can be put in their findings?

8. The faculty of a university school of business decides to begin a core curriculum for all undergraduates and, therefore needs, to study the effectiveness of the proposed curriculum. A faculty research group is asked to study the program for two years. The research group, wishing to have a group with which to compare the core curriculum group, requests that the regular program be continued for two years and that students be allowed to volunteer for the core program or the regular program. In this way, the group feels it will have an experimental group and a control group. Discuss the group's proposal critically. Would you have faith in its findings at the end of the two years? Give reasons for reacting positively or negatively to the proposal. What research design would you like to set up?

9. Imagine that you are a graduate school professor and that you are asked to judge the worth of a proposed doctoral thesis. The doctoral student is a database administrator who is instituting a new type of administration into his company's system. He plans to study the effects of the new administration for a three-year period and then write his thesis. To keep the study objective, he will not study any other company situation. Discuss the candidate's proposal, keeping these questions in mind: If the study is completed, should he be awarded the degree? Is such a study suitable for doctoral work?

40.10 Project

Introduction

Little is known of the strategies that researchers adopt during the process of conducting a project, especially in the computing field. You are to conduct a study that contributes to the knowledge of that process by posing the research question: "What are the influencing factors on the research process and how do researchers harness these factors to enhance their research?" The majority of textbooks on research methods in business, education, and the behavioral sciences follow the template for research discussed in Section 40.4, The Scientific Method. Feel free to use this template. The second stage, identifying the problem, should be expanded to include preliminary data and negotiating access and ethical issues.

A succession of interviews needs to be conducted with eight or more academic professionals in either information systems, computer science, or computer engineering to construct a preliminary model of their perception of the research process.

You are to collect data on this recommended factors and to provide a structured way of assessing an individual's perception of the research process.

The majority of textbooks on business, social science, and educational research detail the research process. The preliminary stage of identifying the research problem should be expanded to incorporate preliminary data gathering and negotiating access and ethical issues.

Little has been written on the issues arising at these different stages of the research process. Factors that affect research projects are knowledge, skills, and personal background.

The following list of factors can be constructed, but feel free to modify:

- Motivation
- IT skills
- Financial resources
- Time availability
- Intellectual skills
- Administrative skills
- Creativity
- Communication skills
- Library resources
- Research knowledge
- Perseverance
- Business knowledge

40.11 Problem Solving

Conduct this survey research study and report the results. Consider selecting interviewees all from the same university but at different stages in their career as an experimental control. All interviewees selected should have had at least one example of their work appearing in conference proceedings or refereed journals within the last two years, a second form of experimental control. The interviewees should represent a variety of computing disciplines: networking, database, information systems, computer graphics, operating systems, etc. Further interviews should be conducted with each participant that examines their experiences and the importance of the various research stages and the impact of the various factors. The participants will then be asked to delete any influencing factors they considered unimportant and to add any influencing factors that they considered important. The aggregate results of the interviews should be compiled and analyzed.

CHAPTER 41

SURVEY AND FIELD RESEARCH

OVERVIEW AND LEARNING OBJECTIVES

In This Chapter

41.1 Introduction

Recall the last time you applied for a job. At one point you probably filled out an employment application. This instrument is the organization's effort to collect information directly from the group of candidates for the job. Notice this process involves obtaining information directly from the participants in the study by asking questions.

But what type of information is collected? When you report your age, gender, years of education, and income on the application these response values are *facts*. *Facts* are values that can independently be verified. In addition to facts, often these employment applications contain questions to prompt the respondent's preference, feeling, or behavioral intention. These type of questions are referred to as *opinions*. For example, How would you react to the scenario X? or If Y responded to X in manner Z, rate your degree of agreement/disagreement. Primarily facts differ from opinions with respect to how they are operationally defined. A birth date field data value can be used to capture the respondent's age with little or no controversy. But how we capture, implement, or interpret opinions can and probably would generate considerable controversy. Finally, a question like: How many times have you attended X? are referred to as *behavior*, or actions completed by the respondent.

Often surveys include all three of these types of content. Opinions can be objectively measured but cannot be verified. At the same time, facts can be verified. Behavior can be verified, although facts are usually easier to verify. The point being made is that, without verification, there is a tendency to apply Garbage-In-Garbage-Out to your research.

This chapter's purpose is to introduce several issues and procedures of survey research. Any research study in which participants are directly asked questions, which might be using questionnaires or interviews, to obtain information is referred to as survey research. Note that, a study is not referred to as survey research simply because it employs questionnaires or interviews. The study is a survey research if information is obtained directly from a group of individuals. Survey research is highly valuable when one recognizes that it accommodates large sample sizes, even with the possibility of being efficient at acceptable "costs."

41.2 Types of Surveys

In this section we will view and examine different types of surveys. Generally, these consist of questionnaires and interviews. Questionnaires can be classified as: mail surveys, e-surveys, and group administered surveys.

A *Mail Survey* is a survey in which the respondent completes the survey instrument without intervention from the researcher.

E-surveys are either E-mail surveys or Web-mail surveys.

A *Group-Administered Survey* is where a group of participants are brought together and are asked to respond to a set of structured questions in a survey instrument handed out to each person to complete within a specific time.

41.2.1 Questionnaires

Most surveys use structured instructions to ensure that the participants understand how to properly complete the instrument. Even though mail surveying is appealing, one must keep in mind that they tend to have low response rates.

The difference between E-mail and Web-mail surveys is due to how the communication links are utilized. E-mail surveys are "pushed" directly to the respondent's computer and Web mail surveys are "pulled" by the respondent from the Web site. As attractive as these types of surveys are the researcher needs to consider whether or not the 'have - and - have - not' syndrome (not all participants might have access to E-mail or Web-mail) applies to the participants in the study.

Questionnaires are often convenient for group settings and thus enable high response rates. When administrating the questionnaire, the researcher, due to being present, can respond to respondents questions about the study and get clarification on completing the questionnaire.

41.2.2 Interviews

Classification for interviews includes: *face-to-face interviews*, *focused-interviewing*, and *telephone interviewing*.

Face-to-face interviews are where the survey instrument is orally administered.

Focused Interviews usually provide the interviewer with highly structured predetermined primary questions but allow flexibility to the interviewer in asking secondary questions.

Telephone interviewing is simply an alternative to face-to-face interviews employing telecommunications technology.

In face-to-face interviews where the instrument itself is orally administered the interview must be highly structured. The interviewer must: provide instructions, record responses, and answer respondents' questions. From this list of activities it becomes apparent that interviewers should probably be trained and certified.

Focused interviews tend to generate research studies with a high degree of internal and external validity. But an unfocused interview, where participants are allowed to openly discuss a topic often obtained by the interviewer probing the participant, thereby rendering questionable internal and external validity for the interview and study. This type of interview tends to be expensive but at the same time enlightening and extremely effective.

To determine which method to use, the researcher needs to address the population, sampling, the question, content, bias, and administrative issues. A overview of survey research by the Public Administration Program at North Carolina State University can be found at *http://faculty.chas.ncsu.edu/garson/PA765/survey.htm*.

Whatever method of research, in this case survey research, the researcher must strive for reliability and validity. The interested reader should consult the following articles supported by Colorado State University (2011) found at *http://writing.colostate.edu/guides/research/relvaal/pop2a.cfm*.

41.2.3 Writing your own survey questions

Reconsider the employment application. What is the basic format of this instrument? It contains item topics [personal data, position applying for, record of education, etc.], instructions [how to respond], item formats [birth date as 1/23/43], and arrangements of items. All of these components can have an effect on the information obtained.

A few helpful question writing considerations include:

1. BE NEAT! This is partially accomplished with spacing.
2. Make sure that interviewers and participants fully understand the purpose of the study and how to correctly complete the survey.
3. Be careful with wording and labels. Avoid slang and jargon.
4. Place sensitive items at the end of sections to reduce response bias.
5. Place nonthreatening and interesting items at the beginning of sections.
6. Start the instrument with a purpose of study statement followed by the general instructions needed to correctly respond. Next include the personal demographic section.
7. Visually set off each section, a group of interrelated questions, with spacing and indentations.
8. Remember that item order can both create and eliminate response bias.
9. Consider required response rate, acceptable costs, item types, and item formats needed during the design phase with a keen eye on analysis to be employed.
10. Fine tune the instrument and prepare interviewers through pretesting. Use pretesting to investigate every aspect of the survey project.

This is only a short list of the considerations required when designing your own instrument or interview. The interested reader should consult *Outline on Research Instruments* found at *http://www2.uvawise.edu/pww8y/Resources/MERes/ResInstruments/00ResInstruments.html*. The University of Central Florida publishes an article, *Writing Your Own Survey Questions*, found at *http://www.fctl.ucf.edu/ResearchAndScholarship/SoTL/creatingSoTLProjects/implementingmanaging/writeyourownsurveyitems.php#general*

It is evident that there are many formats and styles for survey instruments. The researcher needs to choose an instrument by comparing the advantages and disadvantages for the different types of surveys. But the best advice is to remember the old computer slogan GIGO, Garbage-In-Garbage-Out. If you start by immediately designing and implementing your own instrument, then your work is just beginning. You will need to determine the instruments' reliability and validity, a task worthy of a whole new textbook devoted to only that issue. Commercial and professional instruments are available for most research domains, complete with results of their reliability and validity analyzes. Start by searching for an appropriate existing survey instrument. Do not design your own instrument unless absolutely necessary.

41.3 Survey Research Sample

Study *Comparative Study of Graduate Information Systems and MBA Students Cognitive Styles*, by King and Stock found on this book's companion DVD. Problem-solving styles are critical dimensions in corporate decision making. In order to study the comparative problem-solving styles of college graduate students majoring in information systems versus business administration, the Kirton Adaption-Innovation Inventory (KAI) was used to determine the cognitive styles of the students enrolled in the Master of Science in Information Systems (MSIS) and the Master of Business Administration (MBA) programs at the University of Mary Hardin-Baylor (UMHB) during the Spring semester of 2007. The results of the study were also analyzed by demographic variants.[1,2]

ON DVD

41.4 Sampling

In most survey research studies the instrument or interview will be administered to a sample of the population. The analysis is done on a *sample*, or a subset of the *population*. Actually, samples are selected from a *sampling frame*, or a published listing of the elements in the population. Generally, sampling is a valuable tool for obtaining data quickly, cheaply and accurately. The purpose of this section is to define and illustrate some of the more common sampling methods. Also discussed is the difference between two major types of sampling procedures: probability and non-probability sampling.

Suppose a sign manufacturer has produced over 1000 signs and wants to determine average packaging weight for various groupings of signs. To solve this problem, focus on a small scale of the problem by considering a population consisting of 24 signs: 8 octagons, 6 pentagons, 6 triangles, and 4 diamonds.

TABLE 41.1 Hypothetical population of signs.

Octagon		Pentagon		Triangle		Diamond	
Id No	Wt	Id No	Wt	Id No	Wt	Id No	Wt
1	40	4	27	10	15	16	45
2	50	5	33	11	15	17	50
3	45	6	30	12	17	18	52
20	45	7	28	13	19	19	57
21	50	8	32	14	18		
22	52	9	31	15	17		
23	57						
24	42		_		_		_
Total	381		181		101		204
$\overline{Wt}$	47.63		30.16		16.83		51

41.4.1 Simple random sampling

We will simply draw a sample of size 8 from our population of discussion. Note that every sign has the same chance of being selected to be in the sample. There are a total of ${}_{24}C_8 = 1{,}598{,}850$ possible samples of size 8 that can be drawn from the hypothetical population of 24 signs. Using a table of random numbers the following simple random sample was obtained. A *random selection*, or a *simple random sample* includes any method where each member of the population has the same probability of being included in the sample.

TABLE 41.2 Simple Random Sample.

Random Sample of Sign Weights	
No.	**Wt.**
10	15
3	45
16	45
18	52
9	31
15	17
20	45
16	45
Total	295
$\overline{Wt}$	36.88

The mean of the sample is 36.88, while the mean of the population weights is 36.12. Notice that the true value of the average sign weights for the population is "essentially" the same as the average sign weight for the selected sample.

However, every once in a while we may select an extreme sample when using simple random sampling. Then the estimator's value may be considerably different than the population's value for the parameter. Consider the following samples drawn from our hypothetical population:

TABLE 41.3 Extreme samples.

Sample I		**Sample II**	
Id No	**Wt**	**Id No**	**Wt**
10	15	21	50
4	27	2	50
12	17	8	32
13	18	22	52
14	18	23	57
15	17	24	42

(*Continued*)

TABLE 41.3 Continued.

Sample I		Sample II	
Id No	**Wt**	**Id No**	**Wt**
6	30	19	57
7	28	16	45
Total	170		385
$\overline{Wt}$	21.25		48.13

The average sign weight for these extreme samples depart from the population average sign weight.

41.4.2 Stratified sampling

Risk of picking an extreme sample are small, but methods are available for controlling the sample variation in the various research study variates. Controlling these variates lowers the probability of selecting an extreme sample. In the population of signs, use sign shape, the *stratum*, for selecting the sample. To obtain sample of size 8, we can select 3 octagons, 2 pentagons, 2 triangles, and one diamond

TABLE 41.4 Sign weights by stratum [sign shape].

Stratum	Id No	Wt
Octagon	2 22 17	50 52 50
Pentagon	6 8	30 32
Triangle	12 14	17 18
Diamond	17	50
Total		266
$\overline{Wt}$		37.38

Since each sign has the same chance to be selected, 1 out of 4, then weighting is unnecessary in the separate strata. It is evident that one would get a better estimate for the population's average sign weight. Why? Because each stratum is represented in the sample selected. There are ${}_9C_3\,{}_6C_2\,{}_6C_2\,{}_4C_1 = 75{,}600$ stratified samples as compared to ${}_{24}C_8 = 1{,}598{,}850$ simple random samples.

Organizations using stratified sampling often extend the method so that cost of conducting the survey varies from stratum to stratum. Then a fixed budget for the project prompts the research group to determine the most efficient way of dividing the total sample by strata. This process of determining the sample size and its breakdown by strata can yield the lowest possible cost that will achieve the desired accuracy.

41.4.3 Cluster sampling

Cluster sampling uses the construction of non-homogeneous clusters of selected units from the population. For the hypothetical population under study we could make clusters, each from a particular weight class. Use the following weight classes:

TABLE 41.5 Weight class frequencies.

Weight Interval	Id Nos
Wt ≤19	10,11,12,13,14,15
20 ≤ Wt ≤ 29	4,7
30 ≤ Wt ≤ 39	5,6,8,9
Wt ≥ 40	1,2,3,20,21,22,23,24,16,17,18,19

We can choose clusters of size 8, by taking 2 units from Wt ≤19, 1 unit from 20 ≤ Wt ≤ 29, 1 unit from 30 ≤ Wt ≤ 39, and 4 units from Wt ≥ 40.

TABLE 41.6 Cluster Samples.

Cluster 1			Cluster 2		
Id No	**Variety**	**Wt**	**Id No**	**Variety**	**Wt**
11	Triangle	15	10	Triangle	15
13	Triangle	19	12	Triangle	19
4	Pentagon	27	4	Pentagon	27
8	Pentagon	32	9	Pentagon	31
20	Octagon	45	19	Diamond	12
24	Octagon	42	22	Octagon	52
16	Diamond	45	2	Octagon	50
19	Diamond	57	3	Octagon	45
Total		282			251
$\overline{Wt}$		35.25			31.38

Notice that the most precise results under cluster sampling, as suggested by comparing the compositions of clusters 1 and 2, will be obtained when each cluster has a heterogeneous composition, while the clusters tend to be similar. The estimate derived from stratified sampling was near the true value of the population parameter because the units in each cluster were all mixed up while similarity was maintained between clusters. The reverse is true for stratified clustering, strata are internally homogeneous, while the strata themselves are as heterogeneous as possible.

In practice, cluster sampling produces less precise estimates than a simple random sample of the same size, since in actual studies one is often required to accept the clusters as they exist. Usually these clusters will be relatively homogeneous.

Suppose the researcher wants to determine the type of activity and average time spent on computing in the home on a daily basis. Image the potential cost of running this study using a simple random sample. Why not use cluster sampling? Households in a single book are fairly homogeneous with respect to family income but at the same time there is great variety from block to block. Then a sample of a large number of neighborhoods containing thousands of households, cluster sampling; might provide as accurate a result as a widely dispersed simple random sample of half as many households, but at much lower cost. While the simple random sample is more precise, practically with respect to size, the larger cluster sample will be cost efficient.

41.4.4 Alternative sampling methods

There are other sampling methods employed in practice. The following brief list provides operational definitions for the alternative methods.

41.4.4.1 Systematic Sampling

For the population of signs we would simply select a random number between 1 and int (n/N), where n is the sample size, N is the population size, and int(_) is the largest positive whole number less than or equal to the function's argument. Then choose every int (n/N) sampling unit from a starting point, say a value between 1 and n. Systematic random sampling is based upon the concept of selecting elements from a randomly arranged sampling frame according to ordered criteria. For a very large sampling frame and a large sample, systematic random sampling is very inexpensive and direct plus very simple to implement.

41.4.4.2 Double Sampling

In the quest for a more cost effective and efficient sampling method for situations requiring large samples and large sampling frames, a sample of a sample can be employed, or *double sampling* can be used. For instances, a simple brief questionnaire could be utilized for obtaining data on representative household occupations for a large population from a sample of ten thousands of households. This process could then be administered a comprehensive questionnaire, using the proportions of each occupational group in the larger sample, for selecting a sample with sample size in the several orders of magnitude lower than for the first questionnaire.

All of the sampling methods discussed so far rely on obtaining a random sample, therefore, you can rely on the Central Limit Theorem and the associated properties of the unit normal distribution to determine the amount of sampling error in the sample. In order to calculate the standard error, or the standard deviation of the sampling distribution of means, it is necessary to evaluate the formula:

$$S_{\bar{X}} = \frac{S_X}{\sqrt{n}}$$

where $S_{\bar{X}}$ is the standard error, S_X is the standard deviation of the distribution of sample means, and n is the common sample size. *Probability sampling* is any method that ensures a random sample.

When there is the presence of a procedure where elements have unequal chances for being included in the sample, it is called *nonprobability sampling methods.* Without a sampling distribution present, nonprobability sampling methods do not have a way to estimate sampling error. These distributions are commonly used when sample selection is based upon visual impression, participant opinion, and subjective factors.

Think about the person-on-the-street interviews seen on local and national newscasts. In *accidental sampling* sample units are selected on the basis of being available or being easily included upon immediate selection.

Most of us have run into people in a mall carrying clipboards with attached questionnaires stopping various people entering or leaving the mall. Usually, the mall pedestrian is pursued by these people and asked to participate in some study; actually the mall pedestrian has been sized up to see whether or not they are qualified for being selected for membership in the sample. *Purposive Sampling* is useful in situations requiring efficient access to a target group.

Modal sampling is when you sample the most frequent case. Interviews of the 'typical voter' is an example of a modal sampling application. But what is a 'typical voter'?

Panel of experts are common at national and /or professional meetings. This is a good approach to elicit the views of individuals with specific expertise. *Expert Sampling* deals with assembling a sample of individuals possessing demonstrable experience and/or expertise within a domain.

Pollsters use *quota sampling* to predict the outcomes of elections. The sampling units are selected on the basis of perceived categories within the population. In stratified sampling the sampling elements are selected within the stratified groups. On the other hand, in quota sampling the elements are selected within presumed subdivisions of the population.

41.5 Sampling Errors

When sampling ensures that every sampling unit has a chance of being selected and the probability of selection is known, then the sampling errors are controllable. The following list points out some important types that the researcher needs to be concerned about when gathering data:

1. reporting and processing errors,
2. non-response errors,
3. errors in sample selection, and unit sampling errors.

41.6 Field Studies

Imagine being asked to implement research on everyday events. For example, assume you have been hired to study the cognitive activities of professional programmers in their corporate computing center.

How would you begin this study? What are we really trying to accomplish? We are attempting to understand how they resolve and solve problems during their daily activities, both on an individual and team basis. What type of organizational structures formally and informally are in place within the working environment?

It will probably take a considerable amount of effort to simply decide which aspects of the personnel's behavior to investigate, let alone determine hypotheses to test. These behaviors involve individual and corporate team interactions.

But how will we collect all this information? Based upon most of our everyday experiences many of us would probably suspect:

1. use of photography to capture photos that can be immediately shared and analyzed.
2. take photographs every day of a sitting or artifact possibly photographs every day of a team's whiteboard – which could contain a list, or log, of an individual's assignments and other personnel with whom they interact.
3. use of video for recording meetings for future interpretation.
4. maintain written notes – a pocket notebook to use on a moments notice, a word processor to organize your thoughts using an "outline" mode or writing observations under continual recording and maintenance.
5. you could classify the systematically organized observations.

Basically, you would be starting to employ a collection of methods that include direct observation of naturally occurring events. *Direct observation* is where the event is observed as it happens. *Naturally occurring events* are events that are not created, sustained, or discontinued solely for research purposes. This type of research is referred to as *Field Research.* In this section, the practices of field research will be discussed.

Through observation you would begin to collect data on the personnel's activities and eventually formulate new hypotheses. Field research is suitable for exploratory and descriptive goal finding studies.

Most of your time during a field study is bound to be through either recall or direct observation. In the proposed study, you would be examining how people solve problems and make decisions in real work environments. By studying the daily events of the personnel in this work environment, it would be best for you to not only observe but simultaneously, take part in the events. After all, the best approach to understand problem solving, is to 'learn it by doing it'. This approach is called *Participant Observation*. Studies where you observe events without participating in them, thus being as unobtrusive as possible, are called *Systematic Observation*.

During the proposed study, it seems appropriate to collect data on *observations of behavior* on the part of the software development team. However, imagine the insight to be gained by collecting their individual *reports of behavior*.

Systematic information on the personnel's daily activities could be obtained simply by asking them to report their activities for a recent specified period of time. You could ask them *open-ended questions* and record the answers, or create a *time diary*. Clearly, both of these techniques require interviewing and recording skills. The time diary would keep data on when an activity starts, ends, and its sequence with the other daily activities. More than likely for your recording in the time diary, you will need to invent a coding scheme for the activities.

But when do you record the activities? The obvious answer is to record the subject's behavior the moment they are observed. This type of procedure is referred to as *Instantaneous Sampling*.

Just like it is important for students to take notes, participant observers must decide how much detail to include in their *field journals.* If something happens, then you must post it in the field journal. How often should you maintain the field journal? The obvious answer is at every available opportunity. Often the difference between remembering and forgetting an event is dependent on sketch notes in the field journal. Remember that the field journal is the participant observer's primary research tool.

Clearly field research involves the researcher following activities and observing through questionnaires and interviewing. These events must be selected, recorded, coded, and interpreted. The field study is really an understanding process, both about the domain and about the cognitive activities under study. The researcher wants to:

1. form hypotheses about the behavioral patterns observed,
2. make predictions, and
3. finally, verify the predictions.

The following article 'The Observational Study of Behavior', by J. Altmann published in *Behavior* 48: 1–41, illustrates how data analysis of abstractions and the associated interpretations can be the primary effort in a field study. When studying this article it is important to try to identify sources of error in the study. Observations need to be measured reliably, or consistently, and they must be valid, in other words, do they measure what they are supposed to measure. Otherwise you cannot describe events and test hypotheses about them. Inadequate sampling could be present in a field study. Also the researcher should be very careful about changing their minds about what to observe, this leads to inadequate sampling. Other error sources are due to recording, observing, and interpreting observations.

A text on qualitative field research which serves as a manual to follow when conducting an ethnographic project, complete with examples, is *A Guide to Qualitative Field Research* by Carol Bailey published by Pine Forge Press, 2nd Edition (September 14, 2006). Another reference, for the interested reader is *The Observational Study of Behavior* by J. Altmann published in Behavior 48: 1–41.

41.7 Field Research Example

ON DVD

Study *Ethnographic Study of MSIS Student Project Team Dynamics*, by Ronald S. King, found on the textbook's DVD. Problem-solving styles and interpersonal dynamics of project teams are often critical dimensions of the ability of a team to function effectively. In order to study the problem-solving styles of students enrolled in a graduate Database Administration and Development class at the University of Mary Hardin-Baylor (UMHB) and to track their intrateam interactions, the theory established by M.J. Kirton in 1976 was used to determine the cognitive styles of the students. Students were categorized according to adaption-innovation characteristics of the Kirton Adaption-Innovation Inventory. These categorizations were then applied to interpret the results of interviews from informants and the team leaders, and about the abilities of project team members to work and communicate with each other.

1. What type of field research is employed? Why?
2. What kind of interviews were made? Why?
3. Describe the three types of participant-observation studies:

41.8 Survey Research Exercises

Answer the following questions:

1. What types of information can be obtained when conducting survey research?
2. List the components and state the purpose of each component of a survey instrument.
3. What is the most important phase in survey research? Defend your answer.
4. List the advantages and disadvantages for:
   a. Face-to-face interviews
   b. Telephone interviews
   c. Mail surveys
5. Operationally define the types of analyses that can be accomplished with survey research.
6. In reference to the King and Stock survey research study:
   a. How was the internal validity addressed?
   b. Is the study replicable?
   c. How could reliability have been addressed?
   d. What were the main results?
   e. List and discuss any surprising results found.

41.9 Sampling Exercises

1. Suppose that random samples of *nj* individuals are taken from strata containing N_j, j = 1,2,…,k. Let $N = \Sigma_{j=1}^{k} N_j$ and $n = \Sigma_{j=1}^{k} n_j$.

   Also let $\bar{X}_j$ and S_j^2 denote the means and variances in the strata. Let be the population mean.

a. Derive the formula for an unbiased estimator for the population mean using a point estimate.

b. Derive the point estimator formula the standard error for the distribution of the sample mean.

c. Develop the confidence interval formulas for a 100 (1– α)% confidence interval for the population mean for stratified random samples.

2. a. Develop the formula for the population total formed by stratified random samples.

b. Develop the formula for a 100(1-α) confidence interval for the population total for stratified random samples.

3. Redo problem 1 for cluster random samples.

4. Redo problem 2 for cluster random samples.

41.10 HYPOTHETICAL RESEARCH PROJECT

Evaluation of the Shattered Dreams Program

Suppose you wanted to determine whether or not the *Shattered Dreams* program changes attitudes toward binge drinking among students. By presenting them with videos of horrible accidents, death, and dismemberment associated with behaviors that people engage in while drunk, it is presumed that students in high school would not drink and drive. So you start the arrangements for a local high school to implement the *Shattered Dreams* program to the senior class enacted prior to the school's senior prom. For each type of sampling, give a *realistic* means of using that approach. Realistic means that you, as a student, could actually experience what the actors experienced in the enactment.

1. Create a different sampling strategy for getting student participation that involves:
   - Simple random sampling
   - Stratified random sampling
   - Convenience sampling
   - Purposive sampling

   Many instances of Web sites on the *Shattered Dreams* can be found on the internet. The interested reader should start by surfing the net for the Texas Alcoholic Beverage Commission, they did not create the program. The Texas Alcoholic Beverage Commission does support and pay some of the cost associated with hosting a Shattered Dreams program.

2. If you used simple random sampling to select the students on campus who participated in your study, to which of the following populations would you feel comfortable generalizing your results? In each case, there are reasons to believe that generalizing would be appropriate, but there are countering reasons to be suspicious

of generalizing. Present both sides of the argument and draw a conclusion for each of these four groups:

- Other students on your campus
- Nonstudents of the same age as the students who participated in your research
- Adults between the ages of 30 and 50

3. What would be the advantages and disadvantages of simple random sampling versus quota sampling?

41.11 Field Study Exercises

1. Operationally define field research.
2. What is meant by systematic observations? Hint: Discuss methods of selecting participants, methods of recording events, and sources of errors.
3. Through a literature or surfing the internet, operationally define and illustrate the following types of sampling: quota sampling, snowball sampling, deviant case sampling, and the use of informants.
4. Discuss the advantages and disadvantages of participant observations.
5. List the dos and don'ts you would recommend to be used in training the observers in the sample study in section 41.5.
6. What are the advantages and disadvantages of observers intruding into the action being observed?
7. How would you determine the validity of a survey instrument you constructed for a study?
8. How would you determine the reliability of a survey instrument you constructed for a study?
9. How does reliability differ from validity? Can a valid instrument be unreliable? Can a reliable instrument be invalid? Can an instrument be both unreliable and invalid? Can an instrument be both valid and reliable?
10. When conducting a survey research study, should the survey instrument's determination of reliability be mandatory? Should the determination of validity be mandatory? Defend your answers.

41.12 SURVEY RESEARCH PROJECT

Global Internet participation has grown at a very fast pace, both in terms of the number of internet users as well as the amount of time spent online. This has encouraged the steady growth of global e-commerce, despite the earlier setbacks of the dot-com crisis. Online grocery retail has been one aspect of B2C e-commerce that has largely disappointed expectations, managing to capture only a marginal fraction of the overall grocery trade. As a consequence, the worldwide failure rate for e-grocers has been extremely high, leading to the suspicions that there are fundamental weaknesses in the model.

1. Develop a research paper that explores the challenges facing online grocery ventures.
2. Do your results ultimately conclude that these barriers pose a credible challenge to the viability of the pure-play (online only) form of the model. Or is there evidence from the few e-grocers who are profitable today that would suggest that online grocery trade is most effectively implemented in conjunction with physical grocery retail stores (the bricks and clicks form of the model)?

41.13 Sampling References

1. Lohr, Sharon L. *Sampling: Design and Analysis*, Duxbury Press: 2002 Edition (December 9, 2009).
2. Thompson, Steven K. *Sampling: Second Edition*, Wiley Interscience, 2 Edition (April 4, 2002).

CHAPTER 42

A METHODOLOGY FOR MODEL CONSTRUCTION

OVERVIEW AND LEARNING OBJECTIVES

In This Chapter

42.1 Introduction

This chapter is a proposal on how to build and evaluate models for solving problems, especially when applying the multiple regression approach. Basically the proposed steps are:

First specify the problem, complete with identifying and screening of the variables. The chosen variables must be correlated to the problem under study and readily accessible. By carefully studying a correlation matrix for all independent and dependent variables for currently maintained data, the researcher is allowed to run pilot tests on simple regression models. These preliminary tests enable insight into future modeling efforts. This is also a good time to establish goals for the project complete with estimates for project completion requirements such as necessary time and cost.

Next a model, or several models to be followed, needs to be developed by selecting a best representative model, using correlation analysis, the researcher should study the interrelationships between various independent variables. Out of this effort, the researcher will be able to select independent variables in regression equations, as well as transformed independent variables that may be used in the final results. Study of residual plots should also be performed at this time. Finally, once a useful model has been chosen, then procedures for verification of the model and model maintenance should be started.

This chapter and the header article are an attempt to illustrate the model-building process. Special emphasis is on procedures for selecting variables for a regression analysis as well as choosing a best regression model for a specific study. The process follows a personal data analysis methodology recommended by Katrina D. Maxwell in *Applied Statistics for Software Managers* published by Pearson Publications, Inc. (2002). Maxwell's text provides not only a personal data analysis methodology, or a model building process, but illustrates the process using actual corporate data to solve specific software productivity, time-to-market, and maintenance problems.

42.2 Sample Study

Study *Forecasting Gold Prices Using Multiple Linear Regression Method* by Ismail, Yahya, and Shabri, found at *http://www.scipub.org/fulltext/ajas/ajas681509-1514.pdf*. Both this article and the material found in this chapter illustrate and outline the steps for constructing linear models. Linear models tend to crystallize one's thinking with regard to the variables operating in an experimental situation. Models form the basis for the analysis and interpretation of data generated by the experiments. Therefore, the model should undergo a preliminary analysis as well as be validated.

Clearly this article touches on a hot topic of our day. Several models were considered before choosing the appropriate model. When conducting a study, you may not only have to consider past models, but you may also have to construct a brand new one conceptually. The following example of applied research illustrates the modeling construction process in this situation.

42.3 Lessons Learned

It usually takes a number of years to collect enough software project data to analyze. Add the software industry's high job turnover rate, and more often than not, you will be analyzing data that was collected by others. What methods should you use? What should you do first?

42.3.1 Step 1: Validate your data

> *Data validation* is reviewing the data to determine that it is the right data for the study and that the values are correctly specified and fit the variable's scope.

Data validation is the most important step, as it is critical to verify that the specific data is complete, accurate, and fit. Data is often assembled from various sources, and it needs to be evaluated to make sure it was gathered correctly and that it adequately represents that for which it was gathered.

A Judgment Analysis (JAN) study is used throughout this chapter for discussing model construction. Consider the following study, which was generated by a research request from the College of Business Administration at Omega University. The Information Systems Department wanted to find and measure a set of criteria that would be descriptive of the doctoral candidate selection process by the IS faculty. This set of criteria would be descriptive of the candidates' success in the program. Selective procedures for doctoral candidates in IS involve looking at many predictor variables; some of these variables are GRE population, GMAT scores, Miller Analogies Test (MAT) scores, Strong Vocational Interest Blank (SVIB), incoming GPA, and previous major field of study. The MMPI is also required for each incoming doctoral candidate. However, in the past many of the doctoral graduates did not take the MMPI. Because of this, the 13 subscales of the MMPI were not included in the study. To describe the doctoral candidate selection process used by the IS faculty, three questions had to be answered:

- How many distinct judgmental policies are presented among the IS faculty with respect to what they consider the important cognitive and affective skills of an outstanding graduate student?
- Which of the affective and cognitive skills being considered for each graduate do the faculty deem most important?
- Which predictor variables used in determining admission to the program relate to the captured judgmental policy or policies?

42.3.1.1 Statement of Problem

At present there seems to be a wide variety of methodologies available to capture and analyze judgmental policies where either the investigator is looking only at predictor variables alone or there is one criterion variable and several predictor variables. The author has been unable to find in the literature a methodology for capturing and analyzing judgmental policies where the investigator is looking at a set of criterion variables and a set of predictor variables in conjunction. Therefore, the problem was to attempt to capture and analyze judgmental policies of the IS faculty while looking at both a set of criterion variables and a set of predictor variables.

42.3.1.2 Purpose

The basic purpose of this study, which was service-oriented in nature, was to develop a methodology for capturing and analyzing judgmental policies where the investigator is looking at a set of criterion variables and a set of predictors in conjunction. This type of methodology was needed so that the technique can be used to a great degree in judgmental research.

42.3.2 Step 2: Select the variables and model

> *Variable and model selection* determines what variables and relationships to study to ensure all steps taken target the intended purpose of the study.

The purpose for variable and model selection is to determine what data we actually have and the semantics of the data. What relationships could we study? What possible relationships should we study? What relationships will we study first? Your answers will depend on the overall goals of the analysis.

42.3.2.1 Operational Definitions

In this section, we provide definitions of various concepts discussed throughout this chapter.

Ipsative JAN refers to the JAN approach in which the judges were using personal knowledge of the capabilities of each IS doctoral graduate to suggest a rank. It is a way of quantifying subjective opinions that yields more objective results.

The *judges* are only those faculty who have had the student as an advisee or student, or worked on special projects with him and rated the student using the criterion instrument.

The *judgmental evaluation procedure* is the outcome of a rating process by a judge who has been presented a series of variables on which he must base a decision. For example, how does a judge involve effective thought processing when presented with a series of variables on which he must base a decision?

Normative JAN refers to the JAN approach in which the judges were utilizing the predictor variables to rank each IS student into one of seven categories.

A *profile score* is a profile that was defined as the set of ratings assigned by a judge to a specific IS doctoral student on the nine criteria, Y_i, $i = 1, \dots, 9$.

Type A JAN is an analysis in which all the judges give judgments on the same subjects with respect to the same criterion variable and predictor variables.

Type B JAN is an analysis in which the judges do not judge the same subjects (different judges could judge a different number of subjects also) with respect to the same criterion variable and predictor variables.

42.3.2.2 Questions Answered

The questions answered were subdivided into two general populations. The first population of concern pertains to how well the methodology answered the questions of the IS research request. These questions are as follows:

- How many different judgmental policies exist within the IS Department concerning what constitutes an outstanding IS doctoral graduate?
- Which of the affective and cognitive skills being considered for each graduate do the faculty deem most important?
- What kinds of information that they now collect are important indicators of a graduate's ability to succeed as an IS employee?

The second population of concern involves interpretation of computer outputs. The major questions that were considered are as follows:

- What type of judgmental policies are captured?
- Is it possible to determine whether or not each judge is a "good" judge?
- How many judgmental policies are captured?
- What are the most important predictor and criterion variables?
- What is the relationship between the criteria and predictors of importance?
- Is the procedure for analyzing a captured policy an intuitive or a statistical process?

42.3.2.3 Limitations

There are a few limitations to this study:

- The subject to predictor ratio should have been larger.
- This specific study cannot be generalized beyond the IS Department at Omega University.

The Criterion Instrument Used to Obtain the Ratings of the Faculty for the 50 Doctoral Graduates with Whom They Were Familiar

Project—Can future success of doctoral candidates be predicted with a high degree of certainty using only a few predictors?

Directions

On an attached sheet you will find a list of the doctoral graduates from the IS Department. As judges (i.e., faculty of the IS Department), please rate the graduates whom you feel qualified to rate. You will be asked to rank them in nine different categories, plus a general overall rank. When evaluating the graduates, please rate them in comparison to the other graduates with whom you are familiar and in terms of what you think constitutes an outstanding IS graduate.

Please rank the graduates from 1 to 5, where 1 = poor, 2 = below average, 3 = average, 4 = above average, and 5 = outstanding, on the following categories.

Name of Graduate________________ Name of Judge____________________

____

Section 1 (pertains to the graduate while a student in the IS Department).

RANK

1. Academic success (final GPA, comps, orals, and dissertation)
2. Verbal ability or verbal articulation
3. Intellectual ability
4. Interpersonal relationships — faculty and supervisors
5. Interpersonal relationships — peers
6. Interpersonal relationships — to people he is serving (students, inmates, etc.)
7. Leadership initiative in getting things done (projects, meetings, etc.)
8. Personal characteristics (emotionally mature, friendly, enthusiastic)
9. Improvement of self (personally and professionally)
10. General overall rank combining all categories above into a single rank from 1 to 5

42.3.2.4 Judgment Analysis

The Judgment Analysis technique is useful for identifying the rating policies existing in a board or committee of judges. With this technique, each judge makes a judgmental decision on one common criterion variable on which he is given a common predictor set profile on a set of judges to determine how many different evaluation policies exist in the group. Next, a multiple regression analysis can be made to determine the composition of each captured policy.

The JAN procedure consists of two basic stages. In the first stage, a least-squares solution of a multiple regression equation is computed for each judge, which predicts the criterion decision he has made. It answers the question of how consistent the judge is in his use of specific variables in arriving at an overall decision by giving his decision-making equation and the R^2, the square of the multiple correlation coefficient, from his multiple regression analysis. A high R^2 indicates that he has a policy and was consistent in expressing it through his criterion decisions.

Next, a hierarchical grouping procedure is initiated that allows the grouping of individual judges according to some objective function by similarity of their decision-making equations in this case. Groupings are selected that yield the smallest decrease in ability to predict the criterion from knowledge of which individual judges are in a particular group (loss in R^2).

The first step in the grouping procedure reduces the number of equations by one. The two judges whose decision-making equations are most similar are grouped together as a unit and all other judges are left ungrouped. From the set of ungrouped judges, the process then selects a judge whose decision-making equation is most similar to that of the grouped judges and groups him with them as a unit. The number of judges remaining in the ungrouped set is reduced by one at each step of the process until all judges have been grouped into a single unit. At each step of the grouping process, an examination of the loss of predictive efficiency allows the investigator to identify the number of various judgmental policies that exist.

If a single policy is characteristic of the whole collection of judges, it can be determined at the final stage of grouping. If a joint policy does not exist, arbitration of the criterion by the whole group may yield a joint policy.

Judgment analysis has been extensively used by the United States Air Force to analyze individual and group judgments and to formulate common or overall judgmental policies. A large amount of this type of work has been done by Ward, Bottenberg, and Christal. Bottenberg and Christal[1] were originally responsible for the mathematical and computer program developments in the field.[2] Showed that JAN was highly valid.

42.3.3 Step 3: Perform preliminary analyses

> *Preliminary analysis* is a review of the data to become more familiar with its parts, descriptions, and relationships, in order to make sure it will not violate any of the assumptions the statistical tests are built on.

The purpose of the preliminary analysis is to check that the assumptions underlying the statistical tests are true. You need to objectively look at the data and decide on the goal for the study. You must determine which of the variables collected has an influence on achieving this goal. The preliminary analysis will utilize different types of graphs, tables, correlations, and other statistical tools.

42.3.3.1 Predictor Variables

The following hierarchy of predictor variables was used in the study. This information was obtained either from the faculty of the IS Department or the Omega University Graduate Office. As mentioned earlier, this information was used in the selection of IS doctoral students.

The cognitive variables are the following:

X_2 = GRE population score in social sciences

X_3 = GRE population score in the humanities

X_4 = GRE population score in the natural sciences

X_5 = GMAT verbal score

X_6 = GMAT qualitative score

X_7 = GMAT composite score

X_8 = Incoming GPA earned in student's Master's degree program (A = 4.0, B = 3.0, C = 2.0, D = 1.0)

X_9 = Miller Analogies Test score

NOTE *Note: Variables X_2, X_3, … , X_9 are in raw score form.*

The affective variables, from Group 5 of the SVIB, are the following:

X_{10} = Interest score for becoming an IT manager

X_{11} = Interest score for becoming a database manager

X_{12}= Interest score for becoming a systems analyst

X_{13} = Interest score for becoming a programmer

X_{14}= Interest score for becoming a Web site designer

X_{15}= Interest score for becoming a CIO

The biographic variables are the following:

X_{16}= Age when admitted to the IS doctoral program

X_{17}= 1 if the Master's degree program was in IS or MIS; = 0 otherwise

There are eight scores on the previous version of the SVIB Group 5 section, occupational interest. This inventory has been revised three times in the last 10 years and new occupational classifications have been added. Because of this, only those occupational populations of interest that are on all editions were used in the study. Also the scores on the same year edition were not rescaled. This decision was made after consulting with the director of research marketing.

The means, standard deviations, and raw scores for the predictor variables X_2, X_3, … , X_{17} are considered confidential information by the faculty of the IS Department. Because of this, they were not included in the study.

42.3.3.2 Criterion Variables

The following hierarchy of criterion variables was used in the study.

The cognitive variables are the following:

Y_1 = Academic success (final GPA, comprehensives, orals, and dissertation)

Y_2 = Verbal ability or verbal articulation

Y_3 = Intellectual ability

The social or affective variables are the following:

Y_4 = Interpersonal relationships with faculty and supervisors

Y_5 = Interpersonal relationships with his peers

Y_6 = Interpersonal relationship with the people he is serving (students, inmates, patients, etc.)

Y_7 = Leadership initiative in getting things done (projects, etc.)

Y_8 = Personal characteristics (emotionally mature, friendly, enthusiastic)

Y_9 = Improvement of self (personally and professionally)

The composite variable is the following:

Y_{10} = General overall rank combining all of the above nine categories into a single rank

Each of the criteria was scored on a scale from 1 to 5 where 1 = poor, 2 = below average, 3 = average, 4 = above average, and 5 = outstanding.

The criterion instrument included in the faculty questionnaire was developed by three members of the IS faculty. Considerations for the instrument were based on the following question. "If it were my job to hire a Ph.D. with a degree in IS work, what characteristics or attributes would I would expect him to have?" From this question, the nine criterion variables were generated and used in the study. Most of the IS doctorates have graduated in the past three years and so consideration was not given to the number of publications, employment effectiveness in their present job, and professional affiliations.

The criterion instrument was submitted to the 16 members of the IS faculty with the following instructions: Please rate each of the above Ph.D. graduates with whom you are familiar on a scale from 1 to 5 where 1 = poor, 2 = below average, 3 = average, 4 = above average, and 5 = outstanding. The basis for comparison should be your conception of what constitutes an outstanding Ph.D. graduate in IS.

Notice that this is an ipsative type of JAN ranking. Approximately 500 ratings were obtained through the distribution of the questionnaire.

42.3.3.3 Questions Asked

- How many of the judges adhered strictly to the profile scores when ranking the IS doctoral graduates?
- How many different judgmental policies existed?
- Which predictor variables $X_2, X_3, \ldots, X_{17}$ were the most effective in redirecting the obtained judgmental policy?

42.3.3.4 Method Used for Organizing Data

The following output data for the normative JAN procedure was presented and summarized in Tables 42.1 to 42.5:

- The means and standard deviations for the profile scores and the judgments
- The correlation matrix for the profile scores

- The correlation matrix for the judgments
- The correlations between the judgments and the profile scores
- The sequential and cumulative R^2 drops, the judges combined, and the single member systems remaining in each step of the JAN procedure

The means and standard deviations for the profile scores and judgments are listed in Table 42.1. The mean for each of the profile scores was approximately 3. Since each Y_i value, where i=l, ... , 9, only takes on values from 1 to 5, and since the 40 sets of ratings were selected at random, these results seem consistent. The standard deviations also seem consistent with what should be expected. (Note that 12 judges returned the instrument, as discussed later.)

TABLE 42.1 Means and standard deviations for the profile scores and the judgments.

Profile	Mean	Standard Deviation
Y_1 = Academic success (final GPA, comps, orals, dissertation)	3.27	1.18
Y_2 = Verbal ability or verbal articulation	3.15	1.37
Y_3 = Intellectual ability	3.20	1.17
Y_4 = Interpersonal relationships - faculty and supervisor	3.1	1.14
Y_5 == Interpersonal relationships - peers	3.07	1.06
Y_6 = Interpersonal relationships - to people he is serving (students, inmates, etc.)	2.92	1.10
Y_7 = Leadership initiative in getting things done (projects, meetings, etc.)	3.02	1.23
Y_8 = Personal characteristics (emotionally mature, friendly, enthusiastic)	3.27	1.22
Y_9 = Improvement of self (personally and professionally)	3.15	1.31
Judgments		
Judge 1	3.15	1.64
Judge 2	3.10	2.05
Judge 3	3.07	1.69
Judge 4	3.60	1.50
Judge 5	3.90	1.91
Judge 6	3.57	1.41
Judge 7	3.40	1.56
Judge 8	3.97	1.80
Judge 9	3.40	1.60
Judge 10	3.45	1.82
Judge 11	3.82	1.80
Judge 12	3.47	1.39

ON DVD All figures and tables in this chapter appear on the companion DVD.

The means for the judgments all lie between 3 and 4. This indicates that the judges tended to put more of the 40 sets of ratings in groups 1 through 3 rather than in groups 5 through 7. The standard deviations of the judgments were also consistent with what should be expected.

The correlation matrix for the profile scores, the correlation matrix for the judgments, the intercorrelation matrix between the profile scores and the judgments, and the initial R^2 for each of the judgments are included in Tables 42.2 through 42.5, respectively.

The correlation matrix for the profile scores, shown in Table 42.2, indicates that the cognitive variables Y_1, Y_2, and Y_3 all correlate highly with one another. Variables Y_4, Y_5, and Y_6 correlate highly with one another but quite low with variables Y_1, Y_2, and Y_3. Since variables Y_4, Y_5, and Y_6 are affective variables, this was to be expected. Variables Y_7, Y_8, and Y_9 are also affective variables, but their interpretation is not quite as clear. Y_8 appears to correlate the highest with Y_4, Y_5, Y_6, and Y_9. Y_7 and Y_9 correlate highly with each other but also fairly high with variables Y_1 through Y_6 and Y_8. It appears that Y_1 through Y_3, Y_4 through Y_6, and Y_8 through Y_9 should be grouped into three separate clusters. A factor analysis was completed on the profile scores and confirmed the three clusters.

TABLE 42.2 Correlation matrix for the profile scores.

Variable	1	2	3	4	5	6	7	8	9
1									
2	0.79								
3	0.85	0.84							
4	0.01	0.07	0.02	0.01					
5	0.06	0.08	0.03	0.84					
6	0.07	0.37	0.21	0.76	0.75				
7	0.42	0.72	0.53	0.39	0.42	0.62			
8	0.21	0.35	0.22	0.72	0.70	0.66	0.54		
9	0.33	0.54	0.42	0.50	0.51	0.59	0.77	0.75	

The correlation matrix for the judgments, shown in Table 42.3, indicates a high agreement among the judges in the ratings that each gave to the 40 sets of ratings. This will also be indicated later in the analysis of the JAN procedure.

TABLE 42.3 Correlation matrix for the judgments.

Variable	1	2	3	4	5	6	7	8	9	10	11	12
1												
2	0.90											
3	0.92	0.83										
4	0.82	0.37	0.73									

(Continued)

TABLE 42.3 Continued.

Variable	1	2	3	4	5	6	7	8	9	10	11	12
5	0.92	0.92	0.90	0.79								
6	0.87	0.90	0.83	0.73	0.89							
7	0.84	0.86	0.86	0.82	0.88	0.85						
8	0.85	0.85	0.84	0.71	0.85	0.81	0.86					
9	0.84	0.91	0.82	0.72	0.90	0.89	0.87	0.79				
10	0.86	0.89	0.85	.80	.87	.80	0.87	.85	0.81			
11	0.93	0.91	.89	0.88	0.92	0.85	.91	0.87	0.86	0.92		
12	0.92	0.90	0.88	0.85	0.90	0.85	0.88	0.88	0.85	0.91	0.95	

The intercorrelation matrix between the profiles and the judgments, shown in Table 42.4, indicates that Judges 4, 7, and 10 have somewhat higher relative correlations with variables Y_1, Y_2, and Y_3 than do the rest of the judges when the correlations are examined for each judge over all the profiles. Again, this result shows up in the analysis of the JAN procedure.

On the criterion instrument, there were 480 usable sets of ratings obtained that could be used in the study. One of the judges was eliminated from the study. From these 480 sets of ratings, 40 sets of ratings were randomly selected. The modified Q-sort instrument with the necessary instructions was then distributed to the 13 faculty members of the IS faculty and to the 3 past members of the IS faculty. Instruments were returned by 12 of the 16 faculty members.

TABLE 42.4 Intercorrelation matrix between the profiles and the judgments.

	Judgments												
	Variables	1	2	3	4	5	6	7	8	9	10	11	12
Profiles	1	0.45	0.39	0.97	0.62	0.35	0.31	0.47	0.39	0.33	0.37	0.57	0.54
	2	0.58	0.56	0.56	0.71	0.56	0.46	0.65	0.61	0.45	0.74	0.74	0.74
	3	0.50	0.40	0.40	0.73	0.43	0.32	0.54	0.46	0.39	0.64	0.62	0.61
	4	0.75	0.71	0.71	0.47	0.73	0.81	0.61	0.68	0.75	0.59	0.64	0.64
	5	0.69	0.78	0.73	0.43	0.77	0.78	0.64	0.67	0.79	0.59	0.61	0.62
	6	0.82	0.85	0.80	0.57	0.83	0.84	0.77	0.81	0.79	0.75	0.75	0.75
	7	0.76	0.76	0.74	0.75	0.78	0.67	0.78	0.68	0.68	0.81	0.85	0.82
	8	0.72	0.76	0.79	0.60	0.77	0.78	0.75	0.73	0.78	0.66	0.77	0.80
	9	0.76	0.75	0.76	0.68	0.79	0.68	0.70	0.65	0.72	0.73	0.82	0.85

The initial R^2 for each of the judges ranged from 0.9735 to 0.8134. This indicates that each judge paid strict attention to the variables upon which the ratings were based. Because of the excellent job done by each of the judges on the modified Q-sort, they were all included in the study. The sequential and cumulative R^2 drops, the judges combined, and the single-member systems remaining in each step of the Type A normative JAN procedure are listed on Table 42.5.

TABLE 42.5 Initial R^2 for each of the judges.

Judges	RSQ
Judge 1	0.9282
Judge 2	0.9521
Judge 3	0.8708
Judge 4	0.8134
Judge 5	0.9212
Judge 6	0.8860
Judge 7	0.8761
Judge 8	0.8436
Judge 9	0.8896
Judge 10	0.9109
Judge 11	0.9719
Judge 12	0.9735

42.3.4 Step 4: Determine design and methodologies of the study

By *design* and *methodologies* for a study, the designer identifies and enables the control, as well as the interpretation of the sources of variation in the study.

42.3.4.1 Subjects Judged

The IS Department had its first doctoral graduate 10 years earlier. Since then, not including the previous year's graduates, 55 individuals have received their doctoral degree from the IS Department. Of these 55 graduates, it was possible to get complete information on the predictor variables for 50 of them. These 50 were the individuals used in the study.

42.3.4.2 Judges

At present, there are 13 faculty members of the IS Department who have worked with some or all of the 50 doctoral graduates. There were three faculty members on the staff of the IS Department over the 10 years who have left for positions in other universities. Since they are familiar with many of the graduates being studied, they were included in the study. These 16 past or present faculty members were used as judges.

42.3.4.3 Strategy Used for Obtaining Data

Procedure 1

Out of the approximately 500 sets of ratings obtained on the criterion instrument, 40 were randomly selected.

Procedure 2

Each of the 16 members of the IS faculty were asked to do a modified Q-sort on the 40 sets of ratings. Each set of ratings was called a profile. They were asked to compare the 40 profiles against one another and place them in seven groups. Group 7 would represent those profiles they considered the best of the group of 40 and group 1 would represent those profiles they considered the poorest in the group of 40. Groups 2 through 6 lay on an ordinal scale between one and seven. The only restrictions were that each judge must place at least one profile in each of the seven groups. Hopefully, this forced the 12 judges to think along the whole continuum. The instrument and the instructions used for the Q-sort were presented to the IS faculty in a training session before they actually conducted this part of the study.

Procedure 3

After the judges completed the rankings, a Type A JAN was run to determine the number of judgmental policies. The decision as to the number of policies existing was based on the following consideration: Any judge who was unable to identify at least one significant factor is failing to relate any predictor variable set to any criterion variable.

Procedure 4

One judgmental policy was determined and a regression model was obtained where the rankings of the 40 profiles scored by each of the judges was the criterion Y_{11}, and $Y_1, \ldots, Y_9$ were the predictors. The systematic procedure was used to determine which of the nine criteria the IS faculty considered most important in an outstanding doctoral graduate. This process is visually displayed in the regression flowchart shown in Figure 42.1. If the subset of variables removed in the restricted model does not constitute an R^2 drop from the full model of 0.05 or more, then this subset of variables was not considered to make a significant contribution. The rationale as to which subsets of variables were grouped together is explained in section 42.3.6.

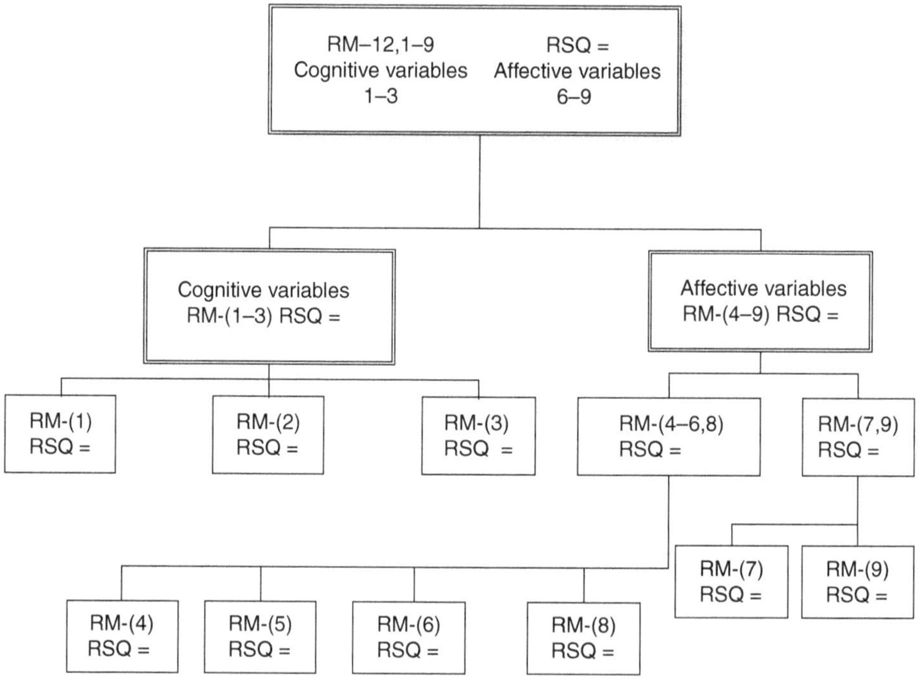

Note: Variables 1, 2,3, … , 9, 12
12, 1–9 means a model using 12 as the criterion and 1–9 as predictors
RM-( ) represents the restricted model formed by dropping variables ( ) out of the full model

FIGURE 42.1 Regression chart for the criterion variables.

Procedure 5

The average score for each IS doctoral graduate was determined for each of the criteria, Y_i, $i = 1, 2, \ldots, 9$. This score on each criterion was determined by averaging all the scores the graduate received from those judges who rated him. Average scores on each of the criteria along with the predetermined weight for the criteria were used to obtain a predicted score, Y_{12}, for each of the IS graduates: $Y_{12} = b_1 Y_1 + b_2 Y_2 + \ldots + b_9 Y_9$.

The second main objective of this methodology was to determine which of the predictors used for admittance to the IS doctoral program are of the greatest importance. Since a judgmental policy was already determined in regard to what constitutes an outstanding IS graduate, this was now possible.

Procedure 6

A regression model was run using the predicted score from Procedure 5 as the criterion, Y_{12}, and $X_2, X_3, \ldots, X_{17}$ were used as predictors. The systematic procedure that was used to determine which of the predictors used for admittance to the IS doctoral program are of greatest importance is given in the regression flowchart in Figure 42.2. If the subset of variables removed in the restricted model does not constitute an R^2 drop from the full model of 0.05 or more, then this subset of variables was not considered to make a significant contribution. The rationale as to which subsets of variables were grouped together is explained in Section 42.3.6.

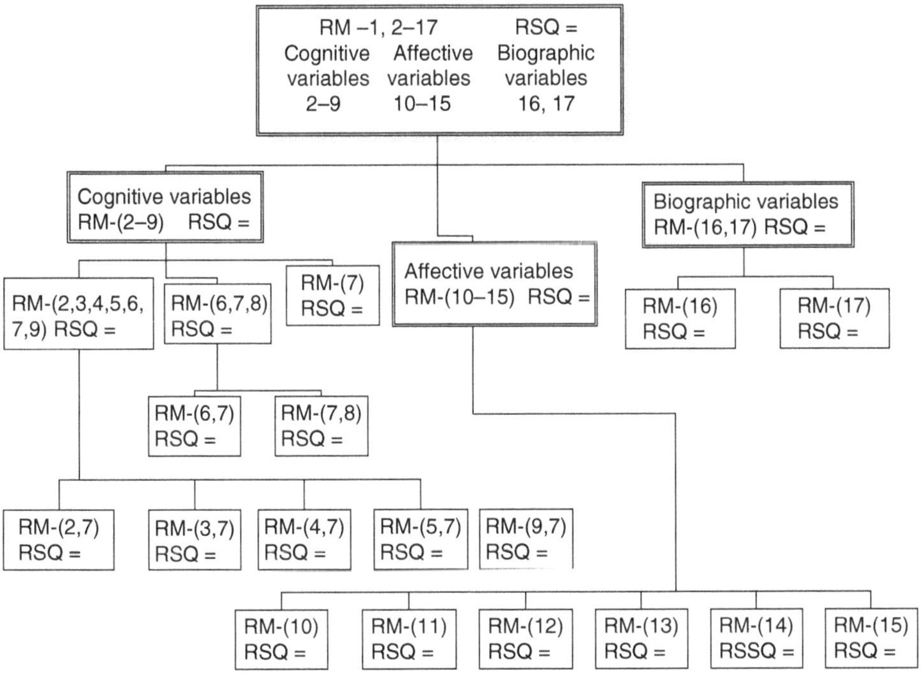

FIGURE 42.2 Regression chart for the predictor variables.

42.3.5 Step 5: Check the model

> *Checking the model* is verifying that the variables of the final model do not violate assumptions upon which the model is based. This involves careful determination that the variables are correctly distributed and standard errors are random, as well as looking at any other stringent variate requirements.

The purpose of checking the model is to make sure that the assumptions underlying the statistical tests used have not been violated In some experimental designs, we may reduce the variance of the estimator and increase the information by either reducing the variance (noise) or increasing the sample size (volume of the signal).

In practice, the assumptions for a linear model are rarely satisfied. The experimenter does not know all of the important variables in a process nor does he know the true functional relationships. Therefore, the function chosen to fit the true relation is only an approximation, and the variables included in the experiment form only a subset of the total. The random error is thus a composite of error caused by the failure to include all of the important process variables as well as the error in approximating the function.

To determine the number of judgmental policies present, the factors previously mentioned were used. Since the original R^2 drop is only 0.0729, this indicates that one judgmental policy provided does not exhibit a large R^2 drop from one grouping to the next. Further examination of the R^2 drop from one grouping to the next, as shown in Table 42.6, indicates a linear trend from one stage to the next. On the basis of this evidence, it was decided that one judgmental policy existed for the 12 judges. Because of this, it was not necessary for the IS faculty to meet and arbitrate a single policy.

TABLE 42.6 Stages of the JAN procedure for the judges.

Methodology for Criterion = 11				
Stages	**Judges**	**R^2**	**Drop in R^2**	**Collective Drop in R^2**
1	1,2,3,4,5,6,7,8,9,10,11,12	0.9084		
2	(3,9),1,2,4,6,7,8,9,10,11,12	0.9065	0.0019	
3	(3,5),(6,9),1,2,4,7,8,10,11,12	0.9041	0.0024	0.0043
4	(3,5),(6,9),(11,12),1,2,4,7,8,10	0.9013	0.0028	0.0071
5	(3,5),(6,9),(7,10),(11,12),1,2,4,8	0.8974	0.0038	0.0109
6	(1,2),(3,5),(6,9),(7,10),(11,12),4,8	0.8930	0.0044	0.0154
7	(1,2),(3,5,8),(6,9),(7,10),(11,12),4	0.8884	0.0046	0.0199
8	(1,2),(3,5,8),(6,9),(4,7,10),(11,12)	0.8831	0.0053	0.0253
9	(1,2),(3,5,8,6,9),(4,7,10),(11,12)	0.8771	0.0060	0.0313
10	(1,2),(3,5,8,6,9,11,12),(4,7,10)	0.8684	0.0087	0.0400
11	(1,2,3,5,8,6,9,11,12),(4,7,10)	0.8565	0.0119	0.0519
12	(1,2,3,5,8,6,9,11,12,4,7,10)	0.8533	0.0210	0.0729

A point of interest that was mentioned previously is again apparent when one examines stages 11 and 12. Judges 4, 7, and 10 are grouped together in stage 11 and when combined with the other nine judges in stage 12, we have the largest R^2 drop. Although this drop was not large enough to warrant two judgmental policies, it was consistent with the results found earlier in the examination of previous tables.

A regression model Y_{11} on $Y_1, Y_2, \ldots, Y_9$ was completed to determine which of the nine criteria judges deemed most important if they were to hire a Ph.D. with a degree in IS. Here we were trying to determine what the faculty considered the highest priority of the nine criteria they considered important for an outstanding doctoral graduate to possess. The schematic for the flowchart Figure 42.2 was developed by an examination of the correlations between the criteria $Y_1, Y_2, \ldots, Y_9$ and factor analysis.

42.3.6 Step 6: Extract the equation

> *Extracting the equation* derives the formula that explains the change in the dependent variable. This formula can be used to predict future values of the dependent variable or to analyze the sources of variation.

In a predictive regression study, the last step is to extract the equation for the purpose of predicting future and current values of the dependent variable.

Type A normative JAN analysis was completed to determine the number of judgmental policies that existed among the IS faculty. Only one judgmental policy existed and arbitration among the judges was not necessary. After this, two regression analyses were made to determine the important criterion and/or predictors as related to the judgmental policy of the IS faculty. A complete discussion of the entire process follows.

The first step in the process was to determine the average score for each IS graduate on each of the criteria, Y_i, $i = 1, 2, \ldots, 9$. This score was determined for each of the criteria by averaging all the scores a graduate received from those judges who rated him on the criterion instrument previously discussed. From the previous regression run of Y_{11} on $Y_1, Y_2, \ldots, Y_9$, the weights assigned to each of the predictor variables $Y_1, Y_2, \ldots, Y_9$ can be used to develop a prediction equation. (These weights were based on the single judgmental policy that existed for the IS department.) These average scores on each of the criteria along with the predetermined weight for the criteria were used to obtain a predicted score Y_{12} for each of the IS graduates. The predictive equation used was the following:

$$y_{12} = 0.3119Y_1 + 0.1243Y_2 + 0.0000Y_3 + 0.2621Y_4 + 0.3292Y_5 + 0.3910Y_6 + 0.2006Y_7 + 0.1200Y_8 + 0.1406Y_9 - 2.4473$$

To determine the number of different policies existing among the IS faculty in regard to which characteristics they deemed most desirable in an IS doctoral graduate, a Type A normative JAN was completed on the data gathered from the 12 judges. The predictor variables used were the nine criterion variables listed in section 42.3.3.1, namely $Y_1, Y_2, \ldots, Y_9$.

The criterion variable Y_{11} refers to the group in which each set of ratings was placed. The range of Y_{11} went from 1 to 7. Group 7 represented those sets of ratings each judge considered to be the most favorable in the 40 sets of ratings, and group 1 represented those sets of ratings that each judge considered least favorable. Groups 2 through 6 lay in the continuum between the two extremes.

The regression flowchart in Figure 42.3 indicates, with respect to the existing judgmental policy of the 12 judges, that the affective variables Y_4, Y_5, ..., Y_9 are more important than the cognitive variables, Y_1, Y_2, and Y_3. Of the affective variables, Y_4, Y_5, Y_6, and Y_8 played the biggest part in determining those doctoral candidates that the IS faculty considered outstanding or whom they would hire.

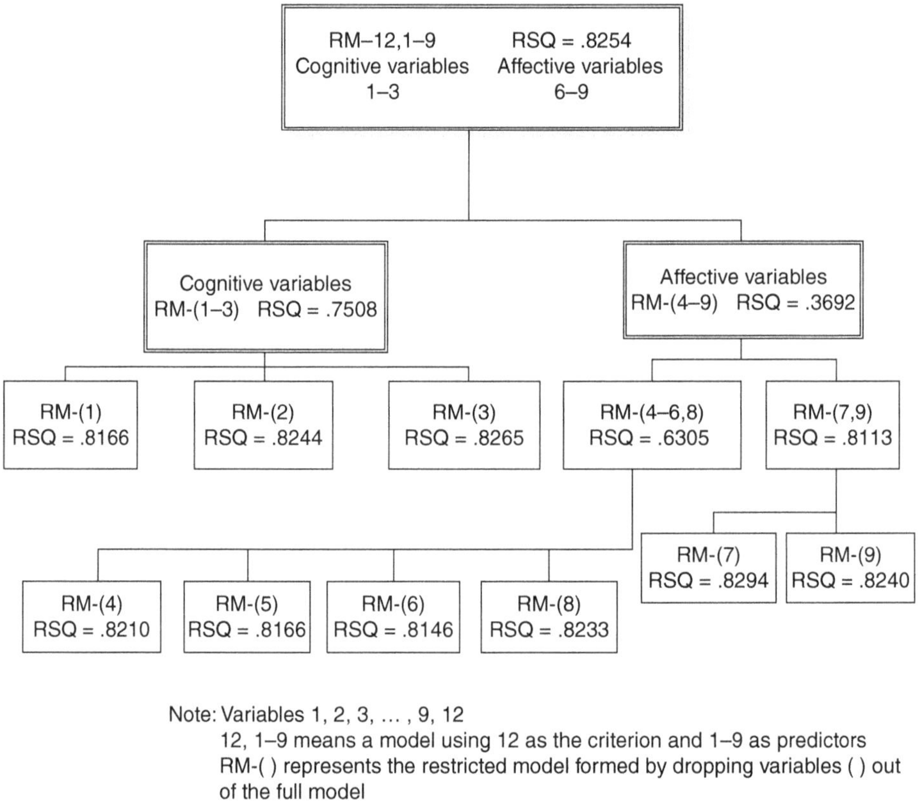

FIGURE 42.3 Final regression chart for the criterion variables.

Next, a regression model was completed using the predicted score Y_{12} as the criterion and the variables X_2, X_3, ... , X_{17}, which are the variables used to determine admittance to the program, as predictors. The regression flowchart in Figure 42.4 was developed through a factor analysis of the correlation matrix of predictors X_2, X_3, ... , X_{17}. The correlation matrix is listed in Table 42.7.

TABLE 42.7 Study correlation matrix for the predictor variables.

JAN Study Correlation Matrix for the Predictor Variables																
Variable	**2**	**3**	**4**	**5**	**6**	**7**	**8**	**9**	**10**	**11**	**12**	**13**	**14**	**15**	**16**	**17**
2																
3	0.44															
4	0.45	0.49														
5	0.75	0.58	0.56													
6	0.28	0.13	0.25	0.41												
7	0.65	0.46	0.49	0.88	0.79											
8	0.12	0.24	−0.08	0.12	0.22	0.2										
9	0.52	0.36	0.55	0.64	0.31	0.59	−0.06									
10	0.05	−0.05	0.01	0.02	−0.04	0	0.11	0.02								
11	0.04	0	0.01	0.05	0.06	0.07	0.04	0.02	0.81							
12	0.24	0.14	0.21	0.1	−0.27	−0.07	0.04	0.1	0.69	0.55						
13	0.11	−0.08	0	0.1	−0.08	0.03	0.17	0.05	0.69	0.49	0.68					
14	0.12	0.13	0.09	0.14	−0.04	0.08	0.26	0.12	0.71	0.69	0.63	0.73				
15	0.01	0.15	0.19	−0.04	−0.37	−0.22	0.099	−0.05	0.41	0.36	0.67	0.51	0.66			
16	0.1	0.11	−0.02	0.02	0.02	0.03	0.05	0.17	0.07	0.09	0.02	0.01	0.25	0.06		
17	−0.03	0.01	−0.06	0.15	0.21	0.2	−0.06	0	−0.34	−0.04	−0.45	−0.3	−0.09	−0.09	−0.23	

The number of factors and the factor loadings suggested the grouping of subsets of variables as indicated on the flowchart. Variable X_7 was dropped from both of the cognitive subsets and individually since it loaded on both of these factors. In Figure 42.4, Y_{12} is denoted by variable 1.

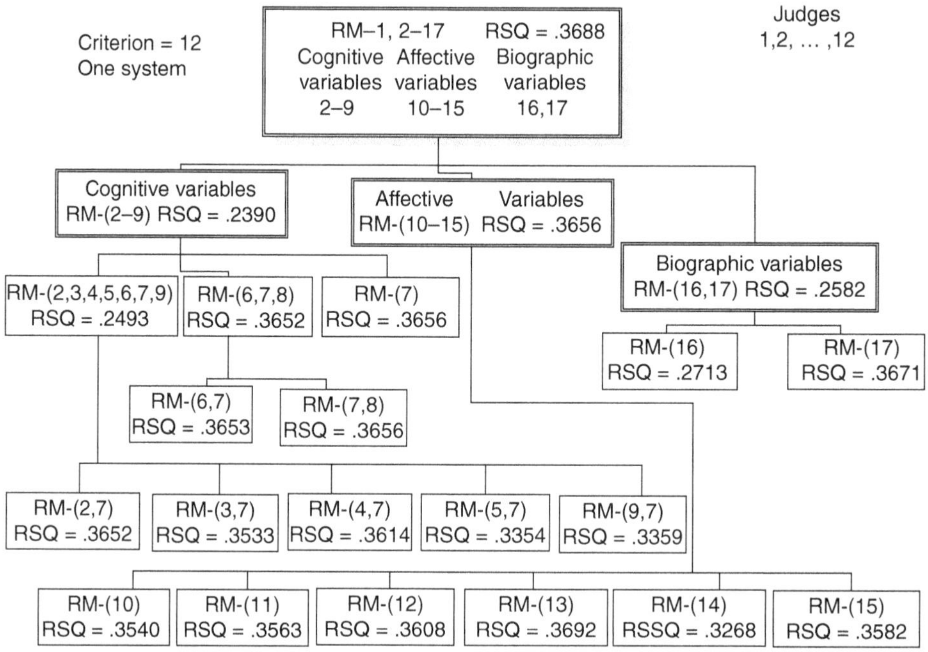

FIGURE 42.4. Final regression chart for the predictor variables.

An examination of the flowchart reveals that variables X_2, X_3, X_4, X_5, and X_9 when considered as a group have the greatest importance in determining admittance to the IS doctoral program based on the judgmental policy of the IS faculty as a single group. It should be noted that because of the multicollinearities existing between these variables, there is virtually no R^2 drop at all when they are dropped out individually. The only other variable that gives a large drop in R^2 is variable X_{16}.

42.3.7 Conclusion

By using the methodology outline in the previous steps, the following results were obtained:

First, one judgmental policy existed, without arbitration, with regard to what constitutes an outstanding graduate.

Second, with regard to this policy, the IS faculty considered the affective variables more important than the cognitive variables. Of these affective variables, Y_4, Y_5, Y_6, and Y_8 were the most important.

Y_4 = Interpersonal relationships–faculty and supervisors

Y_5 = Interpersonal relationships–peers

Y_6 = Interpersonal relationships–with people he is serving (students, inmates, patients, etc.)

Y_8 = Personal characteristics (emotionally mature, friendly, enthusiastic)

Third, with regard to the judgmental policy of the IS faculty, variables X_2, X_3, X_4, X_5, X_9, and X_{16} are the most important in determining admittance to the IS doctoral program.

X_2 = GRE population score in social sciences

X_3 = GRE population score in humanities

X_4 = GRE population score in natural sciences

X_5 = GMAT verbal score

X_9 = Miller Analogies Test score

X_{16} = Age when admitted to the IS doctoral program

The R^2 value of 0.3688 for the full model is somewhat lower than expected. Possibly this is a result of the fact that the predictors, X_2, X_3, ... , X_{17} are either cognitive or interest variables and do not relate to the affective criterion variables Y_4, Y_5, Y_6, and Y_8 that the faculty deemed most important.

42.4 General Modeling Considerations

Both the opening article and the JAN study illustrated several features of modeling that should be present in a study in order to make it possible to achieve success. The remainder of the chapter provides an outline of the necessary components for these features.

The multiple linear regression techniques we have discussed can be very useful but also very dangerous if improperly used and interpreted. Before tackling a large problem by multiple regression methods, it makes sense to plan the project as far ahead as possible, specify the types of work, and provide checkpoints as the work progresses. Planning is the main ingredient.

A more detailed analysis of these steps is provided in the following sections and discussed in (Clauser, B. E. & etc.)[3] and Roeber, E.[4]

42.4.1 Planning the model building process

A good reference on the modeling building process is by N.R. Draper and H. Smith, *Applied Regression Analysis,* John Wiley & Sons, Inc. (1966). The list of necessary steps provided by Draper and Smith are:

Define the problem, select the response, and suggest variables.
Are the proposed variables fundamental to the problem and are they available?
Is the problem potentially solvable?
Correlation matrix and first regression runs.
Establish goals and prepare budget.
Are goals and budget acceptable?

42.5 Development of the Mathematical Model

Plot variables and try various models.
Consult experts for criticism.
Plot new variables and build vectors that explain residuals.
Examine some fitted regression equations.
Are goals met?

42.6 Verification and Maintenance of the Mathematical Model

Are parameters stable over the sample space?
Is there systematic lack of fit?

Practical aspects of the model
Following a template, or outline, enables the researcher to be more efficient and use check points for controlling project costs. A major benefit is placing emphasis on screening procedures, which lead to 'best' choice of regression model for the study under consideration. The reader needs to practice using regression and judgmental analysis to acquire an understanding of these methods. Model building and validation are pertinent to research methods. The following exercises and projects are only a first step in this direction.

42.7 Exercises

1. For the opening study in this chapter, identify the steps that were taken for each step in the modeling construction algorithm.
2. How was this research done?
3. What were the findings? What does it mean?
4. Were the subjects treated in an ethically enlightened manner?
5. Is there a clear explanation of why these particular subjects were sampled?
6. What was the selection process?
7. Was random selection or assignment used?
8. Is there any evidence of bias due to poor sampling or nonresponse (missing data)? Are details of nonparticipants or dropouts given?
9. For a multiple linear regression predictive study, describe the required components for validating the data. Duplicate the process for simple linear regression and for ANOVA.
10. Explain what is meant by *robustness* and what impact this issue has on the responses to Problem 9.

42.8 CLUSTERING PROJECT

Introduction

Clustering techniques fall into two major categories, hierarchical techniques and optimization or partitioned techniques.

Hierarchical Techniques

Hierarchical clustering can be broken down into two major categories agglomerative methods and divisive methods. A procedure for forming agglomerative hierarchical groups of mutually exclusive subsets based on the premise that the greatest amount of information, as indicated by the objective function, is available when a set of n members is ungrouped. The first step is to select and combine the two subsets out of the n possible subsets which would produce the least impairment of the optimum value of the objective function, while reducing the number of subsets to $n-1$. These $n-1$ subsets are then examined to identify which two subsets should be merged in order to obtain the optimum value of the objective function for $n-2$ subsets. This procedure is repeated until the original n members are in a single set or group. Since the number of subsets is reduced by one at each step (n, n–1, ... , 1), the process is referred to as hierarchical grouping. The grouping which occurs at each step usually results in some quantifiable loss. The functional relation used to obtain a value-reflecting number is called the objective function. In general, an objective function may be any functional relationship that an investigator selects to reflect the relative desirability of groupings. The objective function represents loss of information and is reflected by the error sum of squares, (known as Ward's Method). Other examples of objective functions can include:

1. the grand sum of the squared deviations about the means of all measured characteristics, people are clustered to maximize their similarity with respect to measured characteristics,
2. the expected cross-training time in the clustering of jobs to minimize cross-training time when personnel are reassigned according to established policies,
3. the amount of job time incorrectly described to minimize errors. When clustering job descriptions in describing a large number of jobs with a small number of descriptions, and
4. the loss of predictive efficiency when clustering regression equations to minimize the loss of predicted efficiency resulting from reductions in the number of regression equations judgmental analysis (JAN).

An important concept to remember in defining the hierarchical grouping mentioned above is that once two entities are merged, they are joined permanently and become a building block for later merges. This is one of the key differences between the hierarchical techniques and the optimal techniques to be discussed in the next section. Another difference between the two methods is how they arrive at the number of groups. If researchers are not interested in the entire hierarchical tree structure, they must decide at what point the best clustering has been reached. There are many different techniques available to help the user make a selection.

Some of the more common hierarchical techniques are the nearest neighbor (single linkage) method, the farthest neighbor (complete linkage) method, centroid method, and Ward's method (mentioned earlier in the chapter). The basic technique for these methods is similar. However, there are two factors that differentiate between the methods. The first factor is whether the technique can be used with similarity measures, distance measures, or both. The second factor is how these measures are defined, or, in Ward's terminology, the objective function.

In divisive clustering the initial starting point is with all entities in one group. The first step is to split the group into two subsets. Subsequent steps consist of subdividing the subsets until each entity is a separate group. There are two classes of methods for deciding how to split the groups. With the monothetic methods, the data points are based on binary variables and the object is to split the data set on one of these variables so as to minimize the value of some appropriate measure of similarity between the two groups. The splitting process, based on other variables, continues until some criterion is reached. Examples of monothetic techniques include association analysis and the automatic interaction detector method. The polythetic methods are based on the values taken by all of the variables. The divisive techniques have not gained the same popularity as the agglomerative techniques and their use is minimal.

Optimization Techniques

The basic method used by the optimization techniques is to obtain an initial partition of the data units and then alter the cluster membership to get a better partition, i.e., until a local optimum is found. The various methods differ as to how they arrive at an initial partition and how they define "better" (i.e., the objective function). The techniques in this section differ from the hierarchical techniques in two ways. First, members are allowed to change cluster membership. So, a data point poorly classified in an early cycle can be re-classified in a later step. Second, the number of clusters must be specified in advance. If the structure of the data is known a priori, this is not a problem. Otherwise, two possible suggestions are to:

1. try a number of different partitions, or
2. use a hierarchical technique first to help decide upon the number of clusters.

Once the number of clusters has been decided on, the next step is to decide on the initial partition (i.e., to which clusters the data points will initially be assigned). There are a number of different methods being used. In the following discussion let k refer to the number of clusters, and n_i refer to the number of entities in the ith cluster (i=1 to k; $n_1 + n_2 + ... + n_k = n$). The simplest method is to choose the first k data points to act as cluster centers. A second method that tries to account for any nonrandom sequence is to number the entities from 1 to n and choose the integer parts of n/k, 2n/k, ... (k–l)n/k, and n for the initial cluster centers. A third method is to choose the k data points randomly from the data set. In these three methods, once the k seed points have been chosen, the remaining entities are then placed in the cluster that they are "closest" to, usually using the Euclidean squared distance. A fourth method is to partition the data units so that the first n_k in the first partition, the next n_k are in the second cluster, and the last n_k are in the kth cluster. A final methods is to decide on the appropriate cluster for each individual data unit by use of the formula:

$$C = K + \left|\sum_{i=1}^{p} V_i\right|_{/(\text{Max}-\text{Min})}$$

where, C is the cluster number to assign to the data unit (C = 1, 2, ... , K); K is the total number of clusters; V_i is the number of variables being used to cluster (i =1, 2, ... p); Min is the minimum score when all the variables are summed for each data unit; and Max is the maximum score when all the variables arc summed for each data unit.

Once the data points have been allocated to an initial partition, the next step is to search for poor fits. The two most popular types of clustering criteria are variations on the k-means technique and procedures based on the fundamental matrix equation

$$T = W + B$$

where T is the total dispersion matrix and W and B are the matrices of within-cluster and between-cluster variation.

In the k-means approach, once the initial partitions have been established, the cluster centroid can be updated after each addition to the cluster (MacQueen's method), or only after all the points have been allocated (Forgy method and Jancey method).

The basic algorithm for MacQueen's method is as follows:

1. Begin with an initial partition of the data units into clusters (any of the methods mentioned earlier will work).
2. Take each data unit in sequence and compute the distances to all cluster centroids; if the nearest centroid is not that of the data unit's parent cluster, then reassign the data unit and update the centroids of the losing and gaining clusters.
3. Repeat step 2 until convergence is achieved; that is, continue until a full cycle through the data set fails to cause any changes in cluster membership.

The basic algorithm for Forgy's method is to:

1. Begin with any desired initial configuration. Go to step 2 if beginning with a set of seed points; go to step 3 if beginning with a partition of the data units.
2. Allocate each data unit to the cluster with the nearest seed point. The seed points remain fixed for a full cycle through the entire data set.
3. Compute new seed points as the centroids of the clusters of data units.
4. Alternate steps 2 and 3 until the process converges; that is, continue until no data units change their cluster membership at step 2.

At this point a local optimum has been achieved. Unfortunately, there is no way to know whether this is also the global optimum. The solution can either be accepted, or a new initial partition can be obtained and the clusters can be re-analyzed.

Looking at the matrix relationship mentioned earlier (i.e., W = T + B), four possible grouping criteria have been proposed. The first possibility is to minimize the trace W. This is one of the three most popular techniques of cluster analysis. Second, minimize the ratio of determinants $|W|/|T|$.This criterion is known as Wilks' lambda statistic. Third, maximize the largest eigenvalue of W^{-1} B. This criterion is known as Roy's largest root criterion. Fourth is to maximize the trace of W^{-1} B. This criterion is known as Hotelling's trace criterion Each of these methods has its advantages and disadvantages. The major problem with minimizing the trace W is that it is transformation dependent. This means that, in general, different results will be obtained from applying the technique to raw data as opposed to standardized data. A second problem with this criterion is that it tends to find spherical clusters in the data. This is because the trace does not take into account correlations among the variables. The second method, minimize the ratio of determinants $|W|/|T|$, assumes that all the clusters in the data have the same shape. It also tends to divide the data into approximately equal-sized clusters when the separation is not great. Similar problems can be found for the other two criterion.

You are to answer the following questions:

1. What is cluster analysis?
2. How does cluster analysis work?
3. What are the objectives of cluster analysis?
4. How should data be prepared?
5. How can cluster analysis be used to detect outliers?
6. Give an overview of distance measures used in clustering methods.
7. Explain standardizing the data and why it is important in cluster analysis.
8. What are the underlying assumptions in cluster analysis?
9. How are clusters derived and how do you assess them?
10. How do you decide whether hierarchical or nonhierarchical methods should be used?
11. Is interpretation of the clusters interpreting or labeling the clusters?
12. Explain the validation and profiling of the clusters.
13. What is meant by profiling the cluster solution?

42.9 JPC PROJECT

Study the following article: *Determining Component Weights in a Communications Assessment Using Judgmental Policy Capturing* by Leo M. Harvill, Forest Lang, and Ronald S. McCord published in Med Educ Online[serial outline] 2004, 9:12 available from *http://www.med-ed-online.org*. This article utilizes a methodology called judgmental policy capturing (JPC) is described and used to determine component weights for complex assessment components in medical education.

Try to replicate the JAN study in this chapter for a comparable application within an accessible organization. Additionally redo the study using the JPC methodology recommended by Harvill, Lang, and McCord. Finally, run a comparative analysis of the JAN versus JPC methodologies.

References

1. Bottenberg, R.A., and R.E. Christal, *An Iterative Technique for Clustering Criteria which Retains Optimum Predictive Efficiency,* Lackland Air Force Base, Texas: Personnel Laboratory, Wright Air Development Division, March 1961 (WADD-Tn-61-30, ASTIA Document AS-261 615).
2. Clarke, C.M., "An Evaluation of the Validity of Judgmental Analysis," doctoral dissertation, Baylor University, 1989.
3. Clauser B.E., R.G. Subhiyah, R.J. Nungester, D.R Ripkey, S.G. Clyman, D. McKinley, "Scoring a Performance-Based Assessment by Modeling the Judgments of Experts," *Journal of Educational Measurement*: 397–415.
4. Roeber, E, *Setting Standards on Alternate Assessments (Synthesis Report 42),* Minneapolis, MN: University of Minnesota, National Center on Educational Outcomes (April 2002), retrieved January 6, 2003, from *http://education.umn.edU/NCEO/OnlinePubs/Synthesis42.html.*

CHAPTER 43

A GUIDE TO STATISTICAL SOFTWARE

OVERVIEW AND LEARNING OBJECTIVES

In This Chapter

43.1 Introduction

This chapter provides an overview of statistical software. Sketches of answers are provided for responding to: What considerations are important in the design of a statistical software package? What design constructs are possible? How can packages be compared? How can a researcher select the package which best suits their needs?

In many, if not most situations, the researcher will not be able to choose from the entire array of all available packages, but will have to select from a more limited list of the packages available. The questions above remain important, however, for choosing wisely from this limited list and for using the selected package efficiency and effectively.

43.2 Design Constructs

Current statistical software packages designs are dependent upon what type of interface is implemented. Several packages interface with relational database systems such as ORACLE. Other packages interface with mathematical software. For instance MATLAB is one which not only offers statistical capabilities but optimization and symbolic manipulation. Some packages utilize visualization interfaces, which is a very useful feature for exploratory data analysis. User interfaces come in many sizes and shapes as they have evolved from command line to graphical user interfaces to hybrid drag and drop system interfaces.

An applications software package, in statistics or in any other area, may take one of these four forms:

1. A set of independent programs.
2. A set of subroutines to be called by a user-generated program.
3. A single, multipurpose program.
4. An applications compiler.

The design of a package must take into account a number of considerations, including:

1. The intended application of the package (primarily the behavioral sciences, some other area, or a number of fields).
2. The background of users (undergraduates or professional statisticians).
3. The level of experience of the users (familiarity with computers and programming).
4. The hardware the package will run on.

We now consider the four major design constructs in more detail.

43.2.1 Sets of programs

In a package of this kind, each program is complete in itself, and each performs just one statistical operation on the data file: computing descriptive statistics, for example, or performing a t-test. Each program will use the same input formatting, style of documentation, and coding language, and some effort will also be made to standardize the output formats. The advantages of such a package are these: the programs can be easily used by a nonprogrammer, and they generally offer some flexibility of input format. In many cases, however, the user will want to perform many analyses on a particular data file, or the same analysis on several files. A package of individual programs is clumsy and time consuming in such situations.

43.2.2 Sets of subroutines

When the package consists of a set of subroutines, the user must write a main program to read in the data file and call the subroutines to perform statistical analyses. This main program will be written in a high-level language, while the subroutines will be entirely input-free. There are several advantages to such a package

1. Statistical processing can be tailored to each job.
2. Repetitive runs of several programs can be done easily.
3. Routines not in the package can be supplied by the user.

On the other hand, a set of subroutines can be used only by someone familiar with programming in the language of the subroutines; many statistical users will not have such training. Packages of this type generally have less documentation provided than might be required by the general user. Therefore, a high level of sophistication in computing may be required to obtain statistical results.

43.2.3 Large, multiple-use programs

Many packages today are single, large programs and can be divided into two subcategories. In the first, the entire program, which can perform a variety of statistical and processing tasks, is maintained in the computer's memory. Such packages tend to occupy vast amounts of storage, and they can require large amounts of time for *I/O*. They are, however, easily used by the nonprogrammer. In the second, the first section of code loaded into main memory produces a menu of available operations, then generates calls to the parts of the program which perform tasks the user requests, either in a specified code or in some language prose series of statements. Only those parts of the program related to the current tasks are maintained in memory, which can then be used more efficiently. Such systems are well-adapted to interactive use, and some also provide tutorial sessions in the use of the package.

43.2.4 Application compilers

Packages of this type are, in fact, programming languages in which statistical tasks, or large steps in statistical tasks, are initiated by single statements in the language. They are useful to the researcher who wants to perform a nonstandard analysis quickly, but they again require that the user be a competent programmer. Implementation of a statistical applications compiler can be accomplished with a functional paradigm programming language.

43.3 Problem Areas

Many, if not most statistical packages, now in wide use were written by programmers rather than statisticians. Today many package systems are constructed in modules and adapted to current technology but with minimal documentation. These and other conditions have led to the following problem areas in the design and use of statistical packages:

1. *Statistical packages are often misunderstood, inefficiently used, even misused.*

The ease with which some packages can be used may encourage the unwary to apply the analyses in the package to situations and data inconsistent with the techniques available. For example, assume that a group of values has come from a normally distributed population when this has not been established. Documentation should show not only how to cause the analyses to be performed but also the situations to which the analyses are applicable, the assumptions on which the techniques are based, and the conclusions that can be drawn. This information should be available not only in the written documentation of the package but also in online tutorial help sessions.

2. *Overly stringent requirements of system hardware or software.*

Some packages are written for specific architectural platforms, and can be used only with difficulty, if at all, on any other platform. Some require specialized devices, a plotter, for example, that may not be available.

3. *Lack of transportability.*

It should be possible to execute a package on a wide variety of architectures, particularly since hardware/software changes at computer facilities occur with surprising frequency. Acquisition of a new machine should not render all purchased software unusable. Import/export facilities are extremely important components needed in current applications. For instance, Excel has been widely used for entry and management of data.

4. *Inflexible and complicated user interfaces.*

In some earlier packages, the statements used to control the operation of the package seemed almost intentionally obscure, with stringent requirements for the placement of instructions on each line and instructions which, at first glance, seem arbitrary sequences of characters. Graphical user interfaces greatly enhance the usability of a package and should be easily accessible to those who use it. In particular, English-like words should be used, and their placement on the lines should not be overly critical.

5. *Appropriateness of algorithms.*

An algorithm is a series of steps which together describe the operations the computer must perform to accomplish a given task. The algorithms that perform statistical analyses must be appropriate to the demands of the tasks. For example, in computing the variance of a very large list of relatively uniform values by the formula

$$S^2 = \frac{1}{n-1}\sum_{i=1}^{n} X_i^2 - \frac{n}{n-1}\overline{X}^2$$

the difference between the summation and n $\overline{X}^2$ may be very small compared to the value of the summation. In such a case, the variance may be distorted by the limited precision of the computer, and it would be better to use the formula provided by the definition of sample variance to find S^2.

6. *Sophistication of programming language design.*

The language in which the package is written is clearly relevant to its performance. There are many issues in programming language design which relate to package performance, for which we will provide one example. Subroutines may be classified as *open* or *closed.* A closed subroutine (or subprocedure, or subprogram) is compiled and stored in the computer's memory separately from the main program that calls it. A call statement in the main program causes a jump in the flow of execution into the subroutine and at the completion of the subroutine, execution returns to the point in the main program immediately after the call. Such a linking mechanism need not reside in memory simultaneously with its main program except when the subroutine is executed. The code of an open subroutine, in contrast, is a part of the code of the main program, the subroutine is compiled and resides in memory with its main program. Many modern computers are designed to execute such languages efficiently, but may encounter difficulties with closed subroutines. If many users are engaging the computer, there may be insufficient memory remaining to bring in a closed subroutine. Thus, that program will wait until other executions are completed and sufficient space in memory is freed to bring in the subroutine.

7. *Lack of modularity.*

Many early packages are composed of large blocks of code not broken down into modular sections. Such code is difficult to maintain or modify, and often lacks transportability because machine-dependent statements are hard to locate. Object oriented paradigm programming languages alleviate this problem.

8. *Documentation availability.*

Often, in an organizational environment, it is difficult, if not impossible, to obtain duplicate copies of the package manual. When several people are likely to use a given package, each person should be able to have a copy of the manual at reasonable cost.

43.4 Desirable Package Features

It is obvious that the single most desirable characteristic of any package of programs is *ease of use;* the two primary areas here are formatting input of the data file and instructing the package to perform the desired tasks. A package that does not allow direct, easily understood operation in these areas can be more of an impediment than an aid. Other useful characteristics of a statistical program package are:

1. An interactive process to generate successive models, with each model a better approximation of the situation represented by the data. Graphical displays are desirable here.
2. An English-like command structure.
3. Intelligible error messages during execution to avoid dependence on the documentation.
4. Results in English sentences whenever possible, rather than just tables of values.
5. The ability to generate new values based on existing values.
6. The inclusion of a end-user friendly graphical user interface.
7. Hardcopy documentation and online help sessions/tutorials showing examples of data files, runs using the package, and interpretation of results.
8. The ability to manipulate data files.

This last characteristic deserves further comment. In addition to sorting of files, the attachment labels to variables, and the selection of particular cases, it is also often useful to *invert* a file, that is, to rearrange the records in a new order based on the values of a particular variable.

43.5 Evaluation Checklist

Literally hundreds of statistical software packages exist, so to obtain the best value when you are selecting one, you should perform an analysis of competing systems using criteria which can be objectively applied to each system. The criteria listed below have been found useful in such analyses, for both statistical and non-statistical packages.

1. Cost elements

In addition to the cost of the package itself, there are many other factors that may affect the entire cost associated with the package, including

a. Conversion to the new package.

b. Training support personnel.

c. System modifications to accept the new package.

d. New forms, manuals, and other operating aids.

e. Maintenance.

f. Additional charges for multiple machines or locations.

2. Availability

Is the package operational? What is its operational history? Seek references and contact other users of the candidate package.

3. Equipment configuration

Ascertain the minimal and optimal configurations of hardware for the package, and where your machine fits into that range.

4. Software environment

Most installations have standards with respect to programming languages, operating systems, and utilities, and the usages of the package should conform to those standards.

5. Quality of system design

Packages built up of *modules* often employ object-oriented programming languages which can be more easily maintained and modified; expandable packages can be increased in size to handle new processing tasks; and fast, accurate algorithms will handle more tasks more precisely. All these tasks can be accomplished while minimizing module duplications in object-oriented processing.

6. Documentation [hard copy and online help sessions/tutorials]

Documentation should be available for all levels of the package, including:

a. A general overview and introduction to the package.

b. A systems manual and/or online support for describing the package processes, and a broad range of input/output formats.

c. Program-level documentation, with descriptions and flowcharts, record organization and format, and annotated program listings.

d. Operational instructions.

e. User instructions, with example sessions.

7. Installation support

In installing the package, support should be available for preparation, training personnel, and the installation itself.

8. Maintenance

Who assumes responsibility for maintenance, and how much support is provided?

9. Enhancement

The supplier may be planning enhancements of the package. The user should be aware of the cost, availability, and scheduling of these changes.

10. Supplier integrity

Is the supplier committed to support the package, and the maintenance of it? Does the supplier have the personnel and financial resources to assure such a commitment?

11. Application specific criteria

Does the package offer experimental design and graphical processing?

Some packages for the interested reader to check out include:

SAS- a "professional" large multi-purpose program, still used primarily from command line - although menu driven graphical user interface upgrades are becoming common. SAS evolved from a statistical analysis system to a complete data management and analysis system. Extensive experimental analysis and regression analysis as well as analyses of time series are available. (*www.sas.com/technologies/analytic/statistics/index.html*).

SPSS- heavily used in the social sciences offers a system with a gentle learning curve. This is a modular system offering complete graphics, broad data management and reporting capabilities. (*www.spss.com*).

BMDP- originally designed for bio-medical analysis offers extensive regression (stepwise, regression on principal components, ridge regression, and non-linear regression analyses beside the normative offering of regression types) and experimental design analyses. New easy-to-use interfacing that uses point and click coupled with fill-in-the-blank interfacing is offered in newer versions of BMDP.

STATISTICA- a menu-driven system that requires no programming. This system offers simple t–tests to neural networks and experimental designs. (*www.statsoft.com*).

S-PLUS- offers statistical and graphical analysis with extensions in I/O, data management facilities, and system integration capabilities at *www.insightful.com*.

The interested reader should consult the following articles for a more comprehensive review on statistical software by George Mason University:

1. *A Guide to Statistical Software* at *http://www.galaxy.gmu.edu/papers/astrl.html*.
2. *A Guide to Statistical Software: Comparative Statistical Software Review* by John A. Wass, Scientific Computing at *http://scientificcomputing.com/comparative_statistical_software.aspx*.
3. For using Excel for statistical analysis, the reader should consult *Excel for Statistics - Tips and Warnings by* Statistical Services Center at the Applied Statistics at the University of Reading found at *http://www.reading.ac.uk/ssc/publications/guides/topxfs.html*.

43.6 Exercises

1. Describe each of the following software package design constructs:

   a. A set of independent programs

   b. A set of subroutines to be called from a user-generated program

   c. A single, large program

   d. An applications compiler

2. Identify statistical packages in the four categories of Problem 1.

3. What data management facilities should be part of a statistical software package?

4. If the code of a package is accessible, what program maintenance facilities are desirable?

5. Become familiar with a statistical package available at your facility other than the package being used in the course.

43.7 Statistical Computing Exercises

1. Develop pseudocode for the following algorithms:

   a. Ranking numerical data

   b. Finding the maximum value

   c. Constructing a frequency table with data grouped in classes

   d. Outputing frequency distribution

   e. Outputing a stem and leaf plot

   f. Outputing a box and whisker plot

   g. Outputing a scattergram

   h. Performing a multiple linear regression using the normal equations

   i. Performing a one-way ANOVA

   j. Outputing a two-way ANOVA and means table

CHAPTER 44

PRODUCT DEVELOPMENT

OVERVIEW AND LEARNING OBJECTIVES

In This Chapter

44.1 Introduction

A college major in computer science would classify computational complexity as a theoretical or foundational topic. The same student would also classify data structures or design of algorithms as applied computer science studies. What is the difference between theoretical and applied studies? The student's perception probably leads them in this direction based upon operationally defining applied areas of study as those studies being real-world-solutions to real-world-problems. At the same time, the student probably perceives areas of study that expand knowledge in the field as theoretical.

However, appearances can be deceptive. Many of the topics covered in computational complexity classes evolved from real-world problems, such as the need for determining if a problem is solvable with a specific architecture. On the other hand, attempting to solve real-world-problems can help one to discover new knowledge and is, therefore, defined as usually theoretical. There appears to be a non-empty intersection between theoretical and applied domains. Possibly it is better to think of application as a catalyst for expanding knowledge in the field and theoretical studies as preparation for application.

44.2 Sample Study

The study discussed here is a cognitive-map-based approach to the coordination of distributed cooperative agents available online at *http://ieeexplore.ieee.org/*.

Ternary logic is employed, through the use of negative-positive-neural logic (NPN), in this study to allow for response to stimuli by representative states of "agreement, disagreement, or maybe." Using NPN components, the researchers have has constructed a pooled cognitive map that is derived from a network of cognitive maps. Each node in the network represents either an Artificial Intelligent (AI) agent's or a human agent's cognitive map; it represents the agent's positive to negative reactions to stimuli. The network is implemented in the LISP and C programming languages.

A typical application for this system consists of gathering a group of both AI and human experts on a domain area to consider a specific problem. The system, called D-POOL, commences with several of the local nodes, broadcasting the problem statement throughout the network. Next a determination of the identity of valid experts as well as pooling the cognitive maps, is derived from their assigned groups. Finally, all the originally chosen local nodes, or experts, have their cognitive map stored. When the problem of interest is presented to each expert then focuses of attention, not solutions, are generated. Next, a pseudo-cognitive map is constructed, representative of the union of each of the component cognitive maps for all chosen AI and human experts remaining in the study. The foci of attention are passed to the pseudo-cognitive map. This map represents the knowledge mapping derived by coordination of distributed cooperative human and AI experts, still remaining in the study thereby formulating a solution to the problem. DPOOL generates cooperative, or compromised, problem solving through communication and perspective sharing for collections of human and AI experts. By solving two large scale sample problems, where several domain areas of expertise are present, DPOOL's applicability for distributed group decision support has been demonstrated.

44.3 Types of Research

An attempt has been made in the following table to classify in terms of the intended purpose, code words to describe research, and researchers control in their participation of the research. Experimental versus Field Research is the major factor of difference. For each purpose an example study has been stated. The ratings of researcher control have subjectively been made, as illustrated in Table 44.1.

TABLE 44.1 Types of research.

Types of Research	Purpose	Code Words to Look for in Research	Degree of Control over Factors	Example
Nonexperimental (Descriptive)	Describe the characteristics of an existing phenomenon	Describe, interview, review literature, participant observation, focus groups	None or low	King-Stock article
Nonexperimental (Historical)	Relate events that have occurred in the past to current events	Past, describe	None or low	An analysis of software engineering practices
Nonexperimental (Correlational)	Examine the relationships between variables	Relationship, relate to, associated with, predicts	Low to medium	An investigation that focuses on the relationship between software development time and productivity
Experimental	To test for true cause and effect relationships	Function of, cause of, comparison between, effects of	High	The effect of a programming language on software development time
Quasi-experimental	To test for causal relationships without having full control	Function of, cause of, comparison between, effects of	Moderate to high	Classification differences in neural nets and traditional classification methods

All figures and tables in this chapter appear on the companion DVD.

ON DVD

44.4 Adequacy Testing Theory in a Field of Study

For the article presented in the previous section, the hypothesis tested is:

Coherence and coordination can be achieved with a set of autonomous intelligent agents, without centralized control or viewpoints in a distributed environment, through agent actions based upon cognitive integrity, transparency, and simplicity.

The basic need for this hypothesis test includes the following issues:

- Each node, either an expert system or human expert, is represented as a cognitive map system. A systematic approach to problem solving is taken to construct a pooled cognitive map for representing the pooled knowledge and reasoning techniques contributed by the participating experts. Note that this involves fuzzy reasoning.
- Enforcing cognitive integrity enables the agents to derive a sufficient "big picture" of the distributed environment, or the composite and/or derivative view.
- Enforcing both cognitive transparency and simplicity ensures that the "picture" given to each agent is clearly understandable, modifiable, and transmittable.

We are left with asking what criteria could be used to determine whether a theory is an adequate one. A number of such criteria can be found in the literature. However, it must be recognized that their application is a matter of rational analysis and can give only a general indication of theoretical adequacy. This theory from the field of study in distributed artificial intelligence passes the following criteria for testing the adequacy of the latter hypothesis test:

Criteria for adequacy testing theory

- Testability — A theory should lead to deductions that can be credited or discredited by empirical test.
- Consistency — A theory should be free from internal contradiction or be logically consistent.
- Scope — A theory should account for a relatively wide range of data.
- Predictive power — A theory should provide a basis for successful prediction.
- Fruitfulness — A theory should generate a variety of new hypotheses.
- Parsimony — Other things being equal, a simpler theory is preferable to a more complicated one.

Of these, predictive power is perhaps the acid test of a theory. Two points should be made, however. The first is that the theory selected for use here is seen as one way of approaching administrative behavior. Other theories may have greater value and their development and utilization should, of course, be encouraged. On the other hand, and this is the second point, the criteria used appear to be useful in providing a general assessment of theory.

This is especially true when theoretical formulations may be deficient in obvious respects, as has been undeniably the case with several approaches in administration.

44.5 Product Development Methodology

Additionally, the latter lead article, on DPOOL, illustrates the systematic approach to product development.

This development involves two categories of activities called *system analysis* and *system synthesis*. System analysis deals primarily with the first of two problem steps: identifying the problem and analyzing the problem and setting goals. System synthesis techniques are useful for the balance of problem solving, which incorporate the steps in Figure 44.1.

Mission analysis is a determination of where we are going and how we know when we've arrived. Mission analysis includes the steps of identifying an overall mission objective and the specific measurable performance requirements for satisfactory completion of the mission. The mission is what has to be accomplished or is required.

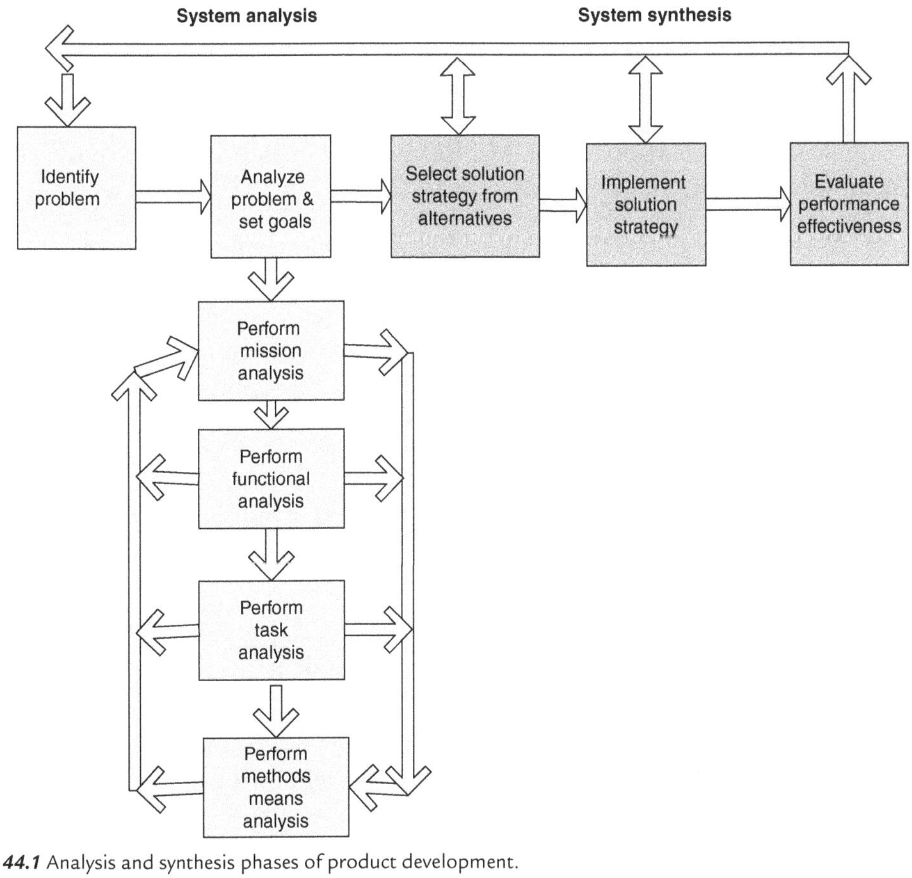

FIGURE 44.1 Analysis and synthesis phases of product development.

Functional analysis and *task analysis* are quite closely related to mission analysis, and consist of breaking down all of the functions identified in the mission profile into its constituent component functions. Functional and task analyses may be viewed as vertical expansions of the mission profile, and continue until units of performance, or tasks, are identified.

The remaining system analysis step is called *methods-means analysis*. This analysis step identifies for each performance requirement or family of performance requirements possible strategies and vehicles for accomplishing the performance requirements. This step only identifies what is available, not how.

The previous diagram can be used to derive a more detailed level diagram as shown in Figure 44.2:

In summary, we obtain the information shown in Table 44.2.

Once the data base, or collection of questions, formed from the feasible *whats* is obtained through the system analysis process, the system synthesis uses this data to:

1. select solution strategies and vehicles for the identified performance requirements,
2. making and/or buying the system components that are required,
3. trying out the components in order to determine their effectiveness, and
4. implementing them as well as evaluating the total system product in the environment for which they were designed.

Walmart has been known for efficiently placing the product in the customer's hands in high quality. How is this feat accomplished? The answer is a composition of the following buzz phrases: Supply Chain Management (SCM), Entity Relationship Program (ERP), and Customer Relationship Management (CRM).

Basically, SCM is a model of the corporation's participating people, other organizations, and technology in the flow from the conceptualization of the product to placing the product in the customer's hands. The flow integrates the following functional and business components: manufacturing, inventory, marketing, sales, and the actual customers. Since the major emphasis is to produce products that are generated by satisfying customers expectations, then the SCM has a feedback loop using the current product information for product ratings of satisfaction and identification for needed modifications.

The ERP system is the software and organizational implementation of the SCM. CRM is a software module that interfaces with the ERP.

Many companies currently are in need of product development systems that integrate data across the company's functional divisions and their associated client base to distill information and business intelligence. This data flows through the complete cycle of product development and business functions. A successful flow demands that each individual in the corporation has a basic understanding of what each functional and business component contributes to the flow of the product from the origin of the system to the customer,

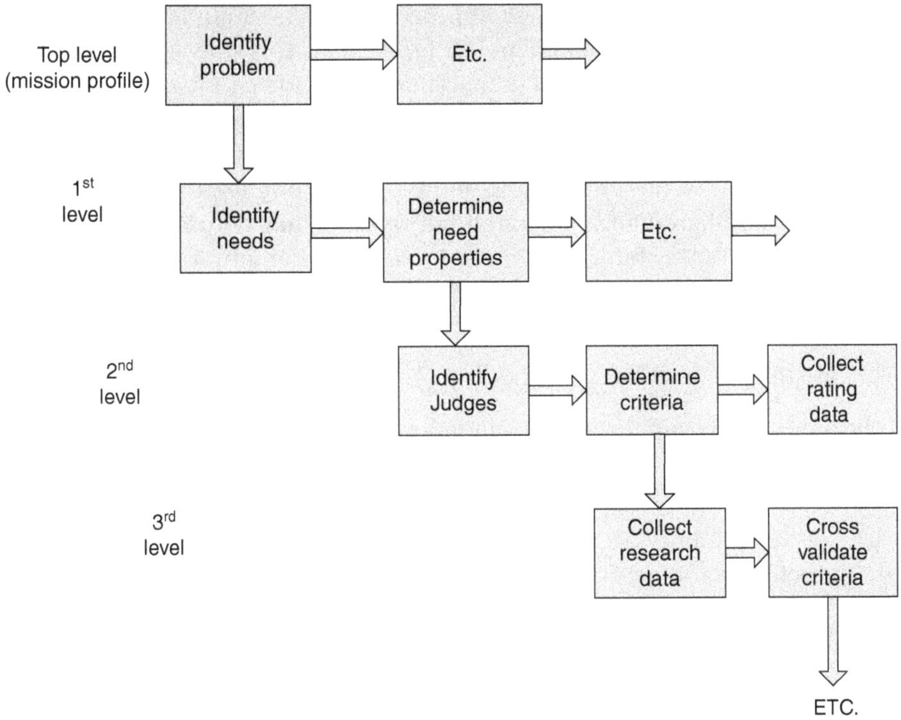

FIGURE 44.2 Problem identification level diagram.

TABLE 44.2 Questions to be answered by a system analysis.

Questions to be Answered in System Analysis	Steps to be Done in System Analysis
Where are we going and how do we know when we've arrived?	Determine mission objectives and performance.
What are the things that will keep us from where we are going and how do we eliminate them?	Determine and reconcile constraints.
What are the major milestones along the way to where we're going?	Determine mission profile.
What are the things that must be done to get to each milestone?	Perform functional analysis.
Of what specific tasks are the things composed?	Perform task analysis.
What are the possible ways of getting the things done?	Perform methods and means analysis.

complete with customer feedback. Multiple layer collaborative Decision Support System (DSS) architectures implement the SCM system. Such a system must allow for communication integration and sharing of information on a corporate-wide basis. This data can enable initiating alerts, assisting with diagnosis of a problem, and analyzing problems.

The use of product design and development processes across multiple companies is referred to as Collaborative Product Development (CPD). Usually this calls for transparent integration across heterogeneous computer center architecture and system/application system platforms. CPD can improve product launch success and, therefore, have an impact on reducing cost and speeding up time to market. Mandatory benefits of a CPD should include sharing specifications, test results, design changes, and the use of online prototyping, and customer feedback processing. Shared and integrated information may realize reduced development costs and time due to tightly integrating and streamlining communication channels.

Every product development process should ask:

1. What are the needs of the customer?
2. How large of a potential market is there for the product?
3. What is the perceived product modification cycle and projected half-life for the product?

Then using the SCM data flows, the following chain of events can be conducted: planning for design, the actual manufacturing process for the product, and for producing the product. Rather than having each of the functional and business components corporately organized as a disconnected network of independent islands of knowledge, CPDs offer an architectural supportive of product design and development, as well as new product development. The end result of the CPDs is the development, processing, and customer delivery that gets the product to the customer cheaply, timely, and with high quality and, simultaneously, maintains low inventory levels.

For more detailed overviews, the interested reader should consult:

1. for SCM: *Supply Chain* by Wikipedia found at *http://en.wikipedia.org/wiki/Supply_chain, and*
2. for Product Development: *Principles of Good Product Development: A Development Methodology* found at *http://www.johnstark.com/prn09.htnl.*

44.6 Exercises

Problems 1–3 apply to the article discussed in section 44.1.

1. Why was this research needed?
2. How was this research done?
3. What are the findings? What does it mean?
4. Operationally define what is meant by a theory. Hint: A theory involves propositions about constructs, presents a systematic view, and explains phenomena.
5. What is a "computation"? What are the objects manipulated in a computation?
6. What is an "algorithm"? What is the relation between algorithms and computations?

7. Discuss the purpose of a theory of algorithms. What is meant by the scope for a theory of algorithms?

8. Describe how you would implement and empirically test the algorithmic approach presented in the chapter's study.

9. Distinguish between the *static* and *dynamic* views of science. Hint: What is the meaning of "heuristic"?

10. Through a background review determine what is meant by a conjoint analysis.

11. Why and when do we use conjoint analysis?

12. How do you use conjoint analysis?

13. Illustrate how conjoint analysis is the application of ANOVA to the analysis of customer preferences.

14. What are the three most common applications of conjoint results?

44.7 Conjoint Analysis Project

Suppose a company is developing a new product and you have been, as a valued customer, requested to respond to a survey. One of the tasks requested in responding would probably be for you to arrange a list of combinations of product attributes in decreasing order of precedence. Using this type of data the company can map a list of product utilities onto the ordered feature precedence list. This results in the company knowing the product utilities needed to encourage customer purchase or satisfaction of the product.

A methodology called *Conjoint Analysis* is a method for determining the product feature profile in customer purchasing through statistically analyzed survey research. Special attention should be paid to the use of possible combinations of features presented to potential customers; this is because:

1. individual attributes have a different impact on precedence rankings than do combination of attributes, and

2. customer visualization with combinations of product features is usually easier for the customer.

Prepare an annotated background review on conjoint analysis. The review should address:

1. the algorithmic steps in a typical conjoint analysis, and

2. identification of the methodologies for conducting a conjoint analysis.

3. find at least 3 to 6 research questions in the conjoint analysis domain.

CHAPTER 45

THE AXIOMATIC RESEARCH METHOD

OVERVIEW AND LEARNING OBJECTIVES

In This Chapter

45.1 Introduction

Whether a research study is labeled as applied or theoretical is immaterial. What is important is that the study be precise and accurate. A research field is not a collection of diverse topical areas; but it is a discipline. Since it is a discipline then researchers should possess a familiarity with axiomatic approaches and a proficiency at formal reasoning. Formalism often enables researchers to convey ideas or results with precision and clarity.

45.2 Sample Study

Study *Discovery of Functional and Approximate Functional Dependencies in Relational Databases* available at *http://www.emis.de/journals/HOA/JAMDS/7/149.pdf*. This study develops the foundation for a simple, yet efficient method for uncovering functional and approximate functional dependencies in relational databases.[1] The technique is based upon the mathematical theory of partitions done over a relation's row identifiers. Using a level-wise algorithm, the minimal nontrivial functional dependencies can be found using computations conducted on integers. Therefore, the required operations on partitions are both simple and fast. Additionally, the row identifiers provide the added advantage of nominally identifying the exceptions to approximate functional dependencies, which can be used effectively in practical data mining applications.

45.3 Axiomatic Development as a Research Method

Assume that you are a new employee for a globally sized company and put in charge of designing the corporate database. How will you convince the management and stakeholders that your database is valid and reliable? Clearly, this cannot be achieved empirically since, for all practical matters, the database appears to us as infinitely large. Thanks to the axiomatic method, this task can be resolved. Relational database design through the use of formal semantics can be rigorously and ruthlessly specified.

If you implement a software package for electronic mail and are unfamiliar with the effects of flow, then your package will never run. Axiomatic development in general, as illustrated by the article discussed, forces the researcher to face and deal with questions as fundamental and diverse as:

- ALGORITHM EXISTENCE
  - Does there exist a solution for a given problem?
- ALGORITHM SPECIFICATION
  - How should the algorithms be specified?
- EXPRESSIVE POWER
  - Does one method of specifying algorithms have greater expressive power than a second; that is, can more algorithms be specified using it?

- ALGORITHM TERMINATION
  - Does a given algorithm always halt or terminate?
- ALGORITHM CORRECTNESS
  - Is a given algorithm correct?
- ALGORITHM CONSTRUCTION
  - How is a correct algorithm obtained?
- ALGORITHM SEMANTICS
  - What is the meaning of a given algorithm?
- ALGORITHM EFFICIENCY
  - Is there a more efficient algorithm for a given problem?
- ALGORITHM IMPLEMENTATION
  - Can a given software program be implemented for the algorithm?
- ALGORITHM COMPARISON
  - Is one algorithm more powerful than a second? Recall that an algorithm has:
- A finite set of instructions
- A computing agent that can carry out the instructions of the algorithm
- A computing agent that must be able to store, retrieve, and make steps in a computation
- A computing agent that must carry out the instructions in discrete steps
- A computing agent that must carry out the instructions deterministically or mechanistically
- The axiomatic development of a database is an excellent example of axiomatic development as a major research method, as is[1] on data mining methodology

The axiomatic development of the relational data model is a perfect example of theoretical research that has had a major impact on real world problem solving and applications. The next section will introduce the foundation for the axiomatic development of the relational data model. A section then follows on strong relations. Finally, there is a section on strong inter-related relations.

45.4 Definition for the Relational Data Model

Reconsider the earlier problem where you have been hired by the airline company in the world. The company's corporate assignment is to design and implement a new world-wide corporate data base. How will you convince the corporate board that your data base is valid and reliable? Many individuals would choose to run a pilot study on query and data manipulation processing using a sub-data base for which the correct answers were known. Then tests are passed if the processing generates the known answers. But this pilot testing is totally inaccurate. What if an extreme sub-data base was chosen? What if the query and/or data manipulation processes were inadequate or misrepresentative? One must remember that any data base, which can be practically viewed as an infinitely large data base, can not be empirically tested.

The relational data model has been axiomatically derived and, as a result, offers the capability to test the criteria of goodness of a proposed relational data base design. This approach clearly enables the designer to defend his data base design to the corporate board members.

A relational data model is defined by the following list [found in *Database Systems the Complete Book*, by Hector Garcia-Molina, Jeffrey D. Ullman, and Jennifer Widom, Prentice Hall, Inc.]:

1. A relation is a uniquely named two dimensional table.
2. An attribute is a uniquely named, within the definition of the relation, column of the relation.
3. The named relation and the set of attributes for a relation is called a schema.
4. The rows of a relation are called tuples. A tuple has one component for each attribute of the relation.
5. A design is composed of one or more relational schema.
6. The set of schemas for the relations in a design is called a database schema.
7. The relational model requires that each component of each tuple be atomic, they must be single valued.
8. Each attribute of a relation is a domain; that is a particular data type.
9. If an assigned value to an attribute X forces another attribute Y to have a specific value, then X functionally determines Y and is denoted as $X \rightarrow Y$. $X \rightarrow Y$ is called a functional dependency.
10. A set of two or more attributes is a key for a relation R if:
    a. those attributes functionally determine all other attributes of the relation, and
    b. no proper subset of the set of attributes functionally determines all other attributes of R, this minimal set of attributes is called a primary key.

Practically speaking the values assigned to tuples in a relation are facts, while the functional dependencies associated with the relation represent information. This becomes readily apparent when one is exposed to Armstrong's Axioms [found in *Database Systems the Complete Book*, by Hector Garcia-Molina, Jeffrey D. Ullman, and Jennifer Widom, Prentice Hall, Inc.], statements that are true without proof.

Reflexivity Axiom: If X is a set of attributes and $Y \subseteq X$, then $Y \rightarrow X$.

Augumentation Axiom: If $X \rightarrow Y$ and Z is a set of attributes, then

XZ → YZ.

Transitivity Axiom: If X → Y and Y → Z, then X → Z.

Using these rules one can formally prove that:

Union Rule: If X → Y holds and X → Z, then X → YZ holds.

Decomposition Rule: If X → YZ holds, then X → Y holds and X → Z holds.

Pseudotransitivity Rule: If X → Y holds and WY → Z holds, then WX → Z holds.

Before further developing this axiomatic data model, investigating some strengths and weaknesses of relations is needed.

45.5 Strong Relations

Consider the following toy corporate workshop data base design:

1. (HOUR, ROOM) → WORKSHOP
2. (WORKSHOP, EMPLOYEE) → RATING
3. (INSTRUCTOR, HOUR) → ROOM
4. (WORKSHOP) → INSTRUCTOR
5. (HOUR, EMPLOYEE) → ROOM

TABLE 45.1 Workshop assignment relation.

Employee	Workshop	Instructor	Hour	Room	Rating
Giatros	SQL	Rainwater	8:00	100	Excellent
King	DB Adm	Jaska	8:00	200	Good
Last	DB Adm	Jaska	8:00	200	Adequate
Green	Algebra	Rainwater	9:00	400	Excellent

The following decomposition of the Workshop Assignment into projected sub-relations can be generated:

- R1(EMPLOYEE, HOUR, WORKSHOP)
- R2(EMPLOYEE, WORKSHOP, RATING)
- R3(WORKSHOP, INSTRUCTOR)
- R4(INSTRUCTOR, HOUR, ROOM)

The goal is to develop a decomposition into projected sub-relations of the original relation. These become a collection of relations that retain the same information as the original relation. One way to ensure this requirement is to enforce that the join of the sub-relations generates the same information as represented by the original relation. This join feature of a decomposition is referred to as a *lossless join decomposition*.

Another approach is to require the join of the projected sub-relations which does not include any non-existent tuples from the original relation. Such a decomposition is referred to as *functionally preserving*. Both of these types of decompositions ensure that answers to any ad hoc inquiries are correct.

Consider two relations R and S. Let $\sigma_{predicate}(R)$ be the selection function (a subset of the rows) and $\pi_{predicate}$(R) be the projection function (a subset of the columns). The *join* of R and S, denoted by R ∞ S is a relation on schema R ∪ S formally defined as:

$$\mathrm{R} \infty \mathrm{S} = \pi_{\mathrm{RUS}}\,(\sigma_1\ R.A_I1 = S.A_I1\ and\ R.A_I2 = S.[.A]\ _12\ and\ \ldots\ and\ R.A_In = S.A_1n)\ R\,x\,S)$$

$$\text{where } \mathrm{R} \cap \mathrm{S} = \{A_I, A_I, \ldots, A_I\}$$

The preceding sets of relation schema can be populated as follows:

TABLE 45.2 Decomposition of the workshop assignment relation.

R1		
Employee	**Hour**	**Workshop**
Giatros	8:00	SQL
King	8:00	DB Adm
Last	8:00	DB Adm
Green	9:00	Algebra

R2		
Employee	**Workshop**	**Rating**
Giatros	SQL	Excellent
King	DB Adm	Good
Last	DB Adm	Adequate
Green	Algebra	Excellent

R3	
Workshop	**Instructor**
SQL	Rainwater
DB Adm	Jaska
Algebra	Rainwater

R4		
Instructor	**Hour**	**Room**
Rainwater	8:00	100
Jaska	8:00	200
Rainwater	9:00	400

Now suppose that a list of Workshops with their corresponding room numbers is required. Relations R1 and R4 contain the necessary information and can be joined using the attribute HOUR. The result of this join is:

TABLE 45.3 Regeneration of the workshop assignment relation.

R1 × R4				
Employee	**Workshop**	**Instructor**	**Hour**	**Room**
Giatros	SQL	Rainwater	8:00	100
Giatros	SQL	Jaska	8:00	200
King	DB Adm	Rainwater	8:00	100
King	DB Adm	Jaska	8:00	200
Last	DB Adm	Rainwater	8:00	100
Last	DB Adm	Jaska	8:00	200
Green	Algebra	Rainwater	9:00	400

This join creates the following invalid information (denoted by the colored rows):

- Giatros, King, and Last are assigned to the same workshop at the same time from two different instructors in two different rooms.
- Rainwater (the SQL teacher) teaches DB Adm.
- Jaska (the DB Adm teacher) teaches SQL.

Both instructors teach different workshops at the same time.

What is meant by a strong relation? Two examples are [found in *Database Systems the Complete Book*, by Hector Garcia-Molina, Jeffrey D. Ullman, and Jennifer Widom, Prentice Hall, Inc.]:

A relation is Boyce-Codd normal form (BCNF) if and only if whenever there is a nontrivial functional dependency $X \rightarrow Y$ where $Y \subseteq R$, then it is the case that X is a superkey for R.

A relation R is in third normal form (3NF) if whenever $X \rightarrow Y$ is a nontrivial functional dependency, either X is a superkey or Y is a member of some key for R.

A relational database design algorithm will be introduced that retains the lossless join property, therefore, ensuring correctly answering unanticipated queries. However, the choice of joins must be evaluated carefully. Many different sequences of joins will recreate an instance of a decomposition which is not a lossless join decomposition. Additionally an algorithm for generating a lossless and functionally preserving decomposition is described.

45.6 Strong Inter-related Relations

Step 1: Form a minimal cover:

- for each set of attributes X and singular attributes, form functional dependencies of the form X--> a

- Eliminate redundant functional dependencies
- Eliminate redundant attributes on the left side of the remaining functional dependencies

Step 2: Each functional dependency in the minimal cover made a relation with left side as a primary key

Step 3: If the collection of relations in step 2 contains the original primary key k then you are done, otherwise, add the all key relation (k}.

Result: a 3NF lossless join and functional dependency preserving collection of relations to replace the original relation.

45.6.1 BCNF algorithm

Step 1: Build a minimal cover.

Step 2: Any X--> a that violates BCNF chosen. Then (R-a) and (Xa) form a lossless join.

Step 3: Repeat step 2 until all violations of BCNF are eliminated or you are left with an {a,b} relation.

Result: new collection of sub-relations is a lossless join replacement for the original relation.

45.6.2 Test for functional dependency preserving

Note: X --> Y if and only if X^+ contains Y.

Therefore, test each original functional dependency on the union of the new sub-relations new functional dependency set.

If this test fails then use the 3NF algorithmic solution, otherwise choose the BCNF algorithmic solution.

45.6.3 To find a key

Note: if X is a key then X -> a for every attribute a.

If X generates every a in R and is minimal it is a candidate key.

EXAMPLE: Given R(abcd) with F={a -->bc, a-->d, b -->d}

then a^+ contains {a, b, c} which contains {a, b, c, d} or a^+ = R. thus (a) is a candidate key.

45.7 Summary

The overall process is similar to assembling a great basketball team. First, the players (the sub-relations) chosen must be coachable and athletic (a high degree of normalization). Second, the chosen players must execute as a stringent defensive team (a lossless join decomposition) and as a powerful offensive team (a functionally preserving decomposition).

With a combination of an outstanding set of coaches, players, and finances, the basketball team might win a championship. In the relational data modeling world, the 3NF ensures that we can always find a lossless join and functionally dependency preserving decomposition. In fact, we might even find a better design (an entire 'team' of super stars who play out of love for the game): a BCNF decomposition which is both a lossless join and functionally dependency preserving.

45.8 The Axiomatic Method as a Tool for Research

The major value of the axiomatic method is its use as a tool for opening up vast new domains for investigation. Use of the axiomatic method in teaching affords the student a chance to experience the thrill of creating concepts and to glimpse the inherent beauty of the field under study. The axiomatic method not only encourages researchers to develop their own creative powers, but sometimes leads to the invention of new methods not previously conceived. A great merit of the axiomatic approach, as illustrated in both the sample study article and the discussion in previous sections, is that it admits the possibility of a mixture of deductive and empirical thinking. Axiomatic development is simply the application of the scientific method to problems classically recognized as purely theoretical.

45.9 Exercises

Questions 1–10 apply to the sample study article.

1. Why was this research needed?
2. How was this research done?
3. What are the findings? What does it mean?
4. How is validity addressed?
5. Discuss the ability to replicate this study.
6. Discuss the ability to generalize this study.
7. Describe how you would implement and test the algorithmic approach.
8. **a.** Does the abstract provide sufficient information for you to make a decision about whether to read the full article?

   b. Is the purpose of the study clearly stated?

   c. Does the introduction make the purpose easy to understand?
9. **a.** Is the topic of the study introduced in terms of what is already known about the subject?

   b. Have the authors described relevant theories? Do they propose any problems with existing theories that they will address in their study?

c. Have conflicting findings from previous studies been discussed? Was any insight provided as to how to explain the conflicting results?

d. Were decisions about the design and procedures of the present study justified in terms of prior studies?

e. Were the study goals or hypotheses clearly stated?

10. a. Do the study's conclusions relate directly to the results, or do they range beyond the results into speculation?

b. Do the study's conclusions relate directly to the purpose and hypotheses of the study?

c. Are the results linked in the discussion to prior knowledge?

d. Are limitations of the study discussed?

e. Have the authors given you an adequate answer to the question, "So what?"

45.10 Semantic Data Models Project

Semantic data models have emerged from a requirement for more expressive conceptual data models. Current generation data models lack direct support for relationships, data abstraction, inheritance, constraints, unstructured objects, and the dynamic properties of an application. Although the need for data models with richer semantics is widely recognized, no single approach has won general acceptance.

a. What are the generic properties of semantic data models?

b. Present a representative selection of models that have been proposed since the mid-1970s.

c. Offer guidelines for the comparison of models.

d. Discuss the future directions in the population of conceptual data modeling.

45.11 Extended Relational Models

Historically, there has been a need within the database community to extend the relational model to accommodate more real-world knowledge. It is interesting that many of the required extensions cannot be accommodated by the model theoretic paradigm for relational databases.

a. Perform a topical search on the large number of proposals that have been conducted in response to this need.

b. How can one begin to compare these proposals? Discuss the limitations you discover in your topical search. Why should one believe that the definitions for these proposals are correct or complete?

c. Claim: The kinds of real-world knowledge that these extended data models attempt to capture have natural representations as first order formulae.
The benefits of this claim would be:

The nonlogical data models can be equivalently formalized by suitably restricted classes of first order theories. As a result, the semantics of the nonlogical data model would be precisely defined by its logical content; two different nonlogical data models could be compared; the definition of an answer to a query remains the same as long as the extension to the data model is first order definable; and integrity constraints remain the same as long as the extension to the data model is first order definable. Verify this claim.

Reference

1. R.S. King and J.J. Legendre, "Discovery of Functional and Approximate Functional Dependencies in Relational Databases," *Journal for Applied Mathematics and Decision Sciences* (2003): 49–59 (available online at *http://www.emis.de/journals/HOA/JAMDS/7/149.pdf*).

UNIT 10

SIMULATION AND RESEARCH ISSUES

Review *Detection of intrusions in information systems by sequential change-point methods* by Tartakovsky and etc. found at *www.sciencedirect.com*.

- Why did the authors run a Monte Carlo simulation?
- Briefly outline the computational steps in this simulation?
- What timing variables were included?
- How was randomization utilized?

Solving a research problem is like solving a detective mystery. The task requires planning and the careful collection of facts. With a little planning, your research experience can be positive and rewarding. This unit illustrates two additional types of research tools. Chapters 46 and 47 deal with the topic of simulation and Chapter 48 provides an outline for performing a Market Basket Analysis with the aid of clustering methods.

Simulation typically is nothing more or less than the technique of performing *sampling experiments* on the model of the system. The experiments are done on the model rather than on the real system itself only because the latter would be too inconvenient, expensive, and time consuming. Otherwise, simulated experiments should be viewed as virtually indistinguishable from ordinary statistical experiments, so that they also should be based upon sound statistical theory. Simulated experiments usually are executed on a computer because of the vast amount of data being generated and processed rather than because of any inherent relationship between the experiment and the computer.

Tartakovsky's article as well as the research study vignette presented in Chapter 48 illustrate the breadth of topics and methods in the realm of computing research methods. A short list of computing research methods includes: field studies, instrument/product development, software experiments, mathematical proofs, simulation, model building, and conceptual analysis.

Information systems focus is on:

- developing an understanding of the process of developing and using information and communication technologies in an organization, and
- filling the gap between the foci of attention by the fields of engineering and the behavioral sciences/economics.

The focus of engineering is upon producing new technological capabilities, while the behavioral sciences/economics focus is to recommend trends on the basis of technological capacity.

In summary, information systems has:

- developed knowledge to explain socio-economic changes associated with information technologies, and
- provided associated relevant guidance to the organizations for necessary action.

Therefore the information system researcher employs a *variety of research methods* and pursues a *variety of theoretical perspectives*.

CHAPTER 46

MONTE CARLO SIMULATION OVERVIEW

OVERVIEW AND LEARNING OBJECTIVES

In This Chapter

46.1 Introduction

The methodology of simulation, initially used by operations research, was created by John von Neumann and Stanislaw Ulam in the late 1940s. Since the advent of digital computing, simulation has been used for countless business applications. Essentially, simulation is the use of a system model that has the desired characteristics of reality in order to reproduce the essence of the actual operations. The Monte Carlo method is simulation by sampling techniques. It involves determining the probability distribution of the variable under consideration and then sampling from this distribution by means of random numbers to obtain data. In effect, a set of random numbers is used to generate a set of values that have the same distributional characteristics as the actual experience it is desired to simulate. In this chapter, we will study Monte Carlo simulation with actual business applications.

46.2 Application: Determination of the Number of Production Units

Suppose that the demand for an advertisement bulletin has the demand distribution shown in Table 46.1:

TABLE 46.1 Unit demand distribution.

Prob.	Demand
0	10,000
0.1	20,000
0.45	40,000
0.75	60,000

ON DVD All tables in this chapter appear on the companion DVD.

The bulletins sell for $4.00, and the variable cost of producing each bulletin is $1.50. Left-over bulletins must be disposed of at a cost of $0.20. The company wants to determine how many bulletins should be printed.

Production quantities of 10,000, 20,000, 40,000, or 60,000 are considered reasonable demands over 1,000 iterations. Basically, this is a production quantity versus demand problem. The number of units sold is the smaller of our production quantity and demand, for which we compute the revenue where:

$$Revenue = MIN(produced, demand) * unit\text{-}price.$$

Additionally, we need to compute the total production cost where:

$$Total\ production\ cost = produced * unit\text{-}prod\text{-}cost.$$

If we produce more bulletins than are demanded, the number of units left over equals production minus demand; otherwise no units are left over. This means we must compute the disposal cost where:

*Disposal cost = unit-disp-cost * produced-demand, if produced > demand*

and

Disposal cost = 0 otherwise.

Using Excel, the first step is to set up the spreadsheet to compute the profit given a specific number of units produced, which utilizes the inverse transformation method applied to the cumulative distribution of the number of units produced. See Table 46.2 a and b:

TABLE 46.2a The basic spreadsheet, in cells A1:H12.

Produced	40000					
Rand#	0.433481				**Prob.**	**Demand**
					0	10000
Demand	20000				0.1	20000
Unit Prod Cost	1.5				0.45	40000
Unit Price	4				0.75	60000
Unit Disp Cost	0.2					
Revenue	80000					
Total Var Cost	60000					
Total Disposal Cost	4000					
Profit	16000					

TABLE 46.2b Spreadsheet formulas for Table 46.2a.

Produced	40000
Rand#	=RAND()
Demand	=VLOOKUP(C2,$G$3:$H$6,2)
Unit Prod Cost	1.5
Unit Price	4
Unit Disp Cost	0.2
Revenue	=MIN(C1,C4)*C6
Total Var Cost	=C1*C5
Total Disposal Cost	=C7*IF(C1>C4,C1-C4,0)
Profit	=C9-C10-C11

RAND(.) is a function to create a uniform random number. This random number is used as the first argument in the VLOOKUP(rand, lookup, 2) function to apply the inverse transformation method using the cumulative distribution table for demand. In Excel, the second column of the table must be the resultant values when executing the VLOOKUP(.) function.

In order to complete the simulation, we need to build a table of 1,000 rows of demand for each production quantity (10,000, 20,000, 40,000, 60,000). In the **Edit** menu, select **Fill** and then **Series**. This selection brings up the Series dialog box where we set Step value to **1**, Stop value to **1000**, where the series resides to **Columns**, and the type of series to **Linear**. Click **OK**. Note that before selecting Fill under the Edit menu, the beginning cell must be selected and have its value set to 1.

Next we set the values of our demand units (10,000, 20,000, 40,000, 60,000) in the row immediately preceding the first row of our demand table. In the left uppermost cell of the table, we establish the value of our profit by copying the value computed in the bottom cell of the spreadsheet shown in Table 46.2b. This action establishes the linkage between the spreadsheet and the demand table, which is shown in Table 46.3.

TABLE 46.3 Partial listing of the demand distribution table, in cells A16:E1016.

16000	10000	20000	40000	60000
1				
2				
3				
4				
5				
6				
7				
8				
9				
10				
11				
12				
13				
14				
15				
16				
17				
18				
19				
20				
21				
22				

TABLE 46.3 Continued.

16000	10000	20000	40000	60000
23				
24				
25				
26				
27				
28				
29				

We are now ready to simulate 1,000 iterations of demand for each production quantity. Select the table range (A16:E1016), and then click **Table** on the **Data** menu. To set up a two-way data table, in the Table dialog box select any blank cell for the column input cell and choose the production quantity in cell C1 as the row input cell. When you click **OK**, Excel simulates 1,000 demand values for each order quantity, as shown in Table 46.4.

TABLE 46.4 Partial demand distribution table for 1,000 demand values for each order quantity.

16000	10000	20000	40000	60000
1	25000	50000	16000	66000
2	25000	50000	100000	150000
3	25000	50000	16000	-18000
4	25000	50000	100000	66000
5	25000	50000	16000	66000
6	25000	50000	16000	-18000
7	25000	50000	100000	-18000
8	25000	50000	100000	-18000
9	25000	50000	-26000	-18000
10	25000	50000	16000	66000
11	25000	50000	100000	66000
12	25000	50000	16000	-18000
13	25000	50000	-26000	150000
14	25000	50000	16000	150000
15	25000	50000	16000	150000
16	25000	50000	-26000	-18000
17	25000	50000	100000	66000
18	25000	50000	100000	-18000
19	25000	50000	16000	-18000
20	25000	50000	100000	150000

(Continued)

TABLE 46.4 Continued.

16000	10000	20000	40000	60000
21	25000	50000	16000	-18000
22	25000	50000	-26000	-18000
23	25000	50000	100000	66000
24	25000	8000	16000	-18000
25	25000	50000	100000	-18000
26	25000	50000	16000	150000
27	25000	50000	16000	150000
28	25000	50000	16000	66000
29	25000	50000	16000	150000

Note that each time any cell is reset in Excel, the RAND(.) function automatically reiterates. Next we use the AVERAGE(B16:B1016) and STDEV(B16:B1016) functions to compute the average simulated profit and the standard deviation of our simulated profits for each order quantity. Each time we press "F," 1,000 iterations of demand are simulated for each order quantity. Producing 40,000 bulletins always yields the largest expected profit.

46.3 Exercises

46.3.1 Games and Simulation Project

1. The use of games and simulations raises many questions about their use, the role of the teacher, the time and space required, which games are appropriate, evaluation procedures, and the benefits and drawbacks of using games. Little research is available on evaluating games and simulations and how teachers can be assisted in making their own games and simulations.

   a. Develop suggestions for creating new games and simulations for future pedagogical uses that emerge almost naturally from the study and use of existing ones.

   b. Present an outline of how to evaluate and design games and simulations in the populations of business, finance, and economics.

46.3.2 Risk Analysis Project

1. RiskAMP is a full-featured Monte Carlo simulation engine for Excel. With this add-in, you can add Risk Analysis to your spreadsheet models quickly and easily. Visit the following Web site to download RiskAMP: *http://www.riskamp.com/home?gclid=CJTx6YDXrp0CFYNX2godGkRAyw*.

2. Using the add-in from Problem 1, calculate the time it takes to commute from home to the office. Suppose your commute to work consists of the following:

   - Drive 2 miles on a highway, with 90% probability you will be able to average 65 MPH the whole way, but with a 10% probability that a traffic jam will result in average speed of 20 MPH.

- Come to an intersection with a traffic light that is red for 90 seconds, then green for 30 seconds.
- Travel 2 more miles on a surface street, averaging 30 MPH with a standard deviation of 10 MPH.

You want to know how much time to allow for your commute in order to have a 75% probability of arriving at work on time. Additionally, when you have an important meeting, you want to know how early you need to leave the house in order to have a 99.5% probability of arriving on time.

3. Look up the following articles and write a literature review for each article:

- Hertz, D.B., "Risk Analysis in Capital Investment," *Harvard Business Review* (January-February 1964): 95–106. This article discusses the application of Monte Carlo methods to corporate finance.
- Boyle, P., "Options: A Monte Carlo Approach," *Journal of Financial Economics* (May 1977): 323–338. This paper pioneered the use of simulation in derivative valuation.

CHAPTER 47

HOW TO CONDUCT A SIMULATION

OVERVIEW AND LEARNING OBJECTIVES

In This Chapter

47.1 Introduction

Chapter 46 presented a brief overview of the Monte Carlo Method. This technique is a computational technique applied to *static models*. Static models provide for the relationship between the system attributes when the system is in equilibrium. If a point of equilibrium is changed by altering any of the attribute values, then the model enables the new values for all the attributes to be derived. However, the model does not show 'how' the attribute values were modified.

Dynamic models follow the changes over time that result from the system attributes. In this chapter, an algorithm for a *discrete event simulation* is presented through applying the dynamic model which maintains a current events chain. A polling and reassignment of attribute values take place each time an event experiences a change of state.

47.2 Manufacturing Example

Suppose that the Easy Rider Company, which manufactures competitive-style bicycles, wants to determine the optimum employment of maintenance personnel in order to reduce production losses caused by poor machine adjustments of the present machine operations and downtime in machinery. The company has already recorded the total number of incidents where equipment needs service each hour and the time required to perform the servicing. Table 47.1 shows the delays per hour and Table 47.2 illustrates the service time for machine downtime in the plant.

TABLE 47.1 Delays per hour on plant machinery.

Delays Per Hour	Number of Occurrences	Relative Frequency (%)	Cumulative Frequency (%)
17 and under	19	4.77%	4.77%
18	35	8.79%	13.57%
19	40	10.05%	23.62%
20	74	18.59%	42.21%
21	95	23.87%	66.08%
22	70	17.59%	83.67%
23	45	11.31%	94.97%
over 23	20	5.03%	100.00%
Total =	398		

TABLE 47.2 Service times for downtime of plant machinery per hour.

Delays Per Hour	Number of Occurrences	Relative Frequency (%)	Cumulative Frequency (%)
under 7	18	5.39%	5.39%
8	82	24.55%	29.94%
9	136	40.72%	70.66%
10	82	24.55%	95.21%
over 10	16	4.79%	100.00%
Total =	334		

ON DVD

All figures and tables in this chapter appear on the companion DVD.

In order to apply the inverse transformation technique, the following cumulative distributions are generated, as shown in Tables 47.3 and 47.4.

TABLE 47.3 Cumulative distribution for delays per hour.

Frequency (%)	Cumulative Delays per Hour
0.00%	17
4.77%	17.5
13.57%	18.5
23.62%	19.5
42.21%	20.5
66.08%	21.5
83.67%	22.5
94.97%	23

TABLE 47.4 Cumulative distribution for servicing.

Cumulative Frequency (%)	Delays Per Hour
0.00%	7
5.39%	7.5
29.94%	8.5
70.66%	9.5
95.21%	10.5

From these tables we can generate the graphs shown in Figures 47.1 to 47.4.

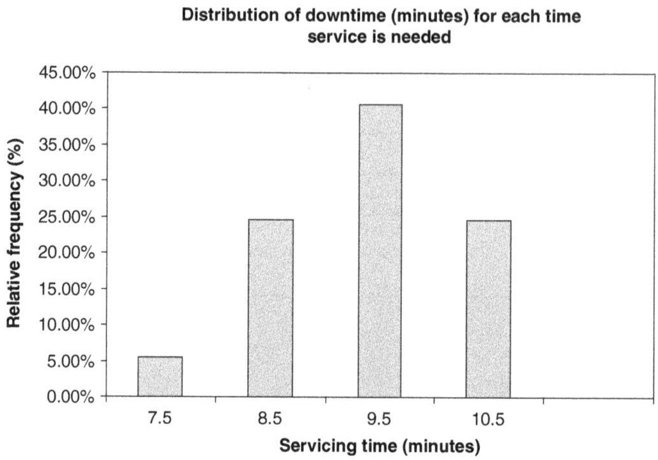

FIGURE 47.1 Distribution of downtime for each time service is needed.

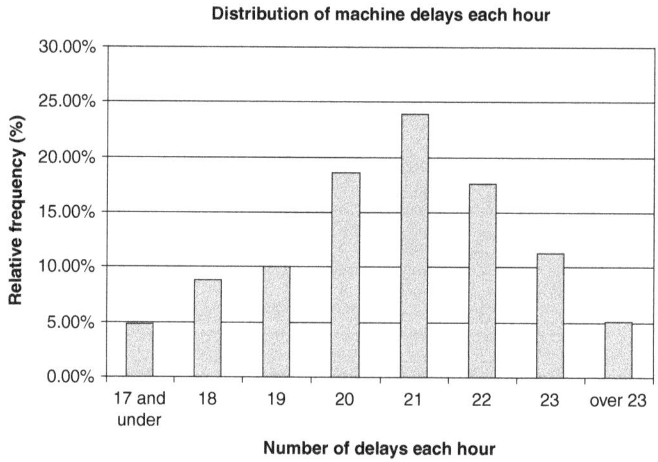

FIGURE 47.2 Distribution of machine delays each hour.

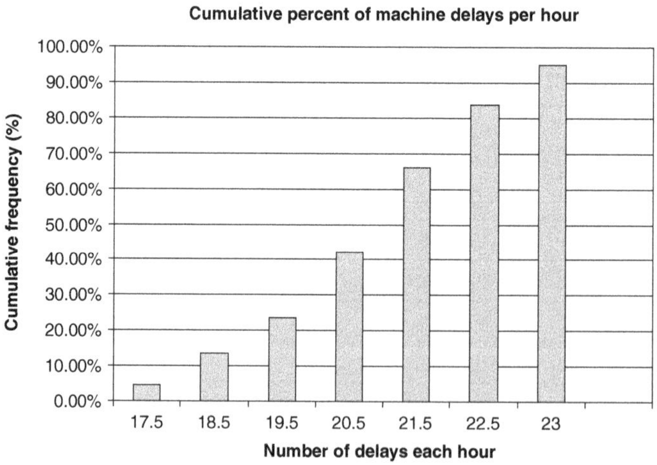

FIGURE 47.3 Cumulative percent of machine delays per hour.

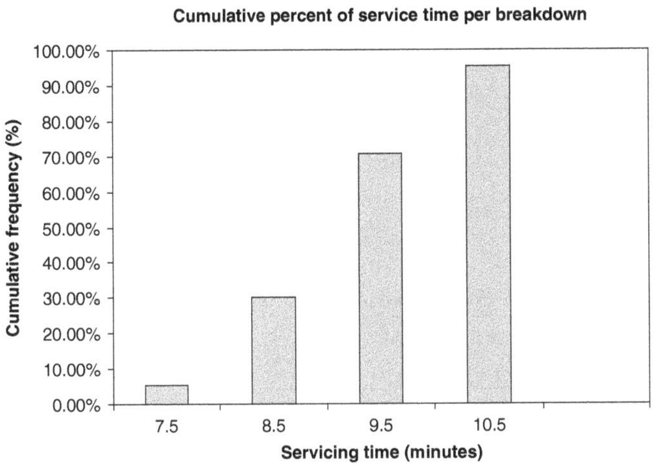

FIGURE 47.4 Cumulative percent of service time per breakdown.

Next, we develop a table to generate random machine servicing time. Table 47.5a illustrates this process for one hour.

TABLE 47.5a Simulation table for one hour of machine servicing time, A40:G50.

Random No	Simulation of Delays in One Hour	Random No	Simulation of Individual Servicing Time Per Hour Ind Service Time		Tot Hrs Servicing Time (Minutes)
34.60%	19.5	19.10%	7.5	62.5	171
		6.31%	7.5		
		79.27%	9.5		
		60.23%	8.5	E63	
		28.40%	7.5		
		36.73%	8.5		
		86.22%	9.5		

First, a random number is generated and used as a search argument in applying the inverse transformation method to obtain the number of one-hour delays. This is the number of individual service times to be obtained, each using a random number as the argument when applying the inverse transformation method to the cumulative distribution of service times. Finally, the sum of all the individual service times is obtained, which represents the total servicing time for the original number of delays. The value E63 stored in the second to last column is the ending address for the random numbers in the third column, the 19 delays. The value 62.5 is an intermediate value used in formulating the address E63, which happens to be the row number affiliated with the first delay random value plus 19.5.

Excel formulas for setting up this table are given in Table 47.5(b). VLOOKUP(.) is used to implement the inverse transformation method, with the first argument as the random value, the second argument as the table, and the third argument indicating that the second column of the table is where the delay value is to be found. The major trick involved in this table is forced upon us by the fact that different initial random numbers will generate a different number of delays. These delays begin in B44 and end at B[44 + no. of delays]. To accomplish the latter task in Excel requires several steps. First, we check to see IF(no. of delays)+(row no. of no. of delays value)–(current row no.) is greater than 0. Then we generate a random number or we simply place a blank in the cell. This formula is posted in the first cell in the column and then copied/pasted into the next 25 cells in the column.

TABLE 47.5b Formulas used in Part A, A40:G50.

Random No	Simulation of Delays in One Hour	Tot Hrs Servicing Time (minutes)
=RAND()	=VLOOKUP(A44,$E$25:$F$33,2)	=SUM($E$44:INDIRECT($F$47))

Next, using the IF(.) and CONCATENATE(.) functions in conjunction with the TRUNC(.) function, the address for the last cell to be summed, when calculating the total service time, is posted to the fifth cell in the second to last column. Finally, the sum is obtained, in the last column, using indirect addressing for the ending address, via the INDIRECT(.) function.

Table 47.6a provides a summary of the calculations based on three, four, and five servicemen. This table is the main table for implementing the simulation. The total servicing time for each hour of the study is adjusted for a personal time allowance (10 minutes for each serviceman per hour) of 30 minutes for three servicemen during the first hour. This servicing time plus 30 minutes for personal affairs is the total number of minutes used for three servicemen. If the total servicing time +30 exceeds 3 hours, then the servicemen experience waiting time; otherwise, they have idle time on their hands. Similar computations are made for the cases of four and five servicemen.

TABLE 47.6a Simulation table for idle and wait times for three, four, and five servicemen.

Three Servicemen				
Hour of Study	Total Service	Busy	Idle	Ops Wait
171	171	180	0	21
1	196.5	180	0	46.5
2	195.5	180	0	45.5
3	175	180	0	25
4	194	180	0	44
5	177.5	180	0	27.5
6	165.5	180	0	15.5
7	166.5	180	0	16.5
8	179	180	0	29
9	148.5	178.5	1.5	0
10	174.5	180	0	24.5

(Continued)

TABLE 47.6a Continued.

Four Servicemen			Five Servicemen		
Busy	**Idle**	**Ops Wait**	**Busy**	**Idle**	**Ops Wait**
211	29	0	221	79	0
236.5	0	26.5	246.5	23.5	0
235.5	0	25.5	245.5	24.5	0
215	0	5	225	45	0
234	0	24	244	26	0
217.5	0	7.5	227.5	42.5	0
205.5	4.5	0	215.5	54.5	0
206.5	3.5	0	216.5	53.5	0
219	0	9	229	41	0
188.5	21.5	0	198.5	71.5	0
214.5	0	4.5	224.5	45.5	0
Hour of Study			**Total Service**		
=SUM($E$44:INDIRECT($F$47))			=SUM($E$44:INDIRECT($F$47))		
1			=TABLE(,G44)		
2			=TABLE(,G44)		
3			=TABLE(,G44)		
4			=TABLE(,G44)		
5			=TABLE(,G44)		
6			=TABLE(,G44)		
7			=TABLE(,G44)		
8			=TABLE(,G44)		
9			=TABLE(,G44)		

TABLE 47.6b Simulation table, with formulas, for idle and wait times for three, four, and five servicemen.

Three Servicemen		
Busy	**Idle**	**Ops Wait**
=IF(K42+30>180,180,K42+30)	=IF(K42+30<180,–K42–30+180,0)	=IF(K42+30>180,K42+30–180,0)
=IF(K43+30>180,180,K43+30)	=IF(K43+30<180,–K43–30+180,0)	=IF(K43+30>180,K43+30–180,0)
=IF(K44+30>180,180,K44+30)	=IF(K44+30<180,–K44–30+180,0)	=IF(K44+30>180,K44+30–180,0)
=IF(K45+30>180,180,K45+30)	=IF(K45+30<180,–K45–30+180,0)	=IF(K45+30>180,K45+30–180,0)
=IF(K46+30>180,180,K46+30)	=IF(K46+30<180,-K46–30+180,0)	=IF(K46+30>180,K46+30–180,0)
=IF(K47+30>180,180,K47+30)	=IF(K47+30<180,–K47–30+180,0)	=IF(K47+30>180,K47+30–180,0)
=IF(K48+30>180,180,K48+30)	=IF(K48+30<180,–K48–30+180,0)	=IF(K48+30>180,K48+30–180,0)
=IF(K49+30>180,180,K49+30)	=IF(K49+30<180,–K49–30+180,0)	=IF(K49+30>180,K49+30–180,0)

(Continued)

TABLE 47.6b Continued.

Three Servicemen		
Busy	**Idle**	**Ops Wait**
=IF(K50+30>180,180,K50+30)	=IF(K50+30<180,–K50–30+180,0)	=IF(K50+30>180,K50+30–180,0)
=IF(K51+30>180,180,K51+30)	=IF(K51+30<180,–K51–30+180,0)	=IF(K51+30>180,K51+30–180,0)
=IF(K52+30>180,180,K52+30)	=IF(K52+30<180,–K52–30+180,0)	=IF(K52+30>180,K52+30–180,0)
Four Servicemen		
Busy	**Idle**	**Ops Wait**
=K42+40	=IF(O42<240,–O42+240,0)	=IF(O42>240,O42–240,0)
=K43+40	=IF(O43+30<240,–O43–30+240,0)	=IF(O43+30>240,O43+30–240,0)
=K44+40	=IF(O44+30<240,–O44–30+240,0)	=IF(O44+30>240,O44+30–240,0)
=K45+40	=IF(O45+30<240,–O45–30+240,0)	=IF(O45+30>240,O45+30–240,0)
=K46+40	=IF(O46+30<240,–O46–30+240,0)	=IF(O46+30>240,O46+30–240,0)
=K47+40	=IF(O47+30<240,–O47–30+240,0)	=IF(O47+30>240,O47+30–240,0)
=K48+40	=IF(O48+30<240,–O48–30+240,0)	=IF(O48+30>240,O48+30–240,0)
=K49+40	=IF(O49+30<240,–O49–30+240,0)	=IF(O49+30>240,O49+30–240,0)
=K50+40	=IF(O50+30<240,–O50–30+240,0)	=IF(O50+30>240,O50+30–240,0)
=K51+40	=IF(O51+30<240,–O51–30+240,0)	=IF(O51+30>240,O51+30–240,0)
=K52+40	=IF(O52+30<240,–O52–30+240,0)	=IF(O52+30>240,O52+30–240,0)
Five Servicemen		
Busy	**Idle**	**Ops Wait**
=K42+50	=IF(R42<300,–R42+300,0)	=IF(R42>300,R42–300,0)
=K43+50	=IF(R43+30<300,–R43–30+300,0)	=IF(R43+30>300,R43+30–300,0)
=K44+50	=IF(R44+30<300,–R44–30+300,0)	=IF(R44+30>300,R44+30–300,0)
=K45+50	=IF(R45+30<300,–R45–30+300,0)	=IF(R45+30>300,R45+30–300,0)
=K46+50	=IF(R46+30<300,–R46–30+300,0)	=IF(R46+30>300,R46+30–300,0)
=K47+50	=IF(R47+30<300,–R47–30+300,0)	=IF(R47+30>300,R47+30–300,0)
=K48+50	=IF(R48+30<300,–R48–30+300,0)	=IF(R48+30>300,R48+30–300,0)
=K49+50	=IF(R49+30<300,–R49–30+300,0)	=IF(R49+30>300,R49+30–300,0)
=K50+50	=IF(R50+30<300,–R50–30+300,0)	=IF(R50+30>300,R50+30–300,0)
=K51+50	=IF(R51+30<300,–R51–30+300,0)	=IF(R51+30>300,R51+30–300,0)
=K52+50	=IF(R52+30<300,–R52–30+300,0)	=IF(R52+30>300,R52+30–300,0)

To perform the simulation, we must first post the total service time from Table 47.5b to the left uppermost cell in Table 47.6a. Next, use the **Fill** selection from the **Edit** menu to fill a series from 1 to 100 in column 1. Then select the first and second columns of the table, followed by selecting **Table** from the **Data** menu. Finally, fill in the options in the Table dialog box and click **OK** to generate the total service times in column 2. We can then fill in the rest of the table.

Based upon an hourly wage including fringe benefits of $4.00 for maintenance personnel and $15.00 per hour for idle automatic machines, the study, executed for 100 one-hour periods, indicates that the employment of five maintenance is best, as shown in Table 47.7.

TABLE 47.7 Simulated costs breakdown.

Idle Machine (Operator) Cost: $15/hr. **Maintenance Personnel Cost: $48/man/12 hours** **Total Cost = Idle Machine Cost + Maintenance Cost** **Cost Analysis of Using Maintenance Men Based on Self-Service Time of Operators**			
Maintenance Personnel	**Three**	**Four**	**Five**
Total Time Lost	2585.5	860	0
Total Pay Value of Ops	646.375	215	0
Main. Per	790.375	407	240

47.3 Discrete Event Simulation

The previous examples of simulation we discussed did not involve time as an element of the simulation. Such simulations are referred to as *static*. However, many systems experience changes in state over time. For example, inventory and queuing systems have the state variables change only at a discrete set of points in time and are called *discrete event simulations*. But consider the water behind a dam. Clearly evaporation and heavy rains make the state variable. The water level experiences changes in state continuously over time. Systems like this are referred to as *continuous simulations*.

Consider simulating a network system. For this type of study, the researcher must:

- Identify the system, which is a collection of entities that interact over time to accomplish the goals of the study.
- Construct a model, or an abstract representation of the system that describes a system in terms of state, entities and related attributes, sets, events, activities, and delays.
    - The *system state* is a collection of variables that contain all of the information to describe the system at any time.
    - A *set* is simply a collection of associated states.
    - An *event* is an occurrence that changes the state of a system.
    - A known time interval, such as for servicing an entity, is an *activity*.
    - A time interval of unknown length until an activity is actually completed, such as a wait time in a queue, is referred to as a *delay*.

When setting up a discrete event simulation, one must operationally define the system state, entities, events, activities, and delays.

The main algorithm involves the use of a future events list. The future events list is entered by the entity activities in a "first-in-first-out" manner. This essentially represents maintenance of the events throughout the simulation. Note that every simulation must have a stopping event, which defines how long the simulation will run.

- Collect, code, and verify the data.
- Validate the model.
- Devise an experimental design if the simulation study will be run on different alternatives.
- Determine if more runs of the simulation are required or whether the sample size is sufficient.
- Document the program, report the results, and implement the simulation.

47.4 Discrete Event Simulation of a Simplified Token Ring

Consider two VPNs connected to a common server on a Token Ring, as shown in Figure 47.5.

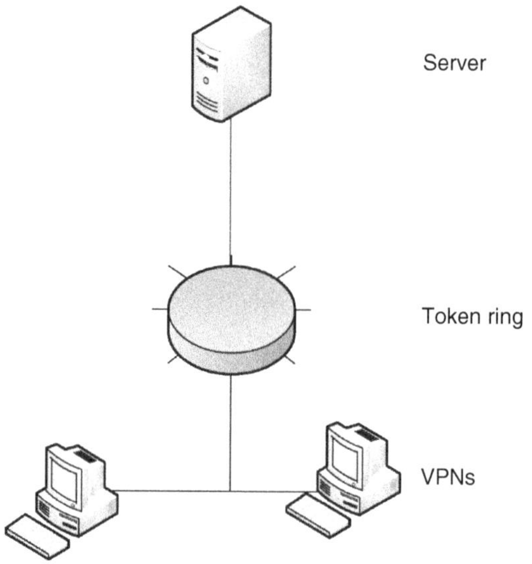

FIGURE 47.5 Token ring architecture.

System State: [VPNQ(t), VPN(t), SQ(t), S(t)]

VPNQ(t) = number of tokens in a VPN queue
VPN(t) = number of tokens being served by a VPN (0, 1, or 2)
SQ(t) = number of tokens at server queue
S(t) = number of tokens (0 or 1) being served by server, all at simulation time t

Events: (AVPN, t, Ti), token Ti arrives at VPN queue at time t
(EVPN, t, Ti), token Ti ends VPN service at time t
(ES, t, Ti), token Ti ends server service at time t

Entities: Six tokens T1, T2, ... T6

Sets: VPN queue, all tokens waiting for final servicing at one of the VPNs, ordered on a FIFO basis
Server queue, all tokens waiting for server processing ordered on a FIFO basis

Activities: VPN servicing time, server processing time, and transfer (Token Ring travel) time

Delays: Delay at VPN queue and delay at server queue

Let's assume for the sake of illustration that at the beginning of the simulation there are five tokens in the VPN queue (T2, T3, T4, T5, T6) and one token at the server at time 0, T1. The following activities will be used to generate an event diagram of the simulation, utilizing a future events list (FEL).

VPN Time:	10	5	5	10	15	10	10
Server Time:	12	12	12	16	12	16	
Transfer Time	60	100	40	40	60		

When the end-VPN servicing (EVPN) event occurs, say for token Ti at time t, other events may be triggered. If the server is idle (S(t) = 0), token Ti begins being serviced by the server and an end-process server event (ES) is scheduled on the FEL; otherwise, token Ti joins the server queue. If at the time there is another token waiting for the VPN for servicing, it will be removed from the queue and will begin being serviced by the VPN through the execution of an end-VPN servicing event on the FEL. These activities and events lead to the event diagram in Table 47.8:

TABLE 47.8 Simulated event diagram.

System State					Sets				
Clock	VPNQ(t)	VPN(t)	SQ(t)	S(t)	VPN Queue	Server Queue	FEL	B_{VPN}	B_{Server}
0	3	2	0	1	T4		(EVPN, 5,T3)	0	0
					T5		(EVPN,10,T2)		
					T6		(ES,12,T1)		
5	2	2	1	1	T5	T3	(EVPN,10,T2)	10	5
					T6		(EVPN,5+5,T4)		
							(ES,12,T1)		
10	1	2	2	1	T6	T3	(EVPN,10,T4)	10	10
						T2	(ES,12,T1)		

(Continued)

TABLE 47.8 Continued.

System State					Sets				
Clock	VPNQ(t)	VPN(t)	SQ(t)	S(t)	VPN Queue	Server Queue	FEL	B_{VPN}	B_{Server}
							(EVPN,10+10,T5)		
10	0	2	3	1		T3	(ES,112,T1)	20	10
						T2	(EVPN,20,T5)		
						T4	(EVPN,10+15,T6)		
12	0	2	2	1		T2	(EVPN,20,T5)	24	12
						T4	(ES,12+12,T3)		
							(EVPN,25,T6)		
							(AVPN,12+60,T1)		
20	1	1	3	1		T2	(ES,24,T3)	40	20
						T4	(EVPN,25,T6)		
						T5	(AVPN,72,T1)		
24	0	1	2	1		T4	(EVPN,25,T6)	44	24
						T5	(ES,24+12,T2)		
							(AVPN,72,T1)		
							(AVPN,24+100,T3)		
25	0	0	3	1		T4	(ES,36,T2)	45	25
						T5	(AVPN,72,T1)		
						T6	(AVPN,124,T3)		
36	0	0	2	1		T5	(ES,36+36,T4)		
						T6	(AVPN,72,T1)		
							(AVPN,36+40,T2)		
							(AVPN,124,T3)		
52	0	0	1	1		T6	(ES,52+12)	45	52
							(AVPN,72,T1)		
							(AVPN,76,T2)		
							(AVPN,52+40,T4)		
							(AVPN,124,T3)		
64	0	0	0	1			(AVPN,72,T1)	45	64
							(AVPN,76,T2)		
							(ES,64+16,T6)		
							(AVPN,92,T4)		
							(AVPN,124,T3)		
							(AVPN,64+80,T5)		
72	0	1	0	1			(AVPN,76,T2)	45	72

TABLE 47.8 Continued.

System State					Sets				
Clock	VPNQ(t)	VPN(t)	SQ(t)	S(t)	VPN Queue	Server Queue	FEL	B_{VPN}	B_{Server}
							(ES,80,T6)		
							(EVPN,72+10,T1)		
							(AVPN,92,T4)		
							(AVPN,124,Y3)		
							(AVPN,144,T5)		
76	0	2	0	1			(ES,80,T6)	49	76
							(EVPN,82,T1)		
							(EVPN,76+10,T2)		
							(AVPN,92,T4)		
							(AVPN,124,T3)		
							(AVPN,144,T5)		

As can be seen from the event diagram, statistics can be gathered through intermediate variables during changes in state. In order to estimate the VPN and server utilization, two cumulative statistics have been kept:

$$B_{VPN} = \text{total busy time of both VPNs from time 0 to time t}$$

$$B_{Server} = \text{total busy time of the server from time 0 to time t}$$

Note both VPNs are busy from time 0 to time 20 and only one VPN is busy from 20 to 24. Therefore, B_{VPN} increases by only 4 over the time interval [20, 24]. From time 25 to time 36 both VPNs are idle. Therefore,

$$\text{average VPN utilization} = \frac{49/2}{76} = 0.32$$

$$\text{average server utilization} = 76/76 = 1.00$$

These estimates cannot be regarded as accurate estimates of the long run "steady state" utilizations of the VPNs and server, as a considerably longer simulation would be needed to reduce the effect of the assumed conditions at time 0 and to realize accurate estimates.

47.5 Summary

Chapter 46 illustrated a static model, which is a snapshot of a system at a particular point in time. This chapter demonstrated how to implement the algorithm for a discrete event simulation. This approach, which is a dynamic model, conveyed the essence of changes made in the system over time. This approach is comparable to an accountant's balance sheet. The balance sheet reflects the stock of certain variables in the firm at the end of an accounting period, as compared with the profit and loss or cash flow statements, which itemize movements among categories during the accounting period. A feature of the dynamic model is that it enables the production of static summaries in addition to dynamic flow information.

As an individual enhances their own computing skills they become acquainted with simulation through management gaming and through introduction to operation research methods, also referred to as management science. At the same time technological improvements (such as parallel processing platforms), advances in decision support systems, new language paradigms (object oriented as well as functional), and simulation extensions or modular add-ons for database management, all indicate that simulation has evolved from a research tool of last resort to a methodology of preference. Simulation studies are becoming more cost effective, with acceptable reliability, more end-user friendly, and more understandable.

47.6 Projects

1. Prepare and turn in a literature review of the following article: "Exploring Monte Carlo Simulation Applications for Project Management," by Young Hoon Kwak and Lisa Ingall, available online at *http://home.gwu.edu/~kwak/Monte_Carlo_Kwak_Ingall.pdf*.

2. Prepare and turn in a literature review of the following article: "A Practical Application of Monte Carlo Simulation in Forecasting," by James D. Whiteside II, AACE International Transactions (2008).

3. Your stockbroker suggests you buy 1,000 shares of International Electronics. The stock currently is selling at $38.75 per share. From research on the Web you discover that the stock is quite volatile and that, based on recent past data, in 21% of the weeks it goes up $1.00, 17% of the weeks it goes up $0.50, 10% of the weeks it remains unchanged, 35% of the weeks it goes down $0.50, and 17% of the weeks it goes down $1.00.

    a. Create a worksheet to simulate the price of the stock and the value of your portfolio for 100 weeks assuming the preceding distribution. You should have 100 rows in the main part of your worksheet, one for each week. You should have separate columns for the week number, a random number, the price change that week, the closing price of the stock that week, and the value of your portfolio. Have separate cells at the top of the worksheet for the resulting High, Low, and Close for the stock price over the 100 weeks. (These are the highest price the stock reaches, the lowest price the stock reaches, and the closing price after 100 weeks.) Have a separate cell that shows the amount of money you would have made (or lost) on the investment. Finally, have a cell containing either "I made money at the end of 100 weeks" or "I lost money at the end of 100 weeks" or "I was even at the end of

100 weeks." Run your simulation several times. Based on the history data and your simulations, is this a good investment?

b. Construct a chart with the number of the week along the x-axis and the price of a share of stock along the y-axis. Press F9 repeatedly to see the result of different simulations.

c. On telling your broker your findings, she suggests that you employ the following strategy: Buy the 1,000 shares of stock now and put in a sell order for the stock at \$43.75 or at 100 weeks, whichever comes first. That way, if the stock reaches \$43.75 you will sell out and achieve a profit of \$5 per share. Otherwise, you will sell out at the end of 100 weeks and take whatever profit or loss results. On inquiry, your broker tells you that there will be a commission cost to you of \$79 for purchasing the stock and another \$79 for selling the stock. Create a worksheet like the worksheet in part a for simulating your broker's strategy. Run the simulation 10 times and record the consequences of using this investment strategy.

d. Prepare a brief report on this investment strategy. Include copies of appropriate worksheets and charts. Would you make money or lose money on average using this strategy? How much on average? Should you invest?

4. Persistent Telemarketing, Inc. has offered you a commissioned part-time job. They will give you 500 names and telephone numbers. You are to telephone these people and sell them disposable cell phones on behalf of a well-known telephone company. The customer will receive a cell phone in the mail. The phone has a fixed number of built-in anywhere, anytime minutes. The customer pays in advance and then throws away the cell phone when the minutes run out. There are four different plans that you will be offering with different amounts of minutes allowed under each plan: \$300, \$400, \$600, and \$1,000. Persistent Telemarketing will pay for all of your telephone calls and will give you a 10 percent commission on all of your sales

You speak to a friend who has done this. She tells you that 20% of the time she reaches a person on the first call, 15% of the time she reaches a person on the second call, and 10% of the time she reaches a person on the third call. She said that she never makes more than three calls to a person, so 55% of the time she never reaches a person at a given number.

If she connects to the person, she said that 85% of the time she doesn't make a sale. When she does connect to a person, 8% of the time she sells the \$300 plan, 4% of the time she sells the \$400 plan, 1% of the time she sells the \$600 plan, and 2% of the time she sells the \$1,000 plan.

Assume that to make a call and not reach someone takes 15 seconds. If someone answers and the person does not buy a plan, assume the call takes an average of 45 seconds. If someone answers and the person does purchase a plan, assume it takes an average of 5 minutes to make the sale and obtain and record all of the information.

You would like to estimate how much money you might make if you take this job, how many phone calls you will make, and how much time you will spend. Create a worksheet to simulate this job. You should have a row for each of the 500 people you'll be given. You should have a cell in each row for at least: the prospect number (1 to 500), how many calls it takes to reach the person (1, 2, 3, never), how much you sell to the person ($0, $300, $400, $600, $1,000), how much you make in commission for the person, and how much time you spend on each person.

At the top of the worksheet you should have well-labeled cells with: the total number of calls made, the total number of sales pitches (answered calls) made, the total number of sales made, the total amount of commission made, the total time spent, and the expected dollars per hour you will make.

CHAPTER 48

A RESEARCH STUDY VIGNETTE

OVERVIEW AND LEARNING OBJECTIVES

In This Chapter

48.1 Introduction

This chapter is a short descriptive tour through a hypothetical research study. The purpose of the vignette is to point out tips of advice on practical issues the researcher should practice while conducting research. The vignette presented is resolved through experimentation. Experimentation has broad applications in the computing field.

48.2 The Vignette Setting

A commercial software design and development firm is concerned about how to ensure product quality efficiently to their customer base. As a result the corporate management board has placed several of the software design and development personnel on a team to identify and determine which aspects of attributes prompt meeting delivery deadlines and produce quality software in their industry. Clearly, the answer can be derived from several diverse viewpoints and motives.

Any good software project manager would take attribute measurements of the process and product in order to forecast delivery time and the expected budget. Customers would perform a feature analysis of the product with respect to their requirements and make an assessment of adequate quality. Corporate management would need to determine if upgrades were needed for meeting customer satisfaction.

Is this project noteworthy?

The effectiveness of research demands depend greatly on the proper selection of projects. In the corporate world top level management should play a part in the project selection process due to the fact that the benefits from research projects cross functional and business function boundaries. Therefore, since this vignette is hypothetical, this project has been determined to be noteworthy and already passed the first hurdle in becoming a research study.

Select Research Projects Carefully.

48.3 Preliminary Research Study Statement

Anyone that has participated in a software development project, whether as a student or a professional, knows that screening and reporting relevant data on pertinent attributes and development processes of the software relate to the answer. The study team's past experiences would make them keenly aware that there must be an assessment of participant's progress in any project as well as assessment of the product's quality. Most of the study team would consider the variety of measurement techniques used by a software engineer.

Integrate your experiences and educational background into the research study.

48.4 Background Review

A thorough background review should be immediately initiated: a literature search, both an internet and web search, commercial software development shop visitations, interview/questionnaire surveys of successful/experienced software development project personnel, and any other creative ideas of searching for information. Identification of academic and professional societies pertinent to the study would be necessary. Also identification of key individuals or experts within the field of the study would be required.

The participants in the background study need answers for:

What are the attributes of concern?

What is the impact of the attributes of concern on software development?

Who are some of the nationally known experts in the domain of software development measurement?

Which professional / academic societies are pertinent to the study?

Are there any available RFPs applicable to the study?

Each team member should list at least 6 researchable questions for future research projects in the study domain.

48.5 Formulating a Project that can be Resolved

For the sake of illustration, let's assume that the study team has settled on using software product metrics and software development process metrics. The team has reached agreement that two major merits for product metrics are that product quality is measurable and product attribute characteristics are quantifiable.

Goodness criteria for the product is a very controllable and well-defined domain, especially in software engineering. However, the process metrics area is overwhelming and as a result, the research team would be aware that it is absolutely necessary to complete both a Joint Application Development (JAD) session and pilot testing of candidate attributes and potential analysis models.

JAD is a technique for including users in the development of a new system. The JAD sessions are generally conducted in specially configured rooms, with the final outcome being general agreement between users and system professionals as to general information requirements of the business. JAD sessions are extremely appropriate during system planning and system analysis phases of the software development life cycle (SDLC). This would be a perfect technique to use in the study since the members of the study team would be so familiar with it. JAD sessions would enable the study team, corporate management, customer representatives, and other stakeholders to reach a consensus on attributes, metrics, and preliminary analysis methods/models to be included in the study. A brief list of attributes would probably contain: productivity cost and resource requirements, effectiveness of quality, effectiveness of development tools, and many others.

To complicate matters more is the fact that current software metrics are specific to a given phase of the SDLC. For example the lines of code (LOC) can be used to measure the amount of work done and predict size, but not before the project implementation phase begins. LOC is commonly used to measure productivity. The study team could even consider using LOC to measure quality. Detection of defaults is relevant only during the specification phase.

By this time the study team would realize that the project could get out of hand; downsizing to make the study realistic as well as ensure that precise and accurate results be generated might be required.

> *Always remember, be patient and strive for simplicity.*
>
> *Be careful that the real problem is uncovered and solved rather than dealing with symptoms of the actual problem.*

Do not be a Robinson Crusoe, the world's first system analyst. At one point of his island story, he set out to build a dugout canoe. He located a fantastic tree in the exact center of the island. A fabulous canoe was manufactured by the obsessive Crusoe, but to no avail. It was literally impossible to transport the final product to water. Another failed project! Researchers must realize that they do not need to construct an eight lane highway for someone's driveway at their private home.

48.6 Attribute Screening

Guidelines and processes that will alleviate the attribute screening could require that:

1. Attribute Measurement is dependent upon the SDLC phases.
2. Scope and cost are among the most pertinent attributes, especially for measuring aspects of product quality.
3. Should internal attributes, which are only directly measurable, or external attributes be selected for the study?

At this juncture, it would be apparent that a pilot study should be implemented to finalize the attribute list and investigate various models for the data analysis. The pilot study charge should be to find out the answers to these questions for any available and accessible data.

Is the data meaningful?

What is the origin of the data?

Is the data complete?

Is the data understandable?

Is the data pertinent to the research?

Does the data meet the requirements for the proposed preliminary analysis methods?

During these past components of the study, team members should not let themselves be sidetracked! During screening, team members may have asked: 'Is there a way to characterize metrics that may be directly related to quality, productivity, and technology?' This would be a great research project.

Several team members might have asked: 'Should the firm invest in implementing efficient fault tolerant software?' Fault tolerance is a software/hardware technique where intervention and prevention for system failures is employed. For example, defensive programming based on redundancy can be used to cope with mistakes in the real environment. All software faults result from design errors. Therefore, to implement fault tolerant computing for coping with design faults, programs need to be tested repeatedly during run time and designed to dynamically recover from errors. The fundamental method for implementing software fault tolerance uses redundancy or alterations.

Consider the typical sort processing example. Assume that the software system allocates one controller block, one recording block, and three sorting blocks to complete the sort. Each request cycle begins with the controller block providing a copy of the original data file to the first sorting block to sort. After this sorting block finishes sorting, a request is sent from the controller block to the recording block to check whether the result is correct based upon pre-established specifications. If this sorting block fails to produce the specified result, the controller block will send another copy of the original data file to a different sorting block, then begins execution. This process is repeated until the data is correctly sorted or no more sorting blocks are available. In the latter case, the system fails.

Care must be taken; clearly, such activity by the study team indicates the team would have become side-tracked.

> *Be patient!*
>
> *Stay focused!*
>
> *In the solution of a research study, new research projects are discovered.*

Do not be discouraged, simply file these questions in your conceptual framework. These research questions could become some of your future research studies. Congratulations would be in order for the study team members!

> *Continuously build and enhance your conceptual domain.*
>
> *Remember the three Ps.*
>
> *Plan, prepare and then produce.*

After completion of a well-defined pilot testing, attribute testing, and successful JAD sessions the study team probably should have completed 90% of the overall study.

48.7 Study Design

Consensus would have been reached obtained in the study by this time. Both product and process metrics could have been chosen for inclusion in the study. Product metrics measure things like lines of code, whereas process metrics measure the efficiency of fault detection during development. The project could have been directed to concentrate on scope and elapsed time.

Each type of metric measurement needing to be measured would be defined either using internal or external attributes. Internal attributes are measured in terms of the entity itself. External attributes represent a measure of the relationship to some external requirement defined by the environment in which the entity is produced or used. For instances, software code maintainability represents the ability of a specific product entity to adhere to accommodating changes in a reasonable manner. Another example is being able to modify the code in one module without having to understand the code in another module.

It is likely that the study team would have decided that the study concentrate on attributes that can be measured directly by examining the process, the product, and the resources applied. Indirect measures can be obtained by establishing an empirical relationship between the measure desired and other countable attributes of the entity. Counting man hours devoted to a task is a direct measurement for the procedural design task. A module complexity attribute can be determined indirectly through measuring other countable characteristics of the software design logic such as the number of simple branch points in the algorithm. Then complexity might simply be a function of the number of simple branches found in the code.

Process improvement through defect measurement and analyzes is facilitated by the nature of the data collected. A modification of the Software Engineering Institute's (SEI) Process Maturity Model can be utilized. The usage of a nationally recognized standard lends validity and reliability to the research study, especially internal validity and external validity.

Metrics can be chosen for each phase of the SDLC: specification, design, implementation and integration, and maintenance. Time spent on specification development, size, effort, and number of pages in the specification document could be used for the specification phase. The number of modules as the size of the product, module cohesion and coupling as a measure of the quality of the design, and fault statistics could be used for the design phase. Total number of test cases which result in a failure as well as other accurately kept fault statistics would probably be utilized in the implementation phase. The metrics used for measurement in all the other phases are pertinent to the metric measurements needed for the maintenance phase.

The discussion on screening for attributes implies that the research study would not have a design but a collection of designs, indicated by the phases of the SDLC. In fact, the implementation phase might help the study team in integrating these phases. Now, the team should consider the screening process for selecting a good design.

What precisely is a research design? First it is a *plan*, *structure*, and *strategy* of the investigation to be conducted. Planning is actually the overall scheme or programmed activities of the research. Such an program of activities includes everything that must be done by the researcher from writing the hypothesis(es) to the analysis(es) of the data. The outline, scheme, and modeling of the operation of the variables specify the structure of the design. Strategy is composed of the methods for data gathering and analyzing the data. *How* research objectives are reached and how to tackle the research problems is the purpose of the strategy.

What is the purpose of a research design? Primarily the purpose is to obtain answers to the research questions. Second, the research is to control variance. The major emphasis in research is to get answers to the questions while controlling for experimental, extraneous, and error variances for the research studied.

A research design enables the researcher to answer research questions: with validity, objectively, accurately, and with economic control. Design does not inform the researcher on what to do, rather the design infers the direction of observation-making and analysis. Acceptable designs must address:

How many observations should be made?

Which variables are active?

Which variables are assigned?

Researchers need to be able to manipulate the active variables and categorize the assigned variables. Additionally, the design tells the researcher what type of statistical analysis, and/or other methodologies, to use. More importantly, an adequate design will outline possible conclusions to be derived from the statistical design.

Note that in the computing field, projects are not just limited to statistical analysis but can involve measurement, simulation, and analytic modeling. Often the emphasis of the research may be dependent upon techniques of performance evaluation, workload characterization, and benchmarking. Many such studies implement queuing network-based simulation models of computer systems. However, stochastic analysis can be used instead of using simulation models of computer systems. Using these models for various computing platforms, the research study goals, probably resolved with an experimental design, could be:

1. to compare a number of systems and decide which system is best for a given set of applications,
2. to determine for a specific system how well it is performing and whether any improvements need to be made.

The Design choice influences which method of analysis are needed.

For each one of the SDLC phases, should the design approached be experimental or nonexperimental? If an independent variable can be manipulated, then an experimental approach can and should be used. The variables in this study are numerous and complex, which indicates to the study team that both experimental and nonexperimental designs are required.

Every adequate design must answer:

Does the design answer the research questions?

Will the design enable testing the hypotheses?

Researchers must remember to solve problems, not just apply techniques.

The main objective of any study is to resolve the answer to the question not just apply techniques. The method or technique is secondary.

Design research to answer the research questions.

Does the design adequately control independent variables?

Randomize whenever possible.

Control independent variables [which limits systematic variance from extraneous and unwanted sources to have presence in the study].

Remember that a perfectly good design can and is often poorly used.

A good design must possess both internal validity and external validity.

If experimental design is required, then other potential questions are:

Should factors be included, if so what are the pertinent factors?

How many levels of each factor should be included?

How should the levels of the factors be spaced?

How should the experimental units be selected?

How many experimental units should be chosen for each treatment combination?

What steps should be taken to control experimental error?

What criterion measures should be used to evaluate the effects of the treatment factors?

What if a modeling approach had been decided upon? Before undertaking more elaborate testing, a good approach is to begin looking for obvious errors or oversights in the model.

When screening and evaluating candidate models, the researcher can:

Determine whether or not the current candidate model reveals any such mistakes as based upon the problem statement.

Determine if all mathematical expressions are consistent with the units the model uses.

By varying the input parameters and/or the predictor variables, decide whether or not the output from the model behaves in a acceptable manner.

Perform retrospective tests on the model to check if good results are produced.

Whatever methodology is used, researchers must address the validity issue. Internal validity asks the question: Does the experimental manipulation really make a statistically significant difference? If the study results and design are representative, or generalizable, then the study possesses external validity.

To what populations can the results be generalized?

Are the variables of this research representative?

48.8 Reporting Results

The team has planned, and then re-planned many times. Additionally the team has prepared. At this point it would be time for DOING: implement the plan, perform the analysis, and convey the results. The study will have: finished data validation, possibly created new variables, selected variables for the final model, removed outliers, chosen the "best" model or models, and introduced knowing how to process categorical variables. Most of these tasks required many decisions to be made for which no clear cut rules existed.

Decisions have an impact on research results and must be justified.

Research must be shared [discussed] with interested communities.

When reporting results, the study team must be sure the reader understands the results.

Tell them what you are going to tell them, tell them again, and then tell them what you told them once more.

Present the most important results first, then present decreasingly important results. For each set of results, remind the reader what research question the result is answering. After all research questions have been presented, present any qualifications or elaborations using a similar order.

Each finding should be presented first in prose, second in numbers, and then again in prose. In the main results section of the teams final report, limited interpretation of results is permissible, it prepares the reader for the conclusion section.

Conclusions should logically follow from results.

The first paragraph of the discussion section must include a restatement of the research question. A discussion should also include the teams efforts to relate the results to previous research on the topic.

Keep the discussion section as brief as possible.

Toward the end of the discussion section the team should present what the team believed to be some questions that remain unanswered despite the efforts made by the research study team. It could be that questions are raised because of the results obtained by the study.

The discussion section should not contain any surprises for the reader.

Don't allow detours.

48.9 Promoting Research Results

This section contains the main points of the complete report but none of the qualifications and elaborations necessary to fully understand those main points. This section is a teaser. Hopefully the reader will use the executive summary to decide whether or not to read and/or study the full report.

Make the executive summary informative and brief.

How? Rewrite, then rewrite, and finally you may have to rewrite again.

48.10 Vignette Closing

As a runner it surprises people when they inform others that being fit or to maintain good health are not why they are compulsive about running. Nor is it the acquisition of trophies and competition that is important. Running is something they enjoy, something that they continuously and perpetually practice on a consistent basis. Being a runner is part of one's identity.

Suppose the study is completed. The reader has just witnessed and participated in extracting knowledge from data. The reader should also have transferred the new knowledge to other groups of individuals. Continue the journey by continuously performing a background review, searching the Web, interacting with other researches in the study domain, and joining a professional society and attending their conferences. Hopefully the reader will consider not only enhancing their conceptual framework, but let research be part of their identity, much as the runner claims identity with the "running lifestyle!"

48.11 Chapter Summary

Research in the computing field evolved from ad hoc research to experimental research. 'Ad hoc' research suffers from problems such as invalidity and irreproducibility. Use of experimental designs, whenever possible, avoid these common pitfalls.

The breadth of the computing field includes: human factors research, operating systems, computer systems organization, and data base management systems is only a short list. This breadth of the field implies that researchers need to place an emphasis on measurement, simulation, and analytical modeling as well as experimental research.

Additionally the reader has probably noticed the emphasis on project management in the vignette. Technical projects are also business projects; they involve business issues such as customer satisfaction, resource utilization, cost, profits, and many other business factors. In the vignette the systems development cycle assumed as a framework for both the study team and the project team. This framework can serve as a general scheme for illustrating commonalities and differences among projects. Often individuals preparing for research are either a technical person with little or no education in management or they are a person primarily involved in product-development, marketing, process improvement, or related projects, with limited background in technology. These individuals need to embark on a study of project management.

Participants in research projects need to develop an appreciation for both the project management and the human and organizational side as well. Successful projects are those where good communications, mutual trusts, and teamwork are present.

48.12 Exercises

Various types of research questions are represented in the following list.

a. For each question you are to determine if the actual computing environment, use of simulation, or an analytic model should be used in the study.
b. Using this model describe how to resolve the research question based upon your answer to Part A. The following list of activities may provide guidance in answering:

- design sets of experiments for measurement or simulation.
- check if the model is adequate.
- make comparative performance analyzes for the alternatives.
- control for measurement errors.

1. A performance evaluation for a personal workstation. To design this workstation answer: Which processor should be chosen? How much memory should be recommended? How many external CD drives should be recommended? Assume the workload can be of three types: secretarial, management, or scientific. Also assume the end-user is either novice or expert. This is a system design problem.
2. Determine the best database management system for a social work office information referral system for seven counties for social services for senior citizens. Consideration should be given to cost, availability, and compatibility. (Hint: Look into required workloads and benchmarks). This is a system selection problem.
3. Find for your organization about the current computing system. Consider a system upgrade study. Either replacing the whole system or parts of the current system with newer but compatible units. Assume the compatibility considerations are dictated by the vendor. You are to recommend quantity, speed, and other pertinent attribute values.
4. For the DBMS application in problem (2) using the required workload using appropriate attributes and the social work office resource management policies, optimize the performance of the query and data manipulation processing. This is a system tuning problem.

5. Your office workstation is unacceptably sluggish. Develop a method to determine whether inadequate resources of poor system management is the reason. You can monitor the system and examine the behavior of various resource management policies under different loading conditions. This is a system analysis problem.

6. Your company is considering the expansion of an existing computing facility. Discuss how measurements under the real workload can be utilized to solve this problem. Consider the case where direct measurements are possible and when only indirect measurements are available. Note that *workload characterization only builds a model of the real workload.*

7. Compare the effect of ORACLE, ACCESS, and SYSBASE on user performance. This is a relational data modeling problem.

8. Determine how data technicians organize data, and if they are capable of organizing data into lists, hierarchies, tables, and networks. This is a data model problem.

9. Explore the difference in novice and expert users' performance in searching, storing, and retrieving information from a database. This is a human factors study.

10. Test the assertion that many of the existing optimizing cluster techniques perform as well or better than JAN in clustering the policies of a group of raters or judges. Hint: use a Monte Carlo simulation. Number of policies, raters, and number of raters per policy should be varied and analyzed with an appropriate experimental design.

APPENDIX A

STATISTICAL TABLES

TABLE A-1 The Normal Distribution.

Areas Under the Normal Curve

Z	Cum p	Tail p	Z	Cum p	Tail p	Z	Cum p	Tail p	Z	Cum p	Tail p	Z	Cum p	Tail p
0.00	0.5000	0.5000	0.40	0.6554	0.3446	0.80	0.7881	0.2119	1.20	0.8849	0.1151	1.60	0.9452	0.0548
0.01	0.5040	0.4960	0.41	0.6591	0.3409	0.81	0.7910	0.2090	1.21	0.8869	0.1131	1.61	0.9463	0.0537
0.02	0.5080	0.4920	0.42	0.6628	0.3372	0.82	0.7939	0.2061	1.22	0.8888	0.1112	1.62	0.9474	0.0526
0.03	0.5120	0.4880	0.43	0.6664	0.3336	0.83	0.7967	0.2033	1.23	0.8907	0.1093	1.63	0.9484	0.0516
0.04	0.5160	0.4840	0.44	0.6700	0.3300	0.84	0.7995	0.2005	1.24	0.8925	0.1075	1.64	0.9495	0.0505
0.05	0.5199	0.4801	0.45	0.6736	0.3264	0.85	0.8023	0.1977	1.25	0.8944	0.1056	**1.65**	**0.9505**	**0.0495**
0.06	0.5239	0.4761	0.46	0.6772	0.3228	0.86	0.8051	0.1949	1.26	0.8962	0.1038	1.66	0.9515	0.0485
0.07	0.5279	0.4721	0.47	0.6808	0.3192	0.87	0.8078	0.1922	1.27	0.8980	0.1020	1.67	0.9525	0.0475
0.08	0.5319	0.4681	0.48	0.6844	0.3156	0.88	0.8106	0.1894	1.28	0.8997	0.1003	1.68	0.9535	0.0465
0.09	0.5359	0.4641	0.49	0.6879	0.3121	0.89	0.8133	0.1867	1.29	0.9015	0.0985	1.69	0.9545	0.0455
0.10	0.5398	0.4602	0.50	0.6915	0.3085	0.90	0.8159	0.1841	1.30	0.9032	0.0968	1.70	0.9554	0.0446
0.11	0.5438	0.4562	0.51	0.6950	0.3050	0.91	0.8186	0.1814	1.31	0.9049	0.0951	1.71	0.9564	0.0436
0.12	0.5478	0.4522	0.52	0.6985	0.3015	0.92	0.8212	0.1788	1.32	0.9066	0.0934	1.72	0.9573	0.0427
0.13	0.5517	0.4483	0.53	0.7019	0.2981	0.93	0.8238	0.1762	1.33	0.9082	0.0918	1.73	0.9582	0.0418
0.14	0.5557	0.4443	0.54	0.7054	0.2946	0.94	0.8264	0.1736	1.34	0.9099	0.0901	1.74	0.9591	0.0409
0.15	0.5596	0.4404	0.55	0.7088	0.2912	0.95	0.8289	0.1711	1.35	0.9115	0.0885	1.75	0.9599	0.0401
0.16	0.5636	0.4364	0.56	0.7123	0.2877	0.96	0.8315	0.1685	1.36	0.9131	0.0869	1.76	0.9608	0.0392
0.17	0.5675	0.4325	0.57	0.7157	0.2843	0.97	0.8340	0.1660	1.37	0.9147	0.0853	1.77	0.9616	0.0384
0.18	0.5714	0.4286	0.58	0.7190	0.2810	0.98	0.8365	0.1635	1.38	0.9162	0.0838	1.78	0.9625	0.0375
0.19	0.5753	0.4247	0.59	0.7224	0.2776	0.99	0.8389	0.1611	1.39	0.9177	0.0823	1.79	0.9633	0.0367
0.20	0.5793	0.4207	0.60	0.7257	0.2743	1.00	0.8413	0.1587	1.40	0.9192	0.0808	1.80	0.9641	0.0359
0.21	0.5832	0.4168	0.61	0.7291	0.2709	1.01	0.8438	0.1562	1.41	0.9207	0.0793	1.81	0.9649	0.0351
0.22	0.5871	0.4129	0.62	0.7324	0.2676	1.02	0.8461	0.1539	1.42	0.9222	0.0778	1.82	0.9656	0.0344
0.23	0.5910	0.4090	0.63	0.7357	0.2643	1.03	0.8485	0.1515	1.43	0.9236	0.0764	1.83	0.9664	0.0336
0.24	0.5948	0.4052	0.64	0.7389	0.2611	1.04	0.8508	0.1492	1.44	0.9251	0.0749	1.84	0.9671	0.0329
0.25	0.5987	0.4013	0.65	0.7422	0.2578	1.05	0.8531	0.1469	1.45	0.9265	0.0735	1.85	0.9678	0.0322
0.26	0.6026	0.3974	0.66	0.7454	0.2546	1.06	0.8554	0.1446	1.46	0.9279	0.0721	1.86	0.9686	0.0314
0.27	0.6064	0.3936	0.67	0.7486	0.2514	1.07	0.8577	0.1423	1.47	0.9292	0.0708	1.87	0.9693	0.0307
0.28	0.6103	0.3897	0.68	0.7517	0.2483	1.08	0.8599	0.1401	1.48	0.9306	0.0694	1.88	0.9699	0.0301
0.29	0.6141	0.3859	0.69	0.7549	0.2451	1.09	0.8621	0.1379	1.49	0.9319	0.0681	1.89	0.9706	0.0294
0.30	0.6179	0.3821	0.70	0.7580	0.2420	1.10	0.8643	0.1357	1.50	0.9332	0.0668	1.90	0.9713	0.0287
0.31	0.6217	0.3783	0.71	0.7611	0.2389	1.11	0.8665	0.1335	1.51	0.9345	0.0655	1.91	0.9719	0.0281
0.32	0.6255	0.3745	0.72	0.7642	0.2358	1.12	0.8686	0.1314	1.52	0.9357	0.0643	1.92	0.9726	0.0274
0.33	0.6293	0.3707	0.73	0.7673	0.2327	1.13	0.8708	0.1292	1.53	0.9370	0.0630	1.93	0.9732	0.0268
0.34	0.6331	0.3669	0.74	0.7704	0.2296	1.14	0.8729	0.1271	1.54	0.9382	0.0618	1.94	0.9738	0.0262
0.35	0.6368	0.3632	0.75	0.7734	0.2266	1.15	0.8749	0.1251	1.55	0.9394	0.0606	1.95	0.9744	0.0256
0.36	0.6406	0.3594	0.76	0.7764	0.2236	1.16	0.8770	0.1230	1.56	0.9406	0.0594	**1.96**	**0.9750**	**0.0250**
0.37	0.6443	0.3557	0.77	0.7794	0.2206	1.17	0.8790	0.1210	1.57	0.9418	0.0582	1.97	0.9756	0.0244
0.38	0.6480	0.3520	0.78	0.7823	0.2177	1.18	0.8810	0.1190	1.58	0.9429	0.0571	1.98	0.9761	0.0239
0.39	0.6517	0.3483	0.79	0.7852	0.2148	1.19	0.8830	0.1170	1.59	0.9441	0.0559	1.99	0.9767	0.0233

Z	Cum p	Tail p
2.00	0.9772	0.0228
2.01	0.9778	0.0222
2.02	0.9783	0.0217
2.03	0.9788	0.0212
2.04	0.9793	0.0207
2.05	0.9798	0.0202
2.06	0.9803	0.0197
2.07	0.9808	0.0192
2.08	0.9812	0.0188
2.09	0.9817	0.0183
2.10	0.9821	0.0179
2.11	0.9826	0.0174
2.12	0.9830	0.0170
2.13	0.9834	0.0166
2.14	0.9838	0.0162
2.15	0.9842	0.0158
2.16	0.9846	0.0154
2.17	0.9850	0.0150
2.18	0.9854	0.0146
2.19	0.9857	0.0143
2.20	0.9861	0.0139
2.21	0.9864	0.0136
2.22	0.9868	0.0132
2.23	0.9871	0.0129
2.24	0.9875	0.0125
2.25	0.9878	0.0122
2.26	0.9881	0.0119
2.27	0.9884	0.0116
2.28	0.9887	0.0113
2.29	0.9890	0.0110
2.30	0.9893	0.0107
2.31	0.9896	0.0104
2.32	0.9898	0.0102
2.33	**0.9901**	**0.0099**
2.34	0.9904	0.0096
2.35	0.9906	0.0094
2.36	0.9909	0.0091
2.37	0.9911	0.0089
2.38	0.9913	0.0087
2.39	0.9916	0.0084

Z	Cum p	Tail p
2.40	0.9918	0.0082
2.41	0.9920	0.0080
2.42	0.9922	0.0078
2.43	0.9925	0.0075
2.44	0.9927	0.0073
2.45	0.9929	0.0071
2.46	0.9931	0.0069
2.47	0.9932	0.0068
2.48	0.9934	0.0066
2.49	0.9936	0.0064
2.50	0.9938	0.0062
2.51	0.9940	0.0060
2.52	0.9941	0.0059
2.53	0.9943	0.0057
2.54	0.9945	0.0055
2.55	0.9946	0.0054
2.56	0.9948	0.0052
2.57	0.9949	0.0051
2.58	**0.9951**	**0.0049**
2.59	0.9952	0.0048
2.60	0.9953	0.0047
2.61	0.9955	0.0045
2.62	0.9956	0.0044
2.63	0.9957	0.0043
2.64	0.9959	0.0041
2.65	0.9960	0.0040
2.66	0.9961	0.0039
2.67	0.9962	0.0038
2.68	0.9963	0.0037
2.69	0.9964	0.0036
2.70	0.9965	0.0035
2.71	0.9966	0.0034
2.72	0.9967	0.0033
2.73	0.9968	0.0032
2.74	0.9969	0.0031
2.75	0.9970	0.0030
2.76	0.9971	0.0029
2.77	0.9972	0.0028
2.78	0.9973	0.0027
2.79	0.9974	0.0026

Z	Cum p	Tail p
2.80	0.9974	0.0026
2.81	0.9975	0.0025
2.82	0.9976	0.0024
2.83	0.9977	0.0023
2.84	0.9977	0.0023
2.85	0.9978	0.0022
2.86	0.9979	0.0021
2.87	0.9979	0.0021
2.88	0.9980	0.0020
2.89	0.9981	0.0019
2.90	0.9981	0.0019
2.91	0.9982	0.0018
2.92	0.9982	0.0018
2.93	0.9983	0.0017
2.94	0.9984	0.0016
2.95	0.9984	0.0016
2.96	0.9985	0.0015
2.97	0.9985	0.0015
2.98	0.9986	0.0014
2.99	0.9986	0.0014
3.00	0.9987	0.0013
3.01	0.9987	0.0013
3.02	0.9987	0.0013
3.03	0.9988	0.0012
3.04	0.9988	0.0012
3.05	0.9989	0.0011
3.06	0.9989	0.0011
3.07	0.9989	0.0011
3.08	0.9990	0.0010
3.09	0.9990	0.0010
3.10	0.9990	0.0010
3.11	0.9991	0.0009
3.12	0.9991	0.0009
3.13	0.9991	0.0009
3.14	0.9992	0.0008
3.15	0.9992	0.0008
3.16	0.9992	0.0008
3.17	0.9992	0.0008
3.18	0.9993	0.0007
3.19	0.9993	0.0007

Z	Cum p	Tail p
3.20	0.9993	0.0007
3.21	0.9993	0.0007
3.22	0.9994	0.0006
3.23	0.9994	0.0006
3.24	0.9994	0.0006
3.25	0.9994	0.0006
3.26	0.9994	0.0006
3.27	0.9995	0.0005
3.28	0.9995	0.0005
3.29	0.9995	0.0005
3.30	0.9995	0.0005
3.31	0.9995	0.0005
3.32	0.9995	0.0005
3.33	0.9996	0.0004
3.34	0.9996	0.0004
3.35	0.9996	0.0004
3.36	0.9996	0.0004
3.37	0.9996	0.0004
3.38	0.9996	0.0004
3.39	0.9997	0.0003
3.40	0.9997	0.0003
3.41	0.9997	0.0003
3.42	0.9997	0.0003
3.43	0.9997	0.0003
3.44	0.9997	0.0003
3.45	0.9997	0.0003
3.46	0.9997	0.0003
3.47	0.9997	0.0003
3.48	0.9997	0.0003
3.49	0.9998	0.0002
3.50	0.9998	0.0002
3.60	0.9998	0.0002
3.70	0.9999	0.0001
3.80	0.9999	0.0001
3.90	1.0000	0.0000

TABLE A-2 Binomial Probabilities.

N	r	p									
		0.05	0.10	0.15	0.20	0.25	0.30	0.35	0.40	0.45	0.50
1	0	0.9500	0.9000	0.8500	0.8000	0.7500	0.7000	0.6500	0.6000	0.5500	0.5000
	1	0.0500	0.1000	0.1500	0.2000	0.2500	0.3000	0.3500	0.4000	0.4500	0.5000
2	0	0.9025	0.8100	0.7225	0.6400	0.5625	0.4900	0.4225	0.3600	0.3025	0.2500
	1	0.0950	0.1800	0.2550	0.3200	0.3750	0.4200	0.4550	0.4800	0.4950	0.5000
	2	0.0025	0.0100	0.0225	0.0400	0.0625	0.0900	0.1225	0.1600	0.2025	0.2500
3	0	0.8574	0.7290	0.6141	0.5120	0.4219	0.3430	0.2746	0.2160	0.1664	0.1250
	1	0.1354	0.2430	0.3251	0.3840	0.4219	0.4410	0.4436	0.4320	0.4084	0.3750
	2	0.0071	0.0270	0.0574	0.0960	0.1406	0.1890	0.2389	0.2880	0.3341	0.3750
	3	0.0001	0.0010	0.0034	0.0080	0.0156	0.0270	0.0429	0.0640	0.0911	0.1250
4	0	0.8145	0.6561	0.5220	0.4096	0.3164	0.2401	0.1785	0.1296	0.0915	0.0625
	1	0.1715	0.2916	0.3685	0.4096	0.4219	0.4116	0.3845	0.3456	0.2995	0.2500
	2	0.0135	0.0486	0.0975	0.1536	0.2109	0.2646	0.3105	0.3456	0.3675	0.3750
	3	0.0005	0.0036	0.0115	0.0256	0.0469	0.0756	0.1115	0.1536	0.2005	0.2500
	4	0.0000	0.0001	0.0005	0.0016	0.0039	0.0081	0.0150	0.0256	0.0410	0.0625
5	0	0.7738	0.5905	0.4437	0.3277	0.2373	0.1681	0.1160	0.0778	0.0503	0.0313
	1	0.2036	0.3281	0.3915	0.4096	0.3955	0.3602	0.3124	0.2592	0.2059	0.1563
	2	0.0214	0.0729	0.1382	0.2048	0.2637	0.3087	0.3364	0.3456	0.3369	0.3125
	3	0.0011	0.0081	0.0244	0.0512	0.0879	0.1323	0.1811	0.2304	0.2757	0.3125
	4	0.0000	0.0005	0.0022	0.0064	0.0146	0.0284	0.0488	0.0768	0.1128	0.1563
	5	0.0000	0.0000	0.0001	0.0003	0.0010	0.0024	0.0053	0.0102	0.0185	0.0313
6	0	0.7351	0.5314	0.3771	0.2621	0.1780	0.1176	0.0754	0.0467	0.0277	0.0156
	1	0.2321	0.3543	0.3993	0.3932	0.3560	0.3025	0.2437	0.1866	0.1359	0.0938
	2	0.0305	0.0984	0.1762	0.2458	0.2966	0.3241	0.3280	0.3110	0.2780	0.2344
	3	0.0021	0.0146	0.0415	0.0819	0.1318	0.1852	0.2355	0.2765	0.3032	0.3125
	4	0.0001	0.0012	0.0055	0.0154	0.0330	0.0595	0.0951	0.1382	0.1861	0.2344
	5	0.0000	0.0001	0.0004	0.0015	0.0044	0.0102	0.0205	0.0369	0.0609	0.0938
	6	0.0000	0.0000	0.0000	0.0001	0.0002	0.0007	0.0018	0.0041	0.0083	0.0156
7	0	0.6983	0.4783	0.3206	0.2097	0.1335	0.0824	0.0490	0.0280	0.0152	0.0078
	1	0.2573	0.3720	0.3960	0.3670	0.3115	0.2471	0.1848	0.1306	0.0872	0.0547
	2	0.0406	0.1240	0.2097	0.2753	0.3115	0.3177	0.2985	0.2613	0.2140	0.1641
	3	0.0036	0.0230	0.0617	0.1147	0.1730	0.2269	0.2679	0.2903	0.2918	0.2734
	4	0.0002	0.0026	0.0109	0.0287	0.0577	0.0972	0.1442	0.1935	0.2388	0.2734
	5	0.0000	0.0002	0.0012	0.0043	0.0115	0.0250	0.0466	0.0774	0.1172	0.1641
	6	0.0000	0.0000	0.0001	0.0004	0.0013	0.0036	0.0084	0.0172	0.0320	0.0547
	7	0.0000	0.0000	0.0000	0.0000	0.0001	0.0002	0.0006	0.0016	0.0037	0.0078
8	0	0.6634	0.4305	0.2725	0.1678	0.1001	0.0576	0.0319	0.0168	0.0084	0.0039
	1	0.2793	0.3826	0.3847	0.3355	0.2670	0.1977	0.1373	0.0896	0.0548	0.0313
	2	0.0515	0.1488	0.2376	0.2936	0.3115	0.2965	0.2587	0.2090	0.1569	0.1094
	3	0.0054	0.0331	0.0839	0.1468	0.2076	0.2541	0.2786	0.2787	0.2568	0.2188
	4	0.0004	0.0046	0.0185	0.0459	0.0865	0.1361	0.1875	0.2322	0.2627	0.2734
	5	0.0000	0.0004	0.0026	0.0092	0.0231	0.0467	0.0808	0.1239	0.1719	0.2188

N	r	p									
		0.05	0.10	0.15	0.20	0.25	0.30	0.35	0.40	0.45	0.50
8	6	0.0000	0.0000	0.0002	0.0011	0.0038	0.0100	0.0217	0.0413	0.0703	0.1094
	7	0.0000	0.0000	0.0000	0.0001	0.0004	0.0012	0.0033	0.0079	0.0164	0.0313
	8	0.0000	0.0000	0.0000	0.0000	0.0000	0.0001	0.0002	0.0007	0.0017	0.0039
9	0	0.6302	0.3874	0.2316	0.1342	0.0751	0.0404	0.0207	0.0101	0.0046	0.0020
	1	0.2985	0.3874	0.3679	0.3020	0.2253	0.1556	0.1004	0.0605	0.0339	0.0176
	2	0.0629	0.1722	0.2597	0.3020	0.3003	0.2668	0.2162	0.1612	0.1110	0.0703
	3	0.0077	0.0446	0.1069	0.1762	0.2336	0.2668	0.2716	0.2508	0.2119	0.1641
	4	0.0006	0.0074	0.0283	0.0661	0.1168	0.1715	0.2194	0.2508	0.2600	0.2461
	5	0.0000	0.0008	0.0050	0.0165	0.0389	0.0735	0.1181	0.1672	0.2128	0.2461
	6	0.0000	0.0001	0.0006	0.0028	0.0087	0.0210	0.0424	0.0743	0.1160	0.1641
	7	0.0000	0.0000	0.0000	0.0003	0.0012	0.0039	0.0098	0.0212	0.0407	0.0703
	8	0.0000	0.0000	0.0000	0.0000	0.0001	0.0004	0.0013	0.0035	0.0083	0.0176
	9	0.0000	0.0000	0.0000	0.0000	0.0000	0.0000	0.0001	0.0003	0.0008	0.0020
10	0	0.5987	0.3487	0.1969	0.1074	0.0563	0.0282	0.0135	0.0060	0.0025	0.0010
	1	0.3151	0.3874	0.3474	0.2684	0.1877	0.1211	0.0725	0.0403	0.0207	0.0098
	2	0.0746	0.1937	0.2759	0.3020	0.2816	0.2335	0.1757	0.1209	0.0763	0.0439
	3	0.0105	0.0574	0.1298	0.2013	0.2503	0.2668	0.2522	0.2150	0.1665	0.1172
	4	0.0010	0.0112	0.0401	0.0881	0.1460	0.2001	0.2377	0.2508	0.2384	0.2051
	5	0.0001	0.0015	0.0085	0.0264	0.0584	0.1029	0.1536	0.2007	0.2340	0.2461
	6	0.0000	0.0001	0.0012	0.0055	0.0162	0.0368	0.0689	0.1115	0.1596	0.2051
	7	0.0000	0.0000	0.0001	0.0008	0.0031	0.0090	0.0212	0.0425	0.0746	0.1172
	8	0.0000	0.0000	0.0000	0.0001	0.0004	0.0014	0.0043	0.0106	0.0229	0.0439
	9	0.0000	0.0000	0.0000	0.0000	0.0000	0.0001	0.0005	0.0016	0.0042	0.0098
	10	0.0000	0.0000	0.0000	0.0000	0.0000	0.0000	0.0000	0.0001	0.0003	0.0010
11	0	0.5688	0.3138	0.1673	0.0859	0.0422	0.0198	0.0088	0.0036	0.0014	0.0005
	1	0.3293	0.3835	0.3248	0.2362	0.1549	0.0932	0.0518	0.0266	0.0125	0.0054
	2	0.0867	0.2131	0.2866	0.2953	0.2581	0.1998	0.1395	0.0887	0.0513	0.0269
	3	0.0137	0.0710	0.1517	0.2215	0.2581	0.2568	0.2254	0.1774	0.1259	0.0806
	4	0.0014	0.0158	0.0536	0.1107	0.1721	0.2201	0.2428	0.2365	0.2060	0.1611
	5	0.0001	0.0025	0.0132	0.0388	0.0803	0.1321	0.1830	0.2207	0.2360	0.2256
	6	0.0000	0.0003	0.0023	0.0097	0.0268	0.0566	0.0985	0.1471	0.1931	0.2256
	7	0.0000	0.0000	0.0003	0.0017	0.0064	0.0173	0.0379	0.0701	0.1128	0.1611
	8	0.0000	0.0000	0.0000	0.0002	0.0011	0.0037	0.0102	0.0234	0.0462	0.0806
	9	0.0000	0.0000	0.0000	0.0000	0.0001	0.0005	0.0018	0.0052	0.0126	0.0269
	10	0.0000	0.0000	0.0000	0.0000	0.0000	0.0000	0.0002	0.0007	0.0021	0.0054
	11	0.0000	0.0000	0.0000	0.0000	0.0000	0.0000	0.0000	0.0000	0.0002	0.0005
12	0	0.5404	0.2824	0.1422	0.0687	0.0317	0.0138	0.0057	0.0022	0.0008	0.0002
	1	0.3413	0.3766	0.3012	0.2062	0.1267	0.0712	0.0368	0.0174	0.0075	0.0029
	2	0.0988	0.2301	0.2924	0.2835	0.2323	0.1678	0.1088	0.0639	0.0339	0.0161
	3	0.0173	0.0852	0.1720	0.2362	0.2581	0.2397	0.1954	0.1419	0.0923	0.0537
	4	0.0021	0.0213	0.0683	0.1329	0.1936	0.2311	0.2367	0.2128	0.1700	0.1208
	5	0.0002	0.0038	0.0193	0.0532	0.1032	0.1585	0.2039	0.2270	0.2225	0.1934

N	r	p									
		0.05	0.10	0.15	0.20	0.25	0.30	0.35	0.40	0.45	0.50
12	6	0.0000	0.0005	0.0040	0.0155	0.0401	0.0792	0.1281	0.1766	0.2124	0.2256
	7	0.0000	0.0000	0.0006	0.0033	0.0115	0.0291	0.0591	0.1009	0.1489	0.1934
	8	0.0000	0.0000	0.0001	0.0005	0.0024	0.0078	0.0199	0.0420	0.0762	0.1208
	9	0.0000	0.0000	0.0000	0.0001	0.0004	0.0015	0.0048	0.0125	0.0277	0.0537
	10	0.0000	0.0000	0.0000	0.0000	0.0000	0.0002	0.0008	0.0025	0.0068	0.0161
	11	0.0000	0.0000	0.0000	0.0000	0.0000	0.0000	0.0001	0.0003	0.0010	0.0029
	12	0.0000	0.0000	0.0000	0.0000	0.0000	0.0000	0.0000	0.0000	0.0001	0.0002
13	0	0.5133	0.2542	0.1209	0.0550	0.0238	0.0097	0.0037	0.0013	0.0004	0.0001
	1	0.3512	0.3672	0.2774	0.1787	0.1029	0.0540	0.0259	0.0113	0.0045	0.0016
	2	0.1109	0.2448	0.2937	0.2680	0.2059	0.1388	0.0836	0.0453	0.0220	0.0095
	3	0.0214	0.0997	0.1900	0.2457	0.2517	0.2181	0.1651	0.1107	0.0660	0.0349
	4	0.0028	0.0277	0.0838	0.1535	0.2097	0.2337	0.2222	0.1845	0.1350	0.0873
	5	0.0003	0.0055	0.0266	0.0691	0.1258	0.1803	0.2154	0.2214	0.1989	0.1571
	6	0.0000	0.0008	0.0063	0.0230	0.0559	0.1030	0.1546	0.1968	0.2169	0.2095
	7	0.0000	0.0001	0.0011	0.0058	0.0186	0.0442	0.0833	0.1312	0.1775	0.2095
	8	0.0000	0.0000	0.0001	0.0011	0.0047	0.0142	0.0336	0.0656	0.1089	0.1571
	9	0.0000	0.0000	0.0000	0.0001	0.0009	0.0034	0.0101	0.0243	0.0495	0.0873
	10	0.0000	0.0000	0.0000	0.0000	0.0001	0.0006	0.0022	0.0065	0.0162	0.0349
	11	0.0000	0.0000	0.0000	0.0000	0.0000	0.0001	0.0003	0.0012	0.0036	0.0095
	12	0.0000	0.0000	0.0000	0.0000	0.0000	0.0000	0.0000	0.0001	0.0005	0.0016
14	0	0.4877	0.2288	0.1028	0.0440	0.0178	0.0068	0.0024	0.0008	0.0002	0.0001
	1	0.3593	0.3559	0.2539	0.1539	0.0832	0.0407	0.0181	0.0073	0.0027	0.0009
	2	0.1229	0.2570	0.2912	0.2501	0.1802	0.1134	0.0634	0.0317	0.0141	0.0056
	3	0.0259	0.1142	0.2056	0.2501	0.2402	0.1943	0.1366	0.0845	0.0462	0.0222
	4	0.0037	0.0349	0.0998	0.1720	0.2202	0.2290	0.2022	0.1549	0.1040	0.0611
	5	0.0004	0.0078	0.0352	0.0860	0.1468	0.1963	0.2178	0.2066	0.1701	0.1222
	6	0.0000	0.0013	0.0093	0.0322	0.0734	0.1262	0.1759	0.2066	0.2088	0.1833
	7	0.0000	0.0002	0.0019	0.0092	0.0280	0.0618	0.1082	0.1574	0.1952	0.2095
	8	0.0000	0.0000	0.0003	0.0020	0.0082	0.0232	0.0510	0.0918	0.1398	0.1833
	9	0.0000	0.0000	0.0000	0.0003	0.0018	0.0066	0.0183	0.0408	0.0762	0.1222
	10	0.0000	0.0000	0.0000	0.0000	0.0003	0.0014	0.0049	0.0136	0.0312	0.0611
	11	0.0000	0.0000	0.0000	0.0000	0.0000	0.0002	0.0010	0.0033	0.0093	0.0222
	12	0.0000	0.0000	0.0000	0.0000	0.0000	0.0000	0.0001	0.0005	0.0019	0.0056
	13	0.0000	0.0000	0.0000	0.0000	0.0000	0.0000	0.0000	0.0001	0.0002	0.0009
	14	0.0000	0.0000	0.0000	0.0000	0.0000	0.0000	0.0000	0.0000	0.0000	0.0001
15	0	0.4633	0.2059	0.0874	0.0352	0.0134	0.0047	0.0016	0.0005	0.0001	0.0000
	1	0.3658	0.3432	0.2312	0.1319	0.0668	0.0305	0.0126	0.0047	0.0016	0.0005
	2	0.1348	0.2669	0.2856	0.2309	0.1559	0.0916	0.0476	0.0219	0.0090	0.0032
	3	0.0307	0.1285	0.2184	0.2501	0.2252	0.1700	0.1110	0.0634	0.0318	0.0139
	4	0.0049	0.0428	0.1156	0.1876	0.2252	0.2186	0.1792	0.1268	0.0780	0.0417
	5	0.0006	0.0105	0.0449	0.1032	0.1651	0.2061	0.2123	0.1859	0.1404	0.0916
	6	0.0000	0.0019	0.0132	0.0430	0.0917	0.1472	0.1906	0.2066	0.1914	0.1527

N	r	p									
		0.05	0.10	0.15	0.20	0.25	0.30	0.35	0.40	0.45	0.50
15	7	0.0000	0.0003	0.0030	0.0138	0.0393	0.0811	0.1319	0.1771	0.2013	0.1964
	8	0.0000	0.0000	0.0005	0.0035	0.0131	0.0348	0.0710	0.1181	0.1647	0.1964
	9	0.0000	0.0000	0.0001	0.0007	0.0034	0.0116	0.0298	0.0612	0.1048	0.1527
	10	0.0000	0.0000	0.0000	0.0001	0.0007	0.0030	0.0096	0.0245	0.0515	0.0916
	11	0.0000	0.0000	0.0000	0.0000	0.0001	0.0006	0.0024	0.0074	0.0191	0.0417
	12	0.0000	0.0000	0.0000	0.0000	0.0000	0.0001	0.0004	0.0016	0.0052	0.0139
	13	0.0000	0.0000	0.0000	0.0000	0.0000	0.0000	0.0001	0.0003	0.0010	0.0032
	14	0.0000	0.0000	0.0000	0.0000	0.0000	0.0000	0.0000	0.0000	0.0001	0.0005
	15	0.0000	0.0000	0.0000	0.0000	0.0000	0.0000	0.0000	0.0000	0.0000	0.0000
16	0	0.4401	0.1853	0.0743	0.0281	0.0100	0.0033	0.0010	0.0003	0.0001	0.0000
	1	0.3706	0.3294	0.2097	0.1126	0.0535	0.0228	0.0087	0.0030	0.0009	0.0002
	2	0.1463	0.2745	0.2775	0.2111	0.1336	0.0732	0.0353	0.0150	0.0056	0.0018
	3	0.0359	0.1423	0.2285	0.2463	0.2079	0.1465	0.0888	0.0468	0.0215	0.0085
	4	0.0061	0.0514	0.1311	0.2001	0.2252	0.2040	0.1553	0.1014	0.0572	0.0278
	5	0.0008	0.0137	0.0555	0.1201	0.1802	0.2099	0.2008	0.1623	0.1123	0.0667
	6	0.0001	0.0028	0.0180	0.0550	0.1101	0.1649	0.1982	0.1983	0.1684	0.1222
	7	0.0000	0.0004	0.0045	0.0197	0.0524	0.1010	0.1524	0.1889	0.1969	0.1746
	8	0.0000	0.0001	0.0009	0.0055	0.0197	0.0487	0.0923	0.1417	0.1812	0.1964
	9	0.0000	0.0000	0.0001	0.0012	0.0058	0.0185	0.0442	0.0840	0.1318	0.1746
	10	0.0000	0.0000	0.0000	0.0002	0.0014	0.0056	0.0167	0.0392	0.0755	0.1222
	11	0.0000	0.0000	0.0000	0.0000	0.0002	0.0013	0.0049	0.0142	0.0337	0.0667
	12	0.0000	0.0000	0.0000	0.0000	0.0000	0.0002	0.0011	0.0040	0.0115	0.0278
	13	0.0000	0.0000	0.0000	0.0000	0.0000	0.0000	0.0002	0.0008	0.0029	0.0085
	14	0.0000	0.0000	0.0000	0.0000	0.0000	0.0000	0.0000	0.0001	0.0005	0.0018
	15	0.0000	0.0000	0.0000	0.0000	0.0000	0.0000	0.0000	0.0000	0.0001	0.0002
	16	0.0000	0.0000	0.0000	0.0000	0.0000	0.0000	0.0000	0.0000	0.0000	0.0000
17	0	0.4181	0.1668	0.0631	0.0225	0.0075	0.0023	0.0007	0.0002	0.0000	0.0000
	1	0.3741	0.3150	0.1893	0.0957	0.0426	0.0169	0.0060	0.0019	0.0005	0.0001
	2	0.1575	0.2800	0.2673	0.1914	0.1136	0.0581	0.0260	0.0102	0.0035	0.0010
	3	0.0415	0.1556	0.2359	0.2393	0.1893	0.1245	0.0701	0.0341	0.0144	0.0052
	4	0.0076	0.0605	0.1457	0.2093	0.2209	0.1868	0.1320	0.0796	0.0411	0.0182
	5	0.0010	0.0175	0.0668	0.1361	0.1914	0.2081	0.1849	0.1379	0.0875	0.0472
	6	0.0001	0.0039	0.0236	0.0680	0.1276	0.1784	0.1991	0.1839	0.1432	0.0944
	7	0.0000	0.0007	0.0065	0.0267	0.0668	0.1201	0.1685	0.1927	0.1841	0.1484
	8	0.0000	0.0001	0.0014	0.0084	0.0279	0.0644	0.1134	0.1606	0.1883	0.1855
17	9	0.0000	0.0000	0.0003	0.0021	0.0093	0.0276	0.0611	0.1070	0.1540	0.1855
	10	0.0000	0.0000	0.0000	0.0004	0.0025	0.0095	0.0263	0.0571	0.1008	0.1484
	11	0.0000	0.0000	0.0000	0.0001	0.0005	0.0026	0.0090	0.0242	0.0525	0.0944
	12	0.0000	0.0000	0.0000	0.0000	0.0001	0.0006	0.0024	0.0081	0.0215	0.0472
	13	0.0000	0.0000	0.0000	0.0000	0.0000	0.0001	0.0005	0.0021	0.0068	0.0182
	14	0.0000	0.0000	0.0000	0.0000	0.0000	0.0000	0.0001	0.0004	0.0016	0.0052
	15	0.0000	0.0000	0.0000	0.0000	0.0000	0.0000	0.0000	0.0001	0.0003	0.0010

N	r	p									
		0.05	0.10	0.15	0.20	0.25	0.30	0.35	0.40	0.45	0.50
17	16	0.0000	0.0000	0.0000	0.0000	0.0000	0.0000	0.0000	0.0000	0.0000	0.0001
	17	0.0000	0.0000	0.0000	0.0000	0.0000	0.0000	0.0000	0.0000	0.0000	0.0000
18	0	0.3972	0.1501	0.0536	0.0180	0.0056	0.0016	0.0004	0.0001	0.0000	0.0000
	1	0.3763	0.3002	0.1704	0.0811	0.0338	0.0126	0.0042	0.0012	0.0003	0.0001
	2	0.1683	0.2835	0.2556	0.1723	0.0958	0.0458	0.0190	0.0069	0.0022	0.0006
	3	0.0473	0.1680	0.2406	0.2297	0.1704	0.1046	0.0547	0.0246	0.0095	0.0031
	4	0.0093	0.0700	0.1592	0.2153	0.2130	0.1681	0.1104	0.0614	0.0291	0.0117
	5	0.0014	0.0218	0.0787	0.1507	0.1988	0.2017	0.1664	0.1146	0.0666	0.0327
	6	0.0002	0.0052	0.0301	0.0816	0.1436	0.1873	0.1941	0.1655	0.1181	0.0708
	7	0.0000	0.0010	0.0091	0.0350	0.0820	0.1376	0.1792	0.1892	0.1657	0.1214
	8	0.0000	0.0002	0.0022	0.0120	0.0376	0.0811	0.1327	0.1734	0.1864	0.1669
	9	0.0000	0.0000	0.0004	0.0033	0.0139	0.0386	0.0794	0.1284	0.1694	0.1855
	10	0.0000	0.0000	0.0001	0.0008	0.0042	0.0149	0.0385	0.0771	0.1248	0.1669
	11	0.0000	0.0000	0.0000	0.0001	0.0010	0.0046	0.0151	0.0374	0.0742	0.1214
	12	0.0000	0.0000	0.0000	0.0000	0.0002	0.0012	0.0047	0.0145	0.0354	0.0708
	13	0.0000	0.0000	0.0000	0.0000	0.0000	0.0002	0.0012	0.0045	0.0134	0.0327
	14	0.0000	0.0000	0.0000	0.0000	0.0000	0.0000	0.0002	0.0011	0.0039	0.0117
	15	0.0000	0.0000	0.0000	0.0000	0.0000	0.0000	0.0000	0.0002	0.0009	0.0031
	16	0.0000	0.0000	0.0000	0.0000	0.0000	0.0000	0.0000	0.0000	0.0001	0.0006
	17	0.0000	0.0000	0.0000	0.0000	0.0000	0.0000	0.0000	0.0000	0.0000	0.0001
	18	0.0000	0.0000	0.0000	0.0000	0.0000	0.0000	0.0000	0.0000	0.0000	0.0000
19	0	0.3774	0.1351	0.0456	0.0144	0.0042	0.0011	0.0003	0.0001	0.0000	0.0000
	1	0.3774	0.2852	0.1529	0.0685	0.0268	0.0093	0.0029	0.0008	0.0002	0.0000
	2	0.1787	0.2852	0.2428	0.1540	0.0803	0.0358	0.0138	0.0046	0.0013	0.0003
	3	0.0533	0.1796	0.2428	0.2182	0.1517	0.0869	0.0422	0.0175	0.0062	0.0018
	4	0.0112	0.0798	0.1714	0.2182	0.2023	0.1491	0.0909	0.0467	0.0203	0.0074
	5	0.0018	0.0266	0.0907	0.1636	0.2023	0.1916	0.1468	0.0933	0.0497	0.0222
	6	0.0002	0.0069	0.0374	0.0955	0.1574	0.1916	0.1844	0.1451	0.0949	0.0518
	7	0.0000	0.0014	0.0122	0.0443	0.0974	0.1525	0.1844	0.1797	0.1443	0.0961
	8	0.0000	0.0002	0.0032	0.0166	0.0487	0.0981	0.1489	0.1797	0.1771	0.1442
	9	0.0000	0.0000	0.0007	0.0051	0.0198	0.0514	0.0980	0.1464	0.1771	0.1762
	10	0.0000	0.0000	0.0001	0.0013	0.0066	0.0220	0.0528	0.0976	0.1449	0.1762
	11	0.0000	0.0000	0.0000	0.0003	0.0018	0.0077	0.0233	0.0532	0.0970	0.1442
	12	0.0000	0.0000	0.0000	0.0000	0.0004	0.0022	0.0083	0.0237	0.0529	0.0961
	13	0.0000	0.0000	0.0000	0.0000	0.0001	0.0005	0.0024	0.0085	0.0233	0.0518
19	14	0.0000	0.0000	0.0000	0.0000	0.0000	0.0001	0.0006	0.0024	0.0082	0.0222
	15	0.0000	0.0000	0.0000	0.0000	0.0000	0.0000	0.0001	0.0005	0.0022	0.0074
	16	0.0000	0.0000	0.0000	0.0000	0.0000	0.0000	0.0000	0.0001	0.0005	0.0018
	17	0.0000	0.0000	0.0000	0.0000	0.0000	0.0000	0.0000	0.0000	0.0001	0.0003
	18	0.0000	0.0000	0.0000	0.0000	0.0000	0.0000	0.0000	0.0000	0.0000	0.0000
	19	0.0000	0.0000	0.0000	0.0000	0.0000	0.0000	0.0000	0.0000	0.0000	0.0000

N	r	p									
		0.05	0.10	0.15	0.20	0.25	0.30	0.35	0.40	0.45	0.50
20	0	0.3585	0.1216	0.0388	0.0115	0.0032	0.0008	0.0002	0.0000	0.0000	0.0000
	1	0.3774	0.2702	0.1368	0.0576	0.0211	0.0068	0.0020	0.0005	0.0001	0.0000
	2	0.1887	0.2852	0.2293	0.1369	0.0669	0.0278	0.0100	0.0031	0.0008	0.0002
	3	0.0596	0.1901	0.2428	0.2054	0.1339	0.0716	0.0323	0.0123	0.0040	0.0011
	4	0.0133	0.0898	0.1821	0.2182	0.1897	0.1304	0.0738	0.0350	0.0139	0.0046
	5	0.0022	0.0319	0.1028	0.1746	0.2023	0.1789	0.1272	0.0746	0.0365	0.0148
	6	0.0003	0.0089	0.0454	0.1091	0.1686	0.1916	0.1712	0.1244	0.0746	0.0370
	7	0.0000	0.0020	0.0160	0.0545	0.1124	0.1643	0.1844	0.1659	0.1221	0.0739
	8	0.0000	0.0004	0.0046	0.0222	0.0609	0.1144	0.1614	0.1797	0.1623	0.1201
	9	0.0000	0.0001	0.0011	0.0074	0.0271	0.0654	0.1158	0.1597	0.1771	0.1602
	10	0.0000	0.0000	0.0002	0.0020	0.0099	0.0308	0.0686	0.1171	0.1593	0.1762
	11	0.0000	0.0000	0.0000	0.0005	0.0030	0.0120	0.0336	0.0710	0.1185	0.1602
	12	0.0000	0.0000	0.0000	0.0001	0.0008	0.0039	0.0136	0.0355	0.0727	0.1201
	13	0.0000	0.0000	0.0000	0.0000	0.0002	0.0010	0.0045	0.0146	0.0366	0.0739
	14	0.0000	0.0000	0.0000	0.0000	0.0000	0.0002	0.0012	0.0049	0.0150	0.0370
	15	0.0000	0.0000	0.0000	0.0000	0.0000	0.0000	0.0003	0.0013	0.0049	0.0148
	16	0.0000	0.0000	0.0000	0.0000	0.0000	0.0000	0.0000	0.0003	0.0013	0.0046
	17	0.0000	0.0000	0.0000	0.0000	0.0000	0.0000	0.0000	0.0000	0.0002	0.0011
	18	0.0000	0.0000	0.0000	0.0000	0.0000	0.0000	0.0000	0.0000	0.0000	0.0002
	19	0.0000	0.0000	0.0000	0.0000	0.0000	0.0000	0.0000	0.0000	0.0000	0.0000
	20	0.0000	0.0000	0.0000	0.0000	0.0000	0.0000	0.0000	0.0000	0.0000	0.0000

TABLE A-4a Critical Values of the t Distribution.

df	2-tailed testing			1-tailed testing		
	• •			• •		
	0.1	0.05	0.01	0.1	0.05	0.01
5	2.015	2.571	4.032	1.476	2.015	3.365
6	1.943	2.447	3.707	1.440	1.943	3.143
7	1.895	2.365	3.499	1.415	1.895	2.998
8	1.860	2.306	3.355	1.397	1.860	2.896
9	1.833	2.262	3.250	1.383	1.833	2.821
10	1.812	2.228	3.169	1.372	1.812	2.764
11	1.796	2.201	3.106	1.363	1.796	2.718
12	1.782	2.179	3.055	1.356	1.782	2.681
13	1.771	2.160	3.012	1.350	1.771	2.650
14	1.761	2.145	2.977	1.345	1.761	2.624
15	1.753	2.131	2.947	1.341	1.753	2.602
16	1.746	2.120	2.921	1.337	1.746	2.583
17	1.740	2.110	2.898	1.333	1.740	2.567
18	1.734	2.101	2.878	1.330	1.734	2.552
19	1.729	2.093	2.861	1.328	1.729	2.539
20	1.725	2.086	2.845	1.325	1.725	2.528
21	1.721	2.080	2.831	1.323	1.721	2.518
22	1.717	2.074	2.819	1.321	1.717	2.508
23	1.714	2.069	2.807	1.319	1.714	2.500
24	1.711	2.064	2.797	1.318	1.711	2.492
25	1.708	2.060	2.787	1.316	1.708	2.485
26	1.706	2.056	2.779	1.315	1.706	2.479
27	1.703	2.052	2.771	1.314	1.703	2.473
28	1.701	2.048	2.763	1.313	1.701	2.467
29	1.699	2.045	2.756	1.311	1.699	2.462
30	1.697	2.042	2.750	1.310	1.697	2.457
40	1.684	2.021	2.704	1.303	1.684	2.423
50	1.676	2.009	2.678	1.299	1.676	2.403
60	1.671	2.000	2.660	1.296	1.671	2.390
80	1.664	1.990	2.639	1.292	1.664	2.374
100	1.660	1.984	2.626	1.290	1.660	2.364
120	1.658	1.980	2.617	1.289	1.658	2.358
• •	1.645	1.960	2.576	1.282	1.645	2.327

p	df																			
	5	6	7	8	9	10	11	12	13	14	15	16	17	18	19	20	21	22	23	24
0.500	0.727	0.718	0.711	0.706	0.703	0.700	0.697	0.695	0.694	0.692	0.691	0.690	0.689	0.688	0.688	0.687	0.686	0.686	0.685	0.685
0.450	0.819	0.808	0.800	0.794	0.790	0.786	0.783	0.781	0.779	0.777	0.776	0.774	0.773	0.772	0.771	0.771	0.770	0.769	0.769	0.768
0.400	0.920	0.906	0.896	0.889	0.883	0.879	0.876	0.873	0.870	0.868	0.866	0.865	0.863	0.862	0.861	0.860	0.859	0.858	0.858	0.857
0.350	1.031	1.013	1.001	0.993	0.986	0.980	0.976	0.972	0.969	0.967	0.965	0.963	0.961	0.960	0.958	0.957	0.956	0.955	0.954	0.953
0.300	1.156	1.134	1.119	1.108	1.100	1.093	1.088	1.083	1.079	1.076	1.074	1.071	1.069	1.067	1.066	1.064	1.063	1.061	1.060	1.059
0.250	1.301	1.273	1.254	1.240	1.230	1.221	1.214	1.209	1.204	1.200	1.197	1.194	1.191	1.189	1.187	1.185	1.183	1.182	1.180	1.179
0.200	1.476	1.440	1.415	1.397	1.383	1.372	1.363	1.356	1.350	1.345	1.341	1.337	1.333	1.330	1.328	1.325	1.323	1.321	1.319	1.318
0.150	1.699	1.650	1.617	1.592	1.574	1.559	1.548	1.538	1.530	1.523	1.517	1.512	1.508	1.504	1.500	1.497	1.494	1.492	1.489	1.487
0.100	2.015	1.943	1.895	1.860	1.833	1.812	1.796	1.782	1.771	1.761	1.753	1.746	1.740	1.734	1.729	1.725	1.721	1.717	1.714	1.711
0.095	2.055	1.980	1.929	1.893	1.865	1.844	1.827	1.812	1.801	1.791	1.782	1.775	1.768	1.762	1.757	1.753	1.748	1.745	1.741	1.738
0.090	2.098	2.019	1.966	1.928	1.899	1.877	1.859	1.844	1.832	1.821	1.812	1.805	1.798	1.792	1.786	1.782	1.777	1.773	1.770	1.767
0.085	2.143	2.060	2.005	1.965	1.935	1.911	1.893	1.877	1.864	1.854	1.844	1.836	1.829	1.823	1.817	1.812	1.808	1.804	1.800	1.797
0.080	2.191	2.104	2.046	2.004	1.973	1.948	1.928	1.912	1.899	1.887	1.878	1.869	1.862	1.855	1.850	1.844	1.840	1.835	1.832	1.828
0.075	2.242	2.151	2.090	2.046	2.013	1.987	1.966	1.949	1.935	1.923	1.913	1.904	1.897	1.890	1.884	1.878	1.873	1.869	1.865	1.861
0.070	2.297	2.201	2.136	2.090	2.055	2.028	2.007	1.989	1.974	1.962	1.951	1.942	1.934	1.926	1.920	1.914	1.909	1.905	1.900	1.896
0.065	2.357	2.255	2.187	2.138	2.101	2.073	2.050	2.031	2.016	2.002	1.991	1.981	1.973	1.965	1.959	1.953	1.947	1.942	1.938	1.934
0.060	2.422	2.313	2.241	2.189	2.150	2.120	2.096	2.076	2.060	2.046	2.034	2.024	2.015	2.007	2.000	1.994	1.988	1.983	1.978	1.974
0.055	2.492	2.377	2.300	2.245	2.204	2.172	2.146	2.125	2.108	2.093	2.081	2.070	2.060	2.052	2.045	2.038	2.032	2.027	2.022	2.017
0.050	2.571	2.447	2.365	2.306	2.262	2.228	2.201	2.179	2.160	2.145	2.131	2.120	2.110	2.101	2.093	2.086	2.080	2.074	2.069	2.064
0.045	2.658	2.525	2.436	2.373	2.327	2.290	2.261	2.237	2.218	2.201	2.187	2.175	2.164	2.154	2.146	2.139	2.132	2.126	2.120	2.115
0.040	2.757	2.612	2.517	2.449	2.398	2.359	2.328	2.303	2.282	2.264	2.249	2.235	2.224	2.214	2.205	2.197	2.189	2.183	2.177	2.172
0.035	2.870	2.712	2.608	2.535	2.480	2.437	2.404	2.376	2.353	2.334	2.318	2.304	2.291	2.280	2.271	2.262	2.254	2.247	2.241	2.235
0.030	3.003	2.829	2.715	2.634	2.574	2.527	2.491	2.461	2.436	2.415	2.397	2.382	2.368	2.356	2.346	2.336	2.328	2.320	2.313	2.307
0.025	3.163	2.969	2.841	2.752	2.685	2.634	2.593	2.560	2.533	2.510	2.490	2.473	2.458	2.445	2.433	2.423	2.414	2.405	2.398	2.391
0.020	3.365	3.143	2.998	2.896	2.821	2.764	2.718	2.681	2.650	2.624	2.602	2.583	2.567	2.552	2.539	2.528	2.518	2.508	2.500	2.492
0.015	3.634	3.372	3.203	3.085	2.998	2.932	2.879	2.836	2.801	2.771	2.746	2.724	2.706	2.689	2.674	2.661	2.649	2.639	2.629	2.620
0.010	4.032	3.707	3.499	3.355	3.250	3.169	3.106	3.055	3.012	2.977	2.947	2.921	2.898	2.878	2.861	2.845	2.831	2.819	2.807	2.797
0.005	4.773	4.317	4.029	3.833	3.690	3.581	3.497	3.428	3.372	3.326	3.286	3.252	3.222	3.197	3.174	3.153	3.135	3.119	3.104	3.091
0.001	6.869	5.959	5.408	5.041	4.781	4.587	4.437	4.318	4.221	4.140	4.073	4.015	3.965	3.922	3.883	3.850	3.819	3.792	3.768	3.745
0.0005	7.976	6.788	6.082	5.617	5.291	5.049	4.863	4.716	4.597	4.499	4.417	4.346	4.286	4.233	4.187	4.146	4.110	4.077	4.047	4.021
0.0001	11.178	9.082	7.885	7.120	6.594	6.211	5.921	5.694	5.513	5.363	5.239	5.134	5.044	4.966	4.897	4.837	4.784	4.736	4.693	4.654
0.00005	12.893	10.261	8.783	7.851	7.215	6.757	6.412	6.143	5.928	5.753	5.607	5.484	5.379	5.288	5.209	5.139	5.077	5.022	4.972	4.927
0.00001	17.897	13.555	11.215	9.783	8.827	8.150	7.647	7.261	6.954	6.706	6.502	6.330	6.184	6.058	5.949	5.854	5.769	5.694	5.626	5.566
0.000005	20.591	15.260	12.437	10.731	9.605	8.812	8.227	7.779	7.426	7.142	6.907	6.711	6.545	6.402	6.278	6.170	6.074	5.989	5.913	5.845
0.000001	28.478	20.048	15.767	13.257	11.637	10.516	9.701	9.085	8.603	8.218	7.903	7.642	7.421	7.232	7.069	6.927	6.802	6.691	6.593	6.504

p	df																			
	5	6	7	8	9	10	11	12	13	14	15	16	17	18	19	20	21	22	23	24
0.500	0.000	0.000	0.000	0.000	0.000	0.000	0.000	0.000	0.000	0.000	0.000	0.000	0.000	0.000	0.000	0.000	0.000	0.000	0.000	0.000
0.450	0.132	0.131	0.130	0.130	0.129	0.129	0.129	0.128	0.128	0.128	0.128	0.128	0.128	0.127	0.127	0.127	0.127	0.127	0.127	0.127
0.400	0.267	0.265	0.263	0.262	0.261	0.260	0.260	0.259	0.259	0.258	0.258	0.258	0.257	0.257	0.257	0.257	0.257	0.256	0.256	0.256
0.350	0.408	0.404	0.402	0.399	0.398	0.397	0.396	0.395	0.394	0.393	0.393	0.392	0.392	0.392	0.391	0.391	0.391	0.390	0.390	0.390
0.300	0.559	0.553	0.549	0.546	0.543	0.542	0.540	0.539	0.538	0.537	0.536	0.535	0.534	0.534	0.533	0.533	0.532	0.532	0.532	0.531
0.250	0.727	0.718	0.711	0.706	0.703	0.700	0.697	0.695	0.694	0.692	0.691	0.690	0.689	0.688	0.688	0.687	0.686	0.686	0.685	0.685
0.200	0.920	0.906	0.896	0.889	0.883	0.879	0.876	0.873	0.870	0.868	0.866	0.865	0.863	0.862	0.861	0.860	0.859	0.858	0.858	0.857
0.150	1.156	1.134	1.119	1.108	1.100	1.093	1.088	1.083	1.079	1.076	1.074	1.071	1.069	1.067	1.066	1.064	1.063	1.061	1.060	1.059
0.100	1.476	1.440	1.415	1.397	1.383	1.372	1.363	1.356	1.350	1.345	1.341	1.337	1.333	1.330	1.328	1.325	1.323	1.321	1.319	1.318
0.095	1.516	1.478	1.451	1.432	1.418	1.406	1.397	1.389	1.383	1.377	1.373	1.369	1.365	1.362	1.359	1.357	1.354	1.352	1.350	1.349
0.090	1.558	1.517	1.489	1.469	1.454	1.442	1.432	1.424	1.417	1.411	1.406	1.402	1.398	1.395	1.392	1.389	1.387	1.385	1.383	1.381
0.085	1.602	1.559	1.529	1.508	1.492	1.479	1.468	1.460	1.453	1.447	1.441	1.437	1.433	1.429	1.426	1.424	1.421	1.419	1.417	1.415
0.080	1.649	1.603	1.572	1.549	1.532	1.518	1.507	1.498	1.490	1.484	1.478	1.474	1.469	1.466	1.462	1.459	1.457	1.454	1.452	1.450
0.075	1.699	1.650	1.617	1.592	1.574	1.559	1.548	1.538	1.530	1.523	1.517	1.512	1.508	1.504	1.500	1.497	1.494	1.492	1.489	1.487
0.070	1.753	1.700	1.664	1.638	1.619	1.603	1.591	1.580	1.572	1.565	1.558	1.553	1.548	1.544	1.540	1.537	1.534	1.531	1.529	1.526
0.065	1.810	1.754	1.715	1.687	1.666	1.650	1.636	1.626	1.616	1.609	1.602	1.596	1.591	1.587	1.583	1.579	1.576	1.573	1.570	1.568
0.060	1.873	1.812	1.770	1.740	1.718	1.700	1.686	1.674	1.664	1.656	1.649	1.642	1.637	1.632	1.628	1.624	1.621	1.618	1.615	1.612
0.055	1.941	1.874	1.830	1.797	1.773	1.754	1.738	1.726	1.715	1.706	1.699	1.692	1.686	1.681	1.677	1.672	1.669	1.665	1.662	1.660
0.050	2.015	1.943	1.895	1.860	1.833	1.812	1.796	1.782	1.771	1.761	1.753	1.746	1.740	1.734	1.729	1.725	1.721	1.717	1.714	1.711
0.045	2.098	2.019	1.966	1.928	1.899	1.877	1.859	1.844	1.832	1.821	1.812	1.805	1.798	1.792	1.786	1.782	1.777	1.773	1.770	1.767
0.040	2.191	2.104	2.046	2.004	1.973	1.948	1.928	1.912	1.899	1.887	1.878	1.869	1.862	1.855	1.850	1.844	1.840	1.835	1.832	1.828
0.035	2.297	2.201	2.136	2.090	2.055	2.028	2.007	1.989	1.974	1.962	1.951	1.942	1.934	1.926	1.920	1.914	1.909	1.905	1.900	1.896
0.030	2.422	2.313	2.241	2.189	2.150	2.120	2.096	2.076	2.060	2.046	2.034	2.024	2.015	2.007	2.000	1.994	1.988	1.983	1.978	1.974
0.025	2.571	2.447	2.365	2.306	2.262	2.228	2.201	2.179	2.160	2.145	2.131	2.120	2.110	2.101	2.093	2.086	2.080	2.074	2.069	2.064
0.020	2.757	2.612	2.517	2.449	2.398	2.359	2.328	2.303	2.282	2.264	2.249	2.235	2.224	2.214	2.205	2.197	2.189	2.183	2.177	2.172
0.015	3.003	2.829	2.715	2.634	2.574	2.527	2.491	2.461	2.436	2.415	2.397	2.382	2.368	2.356	2.346	2.336	2.328	2.320	2.313	2.307
0.010	3.365	3.143	2.998	2.896	2.821	2.764	2.718	2.681	2.650	2.624	2.602	2.583	2.567	2.552	2.539	2.528	2.518	2.508	2.500	2.492
0.005	4.032	3.707	3.499	3.355	3.250	3.169	3.106	3.055	3.012	2.977	2.947	2.921	2.898	2.878	2.861	2.845	2.831	2.819	2.807	2.797
0.001	5.893	5.208	4.785	4.501	4.297	4.144	4.025	3.930	3.852	3.787	3.733	3.686	3.646	3.610	3.579	3.552	3.527	3.505	3.485	3.467
0.0005	6.869	5.959	5.408	5.041	4.781	4.587	4.437	4.318	4.221	4.140	4.073	4.015	3.965	3.922	3.883	3.850	3.819	3.792	3.768	3.745
0.0001	9.678	8.025	7.063	6.442	6.010	5.694	5.453	5.263	5.111	4.985	4.880	4.791	4.714	4.648	4.590	4.539	4.493	4.452	4.415	4.382
0.00005	11.178	9.082	7.885	7.120	6.594	6.211	5.921	5.694	5.513	5.363	5.239	5.134	5.044	4.966	4.897	4.837	4.784	4.736	4.693	4.654
0.00001	15.547	12.032	10.103	8.907	8.102	7.527	7.097	6.765	6.501	6.287	6.109	5.959	5.832	5.722	5.627	5.543	5.469	5.402	5.343	5.290
0.000005	17.897	13.555	11.215	9.783	8.827	8.150	7.647	7.261	6.954	6.706	6.502	6.330	6.184	6.058	5.949	5.854	5.769	5.694	5.626	5.566
0.000001	24.771	17.830	14.241	12.110	10.720	9.752	9.043	8.504	8.082	7.743	7.464	7.233	7.036	6.869	6.723	6.597	6.485	6.386	6.297	6.218

TABLE A-4b Critical Values of the F Distribution.

df within	df between										
	1	2	3	4	5	6	7	8	12	24	∞
5	16.26	13.27	12.06	11.39	10.97	10.67	10.46	10.29	9.89	9.47	9.02
6	13.75	10.92	9.78	9.15	8.75	8.47	8.26	8.10	7.72	7.31	6.88
7	12.25	9.55	8.45	7.85	7.46	7.19	6.99	6.84	6.47	6.07	5.65
8	11.26	8.65	7.59	7.01	6.63	6.37	6.18	6.03	5.67	5.28	4.86
9	10.56	8.02	6.99	6.42	6.06	5.80	5.61	5.47	5.11	4.73	4.31
10	10.04	7.56	6.55	5.99	5.64	5.39	5.20	5.06	4.71	4.33	3.91
11	9.65	7.21	6.22	5.67	5.32	5.07	4.89	4.74	4.40	4.02	3.60
12	9.33	6.93	5.95	5.41	5.06	4.82	4.64	4.50	4.16	3.78	3.36
13	9.07	6.70	5.74	5.21	4.86	4.62	4.44	4.30	3.96	3.59	3.17
14	8.86	6.51	5.56	5.04	4.69	4.46	4.28	4.14	3.80	3.43	3.01
15	8.68	6.36	5.42	4.89	4.56	4.32	4.14	4.00	3.67	3.29	2.87
16	8.53	6.23	5.29	4.77	4.44	4.20	4.03	3.89	3.55	3.18	2.75
17	8.40	6.11	5.18	4.67	4.34	4.10	3.93	3.79	3.46	3.08	2.65
18	8.29	6.01	5.09	4.58	4.25	4.01	3.84	3.71	3.37	3.00	2.57
19	8.18	5.93	5.01	4.50	4.17	3.94	3.77	3.63	3.30	2.92	2.49
20	8.10	5.85	4.94	4.43	4.10	3.87	3.70	3.56	3.23	2.86	2.42
21	8.02	5.78	4.87	4.37	4.04	3.81	3.64	3.51	3.17	2.80	2.36
22	7.95	5.72	4.82	4.31	3.99	3.76	3.59	3.45	3.12	2.75	2.31
23	7.88	5.66	4.76	4.26	3.94	3.71	3.54	3.41	3.07	2.70	2.26
24	7.82	5.61	4.72	4.22	3.90	3.67	3.50	3.36	3.03	2.66	2.21
25	7.77	5.57	4.68	4.18	3.85	3.63	3.46	3.32	2.99	2.62	2.17
26	7.72	5.53	4.64	4.14	3.82	3.59	3.42	3.29	2.96	2.58	2.13
27	7.68	5.49	4.60	4.11	3.78	3.56	3.39	3.26	2.93	2.55	2.10
28	7.64	5.45	4.57	4.07	3.75	3.53	3.36	3.23	2.90	2.52	2.07
29	7.60	5.42	4.54	4.04	3.73	3.50	3.33	3.20	2.87	2.49	2.04
30	7.56	5.39	4.51	4.02	3.70	3.47	3.30	3.17	2.84	2.47	2.01
40	7.31	5.18	4.31	3.83	3.51	3.29	3.12	2.99	2.66	2.29	1.81
60	7.08	4.98	4.13	3.65	3.34	3.12	2.95	2.82	2.50	2.12	1.60
80	6.96	4.88	4.04	3.56	3.26	3.04	2.87	2.74	2.42	2.03	1.50
100	6.90	4.82	3.98	3.51	3.21	2.99	2.82	2.69	2.37	1.98	1.43
120	6.85	4.79	3.95	3.48	3.17	2.96	2.79	2.66	2.34	1.95	1.38
∞	6.64	4.61	3.78	3.32	3.02	2.80	2.64	2.51	2.19	1.79	1.00

df within	df between										
	1	2	3	4	5	6	7	8	12	24	∞
5	6.61	5.79	5.41	5.19	5.05	4.95	4.88	4.82	4.68	4.53	4.37
6	5.99	5.14	4.76	4.53	4.39	4.28	4.21	4.15	4.00	3.84	3.67
7	5.59	4.74	4.35	4.12	3.97	3.87	3.79	3.73	3.57	3.41	3.23
8	5.32	4.46	4.07	3.84	3.69	3.58	3.50	3.44	3.28	3.12	2.93
9	5.12	4.26	3.86	3.63	3.48	3.37	3.29	3.23	3.07	2.90	2.71
10	4.96	4.10	3.71	3.48	3.33	3.22	3.14	3.07	2.91	2.74	2.54
11	4.84	3.98	3.59	3.36	3.20	3.09	3.01	2.95	2.79	2.61	2.41
12	4.75	3.89	3.49	3.26	3.11	3.00	2.91	2.85	2.69	2.51	2.30
13	4.67	3.81	3.41	3.18	3.03	2.92	2.83	2.77	2.60	2.42	2.21
14	4.60	3.74	3.34	3.11	2.96	2.85	2.76	2.70	2.53	2.35	2.13
15	4.54	3.68	3.29	3.06	2.90	2.79	2.71	2.64	2.48	2.29	2.07
16	4.49	3.63	3.24	3.01	2.85	2.74	2.66	2.59	2.42	2.24	2.01
17	4.45	3.59	3.20	2.96	2.81	2.70	2.61	2.55	2.38	2.19	1.96
18	4.41	3.55	3.16	2.93	2.77	2.66	2.58	2.51	2.34	2.15	1.92
19	4.38	3.52	3.13	2.90	2.74	2.63	2.54	2.48	2.31	2.11	1.88
20	4.35	3.49	3.10	2.87	2.71	2.60	2.51	2.45	2.28	2.08	1.84
21	4.32	3.47	3.07	2.84	2.68	2.57	2.49	2.42	2.25	2.05	1.81
22	4.30	3.44	3.05	2.82	2.66	2.55	2.46	2.40	2.23	2.03	1.78
23	4.28	3.42	3.03	2.80	2.64	2.53	2.44	2.37	2.20	2.01	1.76
24	4.26	3.40	3.01	2.78	2.62	2.51	2.42	2.36	2.18	1.98	1.73
25	4.24	3.39	2.99	2.76	2.60	2.49	2.40	2.34	2.16	1.96	1.71
26	4.23	3.37	2.98	2.74	2.59	2.47	2.39	2.32	2.15	1.95	1.69
27	4.21	3.35	2.96	2.73	2.57	2.46	2.37	2.31	2.13	1.93	1.67
28	4.20	3.34	2.95	2.71	2.56	2.45	2.36	2.29	2.12	1.91	1.66
29	4.18	3.33	2.93	2.70	2.55	2.43	2.35	2.28	2.10	1.90	1.64
30	4.17	3.32	2.92	2.69	2.53	2.42	2.33	2.27	2.09	1.89	1.62
40	4.08	3.23	2.84	2.61	2.45	2.34	2.25	2.18	2.00	1.79	1.51
60	4.00	3.15	2.76	2.53	2.37	2.25	2.17	2.10	1.92	1.70	1.39
80	3.96	3.11	2.72	2.49	2.33	2.21	2.13	2.06	1.88	1.65	1.33
100	3.94	3.09	2.70	2.46	2.31	2.19	2.10	2.03	1.85	1.63	1.28
120	3.92	3.07	2.68	2.45	2.29	2.18	2.09	2.02	1.83	1.61	1.26
∞	3.84	3.00	2.61	2.37	2.22	2.10	2.01	1.94	1.75	1.52	1.00

TABLE A-5 Critical Values of the Studentized Range Statistic and Dunnett's Test.

df within	df between										
	1	2	3	4	5	6	7	8	12	24	∞
5	6.61	5.79	5.41	5.19	5.05	4.95	4.88	4.82	4.68	4.53	4.37
6	5.99	5.14	4.76	4.53	4.39	4.28	4.21	4.15	4.00	3.84	3.67
7	5.59	4.74	4.35	4.12	3.97	3.87	3.79	3.73	3.57	3.41	3.23
8	5.32	4.46	4.07	3.84	3.69	3.58	3.50	3.44	3.28	3.12	2.93
9	5.12	4.26	3.86	3.63	3.48	3.37	3.29	3.23	3.07	2.90	2.71
10	4.96	4.10	3.71	3.48	3.33	3.22	3.14	3.07	2.91	2.74	2.54
11	4.84	3.98	3.59	3.36	3.20	3.09	3.01	2.95	2.79	2.61	2.41
12	4.75	3.89	3.49	3.26	3.11	3.00	2.91	2.85	2.69	2.51	2.30
13	4.67	3.81	3.41	3.18	3.03	2.92	2.83	2.77	2.60	2.42	2.21
14	4.60	3.74	3.34	3.11	2.96	2.85	2.76	2.70	2.53	2.35	2.13
15	4.54	3.68	3.29	3.06	2.90	2.79	2.71	2.64	2.48	2.29	2.07
16	4.49	3.63	3.24	3.01	2.85	2.74	2.66	2.59	2.42	2.24	2.01
17	4.45	3.59	3.20	2.96	2.81	2.70	2.61	2.55	2.38	2.19	1.96
18	4.41	3.55	3.16	2.93	2.77	2.66	2.58	2.51	2.34	2.15	1.92
19	4.38	3.52	3.13	2.90	2.74	2.63	2.54	2.48	2.31	2.11	1.88
20	4.35	3.49	3.10	2.87	2.71	2.60	2.51	2.45	2.28	2.08	1.84
21	4.32	3.47	3.07	2.84	2.68	2.57	2.49	2.42	2.25	2.05	1.81
22	4.30	3.44	3.05	2.82	2.66	2.55	2.46	2.40	2.23	2.03	1.78
23	4.28	3.42	3.03	2.80	2.64	2.53	2.44	2.37	2.20	2.01	1.76
24	4.26	3.40	3.01	2.78	2.62	2.51	2.42	2.36	2.18	1.98	1.73
25	4.24	3.39	2.99	2.76	2.60	2.49	2.40	2.34	2.16	1.96	1.71
26	4.23	3.37	2.98	2.74	2.59	2.47	2.39	2.32	2.15	1.95	1.69
27	4.21	3.35	2.96	2.73	2.57	2.46	2.37	2.31	2.13	1.93	1.67
28	4.20	3.34	2.95	2.71	2.56	2.45	2.36	2.29	2.12	1.91	1.66
29	4.18	3.33	2.93	2.70	2.55	2.43	2.35	2.28	2.10	1.90	1.64
30	4.17	3.32	2.92	2.69	2.53	2.42	2.33	2.27	2.09	1.89	1.62
40	4.08	3.23	2.84	2.61	2.45	2.34	2.25	2.18	2.00	1.79	1.51
60	4.00	3.15	2.76	2.53	2.37	2.25	2.17	2.10	1.92	1.70	1.39
80	3.96	3.11	2.72	2.49	2.33	2.21	2.13	2.06	1.88	1.65	1.33
100	3.94	3.09	2.70	2.46	2.31	2.19	2.10	2.03	1.85	1.63	1.28
120	3.92	3.07	2.68	2.45	2.29	2.18	2.09	2.02	1.83	1.61	1.26
∞	3.84	3.00	2.61	2.37	2.22	2.10	2.01	1.94	1.75	1.52	1.00

n	α	Number of Groups, Including Control Group								
		2	3	4	5	6	7	8	9	10
5	.05	2.57	3.03	3.29	3.48	3.62	3.73	3.82	3.90	3.97
	.01	4.03	4.63	4.98	5.22	5.41	5.56	5.69	5.80	5.89
6	.05	2.45	2.86	3.10	3.26	3.39	3.49	3.57	3.64	3.71
	.01	3.71	4.21	4.51	4.71	4.87	5.00	5.10	5.20	5.28
7	.05	2.36	2.75	2.97	3.12	3.24	3.33	3.41	3.47	3.53
	.01	3.50	3.95	4.21	4.39	4.53	4.64	4.74	4.82	4.89
8	.05	2.31	2.67	2.88	3.02	3.13	3.22	3.29	3.35	3.41
	.01	3.36	3.77	4.00	4.17	4.29	4.40	4.48	4.56	4.62
9	.05	2.26	2.61	2.81	2.95	3.05	3.14	3.20	3.26	3.32
	.01	3.25	3.63	3.85	4.01	4.12	4.22	4.30	4.37	4.43
10	.05	2.23	2.57	2.76	2.89	2.99	3.07	3.14	3.19	3.24
	.01	3.17	3.53	3.74	3.88	3.99	4.08	4.16	4.22	4.28
11	.05	2.20	2.53	2.72	2.84	2.94	3.02	3.08	3.14	3.19
	.01	3.11	3.45	3.65	3.79	3.89	3.98	4.05	4.11	4.16
12	.05	2.18	2.50	2.68	2.81	2.90	2.98	3.04	3.09	3.14
	.01	3.05	3.39	3.58	3.71	3.81	3.89	3.96	4.02	4.07
13	.05	2.16	2.48	2.65	2.78	2.87	2.94	3.00	3.06	3.10
	.01	3.01	3.33	3.52	3.65	3.74	3.82	3.89	3.94	3.99
14	.05	2.14	2.46	2.63	2.75	2.84	2.91	2.97	3.02	3.07
	.01	2.98	3.29	3.47	3.59	3.69	3.76	3.83	3.88	3.93
15	.05	2.13	2.44	2.61	2.73	2.82	2.89	2.95	3.00	3.04
	.01	2.95	3.25	3.43	3.55	3.64	3.71	3.78	3.83	3.88
16	.05	2.12	2.42	2.59	2.71	2.80	2.87	2.92	2.97	3.02
	.01	2.92	3.22	3.39	3.51	3.60	3.67	3.73	3.78	3.83
17	.05	2.11	2.41	2.58	2.69	2.78	2.85	2.90	2.95	3.00
	.01	2.90	3.19	3.36	3.47	3.56	3.63	3.69	3.74	3.79
18	.05	2.10	2.40	2.56	2.68	2.76	2.83	2.89	2.94	2.98
	.01	2.88	3.17	3.33	3.44	3.53	3.60	3.66	3.71	3.75
19	.05	2.09	2.39	2.55	2.66	2.75	2.81	2.87	2.92	2.96
	.01	2.86	3.15	3.31	3.42	3.50	3.57	3.63	3.68	3.72
20	.05	2.09	2.38	2.54	2.65	2.73	2.80	2.86	2.90	2.95
	.01	2.85	3.13	3.29	3.40	3.48	3.55	3.60	3.65	3.69
24	.05	2.06	2.35	2.51	2.61	2.70	2.76	2.81	2.86	2.90
	.01	2.80	3.07	3.22	3.32	3.40	3.47	3.52	3.57	3.61
30	.05	2.04	2.32	2.47	2.58	2.66	2.72	2.77	2.82	2.86
	.01	2.75	3.01	3.15	3.25	3.33	3.39	3.44	3.49	3.52
40	.05	2.02	2.29	2.44	2.54	2.62	2.68	2.73	2.77	2.81
	.01	2.70	2.95	3.09	3.19	3.26	3.32	3.37	3.41	3.44
60	.05	2.00	2.27	2.41	2.51	2.58	2.64	2.69	2.73	2.77
	.01	2.66	2.90	3.03	3.12	3.19	3.25	3.29	3.33	3.37

TABLE A-6 Critical Values of Dunn's Test.

df	# of Comparisons										
	2	3	4	5	6	7	8	9	10	11	12
5	3.163	3.534	3.810	4.032	4.219	4.382	4.526	4.655	4.773	4.882	4.983
6	2.969	3.287	3.521	3.707	3.863	3.997	4.115	4.221	4.317	4.405	4.486
7	2.841	3.128	3.335	3.499	3.636	3.753	3.855	3.947	4.029	4.105	4.174
8	2.752	3.016	3.206	3.355	3.479	3.584	3.677	3.759	3.833	3.900	3.962
9	2.685	2.933	3.111	3.250	3.364	3.462	3.547	3.622	3.690	3.751	3.808
10	2.634	2.870	3.038	3.169	3.277	3.368	3.448	3.518	3.581	3.639	3.691
11	2.593	2.820	2.981	3.106	3.208	3.295	3.370	3.437	3.497	3.551	3.600
12	2.560	2.779	2.934	3.055	3.153	3.236	3.308	3.371	3.428	3.480	3.527
13	2.533	2.746	2.896	3.012	3.107	3.187	3.256	3.318	3.372	3.422	3.467
14	2.510	2.718	2.864	2.977	3.069	3.146	3.214	3.273	3.326	3.374	3.417
15	2.490	2.694	2.837	2.947	3.036	3.112	3.177	3.235	3.286	3.333	3.375
16	2.473	2.673	2.813	2.921	3.008	3.082	3.146	3.202	3.252	3.297	3.339
17	2.458	2.655	2.793	2.898	2.984	3.056	3.119	3.173	3.222	3.267	3.307
18	2.445	2.639	2.775	2.878	2.963	3.034	3.095	3.149	3.197	3.240	3.279
19	2.433	2.625	2.759	2.861	2.944	3.014	3.074	3.127	3.174	3.216	3.255
20	2.423	2.613	2.744	2.845	2.927	2.996	3.055	3.107	3.153	3.195	3.233
22	2.405	2.591	2.720	2.819	2.899	2.965	3.023	3.074	3.119	3.159	3.196
24	2.391	2.574	2.700	2.797	2.875	2.941	2.997	3.046	3.091	3.130	3.166
26	2.379	2.559	2.684	2.779	2.856	2.920	2.975	3.024	3.067	3.106	3.141
28	2.368	2.546	2.669	2.763	2.839	2.902	2.957	3.004	3.047	3.085	3.120
30	2.360	2.536	2.657	2.750	2.825	2.887	2.941	2.988	3.030	3.067	3.102
32	2.352	2.526	2.647	2.738	2.812	2.874	2.927	2.974	3.015	3.052	3.086
34	2.345	2.518	2.638	2.728	2.802	2.863	2.915	2.961	3.002	3.039	3.072
36	2.339	2.511	2.629	2.719	2.792	2.853	2.905	2.950	2.990	3.027	3.060
38	2.334	2.505	2.622	2.712	2.783	2.844	2.895	2.940	2.980	3.016	3.049
40	2.329	2.499	2.616	2.704	2.776	2.836	2.887	2.931	2.971	3.007	3.039
45	2.319	2.487	2.602	2.690	2.760	2.819	2.869	2.913	2.952	2.987	3.019
50	2.311	2.477	2.591	2.678	2.747	2.805	2.855	2.898	2.937	2.972	3.003
55	2.304	2.469	2.583	2.668	2.737	2.794	2.844	2.887	2.925	2.959	2.990
60	2.299	2.463	2.575	2.660	2.729	2.785	2.834	2.877	2.915	2.948	2.979
65	2.295	2.458	2.569	2.654	2.721	2.778	2.826	2.869	2.906	2.940	2.970
70	2.291	2.453	2.564	2.648	2.715	2.771	2.820	2.862	2.899	2.932	2.962
75	2.287	2.449	2.559	2.643	2.710	2.766	2.814	2.855	2.892	2.926	2.956
80	2.284	2.445	2.555	2.639	2.705	2.761	2.809	2.850	2.887	2.920	2.950
85	2.282	2.442	2.552	2.635	2.701	2.757	2.804	2.846	2.882	2.915	2.945
90	2.280	2.440	2.549	2.632	2.698	2.753	2.800	2.841	2.878	2.911	2.940
95	2.277	2.437	2.546	2.629	2.695	2.750	2.797	2.838	2.874	2.907	2.936
100	2.276	2.435	2.544	2.626	2.692	2.747	2.793	2.834	2.871	2.903	2.933
120	2.270	2.428	2.536	2.617	2.683	2.737	2.783	2.824	2.860	2.892	2.921
∞	2.245	2.398	2.502	2.581	2.644	2.696	2.740	2.779	2.813	2.844	2.872

df	# of Comparisons										
	2	3	4	5	6	7	8	9	10	11	12
5	4.773	5.247	5.604	5.894	6.139	6.352	6.541	6.713	6.869	7.013	7.147
6	4.317	4.698	4.981	5.208	5.398	5.563	5.709	5.840	5.959	6.068	6.169
7	4.029	4.355	4.595	4.785	4.944	5.082	5.202	5.310	5.408	5.497	5.580
8	3.833	4.122	4.334	4.501	4.640	4.759	4.864	4.957	5.041	5.118	5.189
9	3.690	3.954	4.146	4.297	4.422	4.529	4.623	4.706	4.781	4.849	4.912
10	3.581	3.827	4.005	4.144	4.259	4.357	4.442	4.518	4.587	4.649	4.707
11	3.497	3.728	3.895	4.025	4.132	4.223	4.303	4.373	4.437	4.495	4.548
12	3.428	3.649	3.806	3.930	4.031	4.117	4.192	4.258	4.318	4.372	4.421
13	3.372	3.584	3.735	3.852	3.948	4.030	4.101	4.164	4.221	4.272	4.319
14	3.326	3.530	3.675	3.787	3.880	3.958	4.026	4.086	4.140	4.189	4.234
15	3.286	3.484	3.624	3.733	3.822	3.897	3.963	4.021	4.073	4.120	4.163
16	3.252	3.444	3.580	3.686	3.772	3.845	3.909	3.965	4.015	4.060	4.102
17	3.222	3.410	3.543	3.646	3.730	3.801	3.862	3.917	3.965	4.009	4.049
18	3.197	3.380	3.510	3.610	3.692	3.762	3.821	3.874	3.922	3.965	4.004
19	3.174	3.354	3.481	3.579	3.660	3.727	3.786	3.837	3.883	3.925	3.963
20	3.153	3.331	3.455	3.552	3.630	3.697	3.754	3.804	3.850	3.890	3.928
22	3.119	3.291	3.412	3.505	3.581	3.645	3.700	3.749	3.792	3.832	3.867
24	3.091	3.258	3.376	3.467	3.540	3.602	3.656	3.703	3.745	3.783	3.818
26	3.067	3.231	3.346	3.435	3.507	3.567	3.620	3.666	3.707	3.744	3.777
28	3.047	3.208	3.321	3.408	3.479	3.538	3.589	3.634	3.674	3.710	3.743
30	3.030	3.189	3.300	3.385	3.454	3.513	3.563	3.607	3.646	3.681	3.714
32	3.015	3.172	3.281	3.365	3.433	3.491	3.540	3.583	3.622	3.657	3.688
34	3.002	3.157	3.265	3.348	3.415	3.472	3.520	3.563	3.601	3.635	3.666
36	2.990	3.144	3.251	3.333	3.399	3.455	3.503	3.545	3.582	3.616	3.647
38	2.980	3.132	3.238	3.319	3.385	3.440	3.487	3.529	3.566	3.599	3.630
40	2.971	3.122	3.227	3.307	3.372	3.426	3.473	3.514	3.551	3.584	3.614
45	2.952	3.100	3.203	3.281	3.345	3.398	3.444	3.484	3.520	3.552	3.582
50	2.937	3.083	3.184	3.261	3.324	3.376	3.421	3.461	3.496	3.528	3.556
55	2.925	3.069	3.169	3.245	3.307	3.358	3.403	3.442	3.476	3.508	3.536
60	2.915	3.057	3.156	3.232	3.293	3.344	3.388	3.426	3.460	3.491	3.519
65	2.906	3.048	3.146	3.220	3.281	3.331	3.375	3.413	3.447	3.477	3.505
70	2.899	3.039	3.137	3.211	3.271	3.321	3.364	3.402	3.435	3.465	3.492
75	2.892	3.032	3.129	3.202	3.262	3.312	3.354	3.392	3.425	3.455	3.482
80	2.887	3.026	3.122	3.195	3.254	3.304	3.346	3.383	3.416	3.446	3.473
85	2.882	3.020	3.116	3.189	3.248	3.297	3.339	3.376	3.409	3.438	3.465
90	2.878	3.016	3.111	3.183	3.242	3.291	3.333	3.369	3.402	3.431	3.458
95	2.874	3.011	3.106	3.178	3.236	3.285	3.327	3.363	3.396	3.425	3.452
100	2.871	3.007	3.102	3.174	3.232	3.280	3.322	3.358	3.390	3.420	3.446
120	2.860	2.995	3.088	3.160	3.217	3.265	3.306	3.342	3.373	3.402	3.428
∞	2.813	2.942	3.031	3.098	3.153	3.198	3.236	3.270	3.300	3.327	3.352

TABLE A-7 Critical Values of the Chi-Square Distribution.

df	Area in the Upper Tail					
	0.99	0.95	0.9	0.1	**0.05**	**0.01**
1	0.000	0.004	0.016	2.706	3.841	6.635
2	0.020	0.103	0.211	4.605	5.991	9.210
3	0.115	0.352	0.584	6.251	7.815	11.345
4	0.297	0.711	1.064	7.779	9.488	13.277
5	0.554	1.145	1.610	9.236	11.070	15.086
6	0.872	1.635	2.204	10.645	12.592	16.812
7	1.239	2.167	2.833	12.017	14.067	18.475
8	1.646	2.733	3.490	13.362	15.507	20.090
9	2.088	3.325	4.168	14.684	16.919	21.666
10	2.558	3.940	4.865	15.987	18.307	23.209
11	3.053	4.575	5.578	17.275	19.675	24.725
12	3.571	5.226	6.304	18.549	21.026	26.217
13	4.107	5.892	7.042	19.812	22.362	27.688
14	4.660	6.571	7.790	21.064	23.685	29.141
15	5.229	7.261	8.547	22.307	24.996	30.578
16	5.812	7.962	9.312	23.542	26.296	32.000
17	6.408	8.672	10.085	24.769	27.587	33.409
18	7.015	9.390	10.865	25.989	28.869	34.805
19	7.633	10.117	11.651	27.204	30.144	36.191
20	8.260	10.851	12.443	28.412	31.410	37.566
21	8.897	11.591	13.240	29.615	32.671	38.932
22	9.542	12.338	14.041	30.813	33.924	40.289
23	10.196	13.091	14.848	32.007	35.172	41.638
24	10.856	13.848	15.659	33.196	36.415	42.980
25	11.524	14.611	16.473	34.382	37.652	44.314

TABLE A-8 Critical Values of the Binomial Test.

N	2-Tailed Test		1-Tailed Test	
	α = .05	α = .01	α = .05	α = .01
5			0 / 5	
6	0 / 6		0 / 6	
7	0 / 7		0 / 7	0 / 7
8	0 / 8	0 / 8	1 / 7	0 / 8
9	1 / 8	0 / 9	1 / 8	0 / 9
10	1 / 9	0 / 10	1 / 9	0 / 10
11	1 / 10	0 / 11	2 / 9	1 / 10
12	2 / 10	1 / 11	2 / 10	1 / 11
13	2 / 11	1 / 12	3 / 10	1 / 12
14	2 / 12	1 / 13	3 / 11	2 / 12
15	3 / 12	2 / 13	3 / 12	2 / 13
16	3 / 13	2 / 14	4 / 12	2 / 14
17	4 / 13	2 / 15	4 / 13	3 / 14
18	4 / 14	3 / 15	5 / 13	3 / 15
19	4 / 15	3 / 16	5 / 14	4 / 15
20	5 / 15	3 / 17	5 / 15	4 / 16
21	5 / 16	4 / 17	6 / 15	4 / 17
22	5 / 17	4 / 18	6 / 16	5 / 17
23	6 / 17	4 / 19	7 / 16	5 / 18
24	6 / 18	5 / 19	7 / 17	5 / 19
25	7 / 18	5 / 20	7 / 18	6 / 19
26	7 / 19	6 / 20	8 / 18	6 / 20
27	7 / 20	6 / 21	8 / 19	7 / 20
28	8 / 20	6 / 22	9 / 19	7 / 21
29	8 / 21	7 / 22	9 / 20	7 / 22
30	9 / 21	7 / 23	10 / 20	8 / 22
31	9 / 22	7 / 24	10 / 21	8 / 23
32	9 / 23	8 / 24	10 / 22	8 / 24
33	10 / 23	8 / 25	11 / 22	9 / 24
34	10 / 24	9 / 25	11 / 23	9 / 25
35	11 / 24	9 / 26	12 / 23	10 / 25
36	11 / 25	9 / 27	12 / 24	10 / 26
37	12 / 25	10 / 27	13 / 24	10 / 27
38	12 / 26	10 / 28	13 / 25	11 / 27
39	12 / 27	11 / 28	13 / 26	11 / 28
40	13 / 27	11 / 29	14 / 26	12 / 28

TABLE A-9 Critical Values of the Mann-Whitney U Test.

n_2	α	n_1																	
		3	4	5	6	7	8	9	10	11	12	13	14	15	16	17	18	19	20
3	.05	--	--	0	1	1	2	2	3	3	4	4	5	5	6	6	7	7	8
	.01	--	--	--	--	--	--	0	0	0	1	1	1	2	2	2	2	3	3
4	.05		0	1	2	3	4	4	5	6	7	8	9	10	11	11	12	13	14
	.01		--	--	0	0	1	1	2	2	3	3	4	5	5	6	6	7	8
5	.05			2	3	5	6	7	8	9	11	12	13	14	15	17	18	19	20
	.01			0	1	1	2	3	4	5	6	7	7	8	9	10	11	12	13
6	.05				5	6	8	10	11	13	14	16	17	19	21	22	24	25	27
	.01				2	3	4	5	6	7	9	10	11	12	13	15	16	17	18
7	.05					8	10	12	14	16	18	20	22	24	26	28	30	32	34
	.01					4	6	7	9	10	12	13	15	16	18	19	21	22	24
8	.05						13	15	17	19	22	24	26	29	31	34	36	38	41
	.01						7	9	11	13	15	17	18	20	22	24	26	28	30
9	.05							17	20	23	26	28	31	34	37	39	42	45	48
	.01							11	13	16	18	20	22	24	27	29	31	33	36
10	.05								23	26	29	33	36	39	42	45	48	52	55
	.01								16	18	21	24	26	29	31	34	37	39	42
11	.05									30	33	37	40	44	47	51	55	58	62
	.01									21	24	27	30	33	36	39	42	45	48
12	.05										37	41	45	49	53	57	61	65	69
	.01										27	31	34	37	41	44	47	51	54
13	.05											45	50	54	59	63	67	72	76
	.01											34	38	42	45	49	53	57	60
14	.05												55	59	64	69	74	78	83
	.01												42	46	50	54	58	63	67
15	.05													64	70	75	80	85	90
	.01													51	55	60	64	69	73
16	.05														75	81	86	92	98
	.01														60	65	70	74	79
17	.05															87	93	99	105
	.01															70	75	81	86
18	.05																99	106	112
	.01																81	87	92
19	.05																	113	119
	.01																	93	99
20	.05																		127
	.01																		105

n_2	α	n_1																	
		3	4	5	6	7	8	9	10	11	12	13	14	15	16	17	18	19	20
3	.05	--	0	1	2	2	3	3	4	5	5	6	7	7	8	9	9	10	11
	.01	--	--	--	--	0	0	1	1	1	2	2	2	3	3	4	4	4	5
4	.05		1	2	3	4	5	6	7	8	9	10	11	12	14	15	16	17	18
	.01		--	0	1	1	2	3	3	4	5	5	6	7	7	8	9	9	10
5	.05			4	5	6	8	9	11	12	13	15	16	18	19	20	22	23	25
	.01			1	2	3	4	5	6	7	8	9	10	11	12	13	14	15	16
6	.05				7	8	10	12	14	16	17	19	21	23	25	26	28	30	32
	.01				3	4	6	7	8	9	11	12	13	15	16	18	19	20	22
7	.05					11	13	15	17	19	21	24	26	28	30	33	35	37	39
	.01					6	7	9	11	12	14	16	17	19	21	23	24	26	28
8	.05						15	18	20	23	26	28	31	33	36	39	41	44	47
	.01						9	11	13	15	17	20	22	24	26	28	30	32	34
9	.05							21	24	27	30	33	36	39	42	45	48	51	54
	.01							14	16	18	21	23	26	28	31	33	36	38	40
10	.05								27	31	34	37	41	44	48	51	55	58	62
	.01								19	22	24	27	30	33	36	38	41	44	47
11	.05									34	38	42	46	50	54	57	61	65	69
	.01									25	28	31	34	37	41	44	47	50	53
12	.05										42	47	51	55	60	64	68	72	77
	.01										31	35	38	42	46	49	53	56	60
13	.05											51	56	61	65	70	75	80	84
	.01											39	43	47	51	55	59	63	67
14	.05												61	66	71	77	82	87	92
	.01												47	51	56	60	65	69	73
15	.05													72	77	83	88	94	100
	.01													56	61	66	70	75	80
16	.05														83	89	95	101	107
	.01														66	71	76	82	87
17	.05															96	102	109	115
	.01															77	82	88	93
18	.05																109	116	123
	.01																88	94	100
19	.05																	123	130
	.01																	101	107
20	.05																		138
	.01																		114

TABLE A-10 Critical Values of the Wilcoxon Ranked Sums Test.

n	α	m																	
		3	4	5	6	7	8	9	10	11	12	13	14	15	16	17	18	19	20
3	.05	--	--	6	7	7	8	8	9	9	10	10	11	11	12	12	13	13	14
	.01	--	--	--	--	--	--	6	6	6	7	7	7	8	8	8	8	9	9
4	.05	--	10	11	12	13	14	14	15	16	17	18	19	20	21	21	22	23	24
	.01	--	--	--	10	10	11	11	12	12	13	13	14	15	15	16	16	17	18
5	.05	15	16	17	18	20	21	22	23	24	26	27	28	29	30	32	33	34	35
	.01	--	--	15	16	16	17	18	19	20	21	22	22	23	24	25	26	27	28
6	.05	22	23	24	26	27	29	31	32	34	35	37	38	40	42	43	45	46	48
	.01	--	21	22	23	24	25	26	27	28	30	31	32	33	34	36	37	38	39
7	.05	29	31	33	34	36	38	40	42	44	46	48	50	52	54	56	58	60	62
	.01	--	28	29	31	32	34	35	37	38	40	41	43	44	46	47	49	50	52
8	.05	38	40	42	44	46	49	51	53	55	58	60	62	65	67	70	72	74	77
	.01	--	37	38	40	42	43	45	47	49	51	53	54	56	58	60	62	64	66
9	.05	47	49	52	55	57	60	62	65	68	71	73	76	79	82	84	87	90	93
	.01	45	46	48	50	52	54	56	58	61	63	65	67	69	72	74	76	78	81
10	.05	58	60	63	66	69	72	75	78	81	84	88	91	94	97	100	103	107	110
	.01	55	57	59	61	64	66	68	71	73	76	79	81	84	86	89	92	94	97
11	.05	69	72	75	79	82	85	89	92	96	99	103	106	110	113	117	121	124	128
	.01	66	68	71	73	76	79	82	84	87	90	93	96	99	102	105	108	111	114
12	.05	82	85	89	92	96	100	104	107	111	115	119	123	127	131	135	139	143	147
	.01	79	81	84	87	90	93	96	99	102	105	109	112	115	119	122	125	129	132
13	.05	95	99	103	107	111	115	119	124	128	132	136	141	145	150	154	158	163	167
	.01	92	94	98	101	104	108	111	115	118	122	125	129	133	136	140	144	148	151
14	.05	110	114	118	122	127	131	136	141	145	150	155	160	164	169	174	179	183	188
	.01	106	109	112	116	120	123	127	131	135	139	143	147	151	155	159	163	168	172
15	.05	125	130	134	139	144	149	154	159	164	169	174	179	184	190	195	200	205	210
	.01	122	125	128	132	136	140	144	149	153	157	162	166	171	175	180	184	189	193
16	.05	142	147	151	157	162	167	173	178	183	189	195	200	206	211	217	222	228	234
	.01	138	141	145	149	154	158	163	167	172	177	181	186	191	196	201	206	210	215
17	.05	159	164	170	175	181	187	192	198	204	210	216	222	228	234	240	246	252	258
	.01	155	159	163	168	172	177	182	187	192	197	202	207	213	218	223	228	234	239
18	.05	178	183	189	195	201	207	213	219	226	232	238	245	251	257	264	270	277	283
	.01	173	177	182	187	192	197	202	208	213	218	224	229	235	241	246	252	258	263
19	.05	197	203	209	215	222	228	235	242	248	255	262	268	275	282	289	296	303	309
	.01	193	197	202	207	212	218	223	229	235	241	247	253	259	264	271	277	283	289
20	.05	218	224	230	237	244	251	258	265	272	279	286	293	300	308	315	322	329	337
	.01	213	218	223	228	234	240	246	252	258	264	270	277	283	289	296	302	309	315

Note: n is the number of scores in the group with the smallest sum of ranks;
m is the number of scores in the other group.

n	α	m																	
		3	4	5	6	7	8	9	10	11	12	13	14	15	16	17	18	19	20
3	.05	--	6	7	8	8	9	9	10	11	11	12	13	13	14	15	15	16	17
	.01	--	--	--	--	6	6	7	7	7	8	8	8	9	9	10	10	10	11
4	.05	10	11	12	13	14	15	16	17	18	19	20	21	22	24	25	26	27	28
	.01	--	--	10	11	11	12	13	13	14	15	15	16	17	17	18	19	19	20
5	.05	16	17	19	20	21	23	24	26	27	28	30	31	33	34	35	37	38	40
	.01	--	15	16	17	18	19	20	21	22	23	24	25	26	27	28	29	30	31
6	.05	23	24	26	28	29	31	33	35	37	38	40	42	44	46	47	49	51	53
	.01	--	22	23	24	25	27	28	29	30	32	33	34	36	37	39	40	41	43
7	.05	30	32	34	36	39	41	43	45	47	49	52	54	56	58	61	63	65	67
	.01	28	29	31	32	34	35	37	39	40	42	44	45	47	49	51	52	54	56
8	.05	39	41	44	46	49	51	54	56	59	62	64	67	69	72	75	77	80	83
	.01	36	38	40	42	43	45	47	49	51	53	56	58	60	62	64	66	68	70
9	.05	48	51	54	57	60	63	66	69	72	75	78	81	84	87	90	93	96	99
	.01	46	48	50	52	54	56	59	61	63	66	68	71	73	76	78	81	83	85
10	.05	59	62	66	69	72	75	79	82	86	89	92	96	99	103	106	110	113	117
	.01	56	58	61	63	66	68	71	74	77	79	82	85	88	91	93	96	99	102
11	.05	71	74	78	82	85	89	93	97	100	104	108	112	116	120	123	127	131	135
	.01	67	70	73	75	78	81	84	88	91	94	97	100	103	107	110	113	116	119
12	.05	83	87	91	95	99	104	108	112	116	120	125	129	133	138	142	146	150	155
	.01	80	83	86	89	92	95	99	102	106	109	113	116	120	124	127	131	134	138
13	.05	97	101	106	110	115	119	124	128	133	138	142	147	152	156	161	166	171	175
	.01	93	96	100	103	107	111	114	118	122	126	130	134	138	142	146	150	154	158
14	.05	112	116	121	126	131	136	141	146	151	156	161	166	171	176	182	187	192	197
	.01	107	111	115	118	122	127	131	135	139	143	148	152	156	161	165	170	174	178
15	.05	127	132	138	143	148	153	159	164	170	175	181	186	192	197	203	208	214	220
	.01	123	127	131	135	139	144	148	153	157	162	167	171	176	181	186	190	195	200
16	.05	144	150	155	161	166	172	178	184	190	196	201	207	213	219	225	231	237	243
	.01	139	143	148	152	157	162	167	172	177	182	187	192	197	202	207	212	218	223
17	.05	162	168	173	179	186	192	198	204	210	217	223	230	236	242	249	255	262	268
	.01	157	161	166	171	176	181	186	191	197	202	208	213	219	224	230	235	241	246
18	.05	180	187	193	199	206	212	219	226	232	239	246	253	259	266	273	280	287	294
	.01	175	180	185	190	195	201	207	212	218	224	230	236	241	247	253	259	265	271
19	.05	200	207	213	220	227	234	241	248	255	262	270	277	284	291	299	306	313	320
	.01	194	199	205	210	216	222	228	234	240	246	253	259	265	272	278	284	291	297
20	.05	221	228	235	242	249	257	264	272	279	287	294	302	310	317	325	333	340	348
	.01	215	220	226	232	238	244	250	257	263	270	277	283	290	297	303	310	317	324

Note: <u>n</u> is the number of scores in the group with the smallest sum of ranks;
<u>m</u> is the number of scores in the other group.

TABLE A-11 Critical Values of the Wilcoxon Signed Ranks Test.

n	Two-Tailed Test		One-Tailed Test	
	α = .05	α = .01	α = .05	α = .01
5	--	--	0	--
6	0	--	2	--
7	2	--	3	0
8	3	0	5	1
9	5	1	8	3
10	8	3	10	5
11	10	5	13	7
12	13	7	17	9
13	17	9	21	12
14	21	12	25	15
15	25	15	30	19
16	29	19	35	23
17	34	23	41	27
18	40	27	47	32
19	46	32	53	37
20	52	37	60	43
21	58	42	67	49
22	65	48	75	55
23	73	54	83	62
24	81	61	91	69
25	89	68	100	76
26	98	75	110	84
27	107	83	119	92
28	116	91	130	101
29	126	100	140	110
30	137	109	151	120

TABLE A-12 Critical Values of the Correlation Coefficient.

n	2-tailed testing .1	2-tailed testing .05	2-tailed testing .01	1-tailed testing .1	1-tailed testing .05	1-tailed testing .01
5	0.805	0.878	0.959	0.687	0.805	0.934
6	0.729	0.811	0.917	0.608	0.729	0.882
7	0.669	0.754	0.875	0.551	0.669	0.833
8	0.621	0.707	0.834	0.507	0.621	0.789
9	0.582	0.666	0.798	0.472	0.582	0.750
10	0.549	0.632	0.765	0.443	0.549	0.715
11	0.521	0.602	0.735	0.419	0.521	0.685
12	0.497	0.576	0.708	0.398	0.497	0.658
13	0.476	0.553	0.684	0.380	0.476	0.634
14	0.458	0.532	0.661	0.365	0.458	0.612
15	0.441	0.514	0.641	0.351	0.441	0.592
16	0.426	0.497	0.623	0.338	0.426	0.574
17	0.412	0.482	0.606	0.327	0.412	0.558
18	0.400	0.468	0.590	0.317	0.400	0.543
19	0.389	0.456	0.575	0.308	0.389	0.529
20	0.378	0.444	0.561	0.299	0.378	0.516
21	0.369	0.433	0.549	0.291	0.369	0.503
22	0.360	0.423	0.537	0.284	0.360	0.492
23	0.352	0.413	0.526	0.277	0.352	0.482
24	0.344	0.404	0.515	0.271	0.344	0.472
25	0.337	0.396	0.505	0.265	0.337	0.462
26	0.330	0.388	0.496	0.260	0.330	0.453
27	0.323	0.381	0.487	0.255	0.323	0.445
28	0.317	0.374	0.479	0.250	0.317	0.437
29	0.311	0.367	0.471	0.245	0.311	0.430
30	0.306	0.361	0.463	0.241	0.306	0.423
40	0.264	0.312	0.403	0.207	0.264	0.367
50	0.235	0.279	0.361	0.184	0.235	0.328
60	0.214	0.254	0.330	0.168	0.214	0.300
80	0.185	0.220	0.286	0.145	0.185	0.260
100	0.165	0.197	0.256	0.129	0.165	0.232
120	0.151	0.179	0.234	0.118	0.151	0.212
140	0.140	0.166	0.217	0.109	0.140	0.196
160	0.130	0.155	0.203	0.102	0.130	0.184
180	0.123	0.146	0.192	0.096	0.123	0.173
200	0.117	0.139	0.182	0.091	0.117	0.164
300	0.095	0.113	0.149	0.074	0.095	0.134
400	0.082	0.098	0.129	0.064	0.082	0.116
500	0.074	0.088	0.115	0.057	0.074	0.104

TABLE A-13 Transforming r to Z.

r	Z	r	Z	r	Z	r	Z	r	Z
0.002	0.0020	0.102	0.1024	0.202	0.2048	0.302	0.3117	0.402	0.4260
0.004	0.0040	0.104	0.1044	0.204	0.2069	0.304	0.3139	0.404	0.4284
0.006	0.0060	0.106	0.1064	0.206	0.2090	0.306	0.3161	0.406	0.4308
0.008	0.0080	0.108	0.1084	0.208	0.2111	0.308	0.3183	0.408	0.4332
0.010	0.0100	0.110	0.1104	0.210	0.2132	0.310	0.3205	0.410	0.4356
0.012	0.0120	0.112	0.1125	0.212	0.2153	0.312	0.3228	0.412	0.4380
0.014	0.0140	0.114	0.1145	0.214	0.2174	0.314	0.3250	0.414	0.4404
0.016	0.0160	0.116	0.1165	0.216	0.2195	0.316	0.3272	0.416	0.4428
0.018	0.0180	0.118	0.1186	0.218	0.2216	0.318	0.3294	0.418	0.4453
0.020	0.0200	0.120	0.1206	0.220	0.2237	0.320	0.3316	0.420	0.4477
0.022	0.0220	0.122	0.1226	0.222	0.2258	0.322	0.3339	0.422	0.4501
0.024	0.0240	0.124	0.1246	0.224	0.2279	0.324	0.3361	0.424	0.4526
0.026	0.0260	0.126	0.1267	0.226	0.2300	0.326	0.3383	0.426	0.4550
0.028	0.0280	0.128	0.1287	0.228	0.2321	0.328	0.3406	0.428	0.4574
0.030	0.0300	0.130	0.1307	0.230	0.2342	0.330	0.3428	0.430	0.4599
0.032	0.0320	0.132	0.1328	0.232	0.2363	0.332	0.3451	0.432	0.4624
0.034	0.0340	0.134	0.1348	0.234	0.2384	0.334	0.3473	0.434	0.4648
0.036	0.0360	0.136	0.1368	0.236	0.2405	0.336	0.3496	0.436	0.4673
0.038	0.0380	0.138	0.1389	0.238	0.2427	0.338	0.3518	0.438	0.4698
0.040	0.0400	0.140	0.1409	0.240	0.2448	0.340	0.3541	0.440	0.4722
0.042	0.0420	0.142	0.1430	0.242	0.2469	0.342	0.3564	0.442	0.4747
0.044	0.0440	0.144	0.1450	0.244	0.2490	0.344	0.3586	0.444	0.4772
0.046	0.0460	0.146	0.1471	0.246	0.2512	0.346	0.3609	0.446	0.4797
0.048	0.0480	0.148	0.1491	0.248	0.2533	0.348	0.3632	0.448	0.4822
0.050	0.0500	0.150	0.1511	0.250	0.2554	0.350	0.3654	0.450	0.4847
0.052	0.0520	0.152	0.1532	0.252	0.2575	0.352	0.3677	0.452	0.4872
0.054	0.0541	0.154	0.1552	0.254	0.2597	0.354	0.3700	0.454	0.4897
0.056	0.0561	0.156	0.1573	0.256	0.2618	0.356	0.3723	0.456	0.4922
0.058	0.0581	0.158	0.1593	0.258	0.2640	0.358	0.3746	0.458	0.4948
0.060	0.0601	0.160	0.1614	0.260	0.2661	0.360	0.3769	0.460	0.4973
0.062	0.0621	0.162	0.1634	0.262	0.2683	0.362	0.3792	0.462	0.4999
0.064	0.0641	0.164	0.1655	0.264	0.2704	0.364	0.3815	0.464	0.5024
0.066	0.0661	0.166	0.1676	0.266	0.2726	0.366	0.3838	0.466	0.5049
0.068	0.0681	0.168	0.1696	0.268	0.2747	0.368	0.3861	0.468	0.5075
0.070	0.0701	0.170	0.1717	0.270	0.2769	0.370	0.3884	0.470	0.5101
0.072	0.0721	0.172	0.1737	0.272	0.2790	0.372	0.3907	0.472	0.5126
0.074	0.0741	0.174	0.1758	0.274	0.2812	0.374	0.3931	0.474	0.5152
0.076	0.0761	0.176	0.1779	0.276	0.2833	0.376	0.3954	0.476	0.5178
0.078	0.0782	0.178	0.1799	0.278	0.2855	0.378	0.3977	0.478	0.5204
0.080	0.0802	0.180	0.1820	0.280	0.2877	0.380	0.4001	0.480	0.5230
0.082	0.0822	0.182	0.1841	0.282	0.2899	0.382	0.4024	0.482	0.5256
0.084	0.0842	0.184	0.1861	0.284	0.2920	0.384	0.4047	0.484	0.5282
0.086	0.0862	0.186	0.1882	0.286	0.2942	0.386	0.4071	0.486	0.5308
0.088	0.0882	0.188	0.1903	0.288	0.2964	0.388	0.4094	0.488	0.5334
0.090	0.0902	0.190	0.1923	0.290	0.2986	0.390	0.4118	0.490	0.5361
0.092	0.0923	0.192	0.1944	0.292	0.3008	0.392	0.4142	0.492	0.5387
0.094	0.0943	0.194	0.1965	0.294	0.3029	0.394	0.4165	0.494	0.5413
0.096	0.0963	0.196	0.1986	0.296	0.3051	0.396	0.4189	0.496	0.5440
0.098	0.0983	0.198	0.2007	0.298	0.3073	0.398	0.4213	0.498	0.5466
0.100	0.1003	0.200	0.2027	0.300	0.3095	0.400	0.4236	0.500	0.5493

r	Z	r	Z	r	Z	r	Z	r	Z
0.502	0.5520	0.602	0.6963	0.702	0.8712	0.802	1.1042	0.902	1.4828
0.504	0.5547	0.604	0.6994	0.704	0.8752	0.804	1.1098	0.904	1.4937
0.506	0.5573	0.606	0.7026	0.706	0.8792	0.806	1.1155	0.906	1.5047
0.508	0.5600	0.608	0.7057	0.708	0.8832	0.808	1.1212	0.908	1.5160
0.510	0.5627	0.610	0.7089	0.710	0.8872	0.810	1.1270	0.910	1.5275
0.512	0.5654	0.612	0.7121	0.712	0.8912	0.812	1.1329	0.912	1.5393
0.514	0.5682	0.614	0.7153	0.714	0.8953	0.814	1.1388	0.914	1.5513
0.516	0.5709	0.616	0.7185	0.716	0.8994	0.816	1.1447	0.916	1.5636
0.518	0.5736	0.618	0.7218	0.718	0.9035	0.818	1.1507	0.918	1.5762
0.520	0.5763	0.620	0.7250	0.720	0.9076	0.820	1.1568	0.920	1.5890
0.522	0.5791	0.622	0.7283	0.722	0.9118	0.822	1.1630	0.922	1.6022
0.524	0.5818	0.624	0.7315	0.724	0.9160	0.824	1.1692	0.924	1.6157
0.526	0.5846	0.626	0.7348	0.726	0.9202	0.826	1.1754	0.926	1.6296
0.528	0.5874	0.628	0.7381	0.728	0.9245	0.828	1.1817	0.928	1.6438
0.530	0.5901	0.630	0.7414	0.730	0.9287	0.830	1.1881	0.930	1.6584
0.532	0.5929	0.632	0.7447	0.732	0.9330	0.832	1.1946	0.932	1.6734
0.534	0.5957	0.634	0.7481	0.734	0.9373	0.834	1.2011	0.934	1.6888
0.536	0.5985	0.636	0.7514	0.736	0.9417	0.836	1.2077	0.936	1.7047
0.538	0.6013	0.638	0.7548	0.738	0.9461	0.838	1.2144	0.938	1.7211
0.540	0.6042	0.640	0.7582	0.740	0.9505	0.840	1.2212	0.940	1.7380
0.542	0.6070	0.642	0.7616	0.742	0.9549	0.842	1.2280	0.942	1.7555
0.544	0.6098	0.644	0.7650	0.744	0.9594	0.844	1.2349	0.944	1.7736
0.546	0.6127	0.646	0.7684	0.746	0.9639	0.846	1.2419	0.946	1.7923
0.548	0.6155	0.648	0.7718	0.748	0.9684	0.848	1.2490	0.948	1.8117
0.550	0.6184	0.650	0.7753	0.750	0.9730	0.850	1.2562	0.950	1.8318
0.552	0.6213	0.652	0.7788	0.752	0.9775	0.852	1.2634	0.952	1.8527
0.554	0.6241	0.654	0.7823	0.754	0.9822	0.854	1.2707	0.954	1.8745
0.556	0.6270	0.656	0.7858	0.756	0.9868	0.856	1.2782	0.956	1.8972
0.558	0.6299	0.658	0.7893	0.758	0.9915	0.858	1.2857	0.958	1.9210
0.560	0.6328	0.660	0.7928	0.760	0.9962	0.860	1.2933	0.960	1.9459
0.562	0.6358	0.662	0.7964	0.762	1.0010	0.862	1.3011	0.962	1.9721
0.564	0.6387	0.664	0.7999	0.764	1.0058	0.864	1.3089	0.964	1.9996
0.566	0.6416	0.666	0.8035	0.766	1.0106	0.866	1.3169	0.966	2.0287
0.568	0.6446	0.668	0.8071	0.768	1.0154	0.868	1.3249	0.968	2.0595
0.570	0.6475	0.670	0.8107	0.770	1.0203	0.870	1.3331	0.970	2.0923
0.572	0.6505	0.672	0.8144	0.772	1.0253	0.872	1.3414	0.972	2.1273
0.574	0.6535	0.674	0.8180	0.774	1.0302	0.874	1.3498	0.974	2.1649
0.576	0.6565	0.676	0.8217	0.776	1.0352	0.876	1.3583	0.976	2.2054
0.578	0.6595	0.678	0.8254	0.778	1.0403	0.878	1.3670	0.978	2.2494
0.580	0.6625	0.680	0.8291	0.780	1.0454	0.880	1.3758	0.980	2.2976
0.582	0.6655	0.682	0.8328	0.782	1.0505	0.882	1.3847	0.982	2.3507
0.584	0.6685	0.684	0.8366	0.784	1.0557	0.884	1.3938	0.984	2.4101
0.586	0.6716	0.686	0.8404	0.786	1.0609	0.886	1.4030	0.986	2.4774
0.588	0.6746	0.688	0.8441	0.788	1.0661	0.888	1.4124	0.988	2.5550
0.590	0.6777	0.690	0.8480	0.790	1.0714	0.890	1.4219	0.990	2.6467
0.592	0.6807	0.692	0.8518	0.792	1.0768	0.892	1.4316	0.992	2.7587
0.594	0.6838	0.694	0.8556	0.794	1.0822	0.894	1.4415	0.994	2.9031
0.596	0.6869	0.696	0.8595	0.796	1.0876	0.896	1.4516	0.996	3.1063
0.598	0.6900	0.698	0.8634	0.798	1.0931	0.898	1.4618	0.998	3.4534
0.600	0.6931	0.700	0.8673	0.800	1.0986	0.900	1.4722		

TABLE A-14 Statistical Power of the Z Test.

Z	α = .05		α = .01	
	2-tail	1-tail	2-tail	1-tail
1.0	0.170	0.260	0.058	0.092
1.1	0.196	0.293	0.070	0.110
1.2	0.224	0.328	0.085	0.130
1.3	0.255	0.365	0.101	0.152
1.4	0.288	0.403	0.120	0.177
1.5	0.323	0.442	0.141	0.204
1.6	0.360	0.482	0.165	0.234
1.7	0.398	0.522	0.191	0.266
1.8	0.437	0.562	0.219	0.299
1.9	0.476	0.601	0.250	0.335
2.0	0.516	0.639	0.282	0.372
2.1	0.556	0.675	0.317	0.410
2.2	0.595	0.711	0.354	0.450
2.3	0.633	0.744	0.391	0.489
2.4	0.670	0.775	0.430	0.529
2.5	0.705	0.804	0.470	0.569
2.6	0.739	0.830	0.510	0.608
2.7	0.770	0.854	0.549	0.646
2.8	0.800	0.876	0.589	0.682
2.9	0.826	0.895	0.627	0.717
3.0	0.851	0.912	0.664	0.750
3.1	0.873	0.927	0.700	0.780
3.2	0.893	0.940	0.734	0.809
3.3	0.910	0.951	0.766	0.835
3.4	0.925	0.960	0.795	0.859
3.5	0.938	0.968	0.822	0.880
3.6	0.950	0.975	0.847	0.899
3.7	0.959	0.980	0.870	0.915
3.8	0.967	0.984	0.890	0.930
3.9	0.974	0.988	0.907	0.942
4.0	0.979	0.991	0.923	0.953

n	Cohen's *d*																			
	.20	.25	.30	.35	.40	.45	.50	.55	.60	.65	.70	.75	.80	.90	1.00	1.10	1.20	1.30	1.40	1.50
3	.06	.07	.07	.07	.08	.08	.09	.09	.10	.11	.11	.12	.13	.15	.18	.21	.24	.29	.33	.39
4	.07	.08	.09	.10	.11	.12	.13	.15	.17	.18	.21	.23	.25	.31	.37	.44	.52	.59	.66	.72
5	.08	.10	.11	.12	.14	.16	.18	.21	.24	.27	.30	.34	.37	.46	.54	.62	.69	.76	.81	.86
6	.09	.11	.13	.15	.17	.20	.23	.27	.30	.34	.39	.43	.48	.57	.66	.74	.80	.85	.89	.92
7	.10	.12	.15	.17	.21	.24	.28	.32	.37	.42	.47	.52	.57	.66	.75	.81	.87	.91	.94	.96
8	.11	.14	.17	.20	.24	.28	.32	.37	.42	.48	.53	.59	.64	.73	.81	.87	.91	.94	.96	.97
9	.12	.15	.18	.22	.26	.31	.36	.42	.48	.53	.59	.65	.70	.79	.86	.91	.94	.96	.98	.99
10	.13	.16	.20	.24	.29	.35	.40	.46	.52	.59	.64	.70	.75	.83	.89	.93	.96	.98	.99	.99
11	.14	.17	.22	.26	.32	.38	.44	.50	.57	.63	.69	.74	.79	.87	.92	.95	.97	.98	.99	.99
12	.15	.19	.23	.29	.34	.41	.48	.54	.61	.67	.73	.78	.82	.89	.94	.97	.98	.99	.99	
13	.15	.20	.25	.31	.37	.44	.51	.58	.65	.71	.76	.81	.85	.92	.95	.98	.99	.99		
14	.16	.21	.26	.33	.39	.47	.54	.61	.68	.74	.79	.84	.88	.93	.96	.98	.99			
15	.17	.22	.28	.35	.42	.49	.57	.64	.71	.77	.82	.86	.90	.95	.97	.99	.99			
16	.18	.23	.29	.36	.44	.52	.60	.67	.74	.79	.84	.88	.92	.96	.98	.99				
17	.19	.24	.31	.38	.46	.54	.62	.70	.76	.82	.86	.90	.93	.97	.98	.99				
18	.19	.25	.32	.40	.48	.57	.65	.72	.78	.84	.88	.92	.94	.97	.99					
19	.20	.26	.34	.42	.50	.59	.67	.74	.81	.86	.90	.93	.95	.98	.99					
20	.21	.27	.35	.44	.52	.61	.69	.76	.82	.87	.91	.94	.96	.98	.99					
21	.21	.28	.37	.45	.54	.63	.71	.78	.84	.89	.92	.95	.97	.99						
22	.22	.29	.38	.47	.56	.65	.73	.80	.86	.90	.93	.96	.97	.99						
23	.23	.30	.39	.48	.58	.67	.75	.82	.87	.91	.94	.96	.98	.99						
24	.24	.31	.40	.50	.60	.69	.77	.83	.88	.92	.95	.97	.98	.99						
30	.28	.37	.48	.59	.69	.77	.85	.90	.94	.96	.98	.99	.99							
40	.34	.46	.58	.70	.80	.87	.93	.96	.98	.99										
50	.40	.54	.67	.79	.87	.93	.97	.98	.99											
60	.45	.60	.74	.85	.92	.96	.98	.99												
70	.50	.66	.80	.89	.95	.98	.99													
80	.55	.72	.84	.93	.97	.99														
90	.59	.76	.88	.95	.98	.99														
100	.63	.80	.91	.97	.99															

TABLE A-15 Statistical Power of the t Test for One Sample or Two Related Samples.

n	Cohen's *d*																			
	.20	.25	.30	.35	.40	.45	.50	.55	.60	.65	.70	.75	.80	.90	1.00	1.10	1.20	1.30	1.40	1.50
3	.06	.07	.07	.07	.08	.08	.09	.09	.10	.11	.11	.12	.13	.15	.18	.21	.24	.29	.33	.39
4	.07	.08	.09	.10	.11	.12	.13	.15	.17	.18	.21	.23	.25	.31	.37	.44	.52	.59	.66	.72
5	.08	.10	.11	.12	.14	.16	.18	.21	.24	.27	.30	.34	.37	.46	.54	.62	.69	.76	.81	.86
6	.09	.11	.13	.15	.17	.20	.23	.27	.30	.34	.39	.43	.48	.57	.66	.74	.80	.85	.89	.92
7	.10	.12	.15	.17	.21	.24	.28	.32	.37	.42	.47	.52	.57	.66	.75	.81	.87	.91	.94	.96
8	.11	.14	.17	.20	.24	.28	.32	.37	.42	.48	.53	.59	.64	.73	.81	.87	.91	.94	.96	.97
9	.12	.15	.18	.22	.26	.31	.36	.42	.48	.53	.59	.65	.70	.79	.86	.91	.94	.96	.98	.99
10	.13	.16	.20	.24	.29	.35	.40	.46	.52	.59	.64	.70	.75	.83	.89	.93	.96	.98	.99	.99
11	.14	.17	.22	.26	.32	.38	.44	.50	.57	.63	.69	.74	.79	.87	.92	.95	.97	.98	.99	.99
12	.15	.19	.23	.29	.34	.41	.48	.54	.61	.67	.73	.78	.82	.89	.94	.97	.98	.99	.99	
13	.15	.20	.25	.31	.37	.44	.51	.58	.65	.71	.76	.81	.85	.92	.95	.98	.99	.99		
14	.16	.21	.26	.33	.39	.47	.54	.61	.68	.74	.79	.84	.88	.93	.96	.98	.99			
15	.17	.22	.28	.35	.42	.49	.57	.64	.71	.77	.82	.86	.90	.95	.97	.99	.99			
16	.18	.23	.29	.36	.44	.52	.60	.67	.74	.79	.84	.88	.92	.96	.98	.99				
17	.19	.24	.31	.38	.46	.54	.62	.70	.76	.82	.86	.90	.93	.97	.98	.99				
18	.19	.25	.32	.40	.48	.57	.65	.72	.78	.84	.88	.92	.94	.97	.99					
19	.20	.26	.34	.42	.50	.59	.67	.74	.81	.86	.90	.93	.95	.98	.99					
20	.21	.27	.35	.44	.52	.61	.69	.76	.82	.87	.91	.94	.96	.98	.99					
21	.21	.28	.37	.45	.54	.63	.71	.78	.84	.89	.92	.95	.97	.99						
22	.22	.29	.38	.47	.56	.65	.73	.80	.86	.90	.93	.96	.97	.99						
23	.23	.30	.39	.48	.58	.67	.75	.82	.87	.91	.94	.96	.98	.99						
24	.24	.31	.40	.50	.60	.69	.77	.83	.88	.92	.95	.97	.98	.99						
30	.28	.37	.48	.59	.69	.77	.85	.90	.94	.96	.98	.99	.99							
40	.34	.46	.58	.70	.80	.87	.93	.96	.98	.99										
50	.40	.54	.67	.79	.87	.93	.97	.98	.99											
60	.45	.60	.74	.85	.92	.96	.98	.99												
70	.50	.66	.80	.89	.95	.98	.99													
80	.55	.72	.84	.93	.97	.99														
90	.59	.76	.88	.95	.98	.99														
100	.63	.80	.91	.97	.99															

n	Cohen's *d*																			
	.20	.25	.30	.35	.40	.45	.50	.55	.60	.65	.70	.75	.80	.90	1.00	1.10	1.20	1.30	1.40	1.50
3	.01	.01	.01	.01	.01	.01	.01	.01	.01	.01	.01	.01	.02	.02	.02	.02	.02	.02	.02	.02
4	.01	.01	.01	.02	.02	.02	.02	.02	.02	.02	.03	.03	.03	.04	.04	.05	.06	.07	.09	.11
5	.01	.02	.02	.02	.02	.03	.03	.03	.04	.04	.05	.05	.06	.08	.10	.13	.17	.22	.29	.36
6	.02	.02	.02	.03	.03	.04	.04	.05	.06	.07	.08	.09	.11	.15	.20	.27	.34	.43	.52	.62
7	.02	.02	.03	.03	.04	.05	.06	.07	.09	.10	.12	.15	.17	.24	.32	.41	.51	.61	.70	.78
8	.02	.03	.03	.04	.05	.06	.08	.10	.12	.14	.17	.20	.24	.33	.44	.54	.65	.74	.82	.87
9	.03	.03	.04	.05	.06	.08	.10	.12	.15	.19	.22	.27	.32	.42	.54	.65	.75	.83	.89	.93
10	.03	.04	.05	.06	.08	.10	.12	.15	.19	.23	.28	.33	.39	.51	.63	.74	.82	.89	.93	.96
11	.03	.04	.05	.07	.09	.12	.15	.18	.23	.28	.33	.39	.46	.59	.70	.80	.87	.92	.96	.97
12	.03	.05	.06	.08	.10	.14	.17	.22	.27	.33	.39	.45	.52	.65	.76	.85	.91	.95	.97	.98
13	.04	.05	.07	.09	.12	.16	.20	.25	.31	.37	.44	.51	.58	.71	.81	.89	.94	.97	.98	.99
14	.04	.06	.08	.10	.13	.18	.22	.28	.35	.42	.49	.56	.63	.76	.85	.92	.96	.98	.99	.99
15	.04	.06	.08	.11	.15	.20	.25	.31	.38	.46	.53	.61	.68	.80	.88	.94	.97	.98	.99	
16	.05	.06	.09	.12	.17	.22	.28	.35	.42	.50	.58	.65	.72	.83	.91	.95	.98	.99		
17	.05	.07	.10	.14	.18	.24	.30	.38	.46	.54	.62	.69	.76	.86	.93	.97	.98	.99		
18	.05	.08	.11	.15	.20	.26	.33	.41	.49	.57	.65	.73	.79	.89	.94	.97	.99			
19	.06	.08	.11	.16	.21	.28	.36	.44	.52	.61	.69	.76	.82	.91	.96	.98	.99			
20	.06	.09	.12	.17	.23	.30	.38	.47	.56	.64	.72	.79	.84	.92	.97	.99	.99			
21	.06	.09	.13	.18	.25	.32	.41	.50	.59	.67	.75	.81	.87	.94	.97	.99				
22	.06	.10	.14	.20	.26	.34	.43	.52	.62	.70	.77	.84	.88	.95	.98	.99				
23	.07	.10	.15	.21	.28	.36	.46	.55	.64	.73	.80	.86	.90	.96	.98	.99				
24	.07	.11	.16	.22	.30	.39	.48	.58	.67	.75	.82	.87	.92	.97	.99					
30	.09	.14	.21	.29	.39	.50	.61	.71	.79	.86	.91	.94	.97	.99						
40	.13	.20	.30	.42	.54	.66	.77	.85	.91	.95	.97	.99	.99							
50	.16	.26	.39	.53	.66	.78	.87	.93	.96	.98	.99									
60	.20	.33	.47	.62	.76	.86	.93	.97	.99	.99										
70	.24	.39	.55	.71	.83	.91	.96	.99	.99											
80	.28	.45	.62	.77	.88	.95	.98	.99												
90	.32	.50	.68	.83	.92	.97	.99													
100	.36	.55	.74	.87	.95	.98														

TABLE A-16 Statistical Power of the t-Test for Two Independent Samples.

n	Cohen's *d*																			
	.20	.25	.30	.35	.40	.45	.50	.55	.60	.65	.70	.75	.80	.90	1.00	1.10	1.20	1.30	1.40	1.50
3	.01	.01	.01	.01	.01	.01	.01	.01	.01	.01	.01	.01	.01	.01	.02	.02	.02	.02	.02	.03
4	.01	.01	.01	.01	.01	.01	.01	.02	.02	.02	.02	.02	.02	.03	.03	.04	.05	.06	.07	.08
5	.01	.01	.01	.01	.02	.02	.02	.02	.02	.03	.03	.03	.04	.05	.06	.07	.09	.12	.14	.18
6	.01	.01	.01	.02	.02	.02	.02	.03	.03	.03	.04	.05	.05	.07	.09	.12	.15	.19	.24	.29
7	.01	.01	.02	.02	.02	.02	.03	.03	.04	.05	.05	.06	.07	.10	.13	.17	.22	.27	.34	.40
8	.01	.02	.02	.02	.02	.03	.03	.04	.05	.06	.07	.08	.10	.13	.17	.23	.29	.36	.43	.51
9	.01	.02	.02	.02	.03	.03	.04	.05	.06	.07	.09	.10	.12	.16	.22	.28	.36	.44	.52	.60
10	.01	.02	.02	.03	.03	.04	.05	.06	.07	.09	.10	.12	.15	.20	.26	.34	.42	.51	.60	.68
11	.02	.02	.02	.03	.04	.04	.06	.07	.08	.10	.12	.15	.17	.24	.31	.40	.49	.58	.67	.75
12	.02	.02	.03	.03	.04	.05	.06	.08	.10	.12	.14	.17	.20	.27	.36	.45	.55	.64	.73	.80
13	.02	.02	.03	.04	.04	.06	.07	.09	.11	.13	.16	.19	.23	.31	.40	.50	.60	.70	.78	.84
14	.02	.02	.03	.04	.05	.06	.08	.10	.12	.15	.18	.22	.26	.35	.45	.55	.65	.74	.82	.88
15	.02	.02	.03	.04	.05	.07	.09	.11	.14	.17	.20	.24	.29	.38	.49	.60	.70	.78	.85	.91
16	.02	.03	.03	.04	.06	.08	.10	.12	.15	.18	.22	.27	.31	.42	.53	.64	.74	.82	.88	.93
17	.02	.03	.04	.05	.06	.08	.10	.13	.17	.20	.25	.29	.34	.45	.57	.68	.77	.85	.91	.94
18	.02	.03	.04	.05	.07	.09	.11	.14	.18	.22	.27	.32	.37	.49	.61	.71	.81	.88	.92	.96
19	.02	.03	.04	.06	.07	.10	.12	.16	.20	.24	.29	.34	.40	.52	.64	.75	.83	.90	.94	.97
20	.02	.03	.04	.06	.08	.10	.13	.17	.21	.26	.31	.37	.43	.55	.67	.78	.86	.92	.95	.98
21	.02	.03	.05	.06	.08	.11	.14	.18	.23	.28	.33	.39	.46	.58	.70	.80	.88	.93	.96	.98
22	.02	.03	.05	.07	.09	.12	.15	.19	.24	.30	.35	.42	.48	.61	.73	.83	.90	.94	.97	.99
23	.03	.04	.05	.07	.09	.12	.16	.21	.26	.31	.38	.44	.51	.64	.76	.85	.91	.95	.98	.99
24	.03	.04	.05	.07	.10	.13	.17	.22	.27	.33	.40	.46	.53	.67	.78	.87	.93	.96	.98	.99
30	.03	.05	.07	.10	.13	.18	.24	.30	.37	.44	.52	.59	.67	.79	.88	.94	.97	.99		
40	.04	.07	.10	.14	.20	.27	.34	.43	.52	.60	.69	.76	.82	.91	.96	.99				
50	.05	.09	.13	.19	.27	.35	.45	.55	.65	.73	.81	.87	.91	.97	.99					
60	.07	.11	.17	.24	.33	.44	.55	.65	.75	.83	.89	.93	.96	.99						
70	.08	.13	.20	.29	.40	.52	.64	.74	.82	.89	.94	.96	.98							
80	.09	.15	.24	.35	.47	.59	.71	.81	.88	.93	.96	.98	.99							
90	.10	.18	.28	.40	.53	.66	.77	.86	.92	.96	.98	.99								
100	.12	.20	.32	.45	.59	.72	.82	.90	.95	.98	.99									

n	Cohen's *d*																			
	.20	.25	.30	.35	.40	.45	.50	.55	.60	.65	.70	.75	.80	.90	1.00	1.10	1.20	1.30	1.40	1.50
3	.07	.07	.08	.08	.09	.09	.10	.11	.12	.13	.14	.15	.16	.18	.21	.24	.27	.31	.35	.39
4	.07	.08	.09	.10	.11	.12	.13	.14	.16	.17	.19	.21	.22	.26	.31	.36	.41	.46	.51	.57
5	.08	.09	.10	.11	.13	.14	.16	.18	.19	.21	.24	.26	.28	.34	.39	.45	.51	.58	.63	.69
6	.09	.10	.11	.13	.14	.16	.18	.21	.23	.25	.28	.31	.34	.40	.47	.54	.60	.67	.72	.77
7	.09	.11	.12	.14	.16	.18	.21	.23	.26	.29	.32	.36	.39	.46	.53	.61	.67	.74	.79	.84
8	.10	.11	.13	.15	.18	.20	.23	.26	.29	.33	.36	.40	.44	.52	.59	.67	.73	.79	.84	.88
9	.10	.12	.14	.17	.19	.22	.25	.29	.32	.36	.40	.44	.48	.56	.64	.72	.78	.84	.88	.91
10	.11	.13	.15	.18	.21	.24	.27	.31	.35	.39	.43	.48	.52	.61	.69	.76	.82	.87	.91	.94
11	.11	.13	.16	.19	.22	.26	.29	.33	.38	.42	.47	.51	.56	.65	.73	.80	.86	.90	.93	.96
12	.12	.14	.17	.20	.23	.27	.31	.36	.40	.45	.50	.55	.59	.68	.76	.83	.88	.92	.95	.97
13	.12	.15	.18	.21	.25	.29	.33	.38	.43	.48	.53	.58	.63	.72	.80	.86	.90	.94	.96	.98
14	.13	.15	.19	.22	.26	.31	.35	.40	.45	.51	.56	.61	.66	.75	.82	.88	.92	.95	.97	.98
15	.13	.16	.19	.23	.27	.32	.37	.42	.48	.53	.58	.64	.69	.77	.85	.90	.94	.96	.98	.99
16	.13	.17	.20	.24	.29	.34	.39	.44	.50	.56	.61	.66	.71	.80	.87	.92	.95	.97	.98	.99
17	.14	.17	.21	.25	.30	.35	.41	.46	.52	.58	.63	.69	.74	.82	.88	.93	.96	.98	.99	.99
18	.14	.18	.22	.26	.31	.37	.42	.48	.54	.60	.66	.71	.76	.84	.90	.94	.97	.98	.99	
19	.15	.18	.23	.27	.33	.38	.44	.50	.56	.62	.68	.73	.78	.86	.91	.95	.97	.99	.99	
20	.15	.19	.23	.28	.34	.40	.46	.52	.58	.64	.70	.75	.80	.87	.93	.96	.98	.99		
21	.15	.19	.24	.29	.35	.41	.47	.54	.60	.66	.72	.77	.82	.89	.94	.97	.98	.99		
22	.16	.20	.25	.30	.36	.43	.49	.56	.62	.68	.74	.79	.83	.90	.95	.97	.99	.99		
23	.16	.20	.26	.31	.37	.44	.51	.57	.64	.70	.75	.80	.85	.91	.95	.98	.99			
24	.16	.21	.26	.32	.39	.45	.52	.59	.65	.72	.77	.82	.86	.92	.96	.98	.99			
30	.19	.24	.31	.38	.45	.53	.60	.68	.74	.80	.85	.89	.92	.96	.98	.99				
40	.22	.29	.37	.46	.55	.64	.72	.79	.84	.89	.93	.95	.97	.99						
50	.26	.34	.44	.54	.63	.72	.80	.86	.91	.94	.97	.98	.99							
60	.29	.39	.49	.60	.70	.79	.86	.91	.95	.97	.98	.99								
70	.32	.43	.55	.66	.76	.84	.90	.94	.97	.98	.99									
80	.35	.47	.60	.71	.81	.88	.93	.96	.98	.99										
90	.38	.51	.64	.76	.85	.91	.95	.98	.99											
100	.41	.55	.68	.79	.88	.94	.97	.99	.99											

n	Cohen's *d*																			
	.20	.25	.30	.35	.40	.45	.50	.55	.60	.65	.70	.75	.80	.90	1.00	1.10	1.20	1.30	1.40	1.50
3	.01	.01	.01	.01	.02	.02	.02	.02	.02	.02	.02	.02	.03	.03	.03	.04	.04	.05	.06	.06
4	.01	.02	.02	.02	.02	.02	.03	.03	.03	.03	.04	.04	.05	.06	.07	.08	.10	.12	.14	.17
5	.02	.02	.02	.02	.03	.03	.03	.04	.04	.05	.06	.06	.07	.09	.11	.14	.17	.21	.26	.31
6	.02	.02	.02	.03	.03	.04	.04	.05	.06	.07	.08	.09	.10	.13	.16	.21	.25	.31	.37	.44
7	.02	.02	.03	.03	.04	.05	.05	.06	.07	.08	.10	.11	.13	.17	.22	.27	.34	.40	.48	.55
8	.02	.03	.03	.04	.04	.05	.06	.07	.09	.10	.12	.14	.16	.21	.27	.34	.41	.49	.57	.64
9	.02	.03	.03	.04	.05	.06	.07	.09	.10	.12	.14	.17	.19	.25	.33	.40	.49	.57	.65	.72
10	.02	.03	.04	.05	.06	.07	.08	.10	.12	.14	.17	.20	.23	.30	.38	.46	.55	.64	.71	.78
11	.03	.03	.04	.05	.06	.08	.10	.12	.14	.16	.19	.23	.26	.34	.43	.52	.61	.70	.77	.83
12	.03	.04	.04	.06	.07	.09	.11	.13	.16	.18	.22	.25	.29	.38	.48	.57	.66	.75	.82	.87
13	.03	.04	.05	.06	.08	.10	.12	.14	.17	.21	.24	.28	.33	.42	.52	.62	.71	.79	.85	.90
14	.03	.04	.05	.07	.08	.10	.13	.16	.19	.23	.27	.31	.36	.46	.57	.67	.75	.83	.88	.93
15	.03	.04	.06	.07	.09	.11	.14	.17	.21	.25	.29	.34	.39	.50	.61	.71	.79	.86	.91	.94
16	.03	.05	.06	.08	.10	.12	.15	.19	.23	.27	.32	.37	.42	.53	.64	.74	.82	.88	.93	.96
17	.04	.05	.06	.08	.10	.13	.16	.20	.24	.29	.34	.40	.45	.57	.68	.77	.85	.91	.94	.97
18	.04	.05	.07	.09	.11	.14	.18	.22	.26	.31	.37	.42	.48	.60	.71	.80	.87	.92	.96	.98
19	.04	.05	.07	.09	.12	.15	.19	.23	.28	.33	.39	.45	.51	.63	.74	.83	.89	.94	.97	.98
20	.04	.05	.07	.10	.13	.16	.20	.25	.30	.36	.42	.48	.54	.66	.77	.85	.91	.95	.97	.99
21	.04	.06	.08	.10	.13	.17	.21	.26	.32	.38	.44	.50	.57	.69	.79	.87	.92	.96	.98	.99
22	.04	.06	.08	.11	.14	.18	.23	.28	.34	.40	.46	.53	.59	.71	.81	.89	.94	.97	.98	.99
23	.04	.06	.08	.11	.15	.19	.24	.29	.35	.42	.48	.55	.62	.74	.83	.90	.95	.97	.99	.99
24	.05	.06	.09	.12	.16	.20	.25	.31	.37	.44	.51	.57	.64	.76	.85	.92	.96	.98	.99	
30	.06	.08	.11	.15	.20	.26	.33	.40	.47	.55	.62	.69	.76	.86	.93	.97	.99	.99		
40	.07	.11	.15	.21	.28	.36	.44	.53	.62	.70	.77	.83	.88	.95	.98	.99				
50	.09	.13	.19	.27	.36	.45	.55	.65	.74	.81	.87	.92	.95	.98						
60	.10	.16	.24	.33	.43	.54	.65	.74	.82	.88	.93	.96	.98	.99						
70	.12	.19	.28	.39	.51	.62	.73	.82	.88	.93	.96	.98	.99							
80	.14	.22	.33	.45	.57	.69	.79	.87	.92	.96	.98	.99								
90	.16	.25	.37	.50	.63	.75	.84	.91	.95	.98	.99									
100	.18	.28	.41	.55	.69	.80	.88	.94	.97	.99										

TABLE A-17 Statistical Power of the Analysis of Variance.

n	ω^2																			
	.02	.04	.06	.08	.10	.12	.14	.16	.18	.20	.22	.24	.26	.28	.30	.32	.34	.36	.38	.40
3	.06	.07	.08	.09	.11	.12	.13	.15	.17	.18	.20	.22	.25	.27	.29	.32	.35	.38	.41	.44
4	.06	.08	.10	.12	.14	.16	.19	.21	.24	.27	.30	.33	.37	.40	.44	.48	.52	.56	.60	.64
5	.07	.09	.12	.14	.17	.20	.24	.28	.31	.35	.40	.44	.48	.53	.57	.62	.66	.70	.74	.78
6	.08	.10	.14	.17	.21	.25	.29	.34	.39	.44	.49	.54	.59	.63	.68	.73	.77	.81	.84	.87
7	.08	.12	.15	.20	.25	.30	.35	.40	.46	.51	.57	.62	.68	.72	.77	.81	.85	.88	.91	.93
8	.09	.13	.17	.23	.28	.34	.40	.47	.53	.59	.65	.70	.75	.80	.84	.87	.90	.93	.95	.96
9	.09	.14	.20	.26	.32	.39	.46	.52	.59	.65	.71	.76	.81	.85	.89	.92	.94	.96	.97	.98
10	.10	.15	.22	.29	.36	.43	.51	.58	.65	.71	.77	.82	.86	.89	.92	.95	.96	.98	.98	.99
11	.10	.17	.24	.31	.39	.48	.55	.63	.70	.76	.81	.86	.90	.93	.95	.97	.98	.99	.99	
12	.11	.18	.26	.34	.43	.52	.60	.68	.74	.80	.85	.89	.92	.95	.97	.98	.99	.99		
13	.11	.19	.28	.37	.47	.56	.64	.72	.78	.84	.88	.92	.95	.96	.98	.99	.99			
14	.12	.21	.30	.40	.50	.60	.68	.76	.82	.87	.91	.94	.96	.98	.99	.99				
15	.13	.22	.32	.43	.53	.63	.72	.79	.85	.90	.93	.96	.97	.98	.99					
16	.13	.23	.34	.46	.57	.66	.75	.82	.88	.92	.95	.97	.98	.99	.99					
17	.14	.25	.37	.48	.60	.70	.78	.85	.90	.93	.96	.98	.99	.99						
18	.14	.26	.39	.51	.63	.73	.81	.87	.92	.95	.97	.98	.99							
19	.15	.27	.41	.54	.65	.75	.83	.89	.93	.96	.98	.99	.99							
20	.16	.29	.43	.56	.68	.78	.85	.91	.94	.97	.98	.99								
21	.16	.30	.45	.59	.71	.80	.87	.92	.95	.98	.99	.99								
22	.17	.32	.47	.61	.73	.82	.89	.93	.96	.98	.99									
23	.18	.33	.49	.63	.75	.84	.90	.94	.97	.99	.99									
24	.18	.34	.51	.65	.77	.86	.92	.95	.98	.99	.99									
30	.22	.43	.62	.77	.87	.93	.97	.99	.99											
40	.29	.55	.76	.89	.95	.98	.99													
50	.36	.66	.86	.95	.98															
60	.43	.75	.92	.98																
70	.49	.82	.96	.99																
80	.55	.88	.98																	
90	.61	.91	.99																	
100	.66	.94	.99																	

n	ω^2																			
	.02	.04	.06	.08	.10	.12	.14	.16	.18	.20	.22	.24	.26	.28	.30	.32	.34	.36	.38	.40
3	.01	.01	.02	.02	.02	.03	.03	.03	.04	.04	.05	.06	.06	.07	.08	.09	.10	.11	.12	.13
4	.01	.02	.02	.03	.03	.04	.05	.06	.07	.08	.09	.10	.11	.13	.15	.16	.18	.21	.23	.26
5	.02	.02	.03	.04	.05	.06	.07	.08	.10	.11	.13	.15	.18	.20	.23	.26	.29	.32	.36	.40
6	.02	.03	.04	.05	.06	.08	.09	.11	.13	.16	.19	.22	.25	.28	.32	.36	.40	.45	.49	.54
7	.02	.03	.04	.06	.08	.10	.12	.15	.18	.21	.24	.28	.32	.37	.41	.46	.51	.56	.61	.66
8	.02	.03	.05	.07	.09	.12	.15	.18	.22	.26	.30	.35	.40	.45	.50	.55	.61	.66	.71	.75
9	.02	.04	.06	.08	.11	.14	.18	.22	.27	.31	.37	.42	.47	.53	.59	.64	.69	.74	.79	.83
10	.02	.04	.07	.09	.13	.17	.21	.26	.31	.37	.43	.48	.54	.60	.66	.71	.76	.81	.85	.89
11	.03	.05	.08	.11	.15	.19	.25	.30	.36	.42	.48	.55	.61	.67	.73	.78	.82	.86	.90	.92
12	.03	.05	.08	.12	.17	.22	.28	.34	.41	.47	.54	.61	.67	.73	.78	.83	.87	.90	.93	.95
13	.03	.06	.09	.14	.19	.25	.32	.38	.45	.53	.59	.66	.72	.78	.83	.87	.90	.93	.95	.97
14	.03	.06	.10	.16	.21	.28	.35	.42	.50	.57	.64	.71	.77	.82	.87	.90	.93	.95	.97	.98
15	.03	.07	.12	.17	.24	.31	.39	.46	.54	.62	.69	.75	.81	.86	.90	.93	.95	.97	.98	.99
16	.04	.07	.13	.19	.26	.34	.42	.50	.58	.66	.73	.79	.84	.89	.92	.95	.97	.98	.99	.99
17	.04	.08	.14	.21	.28	.37	.46	.54	.62	.70	.77	.83	.87	.91	.94	.96	.98	.99	.99	
18	.04	.09	.15	.22	.31	.40	.49	.58	.66	.74	.80	.85	.90	.93	.95	.97	.98	.99		
19	.04	.09	.16	.24	.33	.43	.52	.61	.70	.77	.83	.88	.92	.95	.97	.98	.99	.99		
20	.05	.10	.17	.26	.36	.46	.55	.65	.73	.80	.86	.90	.93	.96	.98	.99	.99			
21	.05	.11	.19	.28	.38	.48	.58	.68	.76	.82	.88	.92	.95	.97	.98	.99	.99			
22	.05	.11	.20	.30	.40	.51	.61	.71	.78	.85	.90	.93	.96	.98	.99	.99				
23	.05	.12	.21	.32	.43	.54	.64	.73	.81	.87	.91	.95	.97	.98	.99					
24	.06	.13	.22	.33	.45	.56	.67	.76	.83	.89	.93	.96	.97	.99	.99					
30	.07	.17	.31	.45	.58	.70	.80	.87	.92	.96	.98	.99	.99							
40	.10	.26	.44	.62	.76	.86	.92	.96	.98	.99										
50	.14	.35	.57	.75	.87	.94	.97	.99												
60	.18	.44	.68	.85	.94	.98	.99													
70	.22	.52	.77	.91	.97	.99														
80	.26	.60	.84	.95	.99															
90	.30	.67	.89	.97	.99															
100	.35	.74	.93	.98																

n	ω^2																			
	.02	.04	.06	.08	.10	.12	.14	.16	.18	.20	.22	.24	.26	.28	.30	.32	.34	.36	.38	.40
3	.01	.02	.02	.02	.03	.03	.03	.04	.05	.05	.06	.07	.08	.09	.10	.11	.12	.14	.15	.17
4	.01	.02	.02	.03	.04	.05	.05	.07	.08	.09	.11	.12	.14	.16	.18	.21	.24	.26	.30	.33
5	.02	.02	.03	.04	.05	.07	.08	.10	.12	.14	.16	.19	.22	.25	.29	.33	.37	.41	.46	.50
6	.02	.03	.04	.05	.07	.09	.11	.14	.16	.20	.23	.27	.31	.35	.40	.45	.50	.55	.60	.65
7	.02	.03	.05	.06	.09	.11	.14	.18	.21	.26	.30	.35	.40	.46	.51	.56	.62	.67	.72	.77
8	.02	.04	.06	.08	.11	.14	.18	.22	.27	.32	.38	.43	.49	.55	.61	.67	.72	.77	.82	.86
9	.02	.04	.06	.09	.13	.17	.22	.27	.33	.39	.45	.51	.58	.64	.70	.75	.80	.84	.88	.91
10	.02	.05	.07	.11	.15	.20	.26	.32	.38	.45	.52	.59	.65	.71	.77	.82	.86	.90	.93	.95
11	.03	.05	.09	.13	.18	.24	.30	.37	.44	.51	.59	.66	.72	.78	.83	.87	.91	.93	.96	.97
12	.03	.06	.10	.15	.20	.27	.34	.42	.50	.57	.65	.72	.78	.83	.87	.91	.94	.96	.97	.98
13	.03	.06	.11	.17	.23	.31	.39	.47	.55	.63	.70	.77	.82	.87	.91	.94	.96	.98	.99	.99
14	.03	.07	.12	.19	.26	.34	.43	.52	.60	.68	.75	.81	.86	.90	.94	.96	.97	.99	.99	
15	.04	.08	.13	.21	.29	.38	.47	.56	.65	.73	.79	.85	.90	.93	.96	.97	.98	.99		
16	.04	.08	.15	.23	.32	.41	.51	.60	.69	.77	.83	.88	.92	.95	.97	.98	.99	.99		
17	.04	.09	.16	.25	.35	.45	.55	.65	.73	.80	.86	.91	.94	.96	.98	.99	.99			
18	.04	.10	.18	.27	.37	.48	.59	.68	.77	.83	.89	.93	.96	.97	.99	.99				
19	.05	.11	.19	.29	.40	.52	.62	.72	.80	.86	.91	.94	.97	.98	.99					
20	.05	.12	.21	.32	.43	.55	.66	.75	.83	.89	.93	.96	.98	.99	.99					
21	.05	.12	.22	.34	.46	.58	.69	.78	.85	.91	.94	.97	.98	.99						
22	.06	.13	.24	.36	.49	.61	.72	.81	.87	.92	.96	.98	.99	.99						
23	.06	.14	.25	.38	.52	.64	.75	.83	.89	.94	.96	.98	.99							
24	.06	.15	.27	.41	.54	.67	.77	.85	.91	.95	.97	.99	.99							
30	.08	.21	.37	.54	.69	.80	.89	.94	.97	.99	.99									
40	.12	.31	.53	.72	.85	.93	.97	.99												
50	.16	.42	.67	.84	.94	.98	.99													
60	.21	.53	.78	.92	.98	.99														
70	.26	.62	.86	.96	.99															
80	.31	.71	.91	.98																
90	.37	.78	.95	.99																
100	.42	.83	.97																	

n	ω^2																			
	.02	.04	.06	.08	.10	.12	.14	.16	.18	.20	.22	.24	.26	.28	.30	.32	.34	.36	.38	.40
3	.01	.02	.02	.02	.03	.03	.04	.04	.05	.06	.07	.08	.09	.10	.12	.13	.15	.17	.19	.21
4	.01	.02	.03	.03	.04	.05	.06	.08	.09	.11	.13	.15	.17	.20	.22	.26	.29	.33	.37	.41
5	.02	.02	.03	.04	.06	.07	.09	.11	.14	.17	.20	.23	.27	.31	.35	.40	.45	.49	.55	.60
6	.02	.03	.04	.06	.08	.10	.13	.16	.19	.23	.28	.32	.37	.43	.48	.54	.59	.65	.70	.75
7	.02	.03	.05	.07	.10	.13	.17	.21	.26	.31	.36	.42	.48	.54	.60	.66	.71	.77	.81	.85
8	.02	.04	.06	.09	.12	.17	.21	.26	.32	.38	.45	.51	.58	.64	.70	.76	.81	.85	.89	.92
9	.02	.04	.07	.11	.15	.20	.26	.32	.39	.46	.53	.60	.67	.73	.79	.84	.88	.91	.94	.96
10	.03	.05	.08	.13	.18	.24	.31	.38	.46	.53	.61	.68	.74	.80	.85	.89	.92	.95	.97	.98
11	.03	.06	.10	.15	.21	.28	.36	.44	.52	.60	.68	.74	.81	.86	.90	.93	.95	.97	.98	.99
12	.03	.06	.11	.17	.24	.32	.41	.49	.58	.66	.74	.80	.86	.90	.93	.96	.97	.98	.99	
13	.03	.07	.13	.19	.27	.36	.46	.55	.64	.72	.79	.85	.89	.93	.96	.97	.98	.99		
14	.04	.08	.14	.22	.31	.40	.50	.60	.69	.77	.83	.88	.92	.95	.97	.98	.99			
15	.04	.09	.16	.24	.34	.45	.55	.65	.74	.81	.87	.91	.95	.97	.98	.99				
16	.04	.10	.17	.27	.37	.49	.59	.69	.78	.85	.90	.94	.96	.98	.99	.99				
17	.04	.10	.19	.29	.41	.53	.64	.73	.81	.88	.92	.95	.97	.99	.99					
18	.05	.11	.21	.32	.44	.56	.67	.77	.84	.90	.94	.97	.98	.99						
19	.05	.12	.22	.35	.48	.60	.71	.80	.87	.92	.96	.98	.99	.99						
20	.05	.13	.24	.37	.51	.63	.74	.83	.89	.94	.97	.98	.99							
21	.06	.14	.26	.40	.54	.67	.77	.86	.91	.95	.98	.99	.99							
22	.06	.15	.28	.43	.57	.70	.80	.88	.93	.96	.98	.99								
23	.06	.16	.30	.45	.60	.73	.83	.90	.94	.97	.99	.99								
24	.07	.17	.32	.48	.63	.75	.85	.91	.96	.98	.99									
30	.09	.24	.44	.62	.77	.88	.94	.97	.99											
40	.14	.37	.62	.80	.91	.97	.99													
50	.19	.49	.76	.91	.97	.99														
60	.25	.61	.86	.96	.99															
70	.31	.71	.92	.98																
80	.37	.79	.96	.99																
90	.43	.85	.98																	
100	.49	.90	.99																	

n	ω^2																			
	.02	.04	.06	.08	.10	.12	.14	.16	.18	.20	.22	.24	.26	.28	.30	.32	.34	.36	.38	.40
3	.01	.02	.02	.02	.03	.03	.04	.04	.05	.06	.07	.08	.09	.10	.12	.13	.15	.17	.19	.21
4	.01	.02	.03	.03	.04	.05	.06	.08	.09	.11	.13	.15	.17	.20	.22	.26	.29	.33	.37	.41
5	.02	.02	.03	.04	.06	.07	.09	.11	.14	.17	.20	.23	.27	.31	.35	.40	.45	.49	.55	.60
6	.02	.03	.04	.06	.08	.10	.13	.16	.19	.23	.28	.32	.37	.43	.48	.54	.59	.65	.70	.75
7	.02	.03	.05	.07	.10	.13	.17	.21	.26	.31	.36	.42	.48	.54	.60	.66	.71	.77	.81	.85
8	.02	.04	.06	.09	.12	.17	.21	.26	.32	.38	.45	.51	.58	.64	.70	.76	.81	.85	.89	.92
9	.02	.04	.07	.11	.15	.20	.26	.32	.39	.46	.53	.60	.67	.73	.79	.84	.88	.91	.94	.96
10	.03	.05	.08	.13	.18	.24	.31	.38	.46	.53	.61	.68	.74	.80	.85	.89	.92	.95	.97	.98
11	.03	.06	.10	.15	.21	.28	.36	.44	.52	.60	.68	.74	.81	.86	.90	.93	.95	.97	.98	.99
12	.03	.06	.11	.17	.24	.32	.41	.49	.58	.66	.74	.80	.86	.90	.93	.96	.97	.98	.99	
13	.03	.07	.13	.19	.27	.36	.46	.55	.64	.72	.79	.85	.89	.93	.96	.97	.98	.99		
14	.04	.08	.14	.22	.31	.40	.50	.60	.69	.77	.83	.88	.92	.95	.97	.98	.99			
15	.04	.09	.16	.24	.34	.45	.55	.65	.74	.81	.87	.91	.95	.97	.98	.99				
16	.04	.10	.17	.27	.37	.49	.59	.69	.78	.85	.90	.94	.96	.98	.99	.99				
17	.04	.10	.19	.29	.41	.53	.64	.73	.81	.88	.92	.95	.97	.99	.99					
18	.05	.11	.21	.32	.44	.56	.67	.77	.84	.90	.94	.97	.98	.99						
19	.05	.12	.22	.35	.48	.60	.71	.80	.87	.92	.96	.98	.99	.99						
20	.05	.13	.24	.37	.51	.63	.74	.83	.89	.94	.97	.98	.99							
21	.06	.14	.26	.40	.54	.67	.77	.86	.91	.95	.98	.99	.99							
22	.06	.15	.28	.43	.57	.70	.80	.88	.93	.96	.98	.99								
23	.06	.16	.30	.45	.60	.73	.83	.90	.94	.97	.99	.99								
24	.07	.17	.32	.48	.63	.75	.85	.91	.96	.98	.99									
30	.09	.24	.44	.62	.77	.88	.94	.97	.99											
40	.14	.37	.62	.80	.91	.97	.99													
50	.19	.49	.76	.91	.97	.99														
60	.25	.61	.86	.96	.99															
70	.31	.71	.92	.98																
80	.37	.79	.96	.99																
90	.43	.85	.98																	
100	.49	.90	.99																	

TABLE A-18 Statistical Power of the Correlation Coefficient.

n	*r*														
	.10	.15	.20	.25	.30	.35	.40	.45	.50	.55	.60	.65	.70	.75	.80
3	.05	.05	.05	.05	.05	.05	.05	.05	.05	.06	.06	.06	.06	.06	.06
4	.05	.06	.06	.06	.07	.07	.07	.08	.09	.09	.10	.11	.13	.16	.20
5	.06	.06	.07	.08	.08	.09	.10	.12	.13	.16	.18	.22	.28	.36	.48
6	.06	.07	.08	.09	.10	.12	.14	.16	.19	.23	.28	.35	.44	.55	.69
7	.07	.08	.09	.11	.12	.15	.17	.21	.25	.31	.37	.46	.57	.69	.81
8	.07	.08	.10	.12	.14	.17	.21	.25	.31	.38	.46	.56	.67	.78	.88
9	.07	.09	.11	.13	.16	.20	.24	.30	.36	.44	.53	.64	.75	.85	.93
10	.08	.10	.12	.15	.18	.22	.27	.34	.41	.50	.60	.70	.81	.89	.95
11	.08	.10	.13	.16	.20	.25	.31	.38	.46	.56	.66	.76	.85	.92	.97
12	.08	.11	.14	.17	.22	.27	.34	.42	.51	.60	.71	.80	.89	.95	.98
13	.09	.11	.14	.18	.23	.29	.37	.45	.55	.65	.75	.84	.91	.96	.99
14	.09	.12	.15	.20	.25	.32	.40	.49	.58	.69	.78	.87	.93	.97	.99
15	.09	.12	.16	.21	.27	.34	.42	.52	.62	.72	.82	.89	.95	.98	
16	.09	.13	.17	.22	.28	.36	.45	.55	.65	.75	.84	.91	.96	.99	
17	.10	.13	.18	.23	.30	.38	.48	.58	.68	.78	.87	.93	.97	.99	
18	.10	.14	.18	.24	.32	.40	.50	.60	.71	.81	.89	.94	.98	.99	
19	.10	.14	.19	.25	.33	.42	.52	.63	.73	.83	.90	.95	.98		
20	.10	.14	.20	.27	.35	.44	.55	.65	.76	.85	.92	.96	.99		
21	.11	.15	.21	.28	.36	.46	.57	.68	.78	.87	.93	.97	.99		
22	.11	.15	.21	.29	.38	.48	.59	.70	.80	.88	.94	.98	.99		
23	.11	.16	.22	.30	.39	.50	.61	.72	.82	.90	.95	.98	.99		
24	.11	.16	.23	.31	.41	.51	.63	.74	.83	.91	.96	.98			
25	.12	.17	.23	.32	.42	.53	.65	.76	.85	.92	.96	.99			
26	.12	.17	.24	.33	.43	.55	.66	.77	.86	.93	.97	.99			
27	.12	.18	.25	.34	.45	.56	.68	.79	.88	.94	.97	.99			
28	.12	.18	.26	.35	.46	.58	.70	.80	.89	.94	.98	.99			
29	.12	.18	.26	.36	.47	.59	.71	.82	.90	.95	.98	.99			
30	.13	.19	.27	.37	.49	.61	.73	.83	.91	.96	.98				
40	.15	.23	.34	.46	.60	.73	.84	.92	.97	.99					
50	.17	.27	.40	.54	.69	.82	.91	.96	.99						
60	.18	.30	.45	.62	.76	.88	.95	.98							
70	.20	.34	.51	.68	.82	.92	.97	.99							
80	.22	.37	.55	.73	.87	.95	.98								
90	.24	.41	.60	.78	.90	.97	.99								
100	.25	.44	.64	.81	.93	.98									
120	.29	.50	.71	.87	.96	.99									
140	.32	.55	.77	.91	.98										
160	.35	.60	.82	.94	.99										
180	.38	.64	.86	.96	.99										
200	.41	.68	.89	.98											
300	.53	.83	.97												
400	.64	.92	.99												
500	.72	.96													

n	r														
	.10	.15	.20	.25	.30	.35	.40	.45	.50	.55	.60	.65	.70	.75	.80
3	.05	.05	.05	.05	.05	.05	.05	.05	.05	.05	.05	.05	.05	.05	.05
4	.05	.05	.05	.05	.05	.05	.05	.05	.05	.06	.06	.06	.06	.07	.08
5	.05	.05	.05	.05	.06	.06	.06	.07	.07	.08	.09	.10	.13	.16	.23
6	.05	.05	.06	.06	.06	.07	.08	.09	.10	.12	.14	.18	.23	.32	.46
7	.05	.05	.06	.06	.07	.08	.09	.11	.13	.17	.21	.27	.36	.49	.65
8	.05	.06	.06	.07	.08	.10	.12	.14	.17	.22	.28	.37	.48	.62	.78
9	.05	.06	.07	.08	.09	.11	.14	.17	.22	.28	.36	.46	.59	.73	.86
10	.05	.06	.07	.08	.10	.13	.16	.20	.26	.34	.43	.54	.67	.80	.91
11	.06	.06	.08	.09	.12	.15	.19	.24	.31	.39	.50	.62	.74	.86	.94
12	.06	.07	.08	.10	.13	.16	.21	.27	.35	.44	.56	.68	.80	.90	.96
13	.06	.07	.09	.11	.14	.18	.24	.30	.39	.49	.61	.73	.84	.93	.98
14	.06	.07	.09	.12	.15	.20	.26	.34	.43	.54	.66	.78	.88	.95	.98
15	.06	.07	.10	.12	.16	.22	.28	.37	.47	.58	.70	.81	.90	.96	.99
16	.06	.08	.10	.13	.18	.24	.31	.40	.51	.62	.74	.85	.92	.97	.99
17	.06	.08	.11	.14	.19	.25	.33	.43	.54	.66	.77	.87	.94	.98	
18	.06	.08	.11	.15	.20	.27	.36	.46	.57	.69	.80	.89	.95	.99	
19	.07	.09	.12	.16	.22	.29	.38	.49	.61	.72	.83	.91	.96	.99	
20	.07	.09	.12	.17	.23	.31	.40	.51	.63	.75	.85	.93	.97	.99	
21	.07	.09	.13	.17	.24	.32	.43	.54	.66	.78	.87	.94	.98	.99	
22	.07	.09	.13	.18	.25	.34	.45	.57	.69	.80	.89	.95	.98		
23	.07	.10	.14	.19	.27	.36	.47	.59	.71	.82	.91	.96	.99		
24	.07	.10	.14	.20	.28	.38	.49	.61	.73	.84	.92	.97	.99		
25	.07	.10	.15	.21	.29	.39	.51	.63	.75	.86	.93	.97	.99		
26	.07	.10	.15	.22	.30	.41	.53	.66	.77	.87	.94	.98	.99		
27	.07	.11	.16	.23	.32	.43	.55	.68	.79	.89	.95	.98			
28	.08	.11	.16	.23	.33	.44	.57	.69	.81	.90	.96	.99			
29	.08	.11	.17	.24	.34	.46	.58	.71	.82	.91	.96	.99			
30	.08	.12	.17	.25	.35	.47	.60	.73	.84	.92	.97	.99			
40	.09	.14	.23	.33	.47	.61	.75	.86	.93	.98	.99				
50	.10	.17	.28	.41	.57	.72	.84	.93	.97	.99					
60	.11	.20	.33	.49	.65	.80	.90	.96	.99						
70	.13	.23	.38	.55	.72	.86	.94	.98							
80	.14	.26	.43	.61	.78	.90	.97	.99							
90	.15	.29	.47	.67	.83	.93	.98								
100	.16	.32	.51	.72	.87	.96	.99								
120	.19	.37	.59	.79	.92	.98									
140	.21	.42	.66	.85	.96	.99									
160	.24	.47	.72	.90	.98										
180	.26	.52	.77	.93	.99										
200	.29	.56	.82	.95	.99										
300	.41	.74	.94	.99											
400	.52	.86	.98												
500	.61	.92													

n	*r*														
	.10	.15	.20	.25	.30	.35	.40	.45	.50	.55	.60	.65	.70	.75	.80
3	.01	.01	.01	.01	.01	.01	.01	.01	.01	.01	.01	.01	.01	.01	.01
4	.01	.01	.01	.01	.01	.01	.01	.01	.01	.01	.01	.01	.02	.02	.02
5	.01	.01	.01	.01	.01	.02	.02	.02	.02	.02	.02	.03	.03	.04	.06
6	.01	.01	.01	.02	.02	.02	.02	.03	.03	.04	.04	.06	.07	.11	.17
7	.01	.01	.02	.02	.02	.03	.03	.04	.05	.06	.08	.10	.15	.22	.36
8	.01	.02	.02	.02	.03	.03	.04	.05	.07	.09	.12	.17	.24	.36	.55
9	.01	.02	.02	.03	.03	.04	.05	.07	.09	.12	.17	.24	.35	.50	.69
10	.02	.02	.02	.03	.04	.05	.07	.09	.12	.17	.23	.32	.45	.62	.80
11	.02	.02	.03	.04	.05	.06	.08	.11	.15	.21	.29	.40	.55	.71	.87
12	.02	.02	.03	.04	.05	.07	.10	.13	.19	.26	.35	.48	.63	.78	.91
13	.02	.02	.03	.04	.06	.08	.12	.16	.22	.30	.41	.55	.70	.84	.94
14	.02	.03	.04	.05	.07	.10	.13	.18	.25	.35	.47	.61	.76	.88	.96
15	.02	.03	.04	.05	.08	.11	.15	.21	.29	.39	.52	.66	.80	.91	.97
16	.02	.03	.04	.06	.08	.12	.17	.24	.32	.44	.57	.71	.84	.94	.98
17	.02	.03	.05	.06	.09	.13	.19	.26	.36	.48	.62	.76	.87	.95	.99
18	.02	.03	.05	.07	.10	.15	.21	.29	.39	.52	.66	.79	.90	.97	.99
19	.02	.03	.05	.08	.11	.16	.23	.32	.43	.56	.70	.82	.92	.97	
20	.02	.04	.05	.08	.12	.17	.25	.34	.46	.59	.73	.85	.94	.98	
21	.02	.04	.06	.09	.13	.19	.27	.37	.49	.63	.76	.88	.95	.99	
22	.03	.04	.06	.09	.14	.20	.29	.39	.52	.66	.79	.90	.96	.99	
23	.03	.04	.06	.10	.15	.21	.31	.42	.55	.69	.82	.91	.97	.99	
24	.03	.04	.07	.10	.16	.23	.32	.44	.58	.72	.84	.93	.98	.99	
25	.03	.04	.07	.11	.17	.24	.34	.47	.60	.74	.86	.94	.98		
26	.03	.05	.07	.12	.18	.26	.36	.49	.63	.76	.88	.95	.99		
27	.03	.05	.08	.12	.19	.27	.38	.51	.65	.79	.89	.96	.99		
28	.03	.05	.08	.13	.19	.29	.40	.54	.68	.81	.90	.96	.99		
29	.03	.05	.08	.13	.20	.30	.42	.56	.70	.82	.92	.97	.99		
30	.03	.05	.09	.14	.21	.31	.44	.58	.72	.84	.93	.98	.99		
40	.04	.07	.12	.20	.31	.45	.60	.75	.87	.94	.98				
50	.05	.09	.16	.27	.41	.57	.73	.86	.94	.98					
60	.05	.11	.20	.34	.50	.67	.82	.92	.98	.99					
70	.06	.13	.24	.40	.58	.76	.89	.96	.99						
80	.07	.15	.28	.46	.66	.82	.93	.98							
90	.08	.17	.33	.52	.72	.87	.96	.99							
100	.09	.20	.37	.58	.77	.91	.97	.99							
120	.10	.24	.44	.67	.85	.95	.99								
140	.12	.28	.52	.75	.91	.98									
160	.14	.33	.59	.81	.94	.99									
180	.16	.37	.65	.86	.97										
200	.18	.42	.70	.90	.98										
300	.27	.61	.88	.98											
400	.37	.76	.96												
500	.46	.85	.99												

n	r														
	.10	.15	.20	.25	.30	.35	.40	.45	.50	.55	.60	.65	.70	.75	.80
3	.01	.01	.01	.01	.01	.01	.01	.01	.01	.01	.01	.01	.01	.01	.01
4	.01	.01	.01	.01	.01	.01	.01	.01	.01	.01	.01	.01	.01	.01	.01
5	.01	.01	.01	.01	.01	.01	.01	.01	.01	.01	.01	.01	.02	.02	.02
6	.01	.01	.01	.01	.01	.01	.01	.01	.02	.02	.02	.02	.03	.04	.06
7	.01	.01	.01	.01	.01	.01	.02	.02	.02	.03	.03	.04	.06	.10	.17
8	.01	.01	.01	.01	.02	.02	.02	.03	.03	.04	.06	.08	.12	.19	.34
9	.01	.01	.01	.02	.02	.02	.03	.03	.05	.06	.09	.13	.20	.32	.51
10	.01	.01	.01	.02	.02	.03	.03	.05	.06	.09	.13	.19	.29	.44	.66
11	.01	.01	.02	.02	.02	.03	.04	.06	.08	.12	.17	.26	.38	.56	.76
12	.01	.01	.02	.02	.03	.04	.05	.07	.10	.15	.22	.33	.47	.66	.84
13	.01	.01	.02	.02	.03	.04	.06	.09	.13	.19	.27	.40	.56	.74	.89
14	.01	.02	.02	.03	.04	.05	.07	.11	.16	.23	.33	.46	.63	.80	.93
15	.01	.02	.02	.03	.04	.06	.09	.13	.18	.27	.38	.53	.69	.85	.95
16	.01	.02	.02	.03	.05	.07	.10	.15	.21	.31	.43	.59	.75	.89	.97
17	.01	.02	.02	.04	.05	.08	.11	.17	.24	.35	.48	.64	.80	.92	.98
18	.01	.02	.03	.04	.06	.09	.13	.19	.27	.39	.53	.69	.83	.94	.99
19	.01	.02	.03	.04	.06	.10	.14	.21	.31	.43	.58	.73	.87	.95	.99
20	.01	.02	.03	.05	.07	.11	.16	.23	.34	.47	.62	.77	.89	.97	.99
21	.01	.02	.03	.05	.08	.12	.17	.26	.37	.50	.66	.80	.91	.97	
22	.02	.02	.03	.05	.08	.13	.19	.28	.40	.54	.69	.83	.93	.98	
23	.02	.02	.04	.06	.09	.14	.21	.30	.43	.57	.72	.86	.94	.99	
24	.02	.02	.04	.06	.10	.15	.22	.33	.46	.61	.75	.88	.96	.99	
25	.02	.03	.04	.07	.10	.16	.24	.35	.48	.64	.78	.90	.96	.99	
26	.02	.03	.04	.07	.11	.17	.26	.37	.51	.66	.81	.91	.97	.99	
27	.02	.03	.05	.07	.12	.18	.28	.40	.54	.69	.83	.93	.98		
28	.02	.03	.05	.08	.13	.20	.29	.42	.56	.72	.85	.94	.98		
29	.02	.03	.05	.08	.13	.21	.31	.44	.59	.74	.87	.95	.99		
30	.02	.03	.05	.09	.14	.22	.33	.46	.61	.76	.88	.96	.99		
40	.02	.04	.08	.13	.22	.34	.49	.65	.80	.91	.97	.99			
50	.03	.05	.11	.19	.31	.46	.63	.79	.90	.97	.99				
60	.03	.07	.14	.24	.39	.57	.74	.88	.96	.99					
70	.04	.08	.17	.30	.48	.67	.83	.93	.98						
80	.04	.10	.20	.36	.55	.74	.89	.96	.99						
90	.05	.11	.24	.42	.62	.81	.93	.98							
100	.05	.13	.27	.47	.69	.86	.95	.99							
120	.06	.17	.34	.57	.79	.92	.98								
140	.08	.20	.42	.66	.86	.96	.99								
160	.09	.24	.48	.74	.91	.98									
180	.10	.28	.55	.80	.94	.99									
200	.12	.32	.61	.85	.97										
300	.20	.51	.82	.97											
400	.28	.67	.93	.99											
500	.37	.79	.98												

TABLE A-19 Required Sample Size.

Necessary Sample Size

(α = .05, two-tailed test)

_t_ test for One-Sample or Two Related Samples

1-β	Cohen's _d_													
	.20	.30	.40	.50	.60	.70	.80	.90	1.00	1.10	1.20	1.30	1.40	1.50
.50	99	46	27	18	14	11	9	8	7	6	6	5	5	5
.55	112	51	30	20	15	12	10	8	7	7	6	5	5	5
.60	125	57	33	22	16	13	10	9	8	7	6	6	5	5
.65	140	64	37	25	18	14	11	10	8	7	7	6	6	5
.70	157	71	41	27	20	15	12	10	9	8	7	6	6	5
.75	176	80	46	30	22	17	13	11	10	8	7	7	6	6
.80	199	90	52	34	24	19	15	12	10	9	8	7	7	6
.85	227	102	59	38	27	21	17	14	12	10	9	8	7	7
.90	265	119	68	45	32	24	19	16	13	11	10	9	8	7

_t_ test for Two Independent Samples[1]

1-β	Cohen's _d_													
	.20	.30	.40	.50	.60	.70	.80	.90	1.00	1.10	1.20	1.30	1.40	1.50
.50	194	87	50	32	23	17	14	11	9	8	7	6	6	5
.55	219	98	56	36	26	19	15	12	10	9	8	7	6	6
.60	247	110	63	41	29	22	17	14	11	10	9	8	7	6
.65	277	124	70	46	32	24	19	15	13	11	9	8	7	7
.70	310	139	79	51	36	27	21	17	14	12	10	9	8	7
.75	349	156	88	57	40	30	23	19	15	13	11	10	9	8
.80	394	176	100	64	45	34	26	21	17	15	12	11	10	9
.85	450	201	114	73	51	38	30	24	19	16	14	12	11	10
.90	527	235	133	86	60	44	34	27	23	19	16	14	12	11

Correlation

1-β	_r_														
	.10	.15	.20	.25	.30	.35	.40	.45	.50	.55	.60	.65	.70	.75	.80
.50	385	172	97	63	44	32	25	20	16	14	12	10	9	8	7
.55	435	194	109	70	49	36	28	22	18	15	12	11	9	8	7
.60	490	218	122	78	54	40	30	24	19	16	13	11	10	8	7
.65	549	244	137	87	60	44	33	26	21	17	14	12	10	9	7
.70	616	273	153	97	67	49	37	29	23	19	15	13	11	9	8
.75	692	306	171	109	75	54	41	32	25	20	17	14	12	10	8
.80	781	345	193	122	84	61	46	35	28	23	18	15	13	10	9
.85	893	395	220	139	95	69	52	40	31	25	20	17	14	11	9
.90	1045	461	257	162	111	80	60	46	36	29	23	19	15	13	10

k	1-β	ω^2																			
		.02	.04	.06	.08	.10	.12	.14	.16	.18	.20	.22	.24	.26	.28	.30	.32	.34	.36	.38	.40
3	.50	82	41	27	21	16	14	12	10	9	8	7	7	6	6	5	5	5	5	4	4
	.55	92	46	31	23	18	15	13	11	10	9	8	7	7	6	6	6	5	5	5	4
	.60	103	51	34	25	20	17	14	12	11	10	9	8	7	7	6	6	6	5	5	5
	.65	115	57	38	28	22	18	16	14	12	11	10	9	8	8	7	7	6	6	5	5
	.70	127	63	42	31	25	20	17	15	13	12	11	10	9	8	8	7	7	6	6	5
	.75	142	70	46	34	27	23	19	17	15	13	12	11	10	9	8	8	7	7	6	6
	.80	159	79	52	38	30	25	21	18	16	14	13	12	11	10	9	8	8	7	7	6
	.85	180	89	59	43	34	28	24	21	18	16	14	13	12	11	10	9	9	8	8	7
	.90	208	103	68	50	39	32	27	24	21	18	17	15	14	12	11	11	10	9	8	8
4	.50	72	36	24	18	14	12	10	9	8	7	7	6	6	5	5	5	4	4	4	4
	.55	80	40	27	20	16	13	11	10	9	8	7	7	6	6	5	5	5	4	4	4
	.60	89	44	30	22	18	15	13	11	10	9	8	7	7	6	6	5	5	5	5	4
	.65	99	49	33	24	19	16	14	12	11	9	9	8	7	7	6	6	5	5	5	5
	.70	109	54	36	27	21	18	15	13	12	10	9	9	8	7	7	6	6	5	5	5
	.75	121	60	40	30	23	19	17	14	13	11	10	9	8	8	7	7	6	6	6	5
	.80	135	67	44	33	26	21	18	16	14	12	11	10	9	9	8	7	7	6	6	6
	.85	152	75	50	37	29	24	20	18	16	14	12	11	10	9	9	8	8	7	7	6
	.90	175	87	57	42	33	27	23	20	18	16	14	13	12	11	10	9	8	8	7	7
5	.50	64	32	22	16	13	11	9	8	7	7	6	6	5	5	5	4	4	4	4	4
	.55	72	36	24	18	14	12	10	9	8	7	7	6	6	5	5	5	4	4	4	4
	.60	79	39	26	20	16	13	11	10	9	8	7	7	6	6	5	5	5	4	4	4
	.65	87	44	29	22	17	14	12	11	9	9	8	7	7	6	6	5	5	5	4	4
	.70	96	48	32	24	19	16	13	12	10	9	8	8	7	7	6	6	5	5	5	4
	.75	107	53	35	26	21	17	15	13	11	10	9	8	8	7	7	6	6	5	5	5
	.80	118	59	39	29	23	19	16	14	12	11	10	9	8	8	7	7	6	6	5	5
	.85	133	66	44	32	26	21	18	16	14	12	11	10	9	8	8	7	7	6	6	6
	.90	152	75	50	37	29	24	20	18	16	14	12	11	10	9	9	8	8	7	7	6

TABLE A-20 The Poisson Distribution.

x \ μ	.1	.2	.3	.4	.5	.6	.7	.8	.9	1.0
0	.9048	.8187	.7408	.6703	.6065	.5488	.4966	.4493	.4066	.3679
1	.0905	.1637	.2222	.2681	.3033	.3293	.3476	.3595	.3659	.3679
2	.0045	.0164	.0333	.0536	.0758	.0988	.1217	.1438	.1647	.1839
3	.0002	.0011	.0033	.0072	.0126	.0198	.0284	.0383	.0494	.0613
4	.0000	.0001	.0002	.0007	.0016	.0030	.0050	.0077	.0111	.0153
5	.0000	.0000	.0000	.0001	.0002	.0004	.0007	.0012	.0020	.0031
6	.0000	.0000	.0000	.0000	.0000	.0000	.0001	.0002	.0003	.0005
7	.0000	.0000	.0000	.0000	.0000	.0000	.0000	.0000	.0000	.0001

x \ μ	1.1	1.2	1.3	1.4	1.5	1.6	1.7	1.8	1.9	2.0
0	.3329	.3012	.2725	.2466	.2231	.2019	.1827	.1653	.1496	.1353
1	.3662	.3614	.3543	.3452	.3347	.3230	.3106	.2975	.2842	.2707
2	.2014	.2169	.2303	.2417	.2510	.2584	.2640	.2678	.2700	.2707
3	.0738	.0867	.0998	.1128	.1255	.1378	.1496	.1607	.1710	.1804
4	.0203	.0260	.0324	.0395	.0471	.0551	.0636	.0723	.0812	.0902
5	.0045	.0062	.0084	.0111	.0141	.0176	.0216	.0260	.0309	.0361
6	.0008	.0012	.0018	.0026	.0035	.0047	.0061	.0078	.0098	.0120
7	.0001	.0002	.0003	.0005	.0008	.0011	.0015	.0020	.0027	.0034
8	.0000	.0000	.0001	.0001	.0001	.0002	.0003	.0005	.0006	.0009
9	.0000	.0000	.0000	.0000	.0000	.0000	.0001	.0001	.0001	.0002

x \ μ	2.1	2.2	2.3	2.4	2.5	2.6	2.7	2.8	2.9	3.0
0	.1225	.1108	.1003	.0907	.0821	.0743	.0672	.0608	.0550	.0498
1	.2572	.2438	.2306	.2177	.2052	.1931	.1815	.1703	.1596	.1494
2	.2700	.2681	.2652	.2613	.2565	.2510	.2450	.2384	.2314	.2240
3	.1890	.1966	.2033	.2090	.2138	.2176	.2205	.2225	.2237	.2240
4	.0992	.1082	.1169	.1254	.1336	.1414	.1488	.1557	.1622	.1680
5	.0417	.0476	.0538	.0602	.0668	.0735	.0804	.0872	.0940	.1008
6	.0146	.0174	.0206	.0241	.0278	.0319	.0362	.0407	.0455	.0504
7	.0044	.0055	.0068	.0083	.0099	.0118	.0139	.0163	.0188	.0216
8	.0011	.0015	.0019	.0025	.0031	.0038	.0047	.0057	.0068	.0081
9	.0003	.0004	.0005	.0007	.0009	.0011	.0014	.0018	.0022	.0027
10	.0001	.0001	.0001	.0002	.0002	.0003	.0004	.0005	.0006	.0008
11	.0000	.0000	.0000	.0000	.0000	.0001	.0001	.0001	.0002	.0002
12	.0000	.0000	.0000	.0000	.0000	.0000	.0000	.0000	.0000	.0001

x \ μ	3.1	3.2	3.3	3.4	3.5	3.6	3.7	3.8	3.9	4.0
0	.0450	.0408	.0369	.0334	.0302	.0273	.0247	.0224	.0202	.0183
1	.1397	.1304	.1217	.1135	.1057	.0984	.0915	.0850	.0789	.0733
2	.2165	.2087	.2008	.1929	.1850	.1771	.1692	.1615	.1539	.1465
3	.2237	.2226	.2209	.2186	.2158	.2125	.2087	.2046	.2001	.1954
4	.1734	.1781	.1823	.1858	.1888	.1912	.1931	.1944	.1951	.1954
5	.1075	.1140	.1203	.1264	.1322	.1377	.1429	.1477	.1522	.1563
6	.0555	.0608	.0662	.0716	.0771	.0826	.0881	.0936	.0989	.1042
7	.0246	.0278	.0312	.0348	.0385	.0425	.0466	.0508	.0551	.0595
8	.0095	.0111	.0129	.0148	.0169	.0191	.0215	.0241	.0269	.0298
9	.0033	.0040	.0047	.0056	.0066	.0076	.0089	.0102	.0116	.0132
10	.0010	.0013	.0016	.0019	.0023	.0028	.0033	.0039	.0045	.0053
11	.0003	.0004	.0005	.0006	.0007	.0009	.0011	.0013	.0016	.0019
12	.0001	.0001	.0001	.0002	.0002	.0003	.0003	.0004	.0005	.0006
13	.0000	.0000	.0000	.0000	.0001	.0001	.0001	.0001	.0002	.0002
14	.0000	.0000	.0000	.0000	.0000	.0000	.0000	.0000	.0000	.0001

x	μ = 4.1	4.2	4.3	4.4	4.5	4.6	4.7	4.8	4.9	5.0
0	.0166	.0150	.0136	.0123	.0111	.0101	.0091	.0082	.0074	.0067
1	.0679	.0630	.0583	.0540	.0500	.0462	.0427	.0395	.0365	.0337
2	.1393	.1323	.1254	.1188	.1125	.1063	.1005	.0948	.0894	.0842
3	.1904	.1852	.1798	.1743	.1687	.1631	.1574	.1517	.1460	.1404
4	.1951	.1944	.1933	.1917	.1898	.1875	.1849	.1820	.1789	.1755
5	.1600	.1633	.1662	.1687	.1708	.1725	.1738	.1747	.1753	.1755
6	.1093	.1143	.1191	.1237	.1281	.1323	.1362	.1398	.1432	.1462
7	.0640	.0686	.0732	.0778	.0824	.0869	.0914	.0959	.1002	.1044
8	.0328	.0360	.0393	.0428	.0463	.0500	.0537	.0575	.0614	.0653
9	.0150	.0168	.0188	.0209	.0232	.0255	.0280	.0307	.0334	.0363
10	.0061	.0071	.0081	.0092	.0104	.0118	.0132	.0147	.0164	.0181
11	.0023	.0027	.0032	.0037	.0043	.0049	.0056	.0064	.0073	.0082
12	.0008	.0009	.0011	.0014	.0016	.0019	.0022	.0026	.0030	.0034
13	.0002	.0003	.0004	.0005	.0006	.0007	.0008	.0009	.0011	.0013
14	.0001	.0001	.0001	.0001	.0002	.0002	.0003	.0003	.0004	.0005
15	.0000	.0000	.0000	.0000	.0001	.0001	.0001	.0001	.0001	.0002

x	μ = 5.1	5.2	5.3	5.4	5.5	5.6	5.7	5.8	5.9	6.0
0	.0061	.0055	.0050	.0045	.0041	.0037	.0033	.0030	.0027	.0025
1	.0311	.0287	.0265	.0244	.0225	.0207	.0191	.0176	.0162	.0149
2	.0793	.0746	.0701	.0659	.0618	.0580	.0544	.0509	.0477	.0446
3	.1348	.1293	.1239	.1185	.1133	.1082	.1033	.0985	.0938	.0892
4	.1719	.1681	.1641	.1600	.1558	.1515	.1472	.1428	.1383	.1339
5	.1753	.1748	.1740	.1728	.1714	.1697	.1678	.1656	.1632	.1606
6	.1490	.1515	.1537	.1555	.1571	.1584	.1594	.1601	.1605	.1606
7	.1086	.1125	.1163	.1200	.1234	.1267	.1298	.1326	.1353	.1377
8	.0692	.0731	.0771	.0810	.0849	.0887	.0925	.0962	.0998	.1033
9	.0392	.0423	.0454	.0486	.0519	.0552	.0586	.0620	.0654	.0688
10	.0200	.0220	.0241	.0262	.0285	.0309	.0334	.0359	.0386	.0413
11	.0093	.0104	.0116	.0129	.0143	.0157	.0173	.0190	.0207	.0225
12	.0039	.0045	.0051	.0058	.0065	.0073	.0082	.0092	.0102	.0113
13	.0015	.0018	.0021	.0024	.0028	.0032	.0036	.0041	.0046	.0052
14	.0006	.0007	.0008	.0009	.0011	.0013	.0015	.0017	.0019	.0022
15	.0002	.0002	.0003	.0003	.0004	.0005	.0006	.0007	.0008	.0009
16	.0001	.0001	.0001	.0001	.0001	.0002	.0002	.0002	.0003	.0003
17	.0000	.0000	.0000	.0000	.0000	.0001	.0001	.0001	.0001	.0001

x	μ = 6.1	6.2	6.3	6.4	6.5	6.6	6.7	6.8	6.9	7.0
0	.0022	.0020	.0018	.0017	.0015	.0014	.0012	.0011	.0010	.0009
1	.0137	.0126	.0116	.0106	.0098	.0090	.0082	.0076	.0070	.0064
2	.0417	.0390	.0364	.0340	.0318	.0296	.0276	.0258	.0240	.0223
3	.0848	.0806	.0765	.0726	.0688	.0652	.0617	.0584	.0552	.0521
4	.1294	.1249	.1205	.1162	.1118	.1076	.1034	.0992	.0952	.0912
5	.1579	.1549	.1519	.1487	.1454	.1420	.1385	.1349	.1314	.1277
6	.1605	.1601	.1595	.1586	.1575	.1562	.1546	.1529	.1511	.1490
7	.1399	.1418	.1435	.1450	.1462	.1472	.1480	.1486	.1489	.1490
8	.1066	.1099	.1130	.1160	.1188	.1215	.1240	.1263	.1284	.1304
9	.0723	.0757	.0791	.0825	.0858	.0891	.0923	.0954	.0985	.1014
10	.0441	.0469	.0498	.0528	.0558	.0588	.0618	.0649	.0679	.0710
11	.0245	.0265	.0285	.0307	.0330	.0353	.0377	.0401	.0426	.0452
12	.0124	.0137	.0150	.0164	.0179	.0194	.0210	.0227	.0245	.0264
13	.0058	.0065	.0073	.0081	.0089	.0098	.0108	.0119	.0130	.0142
14	.0025	.0029	.0033	.0037	.0041	.0046	.0052	.0058	.0064	.0071
15	.0010	.0012	.0014	.0016	.0018	.0020	.0023	.0026	.0029	.0033
16	.0004	.0005	.0005	.0006	.0007	.0008	.0010	.0011	.0013	.0014
17	.0001	.0002	.0002	.0002	.0003	.0003	.0004	.0004	.0005	.0006
18	.0000	.0001	.0001	.0001	.0001	.0001	.0001	.0002	.0002	.0002
19	.0000	.0000	.0000	.0000	.0000	.0000	.0000	.0001	.0001	.0001

x	μ 7.1	7.2	7.3	7.4	7.5	7.6	7.7	7.8	7.9	8.0
0	.0008	.0007	.0007	.0006	.0006	.0005	.0005	.0004	.0004	.0003
1	.0059	.0054	.0049	.0045	.0041	.0038	.0035	.0032	.0029	.0027
2	.0208	.0194	.0180	.0167	.0156	.0145	.0134	.0125	.0116	.0107
3	.0492	.0464	.0438	.0413	.0389	.0366	.0345	.0324	.0305	.0286
4	.0874	.0836	.0799	.0764	.0729	.0696	.0663	.0632	.0602	.0573
5	.1241	.1204	.1167	.1130	.1094	.1057	.1021	.0986	.0951	.0916
6	.1468	.1445	.1420	.1394	.1367	.1339	.1311	.1282	.1252	.1221
7	.1489	.1486	.1481	.1474	.1465	.1454	.1442	.1428	.1413	.1396
8	.1321	.1337	.1351	.1363	.1373	.1382	.1388	.1392	.1395	.1396
9	.1042	.1070	.1096	.1121	.1144	.1167	.1187	.1207	.1224	.1241
10	.0740	.0770	.0800	.0829	.0858	.0887	.0914	.0941	.0967	.0993
11	.0478	.0504	.0531	.0558	.0585	.0613	.0640	.0667	.0695	.0722
12	.0283	.0303	.0323	.0344	.0366	.0388	.0411	.0434	.0457	.0481
13	.0154	.0168	.0181	.0196	.0211	.0227	.0243	.0260	.0278	.0296
14	.0078	.0086	.0095	.0104	.0113	.0123	.0134	.0145	.0157	.0169
15	.0037	.0041	.0046	.0051	.0057	.0062	.0069	.0075	.0083	.0090
16	.0016	.0019	.0021	.0024	.0026	.0030	.0033	.0037	.0041	.0045
17	.0007	.0008	.0009	.0010	.0012	.0013	.0015	.0017	.0019	.0021
18	.0003	.0003	.0004	.0004	.0005	.0006	.0006	.0007	.0008	.0009
19	.0001	.0001	.0001	.0002	.0002	.0002	.0003	.0003	.0003	.0004
20	.0000	.0000	.0001	.0001	.0001	.0001	.0001	.0001	.0001	.0002
21	.0000	.0000	.0000	.0000	.0000	.0000	.0000	.0000	.0001	.0001

x	μ 8.1	8.2	8.3	8.4	8.5	8.6	8.7	8.8	8.9	9.0
0	.0003	.0003	.0002	.0002	.0002	.0002	.0002	.0002	.0001	.0001
1	.0025	.0023	.0021	.0019	.0017	.0016	.0014	.0013	.0012	.0011
2	.0100	.0092	.0086	.0079	.0074	.0068	.0063	.0058	.0054	.0050
3	.0269	.0252	.0237	.0222	.0208	.0195	.0183	.0171	.0160	.0150
4	.0544	.0517	.0491	.0466	.0443	.0420	.0398	.0377	.0357	.0337
5	.0882	.0849	.0816	.0784	.0752	.0722	.0692	.0663	.0635	.0607
6	.1191	.1160	.1128	.1097	.1066	.1034	.1003	.0972	.0941	.0911
7	.1378	.1358	.1338	.1317	.1294	.1271	.1247	.1222	.1197	.1171
8	.1395	.1392	.1388	.1382	.1375	.1366	.1356	.1344	.1332	.1318
9	.1256	.1269	.1280	.1290	.1299	.1306	.1311	.1315	.1317	.1318
10	.1017	.1040	.1063	.1084	.1104	.1123	.1140	.1157	.1172	.1186
11	.0749	.0776	.0802	.0828	.0853	.0878	.0902	.0925	.0948	.0970
12	.0505	.0530	.0555	.0579	.0604	.0629	.0654	.0679	.0703	.0728
13	.0315	.0334	.0354	.0374	.0395	.0416	.0438	.0459	.0481	.0504
14	.0182	.0196	.0210	.0225	.0240	.0256	.0272	.0289	.0306	.0324
15	.0098	.0107	.0116	.0126	.0136	.0147	.0158	.0169	.0182	.0194
16	.0050	.0055	.0060	.0066	.0072	.0079	.0086	.0093	.0101	.0109
17	.0024	.0026	.0029	.0033	.0036	.0040	.0044	.0048	.0053	.0058
18	.0011	.0012	.0014	.0015	.0017	.0019	.0021	.0024	.0026	.0029
19	.0005	.0005	.0006	.0007	.0008	.0009	.0010	.0011	.0012	.0014
20	.0002	.0002	.0002	.0003	.0003	.0004	.0004	.0005	.0005	.0006
21	.0001	.0001	.0001	.0001	.0001	.0002	.0002	.0002	.0002	.0003
22	.0000	.0000	.0000	.0000	.0001	.0001	.0001	.0001	.0001	.0001

x	μ 9.1	9.2	9.3	9.4	9.5	9.6	9.7	9.8	9.9	10
0	.0001	.0001	.0001	.0001	.0001	.0001	.0001	.0001	.0001	.0000
1	.0010	.0009	.0009	.0008	.0007	.0007	.0006	.0005	.0005	.0005
2	.0046	.0043	.0040	.0037	.0034	.0031	.0029	.0027	.0025	.0023
3	.0140	.0131	.0123	.0115	.0107	.0100	.0093	.0087	.0081	.0076
4	.0319	.0302	.0285	.0269	.0254	.0240	.0226	.0213	.0201	.0189
5	.0581	.0555	.0530	.0506	.0483	.0460	.0439	.0418	.0398	.0378
6	.0881	.0851	.0822	.0793	.0764	.0736	.0709	.0682	.0656	.0631
7	.1145	.1118	.1091	.1064	.1037	.1010	.0982	.0955	.0928	.0901
8	.1302	.1286	.1269	.1251	.1232	.1212	.1191	.1170	.1148	.1126
9	.1317	.1315	.1311	.1306	.1300	.1293	.1284	.1274	.1263	.1251

x	μ = 9.1	9.2	9.3	9.4	9.5	9.6	9.7	9.8	9.9	10
10	.1198	.1210	.1219	.1228	.1235	.1241	.1245	.1249	.1250	.1251
11	.0991	.1012	.1031	.1049	.1067	.1083	.1098	.1112	.1125	.1137
12	.0752	.0776	.0799	.0822	.0844	.0866	.0888	.0908	.0928	.0948
13	.0526	.0549	.0572	.0594	.0617	.0640	.0662	.0685	.0707	.0729
14	.0342	.0361	.0380	.0399	.0419	.0439	.0459	.0479	.0500	.0521
15	.0208	.0221	.0235	.0250	.0265	.0281	.0297	.0313	.0330	.0347
16	.0118	.0127	.0137	.0147	.0157	.0168	.0180	.0192	.0204	.0217
17	.0063	.0069	.0075	.0081	.0088	.0095	.0103	.0111	.0119	.0128
18	.0032	.0035	.0039	.0042	.0046	.0051	.0055	.0060	.0065	.0071
19	.0015	.0017	.0019	.0021	.0023	.0026	.0028	.0031	.0034	.0037
20	.0007	.0008	.0009	.0010	.0011	.0012	.0014	.0015	.0017	.0019
21	.0003	.0003	.0004	.0004	.0005	.0006	.0006	.0007	.0008	.0009
22	.0001	.0001	.0002	.0002	.0002	.0002	.0003	.0003	.0004	.0004
23	.0000	.0001	.0001	.0001	.0001	.0001	.0001	.0001	.0002	.0002
24	.0000	.0000	.0000	.0000	.0000	.0000	.0000	.0001	.0001	.0001

x	μ = 11	12	13	14	15	16	17	18	19	20
0	.0000	.0000	.0000	.0000	.0000	.0000	.0000	.0000	.0000	.0000
1	.0002	.0001	.0000	.0000	.0000	.0000	.0000	.0000	.0000	.0000
2	.0010	.0004	.0002	.0001	.0000	.0000	.0000	.0000	.0000	.0000
3	.0037	.0018	.0008	.0004	.0002	.0001	.0000	.0000	.0000	.0000
4	.0102	.0053	.0027	.0013	.0006	.0003	.0001	.0001	.0000	.0000
5	.0224	.0127	.0070	.0037	.0019	.0010	.0005	.0002	.0001	.0001
6	.0411	.0255	.0152	.0087	.0048	.0026	.0014	.0007	.0004	.0002
7	.0646	.0437	.0281	.0174	.0104	.0060	.0034	.0018	.0010	.0005
8	.0888	.0655	.0457	.0304	.0194	.0120	.0072	.0042	.0024	.0013
9	.1085	.0874	.0661	.0473	.0324	.0213	.0135	.0083	.0050	.0029
10	.1194	.1048	.0859	.0663	.0486	.0341	.0230	.0150	.0095	.0058
11	.1194	.1144	.1015	.0844	.0663	.0496	.0355	.0245	.0164	.0106
12	.1094	.1144	.1099	.0984	.0829	.0661	.0504	.0368	.0259	.0176
13	.0926	.1056	.1099	.1060	.0956	.0814	.0658	.0509	.0378	.0271
14	.0728	.0905	.1021	.1060	.1024	.0930	.0800	.0655	.0514	.0387
15	.0534	.0724	.0885	.0989	.1024	.0992	.0906	.0786	.0650	.0516
16	.0367	.0543	.0719	.0866	.0960	.0992	.0963	.0884	.0772	.0646
17	.0237	.0383	.0550	.0713	.0847	.0934	.0963	.0936	.0863	.0760
18	.0145	.0256	.0397	.0554	.0706	.0830	.0909	.0936	.0911	.0844
19	.0084	.0161	.0272	.0409	.0557	.0699	.0814	.0887	.0911	.0888
20	.0046	.0097	.0177	.0286	.0418	.0559	.0692	.0798	.0866	.0888
21	.0024	.0055	.0109	.0191	.0299	.0426	.0560	.0684	.0783	.0846
22	.0012	.0030	.0065	.0121	.0204	.0310	.0433	.0560	.0676	.0769
23	.0006	.0016	.0037	.0074	.0133	.0216	.0320	.0438	.0559	.0669
24	.0003	.0008	.0020	.0043	.0083	.0144	.0226	.0328	.0442	.0557
25	.0001	.0004	.0010	.0024	.0050	.0092	.0154	.0237	.0336	.0446
26	.0000	.0002	.0005	.0013	.0029	.0057	.0101	.0164	.0246	.0343
27	.0000	.0001	.0002	.0007	.0016	.0034	.0063	.0109	.0173	.0254
28	.0000	.0000	.0001	.0003	.0009	.0019	.0038	.0070	.0117	.0181
29	.0000	.0000	.0001	.0002	.0004	.0011	.0023	.0044	.0077	.0125
30	.0000	.0000	.0000	.0001	.0002	.0006	.0013	.0026	.0049	.0083
31	.0000	.0000	.0000	.0000	.0001	.0003	.0007	.0015	.0030	.0054
32	.0000	.0000	.0000	.0000	.0001	.0001	.0004	.0009	.0018	.0034
33	.0000	.0000	.0000	.0000	.0000	.0001	.0002	.0005	.0010	.0020
34	.0000	.0000	.0000	.0000	.0000	.0000	.0001	.0002	.0006	.0012
35	.0000	.0000	.0000	.0000	.0000	.0000	.0000	.0001	.0003	.0007
36	.0000	.0000	.0000	.0000	.0000	.0000	.0000	.0001	.0002	.0004
37	.0000	.0000	.0000	.0000	.0000	.0000	.0000	.0000	.0001	.0002
38	.0000	.0000	.0000	.0000	.0000	.0000	.0000	.0000	.0000	.0001
39	.0000	.0000	.0000	.0000	.0000	.0000	.0000	.0000	.0000	.0001

TABLE A-21 Critical Values of the Spearman Correlation Coefficient.*

n	2-tailed testing			1-tailed testing		
	α			α		
	.1	.05	.01	.1	.05	.01
5	0.900	1.000	--	0.800	0.900	1.000
6	0.829	0.886	1.000	0.657	0.829	0.943
7	0.714	0.786	0.929	0.571	0.714	0.893
8	0.643	0.738	0.881	0.524	0.643	0.833
9	0.600	0.700	0.833	0.483	0.600	0.783
10	0.564	0.648	0.794	0.455	0.564	0.745
11	0.527	0.609	0.755	0.418	0.527	0.700
12	0.497	0.580	0.727	0.399	0.497	0.671
13	0.478	0.555	0.698	0.379	0.478	0.643
14	0.459	0.534	0.675	0.363	0.459	0.622
15	0.443	0.518	0.654	0.350	0.443	0.600
16	0.426	0.500	0.632	0.338	0.426	0.582
17	0.412	0.485	0.615	0.326	0.412	0.564
18	0.399	0.470	0.598	0.315	0.399	0.548
19	0.389	0.458	0.582	0.307	0.389	0.533
20	0.379	0.445	0.568	0.298	0.379	0.520
21	0.369	0.435	0.555	0.291	0.369	0.508
22	0.360	0.424	0.543	0.283	0.360	0.496
23	0.352	0.415	0.531	0.277	0.352	0.485
24	0.343	0.406	0.520	0.270	0.343	0.475
25	0.336	0.397	0.510	0.265	0.336	0.465
26	0.330	0.389	0.500	0.259	0.330	0.456
27	0.324	0.382	0.491	0.254	0.324	0.448
28	0.317	0.375	0.483	0.249	0.317	0.440
29	0.311	0.368	0.474	0.244	0.311	0.432
30	0.306	0.362	0.467	0.240	0.306	0.425
32	0.296	0.350	0.452	0.232	0.296	0.411
34	0.287	0.339	0.439	0.225	0.287	0.399
36	0.279	0.330	0.427	0.218	0.279	0.388
38	0.271	0.321	0.415	0.212	0.271	0.378
40	0.264	0.313	0.405	0.207	0.264	0.368
42	0.257	0.305	0.395	0.201	0.257	0.359
44	0.251	0.298	0.386	0.197	0.251	0.351
46	0.246	0.291	0.378	0.192	0.246	0.343
48	0.240	0.285	0.370	0.188	0.240	0.336
50	0.235	0.279	0.363	0.184	0.235	0.329
60	0.214	0.254	0.331	0.168	0.214	0.300
70	0.198	0.235	0.307	0.155	0.198	0.278
80	0.185	0.220	0.287	0.145	0.185	0.260

TABLE A-22 Critical Values for Total Number of Runs (U).

From C. Eisenhart and F. Swed, "Tables for testing randomness of grouping in a sequence of alternatives," *The Annals of Statistics*, Vol. 14 (1943), pp. 66-67. Reprinted by permission.

The smaller of n_1 and n_2 \ The larger of n_1 and n_2	5	6	7	8	9	10	11	12	13	14	15	16	17	18	19	20
2								2 6	2 6	2 6	2 6	2 6	2 6	2 6	2 6	2 6
3		2 8	2 8	2 8	2 8	2 8	2 8	2 8	2 8	2 8	3 8	3 8	3 8	3 8	3 8	3 8
4	2 9	2 9	2 10	3 10	3 10	3 10	3 10	3 10	3 10	3 10	3 10	4 10	4 10	4 10	4 10	4 10
5	2 10	3 10	3 11	3 11	3 12	3 12	4 12	4 12	4 12	4 12	4 12	4 12	4 12	5 12	5 12	5 12
6		3 11	3 12	3 12	4 13	4 13	4 13	4 13	5 14	5 14	5 14	5 14	5 14	5 14	6 14	6 14
7			3 13	4 13	4 14	5 14	5 14	5 14	5 15	5 15	6 15	6 16	6 16	6 16	6 16	6 16
8				4 14	5 14	5 15	5 15	6 16	6 16	6 16	6 16	6 17	7 17	7 17	7 17	7 17
9					5 15	5 16	6 16	6 16	6 17	7 17	7 18	7 18	7 18	8 18	8 18	8 18
10						6 16	6 17	7 17	7 18	7 18	7 18	8 19	8 19	8 19	8 20	9 20
11							7 17	7 18	7 19	8 19	8 19	8 20	9 20	9 20	9 21	9 21
12								7 19	8 19	8 20	8 20	9 21	9 21	9 21	10 22	10 22
13									8 20	9 20	9 21	9 21	10 22	10 22	10 23	10 23
14										9 21	9 22	10 22	10 23	10 23	11 23	11 24
15											10 22	10 23	11 23	11 24	11 24	12 25
16												11 23	11 24	11 25	12 25	12 25
17													11 25	12 25	12 26	13 26
18														12 26	13 26	13 27
19															13 27	13 27
20																14 28

TABLE A-23 Critical Values for the Hartley Test of Homogeniety of Variance.
This table is abridged from Table 31 in *Biometrika Tables for Statisticians*, Vol. 1, 2nd ed. New York: Cambridge, 1958. Edited by E. S. Pearson and H. O. Hartley. Reproduced with the permission of the editors and the trustees of Biometrika.

The Hartley F_{max} Test for Homogeneity of Variances

		k = number of variances										
$df = n - 1$	α	2	3	4	5	6	7	8	9	10	11	12
4	.05	9.60	15.5	20.6	25.2	29.5	33.6	37.5	41.4	44.6	48.0	51.4
	.01	23.2	37.	49.	59.	69.	79.	89.	97.	106.	113.	120.
5	.05	7.15	10.8	13.7	16.3	18.7	20.8	22.9	24.7	26.5	28.2	29.9
	.01	14.9	22.	28.	33.	38.	42.	46.	50.	54.	57.	60.
6	.05	5.82	8.38	10.4	12.1	13.7	15.0	16.3	17.5	18.6	19.7	20.7
	.01	11.1	15.5	19.1	22.	25.	27.	30.	32.	34.	36.	37.
7	.05	4.99	6.94	8.44	9.70	10.8	11.8	12.7	13.5	14.3	15.1	15.8
	.01	8.89	12.1	14.5	16.5	18.4	20.	22.	23.	24.	26.	27.
8	.05	4.43	6.00	7.18	8.12	9.03	9.78	10.5	11.1	11.7	12.2	12.7
	.01	7.50	9.9	11.7	13.2	14.5	15.8	16.9	17.9	18.9	19.8	21.
9	.05	4.03	5.34	6.31	7.11	7.80	8.41	8.95	9.45	9.91	10.3	10.7
	.01	6.54	8.5	9.9	11.1	12.1	13.1	13.9	14.7	15.3	16.0	16.6
10	.05	3.72	4.85	5.67	6.34	6.92	7.42	7.87	8.28	8.66	9.01	9.34
	.01	5.85	7.4	8.6	9.6	10.4	11.1	11.8	12.4	12.9	13.4	13.9
12	.05	3.28	4.16	4.79	5.30	5.72	6.09	6.42	6.72	7.00	7.25	7.48
	.01	4.91	6.1	6.9	7.6	8.2	8.7	9.1	9.5	9.9	10.2	10.6
15	.05	2.86	3.54	4.01	4.37	4.68	4.95	5.19	5.40	5.59	5.77	5.93
	.01	4.07	4.9	5.5	6.0	6.4	6.7	7.1	7.3	7.5	7.8	8.0
20	.05	2.46	2.95	3.29	3.54	3.76	3.94	4.10	4.24	4.37	4.49	4.59
	.01	3.32	3.8	4.3	4.6	4.9	5.1	5.3	5.5	5.6	5.8	5.9
30	.05	2.07	2.40	2.61	2.78	2.91	3.02	3.12	3.21	3.29	3.36	3.39
	.01	2.63	3.0	3.3	3.4	3.6	3.7	3.8	3.9	4.0	4.1	4.2
60	.05	1.67	1.85	1.96	2.04	2.11	2.17	2.22	2.26	2.30	2.33	2.36
	.01	1.96	2.2	2.3	2.4	2.4	2.5	2.5	2.6	2.6	2.7	2.7
∞	.05	1.00	1.00	1.00	1.00	1.00	1.00	1.00	1.00	1.00	1.00	1.00
	.01	1.00	1.00	1.00	1.00	1.00	1.00	1.00	1.00	1.00	1.00	1.00

TABLE A-24 The Cochran Test for Homogeneity of Variances.

	k = *number of variances*										
n	2	3	4	5	6	7	8	9	10	12	15
5	.959	.834	.721	.633	.564	.508	.463	.425	.393	.343	.288
6	.937	.793	.676	.588	.520	.466	.423	.387	.357	.310	.259
7	.917	.761	.641	.553	.487	.435	.393	.359	.331	.286	.239
8	.899	.734	.613	.526	.461	.411	.370	.338	.311	.268	.223
9	.882	.711	.590	.504	.440	.391	.352	.321	.295	.254	.210
10	.867	.691	.570	.485	.423	.375	.337	.307	.281	.242	.200
11	.855	.675	.554	.470	.408	.361	.325	.295	.270	.232	.191
12	.843	.660	.539	.456	.396	.350	.314	.285	.261	.223	.184
13	.831	.647	.527	.445	.385	.340	.305	.276	.253	.216	.178
14	.821	.635	.515	.434	.376	.331	.297	.269	.246	.210	.173
15	.812	.624	.505	.425	.367	.323	.289	.262	.240	.205	.168
16	.803	.615	.496	.417	.360	.317	.283	.256	.234	.200	.164
17	.795	.606	.488	.409	.353	.311	.278	.251	.230	.196	.161
18	.788	.598	.481	.403	.347	.305	.272	.246	.225	.191	.157
19	.781	.591	.474	.396	.341	.300	.267	.242	.220	.188	.154
20	.775	.584	.468	.391	.336	.295	.263	.238	.217	.185	.151
21	.769	.578	.462	.385	.331	.291	.259	.234	.213	.182	.149
22	.763	.572	.457	.381	.327	.287	.255	.231	.210	.179	.147
23	.758	.566	.452	.376	.323	.283	.252	.227	.207	.176	.144
24	.753	.561	.447	.372	.319	.279	.249	.224	.205	.174	.142
25	.748	.556	.443	.368	.315	.276	.246	.222	.202	.172	.140
26	.744	.552	.439	.364	.312	.273	.243	.219	.200	.170	.139
27	.739	.548	.435	.361	.309	.270	.241	.217	.197	.168	.137
28	.735	.544	.431	.358	.306	.268	.238	.215	.195	.166	.136
29	.732	.540	.428	.355	.303	.265	.236	.212	.193	.164	.134
30	.728	.536	.424	.352	.301	.263	.234	.210	.192	.163	.133
32	.721	.530	.418	.346	.296	.258	.230	.207	.188	.160	.130
34	.715	.523	.413	.341	.291	.255	.226	.203	.185	.157	.128
36	.709	.518	.408	.337	.287	.251	.223	.200	.182	.154	.126
38	.704	.513	.403	.333	.284	.248	.220	.198	.180	.152	.124
40	.699	.508	.399	.329	.281	.245	.217	.195	.177	.150	.122
42	.694	.504	.395	.326	.277	.242	.215	.193	.175	.148	.121
44	.690	.500	.392	.323	.275	.239	.212	.191	.173	.147	.119
46	.686	.496	.388	.320	.272	.237	.210	.189	.171	.145	.118
48	.682	.492	.385	.317	.270	.235	.208	.187	.170	.143	.117
50	.679	.489	.382	.314	.267	.234	.206	.185	.168	.142	.116
145	.606	.423	.325	.264	.223	.193	.170	.152	.138	.116	.093
∞	.500	.333	.250	.200	.143	.125	.111	.100	.083	.067	.050

The Cochran Test for Homogeneity of Variances (.05 Level) *(Continued)*

n	k = *number of variances*										
	2	3	4	5	6	7	8	9	10	12	15
5	.906	.746	.629	.544	.480	.431	.391	.358	.331	.288	.242
6	.877	.707	.590	.507	.445	.397	.360	.329	.303	.262	.220
7	.853	.677	.560	.478	.418	.373	.336	.307	.282	.244	.203
8	.833	.653	.537	.456	.398	.354	.319	.290	.267	.230	.191
9	.816	.633	.518	.439	.382	.338	.304	.277	.254	.219	.182
10	.801	.617	.502	.424	.368	.326	.293	.266	.244	.210	.174
11	.789	.603	.489	.412	.357	.315	.283	.257	.235	.202	.167
12	.777	.590	.477	.401	.347	.306	.275	.249	.228	.195	.161
13	.767	.580	.467	.392	.339	.299	.267	.242	.222	.190	.157
14	.757	.570	.458	.384	.331	.292	.261	.237	.216	.185	.153
15	.749	.561	.450	.377	.325	.286	.256	.231	.211	.181	.149
16	.741	.554	.443	.370	.319	.280	.251	.227	.207	.177	.146
17	.734	.547	.437	.365	.314	.276	.246	.223	.203	.174	.143
18	.728	.540	.431	.359	.309	.271	.242	.219	.200	.170	.140
19	.722	.534	.425	.354	.304	.267	.238	.215	.197	.168	.138
20	.717	.529	.421	.350	.300	.264	.235	.212	.194	.165	.135
21	.712	.524	.416	.346	.297	.260	.232	.209	.191	.163	.133
22	.707	.519	.412	.342	.293	.257	.229	.207	.188	.160	.132
23	.702	.515	.408	.339	.290	.254	.226	.204	.186	.158	.130
24	.698	.511	.404	.335	.287	.251	.224	.202	.184	.157	.128
25	.694	.507	.401	.332	.284	.249	.222	.200	.182	.155	.127
26	.691	.504	.398	.330	.282	.247	.219	.199	.180	.153	.125
27	.687	.500	.395	.327	.279	.244	.217	.196	.178	.152	.124
28	.684	.497	.392	.324	.277	.242	.215	.194	.177	.150	.123
29	.681	.494	.389	.322	.275	.240	.214	.193	.175	.149	.122
30	.678	.491	.387	.320	.273	.238	.212	.191	.174	.148	.121
32	.672	.486	.382	.315	.269	.235	.209	.188	.171	.145	.119
34	.667	.481	.378	.312	.266	.232	.206	.185	.169	.143	.117
36	.662	.477	.374	.308	.263	.229	.203	.183	.166	.141	.115
38	.658	.473	.370	.305	.260	.227	.201	.181	.164	.139	.114
40	.654	.469	.367	.302	.257	.224	.199	.179	.163	.138	.112
42	.651	.466	.364	.299	.255	.222	.197	.177	.161	.136	.111
44	.647	.463	.361	.297	.253	.220	.195	.175	.159	.135	.110
46	.644	.460	.359	.295	.251	.218	.193	.174	.158	.134	.109
48	.641	.457	.357	.293	.249	.216	.192	.172	.156	.132	.108
50	.638	.454	.354	.290	.247	.215	.190	.171	.155	.131	.107
145	.581	.403	.309	.251	.212	.183	.162	.145	.131	.110	.089
∞	.500	.333	.250	.167	.143	.125	.111	.100	.083	.067	.050

TABLE A-25 Table of Percentage Points of Kolmogorov Statistics.

One-sided test	**p = 0.90**	**0.95**	**0.975**	**0.99**	**0.995**
Two-sided test	**p = 0.80**	**0.90**	**0.95**	**0.98**	**0.99**
$n = 1$	.900	.950	.975	.990	.995
2	.684	.776	.842	.900	.929
3	.565	.636	.708	.785	.829
4	.493	.565	.624	.689	.734
5	.447	.509	.563	.627	.669
6	.410	.468	.519	.577	.617
7	.381	.436	.483	.538	.576
8	.358	.410	.454	.507	.542
9	.339	.387	.430	.480	.513
10	.323	.369	.409	.457	.489
11	.308	.352	.391	.437	.468
12	.296	.338	.375	.419	.449
13	.285	.325	.361	.404	.432
14	.275	.314	.349	.390	.418
15	.266	.304	.338	.377	.404
16	.258	.295	.327	.366	.392
17	.250	.286	.318	.355	.381
18	.244	.279	.309	.346	.371
19	.237	.271	.301	.337	.361
20	.232	.265	.294	.329	.352
21	.226	.259	.287	.321	.344
22	.221	.253	.281	.314	.337
23	.216	.247	.275	.307	.330
24	.212	.242	.269	.301	.323
25	.208	.238	.264	.295	.317
26	.204	.233	.259	.290	.311
27	.200	.229	.254	.284	.305
28	.197	.225	.250	.279	.300
29	.193	.221	.246	.275	.295
30	.190	.218	.242	.270	.290
31	.187	.214	.238	.266	.285
32	.184	.211	.234	.262	.281
33	.182	.208	.231	.258	.277
34	.179	.205	.227	.254	.273
35	.177	.202	.224	.251	.269
36	.174	.199	.221	.247	.265
37	.172	.196	.218	.244	.262
38	.170	.194	.215	.241	.258
39	.168	.191	.213	.238	.255
40	.165	.189	.210	.235	.252
Approximation for $n > 40$:	$\frac{1.07}{\sqrt{n}}$	$\frac{1.22}{\sqrt{n}}$	$\frac{1.36}{\sqrt{n}}$	$\frac{1.52}{\sqrt{n}}$	$\frac{1.63}{\sqrt{n}}$

TABLE A-26 Quantiles of the Smirnov Test Statistic for Two Samples of Equal Size n. Source Z. W. Birnbaum and R. A. Hall, "Small Sample Distribution for Multi-Sample Statistics of the Sminrov Type." *Ann. Math. Statist.*, 31(1960), 710-720.

One-sided test	$p=0.90$	0.95	0.975	0.99	0.995
Two-sided test	$p=0.80$	0.90	0.95	0.98	0.99
$n=3$	2/3	2/3			
4	3/4	3/4	3/4		
5	3/5	3/5	4/5	4/5	4/5
6	3/6	4/6	4/6	5/6	5/6
7	4/7	4/7	5/7	5/7	5/7
8	4/8	4/8	5/8	5/8	6/8
9	4/9	5/9	5/9	6/9	6/9
10	4/10	5/10	6/10	6/10	7/10
11	5/11	5/11	6/11	7/11	7/11
12	5/12	5/12	6/12	7/12	7/12
13	5/13	6/13	6/13	7/13	8/13
14	5/14	6/14	7/14	7/14	8/14
15	5/15	6/15	7/15	8/15	8/15
16	6/16	6/16	7/16	8/16	9/16
17	6/17	7/17	7/17	8/17	9/17
18	6/18	7/18	8/18	9/18	9/18
19	6/19	7/19	8/19	9/19	9/19
20	6/20	7/20	8/20	9/20	10/20
21	6/21	7/21	8/21	9/21	10/21
22	7/22	8/22	8/22	10/22	10/22
23	7/23	8/23	9/23	10/23	10/23
24	7/24	8/24	9/24	10/24	11/24
25	7/25	8/25	9/25	10/25	11/25
26	7/26	8/26	9/26	10/26	11/26
27	7/27	8/27	9/27	11/27	11/27
28	8/28	9/28	10/28	11/28	12/28
29	8/29	9/29	10/29	11/29	12/29
30	8/30	9/30	10/30	11/30	12/30
31	8/31	9/31	10/31	11/31	12/31
32	8/32	9/32	10/32	12/32	12/32
33	8/33	9/33	11/33	12/33	13/33
34	8/34	10/34	11/34	12/34	13/34
35	8/35	10/35	11/35	12/35	13/35
36	9/36	10/36	11/36	12/36	13/36
37	9/37	10/37	11/37	13/37	13/37
38	9/38	10/38	11/38	13/38	14/38
39	9/39	10/39	11/39	13/39	14/39
40	9/40	10/40	12/40	13/40	14/40
Approximation for $n > 40$:	$\frac{1.52}{\sqrt{n}}$	$\frac{1.73}{\sqrt{n}}$	$\frac{1.92}{\sqrt{n}}$	$\frac{2.15}{\sqrt{n}}$	$\frac{2.30}{\sqrt{n}}$

TABLE A-27 Quantiles of the Smirnov Test Statistic for Two Samples of Different Size. Source Frank J. Massey, Jr., "Distribution Table for the Deviation Between Two Sample Cumulatives," *Ann. Math. Statist.*, 23 (1952), 435-441. This table incorporates the corrections reported in Louis S. Davis, "Table Errata 266," *Math. of Computation*, 12(1958), 262-263.

One-sided test		$p = 0.90$	0.95	0.975	0.99	0.995
Two-sided test		$p = 0.80$	0.90	0.95	0.98	0.99
$N_1 = 1$	$N_2 = 9$	17/18				
	10	9/10				
$N_1 = 2$	$N_2 = 3$	5/6				
	4	3/4				
	5	4/5	4/5			
	6	5/6	5/6			
	7	5/7	6/7			
	8	3/4	7/8	7/8		
	9	7/9	8/9	8/9		
	10	7/10	4/5	9/10		
$N_1 = 3$	$N_2 = 4$	3/4	3/4			
	5	2/3	4/5	4/5		
	6	2/3	2/3	5/6		
	7	2/3	5/7	6/7	6/7	
	8	5/8	3/4	3/4	7/8	
	9	2/3	2/3	7/9	8/9	8/9
	10	3/5	7/10	4/5	9/10	9/10
	12	7/12	2/3	3/4	5/6	11/12
$N_1 = 4$	$N_2 = 5$	3/5	3/4	4/5	4/5	
	6	7/12	2/3	3/4	5/6	5/6
	7	17/28	5/7	3/4	6/7	6/7
	8	5/8	5/8	3/4	7/8	7/8
	9	5/9	2/3	3/4	7/9	8/9
	10	11/20	13/20	7/10	4/5	4/5
	12	7/12	2/3	2/3	3/4	5/6
	16	9/16	5/8	11/16	3/4	13/16
$N_1 = 5$	$N_2 = 6$	3/5	2/3	2/3	5/6	5/6
	7	4/7	23/35	5/7	29/35	6/7
	8	11/20	5/8	27/40	4/5	4/5
	9	5/9	3/5	31/45	7/9	4/5
	10	1/2	3/5	7/10	7/10	4/5
	15	8/15	3/5	2/3	11/15	11/15
	20	1/2	11/20	3/5	7/10	3/4
$N_1 = 6$	$N_2 = 7$	23/42	4/7	29/42	5/7	5/6
	8	1/2	7/12	2/3	3/4	3/4
	9	1/2	5/9	2/3	13/18	7/9
	10	1/2	17/30	19/30	7/10	11/15
	12	1/2	7/12	7/12	2/3	3/4
	18	4/9	5/9	11/18	2/3	13/18
	24	11/24	1/2	7/12	5/8	2/3
$N_1 = 7$	$N_2 = 8$	27/56	33/56	5/8	41/56	3/4
	9	31/63	5/9	40/63	5/7	47/63
	10	33/70	39/70	43/70	7/10	5/7
	14	3/7	1/2	4/7	9/14	5/7
	28	3/7	13/28	15/28	17/28	9/14
$N_1 = 8$	$N_2 = 9$	4/9	13/24	5/8	2/3	3/4
	10	19/40	21/40	23/40	27/40	7/10
	12	11/24	1/2	7/12	5/8	2/3
	16	7/16	1/2	9/16	5/8	5/8
	32	13/32	7/16	1/2	9/16	19/32

One-sided test		$p=0.90$	0.95	0.975	0.99	0.995
Two-sided test		$p=0.80$	0.90	0.95	0.98	0.99
$N_1 = 9$	$N_2 = 10$	7/15	1/2	26/45	2/3	31/45
	12	4/9	1/2	5/9	11/18	2/3
	15	19/45	22/45	8/15	3/5	29/45
	18	7/18	4/9	1/2	5/9	11/18
	36	13/36	5/12	17/36	19/36	5/9
$N_1 = 10$	$N_2 = 15$	2/5	7/15	1/2	17/30	19/30
	20	2/5	9/20	1/2	11/20	3/5
	40	7/20	2/5	9/20	1/2	
$N_1 = 12$	$N_2 = 15$	23/60	9/20	1/2	11/20	7/12
	16	3/8	7/16	23/48	13/24	7/12
	18	13/36	5/12	17/36	19/36	5/9
	20	11/30	5/12	7/15	31/60	17/30
$N_1 = 15$	$N_2 = 20$	7/20	2/5	13/30	29/60	31/60
$N_1 = 16$	$N_2 = 20$	27/80	31/80	17/40	19/40	41/80
Large-sample approximation		$1.07\sqrt{\frac{m+n}{mn}}$	$1.22\sqrt{\frac{m+n}{mn}}$	$1.36\sqrt{\frac{m+n}{mn}}$	$1.52\sqrt{\frac{m+n}{mn}}$	$1.63\sqrt{\frac{m+n}{mn}}$

Statistical Tables

Here are a number of useful statistical tables that I use in my teaching and research. I used Excel, SPSS, and Java to generate the values in most of these tables. Several were adapted from published sources. They are all stored in PDF format. Please feel free to use these tables in your research or teaching. However, I assume no responsibility for the accuracy of the information in these files. Feedback and questions about these tables are welcome.

Statistical tables

- The Normal Distribution
- Binomial Probabilities
- Critical Values of the t Distribution
- Critical Values of the F Distribution
- Critical Values of the Studentized Range Statistic
- Critical Values of Dunnett's Test
- Critical Values of Dunn's Test
- Critical Values of the Chi-Square Distribution
- Critical Values of the Binomial Test
- Critical Values of the Mann-Whitney U Test
- Critical Values of the Wilcoxon Ranked-Sums Test

- Critical Values of the Wilcoxon Signed Ranks Test
- Critical Values of the Correlation Coefficient
- Critical Values of the Spearman Correlation Coefficient
- Converting r to Z
- Statistical Power of the Z Test
- Statistical Power of the t-Test for One Sample or Two Related Samples
- Statistical Power of the t-Test for Two Independent Samples
- Statistical Power of the Analysis of Variance
- Statistical Power of the Correlation Coefficient
- Required Sample Size

http://facultyweb.berry.edu/vbissonnette/tables/tables.html

APPENDIX B

DATA FILES

TABLE B-1 American Cities database.

X1	X2	X3	X4	X5	X6	X7	X8	X9	X10	X11
Atlanta	0.107	0.047	0.032	13432	0.124	0.003	1	0	0	East
Baltimore	0.098		0.011	16178	0.09	0.052	1	0	0	East
Boston	0.109	0.062	0.041	14132	0.057	0.215	1	0	0	East
Bridgeport	0.127	0.055	0.022	15784	0.095	0.136	1	0	0	East
Buffalo	0.071	0.097	0.011	18362	0.025	0.214	1	0	0	East
Burlington, VT		0.064	0.049	13979	0.116	0.274	1	0	0	East
Charleston, S.C.	0.033	0.061	0.04	13002	0.087	0.118	1	0	0	East
Charleston, W.V.	0.119	0.063	0.002	18740	0.138	0.047	1	0	0	East
Charlotte, N.C.	0.133	0.039	0.015	10905	0.076	0.042	1	0	0	East
Greensboro	0.128	0.044	0.018	11229	0.106	-0.061	1	0	0	East
Hartford	0.131	0.048	0.103	16547	0.079	0.202	1	0	0	East
Jacksonville	-0.073	0.051	0.015	13790	0.077	0.033	1	0	0	East
Manchester, N.H.		0.035	0.019	10337	0.089	0.034	1	0	0	East
Miami	0.11	0.048	0.051	10718	0.092	0.225	1	0	0	East
Newark	0.111	0.058	0.016	15456	0.076	0.076	1	0	0	East
New York	0.079	0.088	0.012	12286	0.087	0.049	1	0	0	East
Norfolk	-0.012	0.053	0.01	13658	0.031	-0.051	1	0	0	East
Philadelphia	0.107	0.066	0.036	15073	0.069	0.175	1	0	0	East
Pittsburgh	0.12	0.071	0.004	18483	0.061	-0.015	1	0	0	East

(Continued)

TABLE B-1 Continued.

X1	X2	X3	X4	X5	X6	X7	X8	X9	X10	X11
Portland, Me.	0.101	0.054	0.051	11619	0.095	0.03	1	0	0	East
Providence	0.045	0.081	-0.004	11012	0.076	0.028	1	0	0	East
Richmond	-0.039	0.031	0.018	15848	0.199	-0.029	1	0	0	East
Rochester	0.055	0.065	0.016	17890	0.05	0.011	1	0	0	East
Savannah	0.082	0.055	-0.013	16024	0.135	-0.1	1	0	0	East
Tampa-St Petersburg	0.075	0.049	0.034	12333	0.056	0.022	1	0	0	East
Washington, D.C.	0.073		0.027	17139	0.134	0.064	1	0	0	East
Wilmington	0.19	0.074	0.002	17606	0.114	-0.043	1	0	0	East
Birmingham	0.024	0.068	0.024	15681	0.087	0.03	0	1	0	Central
Chicago	0.198		0.004	15226	0.106	-0.012	0	1	0	Central
Cincinnati	0.098	0.06	0.015	16622	0.067	0.094	0	1	0	Central
Cleveland	-0.017	0.056	0.006	17782	0.003	0.053	0	1	0	Central
Columbus	0.063	0.046	0.021	15261	0.065	0.085	0	1	0	Central
Dallas-Ft. Worth	0.231	0.042	0.056	13590	0.109	0.134	0	1	0	Central
Des Molnes	0.224	0.055	0.011	17819	0.134	-0.029	0	1	0	Central
Detroit	0.033	0.12	-0.044	20378	-0.031	-0.013	0	1	0	Central
Fargo	0.134	0.05	0.036	13412	0.114	-0.029	0	1	0	Central
Grand Rapids	0.143	0.067	-0.001	15917	0.032	0.044	0	1	0	Central
Houston	0.162	0.039	0.042	18274	0.079	0.014	0	1	0	Central
Indianapolis	0.095	0.069	0.019	17221	0.05	0.115	0	1	0	Central
Jackson, Miss.	0.053	0.042	0.025	12387	0.092	0	0	1	0	Central
Kansas City, Mo.-Kans.	0.135	0.054	0	16577	0.096	-0.06	0	1	0	Central
Knoxville	0.084	0.055	-0.007	14298	0.096	-0.036	0	1	0	Central
Little Rock	0.025	0.043	0.019	12846	0.088	-0.024	0	1	0	Central
Louisville	0.131	0.074	-0.021	16224	0.041	-0.019	0	1	0	Central
Memphis	0.142	0.059	0.005	13351	0.072	0.085	0	1	0	Central
Milwaukee	0.188	0.043	0.029	18062	0.073	0.059	0	1	0	Central
Minneapolis-St. Paul	0.077	0.042	0.037	16298	0.086	0.131	0	1	0	Central
Mobile	0.243	0.072	0.027	17171	0.133	0.163	0	1	0	Central
New Orleans	0.113	0.066	0.019	16337	0.129	0.024	0	1	0	Central
Oklahoma City	0.204	0.03	0.087	15918	0.283	0.103	0	1	0	Central
Omaha	0.141	0.052	0.018	15053	0.06	-0.021	0	1	0	Central

TABLE B-1 Continued.

X1	X2	X3	X4	X5	X6	X7	X8	X9	X10	X11
San Antonio	0.043	0.064	0.028	10525	0.128	0.12	0	1	0	Central
St. Louis	0.191		-0.015	16349	0.051	0.072	0	1	0	Central
Sioux Falls, S.D.	0.2	0.038	-0.002	17372	0.192	-0.083	0	1	0	Central
Topeka	0.069	0.053	0.018	14702	-0.057	0.222	0	1	0	Central
Tulsa	0.21	0.033	0.038	15238	0.073	0.056	0	1	0	Central
Wichita	0.169	0.033	0.053	15918	0.076	0.125	0	1	0	Central
Albuquerque	0.066	0.071	0.05	12871	0.191	0.041	0	0	1	West
Boise	0.031	0.057	-0.009	11275	0.005	-0.038	0	0	1	West
Casper		0.041	0.139	14768	-0.109	0.406	0	0	1	West
Denver-Boulder	0.049	0.04	0.051	14818	0.113	0.095	0	0	1	West
Fresno	0.108	0.102	0.054	14410	0.152	0.145	0	0	1	West
Great Falls	0.014	0.075	-0.003	19307	0.097	-0.154	0	0	1	West
Honolulu	0.043	0.044	0.039	14524	0.197	-0.011	0	0	1	West
Las Vegas	0.221	0.059	0.07	17985	0.045	0.041	0	0	1	West
Los Angeles-Long Beach	0.109	0.056	0.036	14291	0.077	0.013	0	0	1	West
Phoenix	0.068	0.049	0.048	14083	0.076	-0.047	0	0	1	West
Portland, Oreg.		0.053	0.042	14142	0.069	0.139	0	0	1	West
Sacramento	0.128	0.077	0.054	15953	0.1	0.108	0	0	1	West
Salt Lake City-Ogden	-0.03	0.053	0.058	12708	0.1	0.051	0	0	1	West
San Diego	0.133	0.064	0.045	14228	0.082	-0.018	0	0	1	West
San Francisco-Oakland	0.111	0.054	0.02	17639	0.074	0.033	0	0	1	West
Seattle	0.081	0.058	0.061			0.099	0	0	1	West
Tucson	0.108	0.047	0.057	13459	0.101	0.028	0	0	1	West
U.S. Average	0.1018	0.0569	0.027	15066.19	0.0867	0.0552				

TABLE B-2 American Cities database- Version 2.

City Region	Change in Dept Store Sales	Unempl Rate	Change in Nonfarm Empl	Change in Income Factory Workers	Change in Factory Workers Income	Change in Construction Activity
X1	X2	X3	X4	X5	X6	X7
E	0.107	0.047	0.032	$43,432.00	0.124	0.005
E	0.088		0.011	$46,178.00	0.09	0.052
E	0.109	0.062	0.041	$44,132.00	0.057	0.215
E	0.127	0.065	0.022	$45,784.00	0.095	0.136
E	0.071	0.097	0.011	$48,362.00	0.025	0.214
E		0.064	0.049	$43,979.00	0.116	0.274
E	0.033	0.061	0.04	$43,002.00	0.087	0.118
E	0.119	0.063	0.002	$48,740.00	0.138	0.047
E	0.133	0.039	0.015	$40,905.00	0.076	0.042
E	0.128	0.044	0.018	$41,299.00	0.106	-0.061
E	0.131	0.048	0.103	$46,547.00	0.079	0.202
E	-0.073	0.051	0.015	$43,790.00	0.077	0.033
E		0.035	0.019	$40,337.00	0.089	0.034
E	0.11	0.048	0.051	$40,718.00	0.092	0.225
E	0.111	0.058	0.016	$45,456.00	0.076	0.076
E	0.079	0.088	0.012	$42,285.00	0.087	0.049
E	-0.012	0.053	0.01	$43,658.00	0.031	-0.051
E	0.107	0.066	0.036	$45,073.00	0.069	0.175
E	0.12	0.071	0.004	$48,483.00	0.061	-0.015
E	0.101	0.054	0.051	$41,619.00	0.095	0.03
E	0.045	0.081	-0.004	$41,012.00	0.076	0.028
E	-0.039	0.031	0.018	$45,848.00	0.199	-0.029
E	0.055	0.065	0.016	$47,890.00	0.05	0.011
E	0.082	0.055	0.013	$46,024.00	0.135	-0.1
E	0.075	0.049	0.034	$42,333.00	0.056	0.022
E	0.073		0.027	$47,139.00	0.134	0.064
E	0.19	0.074	0.002	$47,606.00	0.114	-0.043
C	0.024	0.068	0.024	$45,681.00	0.087	0.003
C	0.198		0.004	$45,226.00	0.106	-0.012
C	0.098	0.06	0.015	$46,622.00	0.067	0.094
C	-0.017	0.056	0.006	$47,782.00	0.003	0.053
C	0.063	0.048	0.021	$45,261.00	0.065	0.085

TABLE B-2 Continued.

City Region	Change in Dept Store Sales	Unempl Rate	Change in Nonfarm Empl	Change in Income Factory Workers	Change in Factory Workers Income	Change in Construction Activity
X1	X2	X3	X4	X5	X6	X7
C	0.231	0.042	0.056	$43,590.00	0.109	0.134
C	0.224	0.055	0.011	$47,819.00	0.134	-0.029
C	0.033	0.12	-0.044	$50,378.00	-0.031	-0.013
C	0.134	0.05	0.036	$43,412.00	0.114	-0.029
C	0.143	0.067	-0.001	$45,917.00	0.032	0.044
C	0.162	0.039	0.042	$48,274.00	0.079	0.014
C	0.095	0.069	0.019	$47,122.00	0.05	0.115
C	0.053	0.042	0.025	$42,387.00	0.092	0
C	0.135	0.054	0	$46,577.00	0.096	-0.06
C	0.084	0.055	-0.007	$44,298.00	0.096	-0.036
C	0.025	0.043	0.019	$42,846.00	0.088	-0.024
C	0.131	0.074	-0.021	$46,224.00	0.041	-0.019
C	0.142	0.059	0.005	$43,351.00	0.072	0.085
C	0.188	0.043	0.029	$48,082.00	0.073	0.059
C	0.077	0.042	0.037	$46,298.00	0.086	0.131
C	0.243	0.072	0.027	$47,711.00	0.133	0.163
C	0.113	0.066	0.019	$46,337.00	0.129	0.024
C	0.204	0.03	0.087	$45,918.00	0.283	0.103
C	0.141	0.052	0.018	$45,053.00	0.06	-0.021
C	0.043	0.064	0.028	$40,525.00	0.128	0.12
C	0.191		-0.015	$46,349.00	0.051	0.072
C	0.2	0.038	-0.002	$47,372.00	0.192	-0.083
C	0.069	0.053	0.018	$44,702.00	-0.057	0.222
C	0.21	0.033	0.038	$45,238.00	0.073	0.056
C	0.169	0.033	0.053	$45,918.00	0.076	0.125
W	0.066	0.071	0.05	$42,871.00	0.101	0.041
W	0.031	0.057	-0.009	$41,275.00	0.005	-0.038
W		0.041	0.139	$44,768.00	-0.109	0.406
W	0.049	0.04	0.051	$44,818.00	0.113	0.095
W	0.108	0.102	0.054	$44,410.00	0.152	0.145
W	0.014	0.075	-0.003	$49,307.00	0.097	-0.154
W	0.043	0.044	0.039	$44,542.00	0.197	-0.011

TABLE B-2 Continued.

City Region	Change in Dept Store Sales	Unempl Rate	Change in Nonfarm Empl	Change in Income Factory Workers	Change in Factory Workers Income	Change in Construction Activity
X1	**X2**	**X3**	**X4**	**X5**	**X6**	**X7**
W	0.221	0.059	0.07	$47,985.00	0.045	0.041
W	0.109	0.056	0.036	$44,291.00	0.077	0.013
W	0.068	0.049	0.048	$44,083.00	0.076	-0.047
W		0.053	0.042	$44,142.00	0.069	0.139
W	0.128	0.077	0.054	$45,953.00	0.1	0.108
W	-0.03	0.053	0.058	$42,708.00	0.1	0.051
W	0.133	0.064	0.045	$44,228.00	0.082	-0.018
W	0.111	0.054	0.02	$47,639.00	0.074	0.033
W	0.081	0.058	0.061			0.099
W	0.108	0.047	0.057	$43,459.00	0.101	0.028
AVE(s)	**0.102**	**0.057**	**0.027**	**$45,073.70**	**0.085**	**0.055**
Count	70	70	74	73	73	74

TABLE B-3 Auto file.

Car	Basic Price	Displacement cc	Horsepower	Curb Weight lb.	MPG
Maserati Merak	$31,000	2965	182	3185	14
Maserati Bore	$39,927	4931	315	3540	12
Maserati Khamsin	$43,587	4931	315	3800	12
Mazda GLC	$3,895	1415	65	1995	30
Mazda RX-7	$6,395	1146	100	2350	17
Mercedes-Benz 240D	$20,775	2746	142	3560	14
Mercedes-Benz 300TD	$25,000	2998	77	3780	23
Mercedes-Benz 280CE	$22,481	2998	77	3475	23
Mercedes-Benz 450SL	$34,760	4520	180	3795	12
MG Midget	$5,200	1493	50	1850	23
MGB	$6,550	1798	67	2415	16
Peugeot 504	$7,922	1971	88	2905	22

TABLE B-4 Cost of living.

Component Index Weights	100%	13%	29%	10%	9%	4%	35%
	Composite Index[2]	Grocery Items	Housing	Utilities	Trans-portation	Health care	Misc. goods and Services[3]
Montgomery, Ala.	96.1	96.5	92.6	99.4	98.1	86.5	98.5
Juneau, Alaska	131.6	139.7	136.6	137.6	117.5	148.3	124.3
Phoenix, Ariz.	98.2	100	92.1	91.8	107.6	99.5	101.8
Los Angeles, Calif.	153.1	120.2	245.4	121.3	116.5	114.2	111.7
San Diego, Calif.	141	116.6	212.4	93	119.7	119.4	112.5
San Francisco, Calif.	177	137.9	308.2	107	125.9	126.6	121.8
Colorado Springs, Colo.	96.5	103.1	99	79.7	103.6	102.2	94.4
Denver, Colo.	103.5	107.2	110.6	88.5	96.7	104.6	102.3
Washington, D.C.	137.8	114.5	201.2	106.1	112.8	122	111.3
Jacksonville, Fla.	92.4	103	86.7	80.8	100.2	96.6	93.9
Atlanta, Ga.	97.2	96.5	93.6	88.3	101.4	108.4	100.5
Honolulu, Hawaii	162.4	153.6	243.8	143.7	130.1	115.1	117.3
Chicago, Ill.	128.6	117.8	169.5	109	107.7	121	110.7
Springfield, Ill.	92.1	90.7	85.9	102.1	99.7	92.9	92.8
Des Moines, Iowa	94.4	86.9	95.5	99.9	94.2	92.7	94.9
Topeka, Kans.	92	87.9	81.6	97.9	100.1	92.3	98.2
Baton Rouge, La.	104.1	113.1	95.4	113.1	102.8	100.5	106

(Continued)

TABLE B-4 Continued.

Component Index Weights	**100%**	**13%**	**29%**	**10%**	**9%**	**4%**	**35%**
	Composite index[2]	**Grocery items**	**Housing**	**Utilities**	**Trans-portation**	**Health care**	**Misc. goods and Services[3]**
Baltimore, Md.	107.6	105.7	115	110.5	101.1	105.3	103.1
Boston, Mass.	136.8	115.7	180.8	144.8	109.7	129.4	113.7
Detroit, Mich.	106.6	100	121.7	102.8	101.5	100.6	99.5
Minneapolis, Minn.	102.6	97.3	103	102.2	95.5	95.1	107
Jackson, Miss.	92.3	86.5	85.2	109.5	93.4	87	95.7
St. Louis, Mo.	99.9	103.4	96.5	94.6	96.9	100.9	103.7
Billings, Mont.	97.5	104.2	85.8	103.1	99.5	103.4	101.8
Omaha, Neb.	92.3	89.6	86	104.9	99.4	94.1	93
Las Vegas, Nev.	109.8	110.8	121.7	91.6	109.4	112.7	104.6
Albuquerque, N.M.	103.3	100.2	107.9	122.3	100	96.8	96.8
Buffalo, N.Y.	98.8	102.9	90.8	130.8	103.6	95.6	94
New York (Manhattan), N.Y.	212.1	137.1	399.2	146.6	117.4	139.2	136.2
Charlotte, N.C.	93.5	94.3	84.7	85.3	99.4	104.7	100
Cincinnati, Ohio	95.1	92.5	85.1	110.4	101	101.5	97.8
Cleveland, Ohio	102.7	112.9	100	115.2	100.8	101.5	98.3
Oklahoma City, Okla.	92	89	86	96.4	97.4	96.2	95.1
Eugene, Ore.	109.5	104.2	120.4	85.1	102.3	112.1	111

Portland, Ore.	110.9	116.8	109	117.1	107.1	111.6	109.5
Philadelphia, Pa	119	116.4	126.5	125.6	103.5	110.4	117
Memphis, Tenn.	89.9	92.8	77.6	89.7	94	99.3	97
Dallas, Tex.	95.2	93.6	84.4	97.1	102.5	102.3	101.5
El Paso, Tex.	91.5	105.6	80.3	98.7	95	100	91.7
San Antonio, Tex.	94.2	83.5	97.2	80.4	89.1	97.3	100.5
Salt Lake City, Utah	97.3	103.8	93.2	88.4	97.5	108	99.4
Richmond, Va.	101	108.4	97.9	102.3	95	98	102.4
Seattle, Wash.	118.6	111.8	132.5	109.2	110.2	123.4	114
Milwaukee, Wis.	101	95.1	110.1	93.7	103.4	92.6	98.2
Cheyenne, Wyo.	107.7	114.8	115.6	111.9	95.2	96.9	101.7

TABLE B-5 Fast Food.

Gender	Frequency	Income
Man	4	30
Man	4	58
Man	6	48
Woman	6	58
Woman	5	53
Man	3	49
Man	5	48
Man	6	52
Woman	4	47
Man	4	63
Woman	5	48
Man	6	45
Woman	4	43
Man	4	55
Woman	4	32
Man	6	52
Man	3	44
Woman	4	75
Man	6	53
Woman	5	54
Woman	1	68
Woman	3	50
Man	3	39
Man	3	55
Woman	2	60
Man	2	39
Man	6	55
Man	4	64
Woman	2	30
Man	6	59
Man	4	36
Woman	3	58
Woman	5	46
Man	1	71
Man	2	61
Woman	6	49
Man	2	20
Woman	6	47
Woman	2	61
Man	3	53
Man	6	49
Man	1	33
Man	4	59
Woman	4	60
Man	3	40
Woman	6	49
Woman	2	32
Woman	3	69
Man	5	49
Man	2	61
Man	4	64
Woman	4	78
Woman	3	66
Man	1	24
Man	4	34
Man	4	65
Woman	6	46
Man	5	53
Woman	2	78
Man	4	36
Man	5	49
Man	3	22
Man	2	65
Woman	2	69
Man	6	50
Man	4	44
Woman	2	60
Woman	4	58
Man	4	48
Man	4	62
Woman	5	49
Woman	4	51
Man	4	56

(Continued)

TABLE B-5 Continued.

Gender	Frequency	Income
Man	2	67
Man	3	65
Man	2	71
Woman	5	58
Woman	3	38
Woman	4	52
Woman	3	74
Man	2	70
Woman	2	30
Woman	3	66
Man	3	78
Woman	4	45
Woman	2	34
Man	3	69
Woman	3	53
Man	4	54
Woman	6	51
Man	2	62
Man	3	52
Woman	3	85
Woman	3	86
Woman	3	36
Man	2	78
Woman	6	56
Woman	1	68
Woman	2	72
Woman	6	47
Man	2	65
Man	3	76
Man	4	51
Woman	3	53
Woman	1	68
Woman	3	62
Man	2	39
Man	4	70
Man	4	65
Man	3	93
Woman	2	75
Man	3	69
Woman	2	84
Man	4	56
Woman	4	51
Man	2	76
Man	4	27
Man	4	66
Woman	4	64
Woman	4	52
Man	3	63
Man	2	69
Man	4	63
Man	3	26
Man	3	77
Woman	6	52
Woman	1	94
Man	2	33
Man	2	84
Man	1	32
Man	1	63
Woman	3	65
Woman	6	50
Man	6	52
Man	1	86
Woman	3	22
Man	4	33
Woman	3	44
Woman	3	56
Woman	6	48
Woman	2	30
Man	3	31
Woman	2	64
Woman	4	32
Woman	2	62
Woman	4	71
Woman	1	64
Man	1	79

TABLE B-6 Health file.

Cases	Drink	Health
0	2	2
0	1	1
0	1	2
0	2	1
0	1	1
0	1	1
0	2	3
0	2	1
1	1	4
0	1	1
1	1	2
0	2	2
0	1	3
0	1	1
0	1	1
1	1	3
1	1	1
0	2	3
0	2	2
0	1	2
0	1	1
0	1	2
0	1	2
0	2	3
0	2	3
0	1	3
0	2	2
0	1	3
1	1	2
0	1	2
0	1	2
0	1	2
0	1	2
0	2	1
0	1	1
0	1	1
0	1	2
1	1	3
0	1	2
0	1	1
0	1	1
0	1	2
0	1	1
0	1	3
0	1	1
0	1	2
1	1	1
0	1	2
0	1	1
0	2	3
0	1	2
0	1	2
0	1	1
0	1	3
0	1	3
0	1	2
0	2	1
1	2	3
1	2	4
1	1	2
0	1	2
0	1	2
0	1	1
0	1	4
0	2	1
0	2	1
0	2	2
1	1	2
1	1	4
0	1	2
0	1	1
0	1	1
1	2	3
1	1	1
0	1	2

(Continued)

TABLE B-6 Continued.

Cases	Drink	Health
1	1	2
0	1	1
0	1	2
0	2	2
1	1	1
0	2	4
0	1	1
0	1	2
0	1	2
0	2	2
0	2	1
0	1	1
0	1	1
0	1	2
0	1	2
0	1	3
0	2	1
0	1	1
0	1	2
0	1	2
0	1	2
0	1	3
0	1	2
1	1	2
0	1	1
0	2	1
0	1	2
0	1	2
1	1	3
0	1	1
1	1	2
0	1	1
0	1	2
0	1	2
0	1	2
1	1	1
1	1	4
1	1	2
1	2	2
0	1	2
0	1	1
1	2	2
0	1	1
0	1	2
0	1	2
0	1	3
0	1	2
0	1	1
1	1	3
1	1	1
1	2	2
0	1	3
0	2	4
0	1	2
0	2	2
1	1	1
1	1	2
0	1	1
0	1	1
0	1	1
0	1	1
0	1	1
0	1	1
0	1	1
1	1	1
0	1	2
1	1	1
0	1	1
1	1	2
0	1	1
0	1	1
1	1	1
0	1	1
0	1	2
0	1	2

(*Continued*)

TABLE B-6 Continued.

Cases	Drink	Health
1	1	3
0	1	2
0	2	4
0	1	2
0	1	1
0	1	2
0	1	2
0	1	2
0	1	2
0	2	1
0	1	1
0	1	1
0	1	1
0	1	2
0	1	1
0	1	1
0	1	1
0	2	1
0	2	1
0	2	1
0	1	1
0	2	3
0	1	1
1	1	1
0	1	3
0	1	3
1	1	1
0	1	3
0	1	1
0	1	1
0	1	2
1	1	3
0	1	1
0	1	1
0	2	4
1	1	1
0	1	1
1	1	2
1	1	2
0	1	2
0	1	2
0	1	2
0	1	2
0	1	2
0	1	2
0	1	1
0	1	1
0	1	1
0	1	2
0	1	1
1	1	3
0	2	2
0	2	2
0	1	2
0	1	1
0	1	1
0	2	1
0	1	2
0	1	1
0	1	2
1	2	3
0	1	1
0	2	4
0	1	2
0	1	2
0	1	2
0	1	2
0	1	1
0	1	2
0	2	2
0	1	1
0	1	1
0	2	4
0	1	2
1	1	4

(Continued)

TABLE B-6 Continued.

Cases	Drink	Health
0	2	1
0	1	1
0	1	2
0	1	1
0	2	2
0	2	1
0	1	1
0	1	1
0	1	1
1	1	1
0	1	2
0	1	3
0	2	3
0	2	1
0	2	4
0	1	1
0	2	1
0	1	2
0	1	1
0	1	2
0	1	1
0	2	2
0	1	2
0	1	2
0	1	2
1	1	2
0	2	3
0	1	2
0	1	1
0	1	2
1	1	1
1	2	1
1	1	2
0	1	2
0	1	1
0	1	1
0	2	1
0	1	2
0	1	1
0	1	1
0	1	3
0	1	1
0	1	1
0	2	1
0	1	1
0	1	1
0	1	1
0	1	1
0	1	4
0	1	2
0	1	1
0	1	1
0	1	1
0	1	3
0	1	2
0	1	1
0	1	1
0	1	2
0	2	2
0	1	1
0	1	1
0	1	2
1	1	2
1	2	1
0	1	2
0	1	1
0	1	2
0	2	1
0	2	3

TABLE B-7 Interest rate voltality.

Date	FYGT1	Change
8501	9.02	
8502	9.29	0.27
8503	9.86	0.57
8504	9.14	-0.72
8505	8.46	-0.68
8506	7.8	-0.66
8507	7.86	0.06
8508	8.05	0.19
8509	8.07	0.02
8510	8.01	-0.06
8511	7.88	-0.13
8512	7.67	-0.21
8601	7.73	0.06
8602	7.61	-0.12
8603	7.03	-0.58
8604	6.44	-0.59
8605	6.65	0.21
8606	6.73	0.08
8607	6.27	-0.46
8608	5.93	-0.34
8609	5.77	-0.16
8610	5.72	-0.05
8611	5.8	0.08
8612	5.87	0.07
8701	5.78	-0.09
8702	5.96	0.18
8703	6.03	0.07
8704	6.5	0.47
8705	7	0.5
8706	6.8	-0.2
8707	6.68	-0.12
8708	7.03	0.35
8709	7.67	0.64
8710	7.59	-0.08
8711	6.96	-0.63
8712	7.17	0.21
8801	6.99	-0.18
8802	6.64	-0.35
8803	6.71	0.07
8804	7.01	0.3
8805	7.4	0.39
8806	7.49	0.09
8807	7.75	0.26
8808	8.17	0.42
8809	8.09	-0.08
8810	8.11	0.02
8811	8.48	0.37
8812	8.99	0.51
8901	9.05	0.06
8902	9.25	0.2
8903	9.57	0.32
8904	9.36	-0.21
8905	8.98	-0.38
8906	8.44	-0.54
8907	7.89	-0.55
8908	8.18	0.29
8909	8.22	0.04
8910	7.99	-0.23
8911	7.77	-0.22
8912	7.72	-0.05
9001	7.92	0.2
9002	8.11	0.19
9003	8.35	0.24
9004	8.4	0.05
9005	8.32	-0.08
9006	8.1	-0.22
9007	7.94	-0.16
9008	7.78	-0.16
9009	7.76	-0.02
9010	7.55	-0.21
9011	7.31	-0.24

(Continued)

TABLE B-7 Continued.

Date	FYGT1	Change
9012	7.05	-0.26
9101	6.64	-0.41
9102	6.27	-0.37
9103	6.4	0.13
9104	6.24	-0.16
9105	6.13	-0.11
9106	6.36	0.23
9107	6.31	-0.05
9108	5.78	-0.53
9109	5.57	-0.21
9110	5.33	-0.24
9111	4.89	-0.44
9112	4.38	-0.51
9201	4.15	-0.23
9202	4.29	0.14
9203	4.63	0.34
9204	4.3	-0.33
9205	4.19	-0.11
9206	4.17	-0.02
9207	3.6	-0.57
9208	3.47	-0.13
9209	3.18	-0.29
9210	3.3	0.12
9211	3.68	0.38
9212	3.71	0.03
9301	3.5	-0.21

TABLE B-8 Stock prices.

Stock Prices					
Date	**MSFT**	**GE**	**INTC**	**GM**	**CSCO**
30-Mar-90	0.12152	0.04049	0.03727	0.02228	0.01075
30-Apr-90	0.0474	-0.0039	-0.0539	-0.0354	0.01064
31-May-90	0.25862	0.08352	0.22152	0.11582	0.04211
29-Jun-90	0.0411	0.00544	-0.0259	-0.0206	0.07071
31-Jul-90	-0.125	0.0343	-0.0532	-0.021	-0.0377
31-Aug-90	-0.0752	-0.1344	-0.25	-0.1314	-0.0294
28-Sep-90	0.02439	-0.1134	-0.0037	-0.0881	-0.0909
31-Oct-90	0.0119	-0.0459	0.00752	0.01379	0.31111
30-Nov-90	0.13333	0.05288	0.1194	0.01361	0.33898
31-Dec-90	0.04152	0.05726	0.02667	-0.0582	0.13608
31-Jan-91	0.30399	0.11547	0.18831	0.05455	0.30362
28-Feb-91	0.05732	0.07047	0.04372	0.10069	-0.0427
28-Mar-91	0.02289	0.0239	-0.0209	-0.0443	-0.1295
30-Apr-91	-0.0671	0.01616	0.05348	-0.053	0.22051
31-May-91	0.10859	0.09908	0.13198	0.21748	0.08403
28-Jun-91	-0.0689	-0.0421	-0.1659	-0.0551	-0.0543
31-Jul-91	0.0789	-0.0101	0.01075	-0.0245	0.28689
30-Aug-91	0.15986	0.02218	0.05319	-0.034	0.15605
30-Sep-91	0.04399	-0.0666	-0.1465	-0.0164	-0.0964
31-Oct-91	0.05478	-0.0054	-0.0385	-0.0602	0.18902
29-Nov-91	0.03595	-0.0616	0.00923	-0.1132	0.01539
31-Dec-91	0.14396	0.18996	0.19512	-0.061	0.33838
31-Jan-92	0.0809	-0.0163	0.22194	0.12121	0.13396
28-Feb-92	0.02703	0.04485	0.06054	0.17066	0.08486
31-Mar-92	-0.0405	-0.0296	-0.1299	-0.0233	-0.0307
30-Apr-92	-0.0696	0.01155	-0.0317	0.13311	-0.0759
29-May-92	0.09751	-0.0033	-0.0654	-0.0392	0.25343
30-Jun-92	-0.1322	0.0252	0.14	0.11348	0.02732
31-Jul-92	0.03929	-0.0161	0.04825	-0.054	0.13564
31-Aug-92	0.02405	-0.0327	-0.0293	-0.1586	-0.1218
30-Sep-92	0.08054	0.06541	0.13362	-0.0722	0.128
30-Oct-92	0.10248	-0.0192	0.02814	-0.0428	0.13475
30-Nov-92	0.0493	0.08469	0.05926	0.05528	0.24583
31-Dec-92	-0.0832	0.03459	0.21678	0	0.05184

TABLE B-8 Continued.

Stock Prices					
Date	MSFT	GE	INTC	GM	CSCO
29-Jan-93	0.01318	0.00731	0.22816	0.17054	0.1256
26-Feb-93	-0.0361	-0.0232	0.09133	-0.0079	-0.0099
31-Mar-93	0.10945	0.06692	-0.0129	0.01007	0.0214
30-Apr-93	-0.0757	0.01683	-0.172	0.08638	-0.0782
28-May-93	0.08333	0.02345	0.16557	-0.0135	0.30909
30-Jun-93	-0.0499	0.03914	-0.0079	0.10903	0.01389
30-Jul-93	-0.1591	0.02872	-0.0491	0.08989	-0.0525
31-Aug-93	0.0152	-0.0025	0.22967	-0.0294	-0.0795
30-Sep-93	0.09817	-0.0178	0.10117	-0.1093	0.04712
29-Oct-93	-0.0288	0.01173	-0.1053	0.14072	0.025
30-Nov-93	-0.0016	0.01418	-0.0277	0.11181	0.09756
31-Dec-93	0.00781	0.07339	0.00813	0.04028	0.14889
31-Jan-94	0.05581	0.02741	0.05323	0.11617	0.12186
28-Feb-94	-0.0308	-0.022	0.05364	-0.0457	0.01724
31-Mar-94	0.02727	-0.0442	-0.0182	-0.0751	-0.0712
29-Apr-94	0.09145	-0.0475	-0.0956	0.05336	-0.115
31-May-94	0.16216	0.04462	0.02459	-0.0493	-0.1835
30-Jun-94	-0.0395	-0.0556	-0.064	-0.0651	-0.0556
29-Jul-94	-0.0024	0.08043	0.01393	0.02239	-0.1016
31-Aug-94	0.12864	-0.0124	0.1097	-0.018	0.18155
30-Sep-94	-0.0344	-0.0254	-0.0646	-0.0672	0.10328
31-Oct-94	0.12249	0.01558	0.01114	-0.1573	0.10046
30-Nov-94	-0.002	-0.0588	0.0161	-0.0297	0.07054
30-Dec-94	-0.0278	0.11761	0.01188	0.10492	0.08915
31-Jan-95	-0.0286	0.0098	0.08705	-0.0772	-0.0498
28-Feb-95	0.06105	0.06311	0.14955	0.10161	0.01124
31-Mar-95	0.12897	-0.0062	0.06426	0.03226	0.12963
28-Apr-95	0.14938	0.03704	0.20689	0.02557	0.0459
31-May-95	0.03593	0.03571	0.09646	0.07036	0.09718
30-Jun-95	0.06716	-0.0209	0.12806	-0.0234	0.15571
31-Jul-95	0.00138	0.04656	0.02729	0.04	0.1026
31-Aug-95	0.0221	-0.0021	-0.0558	-0.0272	0.17713
29-Sep-95	-0.0216	0.08977	-0.0183	-0.0053	0.05143
31-Oct-95	0.10497	-0.0078	0.16041	-0.0667	0.12319

(*Continued*)

TABLE B-8 Stock prices.

Stock Prices					
Date	**MSFT**	**GE**	**INTC**	**GM**	**CSCO**
30-Nov-95	-0.1288	0.06126	-0.1288	0.11543	0.08548
29-Dec-95	0.00717	0.07948	-0.0678	0.09021	-0.1129
31-Jan-96	0.05413	0.06597	-0.026	-0.0047	0.11558
29-Feb-96	0.06689	-0.0163	0.06478	-0.0185	0.14114
29-Mar-96	0.04497	0.03755	-0.0329	0.03902	-0.0237
30-Apr-96	0.09818	-0.008	0.19191	0.01878	0.1186
31-May-96	0.04857	0.0712	0.11439	0.0235	0.05542
28-Jun-96	0.01158	0.04834	-0.0273	-0.0499	0.03425
31-Jul-96	-0.0187	-0.0466	0.02366	-0.0692	-0.0861
30-Aug-96	0.03924	0.01064	0.0624	0.02615	0.01932
30-Sep-96	0.07653	0.10027	0.19577	-0.0327	0.17654
31-Oct-96	0.04076	0.06319	0.1518	0.11719	-0.003
29-Nov-96	0.14299	0.07494	0.15472	0.08205	0.09697
31-Dec-96	0.05339	-0.0443	0.03202	-0.0325	-0.0626
31-Jan-97	0.23449	0.04678	0.23952	0.0583	0.09627
28-Feb-97	-0.0441	-0.006	-0.1256	-0.0106	-0.2025
31-Mar-97	-0.0596	-0.0302	-0.0194	-0.0432	-0.1348
30-Apr-97	0.32515	0.11839	0.10099	0.04515	0.07533
30-May-97	0.02058	0.08784	-0.0106	2.31E-10	0.30918
30-Jun-97	0.01915	0.0766	-0.0639	-0.0283	-0.0092
31-Jul-97	0.11968	0.08285	0.29527	0.10987	0.18529
29-Aug-97	-0.0658	-0.1078	0.0034	0.02222	-0.0526
30-Sep-97	0.00095	0.09207	0.00204	0.06673	-0.0307
31-Oct-97	-0.0175	-0.0505	-0.1656	-0.0411	0.12275
28-Nov-97	0.08846	0.14313	0.00812	-0.0428	0.05143
31-Dec-97	-0.0866	-0.0027	-0.095	0.05804	-0.0304
30-Jan-98	0.15426	0.05622	0.15345	-0.0463	0.13117
27-Feb-98	0.13615	0.00323	0.10725	0.19849	0.0446
31-Mar-98	0.05605	0.11238	-0.1296	-0.0172	0.03795
30-Apr-98	0.00698	-0.0116	0.03561	-0.0055	0.0713
29-May-98	-0.0589	-0.0213	-0.116	0.07421	0.03242
30-Jun-98	0.27782	0.08996	0.03762	-0.0704	0.21736
31-Jul-98	0.01442	-0.0125	0.13912	0.08232	0.04005

TABLE B-8 Continued.

	Stock Prices				
Date	**MSFT**	**GE**	**INTC**	**GM**	**CSCO**
31-Aug-98	-0.1273	-0.1055	-0.1566	-0.1893	-0.1449
30-Sep-98	0.14723	-0.0017	0.20457	-0.0559	0.13244
30-Oct-98	-0.038	0.09976	0.04009	0.15148	0.01921
30-Nov-98	0.1523	0.03286	0.20718	0.11375	0.19643
31-Dec-98	0.13678	0.1325	0.10163	0.02415	0.23134
29-Jan-99	0.26183	0.02819	0.18872	0.25415	0.20202
26-Feb-99	-0.1421	-0.0435	-0.1487	-0.0745	-0.1232
31-Mar-99	0.194	0.10629	-0.0089	0.05435	0.12013
30-Apr-99	-0.0927	-0.0475	0.02944	0.02371	0.04107
28-May-99	-0.0077	-0.035	-0.116	-0.0838	-0.0444
30-Jun-99	0.11774	0.11125	0.10058	-0.0435	0.18234
30-Jul-99	-0.0485	-0.0323	0.15966	-0.0739	-0.0359
31-Aug-99	0.07866	0.03039	0.19156	0.09202	0.09155
30-Sep-99	-0.0216	0.05876	-0.0958	-0.05	0.01106
29-Oct-99	0.02208	0.14286	0.04205	0.11917	0.07931
30-Nov-99	-0.0164	-0.0397	-0.0093	0.02928	0.20524
31-Dec-99	0.28231	0.19239	0.07335	0.00955	0.20112
31-Jan-00	-0.1617	-0.1365	0.20197	0.10834	0.02217
29-Feb-00	-0.0868	-0.0094	0.14244	-0.0497	0.20719
31-Mar-00	0.18881	0.17873	0.16759	0.08874	0.16974
28-Apr-00	-0.3435	0.01044	-0.0388	0.13057	-0.1033
31-May-00	-0.103	0.00517	-0.0165	-0.2403	-0.1787
30-Jun-00	0.27872	0.00593	0.07218	-0.1779	0.11636
31-Jul-00	-0.1273	-0.0269	-0.0014	-0.0194	0.0295
31-Aug-00	0	0.14095	0.12195	0.27662	0.04871
29-Sep-00	-0.1361	-0.0147	-0.4449	-0.0996	-0.1949
31-Oct-00	0.14197	-0.0498	0.08271	-0.0442	-0.0249
30-Nov-00	-0.167	-0.0958	-0.1537	-0.1952	-0.1114
29-Dec-00	-0.244	-0.0296	-0.2102	0.02904	-0.201

APPENDIX C

ARTICLES

ARTICLE C-1

Permission to Reprint: International Academy of Business and Public Administration Disciplines (IAPAD). Conference, Dallas, Texas, May 3–6, Conference Proceedings, Vol. 4, No. 2, www.iabpad.com.

Comparative Study of Graduate Information Systems and MBA Students Cognitive Styles

Paul A. Stock
University of Mary Hardin-Baylor

Ronald S. King
University of Mary Hardin-Baylor

ABSTRACT

Problem solving styles are critical dimensions in corporate decision making. In order to study the comparative problem-solving styles of college graduate students majoring in information systems versus business administration, the Kirton Adaption-Innovation Inventory (KAI) was used to determine the cognitive styles of the students enrolled in the Master of Science in Information Systems (MSIS) and the Master of Business Administration (MBA) programs at the University of Mary-Hardin Baylor (UMHB) during the spring semester of 2007. The results of the study were also analyzed by demographic variants.

Introduction

Dr. Michael J. Kirton, an eminent British organizational psychologist, developed a theory on the assumption that all people solve problems and are creative, since both activities are the results of the same brain function. His theory is well established and has been highly validated [Kirton, 1976; Kirton, 1994; Kirton, 1998]. A distinction is made between the level and style of problem solving and creativity. Cognitive level, not measured by the Kirton Adaptive-Inventory (A-I) Inventory [Kirton, 1998], refers to an individual's inherent potential capacity, such as intelligence. Cognitive style, measured by the Kirton A-I Inventory, is defined as the "strategic, stable, characteristic, preferred manner in which people respond and seek to bring about change."

A study, utilizing the Kirton A-I Inventory that compared the problem-solving style, in terms of adaption and innovation, of Information and Computer Science majors was conducted by [Sim, E. R. and Wright, G., 2002]. This study concluded that there is no difference in programming styles between information systems and computer science students as measured by the Kirton's A-I Inventory. The study did conclude, however, there was a gender difference in the Kirton A-I Inventory scores; women scored slightly more adaptively then men.

[Kaufmann, G., Isaken, S. G., and Lauer, K., 1996] developed a regression plane showing the distribution of Kirton A-I Inventory scores across managerial levels by gender. These authors found a striking gender difference when observing Kirton A-I Inventory scores for males and females analyzed across managerial levels. Males were found to be more innovatively inclined than females at the lowest managerial level, whereas females obtained almost a full standard deviation higher innovator score at the top, executive level of leadership.

Cognitive style differences, as measured by the Kirton A-I Inventory, lie on a continuum and range from strong adaption to strong innovation. For large general populations, the distribution of scores forms a normal curve. Smaller groups can be predictably different from general populations, depending on their problem-solving orientation, and may exhibit skewed distributions about different means.

Adaptors prefer to operate with more structure than do innovators. Innovators prefer solving problems by "thinking outside of the box". Adaptors approach problems from within the given frame of reference and strive to produce solutions that are "better" rather than "different." Adaptors are valued in a business environment since they tend to be the system experts and are dedicated to its maintenance and efficiency. Adaptors are especially good at fine-tuning the current rules and procedures in order to make them operate as effectively as possible. Innovators, on the other hand, tend to detach a given problem from the customary frame of reference and search for solutions that are typically seen as "different." In summary, adaptors prefer to solve problems using the rules and innovators tend to solve problems despite the rules.

During the stages of problem solving, innovators and adaptors have distinct patterns of behavior as a result of their distinct cognitive styles. Adaptors generally accept problems as they have been defined, including agreed-upon constraints. During data collection, they tend to exhaustively collect and favor information and perspectives that are closely related to the original problem structure. When generating ideas, adaptors prefer to generate only a few novel and creative solutions that are relevant, readily-acceptable, and aimed at improving the current situation. Such solutions are often easy to implement, which is not necessarily the case for solutions obtained by innovators. Adaptors look for a quick resolution to problems, which leads to increased efficiency.

Innovators tend to reject the original and accepted definition of a problem. Instead, the innovators will redefine the problem. Sometimes, on project teams, this leads to a breakdown in communication. In collecting data, innovators tend to look outside the original problem structure for different perspectives. During the generation of ideas, innovators generally produce numerous novel and creative ideas, and some of these ideas are view as acceptable by others. When evaluating and implementing solutions, innovators are less concerned with immediate efficiency and potential disruption of their project team, and therefore tend to look ahead to potential long-term gains [Kirton, 1994].

During problem definition, an adaptor's chief aim is to define the problem as explicitly as possible in a way that is acceptable to authority, and hopefully to other members of their project team. If a problem definition is supplied at the beginning of a project, adaptors will accept this definition as fixed, clarify any questions that are not well-defined, and move forward to a quick solution. Any ambiguity in the problem definition will cause the adaptor to become frustrated and moving forward becomes difficult. Innovators approach the problem with some freedom to explore different problem formulations. They delight in defining and redefining, even when a problem statement is supplied [Isaksen, S. G. and Kaufmann, G., 1990].

Application

The Kirton A-I Inventory [Kirton, 1998] was introduced in 1976 and measures preferred thinking. Respondents answer 32 questions that focus on how easy or difficult it is for a person to present himself consistently, over a long period of time, in particular ways {refer to Appendix A}. Each answer is assigned a value using a 5-point scale. The inventory is easily understood and can be completed in less than 15 minutes. A person's overall Kirton A-I score will fall between 32 and 160, with a score of 32 representing the theoretical limit of highest adaption, and a score of 160 representing the theoretical limit of highest innovation. In practice, scores typically fall between 40 and 150. For large general populations the distribution of scores forms a normal curve with a theoretical mean of 96. In the United States, the observed mean for the general population is 95, while the observed mean for both U.S. engineers and managers (measured separately) is 97 [Kirton, 1998]. Table C-1 displays the categories of adaptors and innovators for the theoretical distribution of Kirton A-I Inventory scores:

TABLE C-1 Kirton A-I Inventory Score Categories.

Innovators		Adaptors
96–110	Mild	95–80
111–124	Medium	79–65
125–139	High	64–50
140 or more	Very High	49 or less

It is important to note that high scores are not "good" and low scores are not "bad"; it is the relative difference between the scores of the two individuals or between an individual and the mean of a group that is important. In general, a difference of 10 points between individuals is noticeable over time. A difference of 20 points or more can lead to difficulties in communication and may require considerable coping behavior [Kirton, 1998].

The Kirton A-I Inventory was administered to graduate students enrolled in both the Master of Science in Information Systems (MSIS) program and the Master of Business Administration (MBA) program at the University of Mary Hardin-Baylor during the Spring 2007 semester. The Kirton A-I Inventory [Kirton, 1976] found in Appendix A was distributed to the students during a typical class period in the middle of their semester term.

Results

A summary of the Kirton A-I scores is shown in Table C-2 through Table C-7 below.

Conclusion

Our research hypothesis is there are no differences in the problem solving styles between college graduate students who are in the MSIS program and students who are in the MBA program as measured by the Kirton A-I Inventory. Therefore, we compared the mean scores of the students in the MSIS program and the students in the MBA program, as shown in Table C-2 below. Both groups of graduate students scored in the "mild innovator" category. This supports our hypothesis that there is no difference in their problem solving styles. However, the sample size is too small to warrant a high confidence level and more research in this area is needed.

Since earlier research indicated a difference in problem solving styles based on gender, we included this demographic in Table C-2 below. However, the data did not suggest a difference in the overall score based solely on gender. Both male and female graduate students scored at the "mild innovator" level.

TABLE C-2 Demographic Decomposition of Kirton A-I Inventory Scores.

Sample	Size (N)	Range	Mean	Category
All Students	14	85–129	103.86	Mild Innovator
MBA students only	7	85–115	101.00	Mild Innovator
MSIS students only	7	89–129	108.14	Mild Innovator
Male students only	8	85–129	105.25	Mild Innovator
Female students only	6	98–122	103.67	Mild Innovator

We also analyzed the range of scores among all of the students in the sample as shown in Table C-3 below. The KAI scores range from 85 (mild adaptor) to 129 (high innovator). However, the scores show over one-half (57%) of the students fall into the categories of mild adaptor to mild innovator. Furthermore, most of the graduate students (93%) fall in the range of mild adaptor to medium innovator. There were no medium, high, or very high adaptors in the sample. There also were no very high innovators in the sample. There was one high innovator in the sample.

TABLE C-3 Adaptors and Innovators among all Students.

Sample	Frequency	A-I inventory scores	Number of MSIS students	Number of MBA students
VeryHigh Innovators	0	–	–	–
High Innovators	1	129	1 [129]	
Medium Innovators	5	111, 112, 115, 115, 122	2 [115,122]	3 [111, 112, 115]
Mild Innovators	4	98,100,102,105	3 [98, 102, 105]	1 [100]
Mild Adaptors	4	85, 86, 89, 95	2 [89, 95]	2 [85, 86]
Medium Adaptors	0	–	–	–
High Adaptors	0	–	–	–
Very High Adaptors	0	–	–	–

An interesting difference emerges when the KAI scores are divided by the student's undergraduate academic major, as shown in Table C-4 and Table C-5 below. The students who scored at the "mild adaptor" level had an undergraduate emphasis in accounting, computer information systems, and computer science. The students who scored at the "medium innovator" level had an undergraduate emphasis in management, computer graphics design, biology, and church music. It seems logical to have accounting majors and computer science majors who are adaptors, which means they prefer a more structured environment. It seems equally consistent to have management, computer graphics, biology, and music majors who are innovators, which means they produce creative ideas during the problem solving process. This outcome may help college advisors, parents, and prospective students to choose an academic major.

TABLE C-4 Kirton A-I Inventory Scores by Undergraduate BBA Emphasis.

Sample	**Size (N)**	**Scores**	**Mean**	**Category**
Management	1	115	115.00	Medium Innovator
Accounting	3	100, 98, 86	94.67	Mild Adaptor

TABLE C-5 Kirton A-I Inventory Scores by Undergraduate non-BBA majors.

Sample	**Size (N)**	**Scores**	**Mean**	**Category**
English	2	115, 105	110.00	Mild Innovator
Computer Graphics Design	2	122, 102	112.00	Medium Innovator
Computer Information Systems	2	89, 85	87.00	Mild Adaptor
Computer Science	1	95	95.00	Mild Adaptor
Biology	2	129, 112	120.50	Medium Innovator
Church Music	1	111	111.00	Medium Innovator

Another interesting outcome from this study is revealed when the KAI scores are divided based on the number of years of work experience, as shown in Table C-6 below. The graduate students with less than 5 years of work experience scored at the "mild innovator" level. Those students who have between 5 and 10 years of work experience scored in the "medium innovator" level. Finally, those students with 15 or more years of work experience scored in the "mild adaptor" level. Since prior research suggests a person's KAI scores do not tend to change over time, this data suggests that graduate students with less than 10 years of work experience tend to be innovators, while graduate students with over 10 years of experience tend to be adaptors. However, the sample size is too small to place a high level of confidence in this area and more research is warranted in the relationship between the work experience of graduate students and their KAI scores.

TABLE C-6 Kirton A-I Inventory Scores by Years of Working in Field.

Sample	Size (n)	Range	Mean	Category
0 to < 5 years	7	85–122	101.14	Mild Innovator
5 >= and < 10 years	5	109–129	115.00	Medium Innovator
10 >= and < 15 years	0	-	-	-
15 years or more	2	86–95	90.50	Mild Adaptor

Kirton identified three "sub-factors" in the A-I Inventory questions. These three factors are the sufficiency of originality (SO), efficiency (E), and rule conformity (R). We analyzed the data according to these three sub-factors in Table C-7 below. The SO factor addresses the difference people have when developing new ideas. Adaptors tend to produce fewer ideas, which are also more sound and useful to the problem solving process. Innovators tend to produce more ideas, with many of them being outside the box and more risky. The mean SO score for the general population is 40.8 points and questions 1 through 12 on the KAI measure a person's SO factor. Our sample data did not indicate a difference in the SO factor based on the graduate program or gender.

TABLE C-7 Kirton A-I Inventory Factor Scores.

	Sufficiency of Originality (SO) Mean = 40.8*	Conformity to Group Norms and Rules (R) Mean = 35.4*	Efficiency in Regard to Detail (E) Mean = 18.8*
All Students	N=14;MEAN=44.2	N=14; MEAN= 39.2	N=14;MEAN= 20.5
MSIS	N=7; MEAN=44.9	N=7; MEAN= 41.7	N=7; MEAN= 21.6
MBA	N=7; MEAN=43.6	N=7; MEAN= 44.0	N=7; MEAN= 19.4
Female	N=6; MEAN=44.3	N=6; MEAN= 37.7	N=6; MEAN= 21.7
Male	N=8; MEAN=44.1	N=8; MEAN= 40.3	N=8; MEAN= 19.6

SO = sum of values on items 1–12
R = sum of values on items 13–18
E = sum of values on items 19–32
* General population mean from KAI Manual

The rule conformity (R) sub-factor measures differences in people's management structure during the problem solving process. Adaptors tend to accept the rules and even view the rules as an aid to the problem solving process. They also place more emphasis on group cohesion. Innovators view rules and group cohesion as limitations on the problem solving process. They often ignore the rules and pursue their own ideas. The mean R score for the general population is 35.4 and questions 13 through 18 on the KAI measure a person's R factor. Our sample data indicated a difference in the student's R factor scores. Students in the MBA program had higher R factor scores, which indicate they tend to follow the rules more and value group cohesion. Students in the MSIS program had lower R factor scores, which indicate they tend to ignore the rules and develop their own ideas. Our data also showed a difference in the R factor scores based on gender. Our data suggests male students tend to have a higher R factor score, which indicates they tend to follow the rules more and even view rules as an aid. The female students in our sample had lower R factor scores, which indicate they view rules and group cohesion as limits on the decision making process. More research is warranted in the R sub-factor of graduate students based on our results.

The efficiency (E) sub-factor measures the attention people give to detail when solving a problem. Adaptors tend to be more organized, arrange facts in an orderly method, and search for information that is relevant to the problem in a logical manner. Innovators tend to pay less attention to detail, are less thorough, and prefer to develop new ideas rather than search for relevant information. The mean E score for the general population is 18.8 and questions 19 through 32 on the KAI measure a person's E factor. Our data did not indicate any significant difference in the E factor among the graduate students, although students in the MSIS program scored slightly higher than students in the MBA program. Also, female students scored slightly higher than male students.

References

1. Isaksen, S. G. & Kaufmann, G. (1990). Adaptors and Innovators: Different Perceptions of the Psychological Climate for Creativity, *Studia Psychologica*, 32, 3.
2. Kaufmann, G., Isaksen, S. G., & Lauer, K. (1996). Testing the "Glass Ceiling" Effect on Gender Differences in Upper Level Management: The Case of Innovator Orientation, *European Journal of Work and Organizational Psychology*, 5(1), 29–41.
3. Kirton, M. J. (1976). Adaptors & Innovators: A Description and Measure, *Journal of Applied Psychology*, Vol. 61, No.5, 622–629.
4. Kirton, M. J. (1994). A Theory of Cognitive Style. *Adaptors and Innovators: Styles of Creativity and Problem Solving* (Ed.), 1994, Rroutledge, New York, 1–33.
5. Kirton, M. J. (1998). *Kirton Adaption-Innovation Inventory (KAI) Manual* (3rd Edition), Occupational Research Centre: Hertfordshire, UK.
6. Sim, E. R. & Wright, G. (2002) A Comparison of Adaption-Innovation Styles Between Information Systems Majors and Computer Science Majors, *Journal of Information Systems Education.*

PAUL A. STOCK is an Assistant Professor of Economics at the University of Mary Hardin-Baylor. He received his B.S. in Business Administration from Clarion University of Pennsylvania in 1982, his MBA from Oklahoma City University in 1989, and Ph.D. in Economic Education from Ohio University in 1995. Dr. Stock teaches and conducts research in macroeconomics and economic education.

RONALD S. KING is a Professor of Business Information and Technology Systems at the University of Mary Hardin-Baylor. He earned his B.S. in Mathematics from Lamar University in 1966, his M.S. in Mathematics from the University of North Texas in 1968, and the Ph.D. in Statistics from the University of Northern Colorado in 1973. His primary interests are in data mining, database design, and knowledge representation.

Appendix A

Kirton Adaption-Innovation Inventory (KAI)

From a managerial point of view rate the following traits of a successful manager on a scale of 1 to 5 (1 being extremely undesirable and 5 being extremely desirable):

ITEM	RATING
1. Has original ideas	________
2. Proliferates ideas.	________
3. Is stimulating.	________
4. Copes with several new ideas at the same time.	________
5. Will always think of something when stuck	________
6. Would sooner create than improve	________
7. Has fresh perspectives on old problems	________
8. Often risks doing things differently	________
9. Likes to vary set routines at a moment's notice	________
10. Prefers to work on one problem at a time	________
11. Can stand out in disagreement against group	________
12. Needs the stimulation of frequent change	________
13. Prefers changes to occur gradually	________
14. Is thorough	________
15. Masters all details painstakingly	________
16. Is methodical and symmetric	________
17. Enjoys detailed work	________
18. Is (not) a steady plodder	________
19. Is consistent	________

ITEM	RATING
20. Imposes strict order on matters within own control	________
21. Fits readily into "the system"	________
22. Conforms.	________
23. Readily agrees with the team at work	________
24. Never seeks to bend or break the rules	________
25. Never acts without proper authority	________
26. Is prudent when dealing with authority.	________
27. Likes the protection of precise instructions	________
28. Is predictable	________
29. Prefers colleagues who never "rock the boat.	________
30. Likes bosses and work patterns which are consistent	________
31. Works without deviation in a prescribed way	________
32. Holds back ideas until obviously needed.	________

ARTICLE C-2

Permission to Reprint: International Academy of Business and Public Administration Disciplines (IAPAD). Conference, Dallas, Texas, May 3–6, Conference Proceedings, Vol. 4, No. 2, www.iabpad.com

ETHNOGRAPHIC STUDY OF MSIS STUDENT PROJECT TEAM DYNAMICS

Ronald S. King
University of Mary Hardin-Baylor

ABSTRACT

Problem solving styles and interpersonal dynamics of project teams are often critical dimensions of the ability of a team to function effectively. In order to study the problem solving styles of students enrolled in a graduate Database Administration and Development class at the University of Mary Hardin-Baylor (UMHB) and to track their intra-team interactions, the theory established by M. J. Kirton in 1976, was used to determine the cognitive styles of the students. The author categorized students according to Adaption-Innovation characteristics by the Kirton AI Inventory. These categorizations were then applied to interpret the results of interviews from informants, the team leaders, and about the abilities of project team members to work and communicate with each other.

Introduction

Seven students were divided into one team of four students and one team of three students. The teams were asked to design, implement, and evaluate a commercial application database with a Web front-end. Each team leader had completed an undergraduate science degree (one in computer science and the other in biology), had a high GMAT score, a graduate GPA greater than 3.5, and at least seven years of professional experience in the information systems field. Each team was assigned a student who completed an undergraduate computer graphics design degree program, and each team was assigned one military personnel who is currently enrolled in the MSIS program at UMHB. The goal of each project is to expose the students to a simulated and significant corporate initiative that has both technological and managerial dimensions, as well as to allow them to experience the conflicts and management issues that can arise in an intensive team environment.

The Kirton KAI inventory, refer to Appendix B, was administered and each student was classified using the resultant scores. After interviewing the MSIS degree program director and using the list of characteristics for an adaptor and an innovator for each member in the class, the author qualitatively cross validated the classifications assigned to the students as either an adaptor or an innovator. The director and the author have known every team member from previous classes. Additionally, this is a degree program that requires each student applicant to the program to complete an extensive interview with the MSIS degree program director upon admission to graduate studies. Kirton's Adaption-Innovation Inventory is easy to score and takes about 15 minutes for each student to complete. The KAI is a highly validated psychological instrument in existence today [Kirton, M. J., 1976].

Kirton demonstrated, for the U.S. population of managers and engineers [Kirton, M. J., 1998], that the Kirton A-I performance scores were normally distributed with a mean of 96 and standard deviation of 10. The minimal score is 32, one per item, and the maximum score is 160, or 5 on all items. A score above 96 indicates the individual evaluated is innovative and a score below 96 that the individual tends to be adaptive. The following table further breaks down the categories of classification.

TABLE 1 Categories of Innovators and Adaptors by Kirton's A-I Inventory Performance Scores.

Innovators		**Adaptors**
96–110	Mild	95–80
111–124	Medium	79–65
125–139	High	64–50
140 or more	Very High	49 or less

The Kirton A-I Inventory total score can be broken down into three interrelated sub-scores. Sufficiency of Originality (SO), Efficiency for Details (E), and Conformity to Norm Group or Rule (R) are the sub-scores or factors [Kirton, M. J., 1994, 1998]. The SO sub-score relates to a person's style of idea generation, the E sub-score relates to one's method of problem solving, and the R sub-score relates to a person's style when dealing with structure, both impersonal, in a group, and personal, or individually [Kirton, M. J., 1998].

The Sufficiency of Originality (SO) sub-factor deals with the difference people have in their preferred handling of ideas. Adaptors prefer to produce fewer ideas when the consensus of existing ideas is generally agreed upon in terms of soundness and usefulness. Thinking out of the box and taking risks is perceived as valuable by innovators. Assuming a normal distribution of scores, over the sum of the items 1 through 12, implies that SO scores range from 12 to 60 with the general population mean of 36.

The Efficiency for Detail (E) sub-factor deals with the attention people give to detail when problem solving or during decision making. Searching methodically and organizing data in an orderly fashion are preferred methodologies of adaptors. On the other hand, innovators pay less attention to details and prefer to initiate fresh and new approaches to problem solving. Assuming a normal distribution of scores, over the sum of the items 13 through 18, implies that the E scale ranges from 6 to 30 with the general population mean of 18.

[Kirton, M. J., 1994] explains that adaptors and innovators do not naturally work well together. However, bringing together adaptors and innovators on the same team would benefit the team by enabling them to consider a greater variety of alternatives during the problem solving process. As the difference between adaptor and innovator Kirton A-I Inventory performance scores widen, then teamwork becomes more difficult [Kirton, M. J., 1998]. This latter fact is simply due to the natural tendency that people often prefer to work with others like themselves. The Kirton A-I Inventory performance scores can make us aware of differences in team members cognitive styles thus allowing for the accommodation of those style differences.

The Conformity to Group Norm or Rule (R) sub-factor describes differences in the management structure within which problem-solving and decision making take place. Adaptors accept rules as a structure within which to aid the efficiency of deriving solutions in problem-solving while emphasizing group consensus and cohesion. Rather than play by the rules and seek group approval, the innovators tend to break loose or ignore rules in their individual pursuit for solutions. Again assuming a normal distribution of scores, over the sum of the items 19 through 33, implies that the R scales ranges from 15 to 75 with the general population mean of 45.

A study, utilizing the Kirton A-I Inventory, that compared the problem-solving style, in terms of adaption and innovation, of Information and Computer Science majors was conducted by [Sim, E. R. and Wright, G., 2002]. This study concluded that there is no difference in programming styles between information systems and computer science students as measured by the Kirton's A-I Inventory. The study did conclude, however, there was a gender difference in the Kirton A-I Inventory scores; women scored slightly more adaptively then men.

[Kaufmann, G., Isaken, S. G., and Lauer, K., 1996] developed a regression plane showing the distribution of Kirton A-I Inventory scores across managerial levels by gender. These authors found a striking gender difference when observing Kirton A-I Inventory scores for males and females analyzed across managerial levels. Males were found to be more innovatively inclined than females at the lowest managerial level, whereas females obtained almost a full standard deviation higher innovator score at the top, executive, level of leadership.

Characteristics of an Adaptor

1. Produces a few manageable ideas that are relevant, sound, and safe
2. Precise, reliable, master of detail
3. Methodical, prudent, rarely challenges rules except when fully supported
4. Conforming, maintains continuity, stability, and group cohesion
5. Attempts to improve things already done
6. Is cautious about doing new things

Characteristics of an Innovator

1. Proliferates ideas, however relevant they may seem
2. Thinks "outside the box;" interested in the "big picture"
3. Approaches tasks from unsuspected angles
4. Questions the basic assumptions, attempts to manipulate problems
5. Breaks from the mold; challenges rules
6. Prefers to work alone

Informant Training

During a one-time period, the informants were trained, by the author, using the document in Appendix A. Observations gleaned from the informant interviews were then organized into five categories by the author; each focused on issues surrounding a particular theme as follows:

1. Structure
2. Conformity and Consensus
3. Relevance
4. Conflict
5. Positive Value

Results

Tables 2 and 3 present the decomposition of the Kirton A-I performance scores for both teams evaluated in this study.

TABLE 2 Demographic Decomposition of Kirton A-I Inventory Scores.

Sample	Size (N)	Range	Mean	Category
All Students	7	89–129	108.14	Mild Innovator
Male students only	5	89–129	106.6	Mild Innovator
Female students only	2	98–122	110.0	Mild Innovator

TABLE 3 Adaptors and Innovators among all Students.

Sample	Frequency	A-I inventory scores	Number of MSIS students
Mild Innovators	2	102,105	2 [102, 105]
Medium Innovators	2	115, 122	2 [115,122]
High Innovators	1	129	1 [129]
Very High Innovators			
Mild Adaptors	2	89, 95	2 [89, 95]
Medium Adaptors			
High Adaptors			
Very High Adaptors			

As can be seen from these tables the author was limited to a sample size of seven and formed one team of all innovators and a team of two adaptors with one innovator. Only the demographic variables of military service, GPA, undergraduate major, and length of professional field experience were employed by the author in forming the project teams.

For purposes of analyzing the particular themes of interest in the study the author chose to utilize the sub-factors of Kirton' A-I Inventory. The sub-factor scores for the student teams are given in the following.

TABLE 4 Kirton AI Team Composition.

	KAI	SO	E	R	Category
Team I					
Leader	129	45	26	58	High Innovator
Std1	122	54	28	40	Medium Innovator
Std2	115	46	22	47	Medium Innovator
Std3	102	43	24	35	Mild Innovator
Team Template	117	47	25	45	Mild Innovator
Team II					
Leader	95	44	14	37	Mild Adaptor
Std1	89	44	15	30	Mild Adaptor
Std2	105	38	22	45	Mild Innovator
Team Template	96.3	42	17	37.3	Mild Adaptor
All Students	108.1	44.9	21.6	41.7	Mild Innovator

Structure

In general, the author observed that the more adaptive students, on Team II, were more frustrated when they expressed that they had to deal with issues such as open-ended tasks, ambiguous tasks, and/or the inability to contact their team leader or the teacher. Team I, all innovative members, was only slightly bothered by a sudden switch they experienced in their project development direction. The SO sub-factor scores in Table C-4 effect the latter behaviors, where a larger score indicates less sufficiency of originality. Team I with a score of 47 and Team II with a score of 44.9 were both above the mean of 36.

The Team II leader/informant noticed that the adaption student on his team wanted tighter time management, including planning ahead and being on time at team meetings. On the other hand, the innovative student team member tended to be resiliant with his time management. The Team II leader also noted that, without knowing the Adaption-Innovation classification of the other students, the more adaptive student team member tended to bring structure to team meetings through note-taking and sequential processing of information. The E sub-score for Team II, with two adaptors, was 17 which is close to the mean of 18, thus leaning toward preferring an efficiency in regard to detail. At the same time, Team I, all innovators, with an R score of 25, indicated less adherence to efficiency of detail. Both team leaders noticed that the innovative students reported less note-taking and were more inclined to "brainstorm" topics or use nonlinear processes.

Conformity and Consensus

The Team II leader also reported several instances in the area of conformity and consensus. The adaptive team member was more concerned about their team moving forward together and/or getting along, and they expressed the most frustration when this did not happen. The innovative student team member was more likely to work individually at first, bringing their results back to the team later for discussion; whereas the adaptive student tended to work with the full group. Again these results are also indicated by the R sub-factor performance scores for the two teams. Team I with a score of 45 above the mean of 36 indicated a preference to "think out of the box." Whereas Team II with a score of 37.3 tended to prefer group work rather than to work individually. On both teams the Innovative students were perceived as being abrasive and occasionally even offensive in word and/or action. The adaptive member on Team II was perceived as offensive due to their "condescending" attitudes and tendency to reject the ideas of the innovator student team member. This last tendency included one team informant, as related by the team members to the teacher privately, and with one other individual adaptive student as reported by their team informant.

Relevance

Both teams had issues surrounding the perceived relevance of the input from their innovative team members. Adaptive team members did not always see value in the ideas of their innovative teammates. One Innovative teammate felt that their work was disregarded and/or omitted with no good excuse.

Conflict

Team I was composed all innovators. Team II was composed of one innovator and two adaptive members. Team II experienced the greatest amount of conflict overall, as reported by the team informants. The conflict, on Team II, as stated earlier, is related to a composition of adaptors and innovators whereas Team I was composed of all innovators. It would have been interesting to have assigned the Team I leader as the Team II leader. This assignment would have made the Kirton A-I Inventory Performance scores on Team II have a 40 point range, which might have led to team disruption and major conflict [Kirton, M. J., 1994].

Value

In general the students expressed appreciation of each other's work by the end of the semester. Adaptive members seem to recognize value in the innovative teammates bringing ideas to the table, ideas which they might not have devised on their own.

Conclusion

This preliminary study, in spite of small sample size, has demonstrated that applying cognitive style theory to characterize and better understand the personal dynamics of students working in teams is both useful and appropriate. Analysis of the data in this study indicates that their exist relationships between cognitive styles and student's team project behaviors, but further and more rigorous investigations must be made relative to this hypothesis.

A major application of the Kirton A-I performance scores is apparent from the observed student behaviors in this study. In the future the author will not only utilize major, GPA, and professional experience to develop project teams, but Kirton A-I performance scores will be incorporated into the assignment process. The Kirton A-I performance scores will be used to form groups with either homogeneous or heterogeneous cognitive styles for group activities.

Bibliography

1. Isaksen, S. G. and Kaufmann, G. (1990). Adaptors and Innovators: Different Perceptions of the Psychological Climate for Creativity, *Studia Psychologica*, 32, 3.
2. Kaufmann, G., Isaksen, S. G., and Lauer, K. (1996). Testing the "Glass Ceiling" Effect on Gender Differences in Upper Level Management: The Case of Innovator Orientation, *European Journal of Work and Organizational Psychology*, 5(1), 29–41.
3. Kirton, M. J. (1976). Adaptors and Innovators: A Description and Measure, *Journal of Applied Psychology*, 61(5), 622–629.
4. Kirton, M. J. (1994). A Theory of Cognitive Style. In *Adaptors and Innovators: Styles of Creativity and Problem Solving* (Ed.), Rroutledge, New York, 1–33.
5. Kirton, M. J. (1998). *Kirton Adaption-Innovation Inventory (KAI) Manual* (3rd Edition), Occupational Research Centre: Hertfordshire, UK.
6. Sim, E. R. and Wright, G. (2002). A comparison of adaption-innovation styles between information systems majors and computer science majors, *Journal of Information Systems Education*.

RONALD S. KING is a Professor of Business Information and Technology Systems at the University of Mary Hardin-Baylor. He earned his B.S. in Mathematics from Lamar University in 1966, his M.S. in Mathematics from the University of North Texas in 1968, and the Ph.D. in Statistics from the University of Northern Colorado in 1973. His primary interests are in data mining, database design, and knowledge representation.

APPENDIX A

An Ethnographic Study of MSIS Student e-Learning Cultures

Project: An ethnographic study designed to investigate remote learning enhancements in the context of a graduate database class at the University of Mary Hardin-Baylor. The emphasis of the study is to gain an understanding of the e-learning culture of the graduate MSIS student body at UMHB. All results would maintain anonymity.

The Group: The graduate students in the MSIS degree program are required to have either an undergraduate degree in Information Systems or an undergraduate degree in another major with adequate relevant work experience. The current class, Database Development and Administration, has eight students enrolled. I suspect that student preference for class environments swings from totally online to only in a traditional classroom environment. Similarly, I suspect that the student backgrounds vary from having taken only traditional classes to a desire to take only Internet-based courses.

The Request: As a term project team leader, I am requesting that you serve as a concealed observer of your team members. With your help we will complete an observation of their behaviors and the values and beliefs causing these behaviors.

Background: There are two kinds of cultures: objective and subjective culture. We are interested in discovering both types of culture present on your team. The visible part of culture is the objective part. It includes: eye contact, greetings, initiative, food, clothing, and time consciousness. Objective culture is supported by the subjective culture. This is the internal part of culture that drives or motivates the visible, objective culture. It consists of motivations, beliefs, and worldview.

Methodology: Gather data through observation which might be categorized by the following list of things to look for in the individual under observation. Next week I will interview you in order to submit a written report on each of your team member observations. This information will be kept just between you and me to preserve confidentially.

Think of the adjectives which describe the typical business person. Here is a list that comes to my mind: hard driving, self-confident, self-reliant, tough negotiator, decisive, fair, motivated, fact-based, and unemotional, uncaring, scientific, unfeeling, well-educated, rational, closed, control.

Now think of the adjectives describing positive relationships between team members: trusting, caring, emotional, friendship, win-win, self-expression, comfort, and exposed.

But for the negative relationships between team members we might have: vague, dependent, unknown.

In some cases as the leader you could possibly be asking a person to neglect/ignore everything they have been taught all of their life, to give up **control**, and to open themselves up to potential misfortune or failure based on feelings.

Think along the following dimensions:

- Idea generation
  - Likes quality – very focused ideas, highly relevant
  - Likes quantity – large number of ideas, however relevant
- Efficiency
  - Likes lots of detail
  - Prefers the "Big Picture"
- Rule/Group Conformity
  - Likes structure, group work
  - Likes no rules, prefers to work alone
- Coping Ability
- Passive versus Aggressive Behavior
- Introvert/Extrovert
- Female/Male
- Laziness/High Achievers
- Interest/Motivation Level
- Sensing/Intuition
- Thinking/Feeling
- Judging/Perceiving

How do the team members react to?

- Open-ended descriptions
- Ambiguous tasks
- When unable to interact with the teacher or team leader
- Time management
- Team meeting structure
- Group conformity
- Perceived relevance of the input from their teammates

APPENDIX B

Kirton Adaption-Innovation Inventory (KAI)

From a managerial point of view rate the following traits of a successful manager on a scale of 1 to 5, 1 being extremely undesirable and 5 being extremely desirable:

ITEM	RATING
1. Has original ideas	________
2. Proliferates ideas.	________
3. Is stimulating.	________
4. Copes with several new ideas at the same time.	________
5. Will always think of something when stuck	________
6. Would sooner create than improve	________
7. Has fresh perspectives on old problems	________
8. Often risks doing things differently	________
9. Likes to vary set routines at a moment's notice	________
10. Prefers to work on one problem at a time	________
11. Can stand out in disagreement against group	________
12. Needs the stimulation of frequent change	________
13. Prefers changes to occur gradually	________
14. Is thorough	________
15. Masters all details painstakingly.	________
16. Is methodical and symmetric	________
17. Enjoys detailed work	________
18. Is (not) a steady plodder.	________
19. Is consistent	________
20. Imposes strict order on matters within own control	________
21. Fits readily into "the system.	________
22. Conforms.	________
23. Readily agrees with the team at work	________

24. Never seeks to bend or break the rules . . . . ________

25. Never acts without proper authority . . . . ________

26. Is prudent when dealing with authority. . . . . ________

27. Likes the protection of precise instructions . . . . ________

28. Is predictable . . . . ________

29. Prefers colleagues who never "rock the boat". ________

30. Likes bosses and work patterns which are consistent . . . . ________

31. Works without deviation in a prescribed way . . . . ________

32. Holds back ideas until obviously needed. . . . . ________

APPENDIX D

SOLUTIONS TO SELECTED EXERCISES

Solutions (including figures) to selected exercises in the text may be found on the companion DVD.

INDEX

D

E

F

G

H

I

J

K

L

M

N

O

P

Q

R

S

T

U

V

W

Y

Z

www.ingramcontent.com/pod-product-compliance
Lightning Source LLC
LaVergne TN
LVHW062158200926
848484LV00006B/1280
* 9 7 8 1 9 3 6 4 2 0 1 2 4 *